Putnam's Geology

Overleaf: Erosion in Death Valley
William E. Garnett

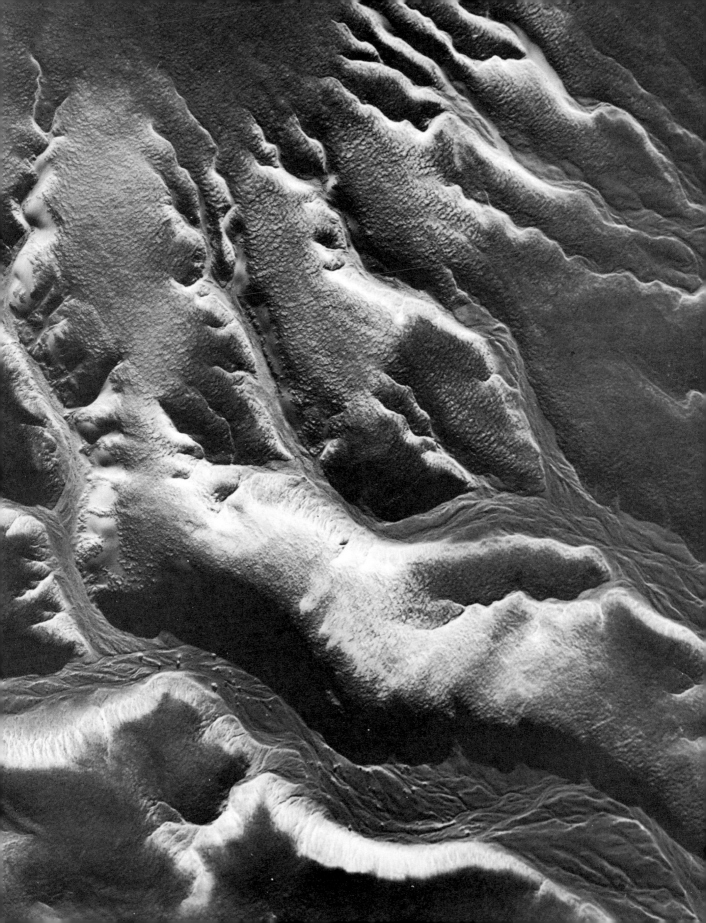

Professor William Clement Putnam, the original author of this book, died on March 16, 1963, at the age of fifty-four. He had taught geology for thirty-two years. This book is the result of his experience as a teacher, and of his love of teaching beginning students.

PUTNAM'S GEOLOGY

THIRD EDITION

PETER W. BIRKELAND   EDWIN E. LARSON

New York ♕ Oxford University Press ♕ 1978

Cover Photo: Sand Dune Arch, Arches National Park, Utah. *Bill Ratcliffe*

Third Printing, 1979

Copyright © 1964, 1971, 1978 by Oxford University Press, Inc.

Library of Congress Cataloging in Publication Data

Putnam, William Clement, 1908–1963.   Putnam's Geology.
  Bibliography: p.
  Includes index.
  1. Geology. I. Birkeland, Peter W. II. Larson, Edwin E., 1931—
III. Title. IV. Title: Geology. QE28.P98 1978   550   77-2789
ISBN 0-19-502285-8

Printed in the United States of America

PREFACE

When we were asked to revise Putnam's *Geology*, we looked upon it as an opportunity to bring an outstanding textbook up to date. Bill Putnam had a clear and lucid style and an ability to explain complex geological matters in a very readable and understandable way. We have tried to maintain that style while adding new data on developments and discoveries that have taken place in the fifteen years since the first edition was published. Perhaps the most spectacular development during those years has been the renewed interest in the theory of continental drift—the basis for the more recent theory of plate tectonics. If ever there was a revolution in geological thinking, it is embodied in plate tectonics, a wide-ranging and unifying theory whose effects reach into many aspects of geology as well as into other sciences. The phenomenon of plate tectonics is still far from being perfectly understood, but the theory has for the most part been accepted by geologists.

Introductory textbooks in physical geology show no one common pattern in their sequence of topics. Nor is there general agreement on the order of presentation in the classroom. The arrangement of this edition is basically that preferred by Bill Putnam, one devel-

oped during the course of his thirty two-years of teaching. Each chapter is self-contained and can be taught in nearly any sequence that an instructor may wish. It does seem logical to us to proceed from a discussion of the materials that make up the visible earth to the various processes responsible for wearing down and shaping the landscape, concluding with a panorama of earthquakes, mountain building, and plate tectonics—in general, the processes that continue to elevate portions of the earth's crust. We find that this ordering of chapters allows for the introduction of essential terminology early in the term.

Plate tectonics has become a guiding theme in the present edition. The theory is briefly introduced in the second chapter and is considered in other chapters in relation to the processes that are described. With the foregoing as background, plate tectonics is explored in detail in a separate and final chapter. As in past editions, environmental issues form an integral part of the text. Chapter summaries and an extensive glossary have been added. The metric system is employed throughout the text with conversions to the English system. A separate student study guide by Dr. Helen A. Biren and Dr. E. Lynn Savage, both of Brooklyn

College of The City University of New York, is also available.

Even in limited travels, one inevitably encounters geological phenomena. We hope that this book will nurture an interest in the landscape that will become a permanent source of pleasure and satisfaction to the reader.

Many people have influenced our thinking during the course of this revision. Our fellow instructors and graduate teaching assistants in the introductory geology course here at the University of Colorado have been helpful in their comments over the years, especially Frank Beck, William C. Bradley, James L. Munoz, and Donald C. Runnells.

We have also benefited from readings of the final manuscript by James L. Baer of Brigham Young University, Douglas G. Brookins of the University of New Mexico, Duncan Heron of Duke University, and Bruce Molnia of the U. S. Geological Survey. In addition, Bruce Bartleson, of the Western State College of Colorado, John T. Andrews and Frank R. Beck, both of the University of Colorado, Bruce E. Brown of the University of Wisconsin, Milwaukee, Paul Catacosinos of Delta College (Michigan), Philip H. Heckel of the University of Iowa, James P. Morgan of Louisiana State University, Dennis I. Netoff of Lincoln College (Illinois), and Theodore W. Ross of Lawrence University provided advice and guidance on materials for us

to consider in this revision. Joyce Berry of Oxford University Press, however, did most of the editorial work, for which we are extremely grateful. We alone, though, are responsible for errors or omissions.

Putnam's original text was known for its excellent photographs. We have tried to maintain those high standards by retaining the best photographs from the previous editions and by adding some of the best work, both past and present, from the U.S. Geological Survey collection as well as the work of many friends, all of whom are duly acknowledged in the captions. We are especially grateful to William C. Bradley for access to his extensive collection, Jan Robertson, whom we commissioned to photograph features in the local area, and Deborah Bowen of Oxford University Press, who did a splendid job of photo research.

We also want to thank Edith Ellis and Paulina Franz, who typed the manuscript drafts, and John Lukens, Richard Carter, Norman Nielsen, and David Tewksbury, who helped with a variety of tasks associated with preparation of the manuscript and the photographs. Finally, we wish to acknowledge the aid of the students in our courses for helping us to formulate many of the ideas presented.

EDWIN E. LARSON
PETER W. BIRKELAND

CONTENTS

I shall lift up mine eyes unto the hills
From whence cometh my help.

Psalms: cxxi, 1.

Fig. 1-1. Monument Valley, Arizona. These monuments are all that remain of a once-continuous layer of sandstone. The rest has been removed by erosional processes acting over long periods of time. *William A. Garnett*

1

GEOLOGY

THE FOUR-DIMENSIONAL SCIENCE

Geology—the study of the earth—is not strictly speaking a fundamental science. Rather, it involves the application of principles from the basic fields of physics, chemistry, biology, and mathematics with those from other secondary-level sciences like astronomy. Not uncommonly, geologists of today find their science to be a kind of natural philosophy encompassing all aspects of the physical and biologic world that surrounds them. One of the ways in which geology is set apart from other sciences is that it was, classically, and still is a field-oriented science. In fact, that aspect is responsible for enticing many a person into the study of geology. In order to make observations of the rocky crust a geologist will often journey to what appears to be the "end of the earth"—at one time struggling up the precipitous slopes of a mountain range in Alaska; at another, wading up a stream valley in a tropical rain forest in Brazil. As often as not, in stopping to rest—to jot down notes, or to compile a map—geologists find themselves in two worlds at once—one the here and now, surrounded by the sights and sounds of the present; the other, the ancient world contained in the record of the rocks and landscape over which they trod.

But the science today is more than the world of a field geologist. Observation of the rocks is only one part of the total attempt to unravel the mysteries of the ages that are contained in those rocks.

Geologists of all kinds, each in a specialized way, contribute to our knowledge of the earth. Some, called *geophysicists*, use the principles of physics to "look" deep inside the earth. Others use conventional and modern techniques to determine the chemical makeup of the rocks and waters, and are known as *geochemists*. Some geologists specialize in the creation and study of models of the earth's processes, and still others approach the understanding of the earth from a theoretical viewpoint. Regardless of degree of specialization, however, each must be aware of how his or her work fits into the understanding of the earth as a whole.

To be sure, there is an overriding historical slant to geology. Most landforms and rocks (Fig. 1-1) we find today at the earth's surface are relics of past ages when the world and the living organisms on it were different. All we have left to tell us of those distant times is the landscape, standing in muted silence, and the

Fig. 1-2. Ancient ripple-marked sandstone at the bottom of the Grand Canyon, Arizona. The sediments are about 1.3 billion years old. *Edwin E. Larson*

fragmentary rock record. Through careful scrutiny of the landforms and the rocks themselves, and through a knowledge of the way things happen on earth today, geologists are able to recreate scenes from the past. Underlying their ability to draw conclusions and to make hypotheses about bygone ages is the idea that the present is the key to the past, a viewpoint called *uniformitarianism*. The term means only that processes we find in evidence in the world today probably existed in the past. Perhaps the rates of those processes were different but, by and large, the processes themselves were the same. That is, rivers under the influence of gravity probably flowed in earlier days much as they do today; the ancient winds probably blew in response to atmospheric disturbances much as they do now, and so on. When one travels to the bottom of the Grand Canyon and observes that not far from a fossil ripple-marked sandstone (Fig. 1-2) the same type of ripple marking is being formed in the soft sand deposited along the banks of the Colorado River, it is difficult to escape that conclusion. The appeal to uniformitarianism is only a re-affirmation that the natural laws have prevailed on our planet continuously and in an unchanging way through the ages. The idea is basic not only to geology but to all the other sciences as well.

The alternative view is that all that surrounds us has no connection with the past. Such a view leads to chaos and confusion, a world in which there is no continuity, each age characterized by processes and events never seen before and probably never to be seen again. All evidence seems to indicate, fortunately, that such is not the case.

In the past, the concept of uniformitarianism was not readily accepted. Before about 1700 many people, particularly those steeped in the dogma of the Christian Church, felt that the earth had been the scene of repeated catastrophes, each of which had wrought large-scale changes both in physical features of the surface and in plants and animals. In between catastrophic events, the earth remained in limbo. By the mid-1700s, however, it became apparent to some that geologic features, even those on a grandiose scale, could result from the relentless action of natural processes. The validity of that view, which was championed first by a Scotsman, James Hutton (1726–97), and subsequently by his friend and countryman, John Playfair (1748–1819), was still being discussed almost a century later. It was not until about 1830 that Charles Lyell (1797–1875), an Englishman familiar with and sympathetic to the views of Hutton, forcefully demonstrated the reality of uniformitarianism through many careful observations of rocks and landforms in western Europe. For his efforts in the field Lyell is generally considered to be the "Father of Geology" (Fig. 1-3).

In 1859 Charles Darwin (1809–82) did for biology what Hutton and Lyell did for geology—he established the idea of biologic uniformitarianism. Once the ideas of physical, chemical, and biologic uniformitarianism were accepted, earth scientists could start to unravel the tangled web of the past.

Sciences such as physics and chemistry are considered to be exact sciences, ones in which the conventional scientific method can be applied. First, experiments are performed, and the various physical and chemical parameters are measured. Based on such observations, a plausible explanation or hypothesis is formed. Subsequent experiments test the hypothesis, and, if it is shown to be in error, it is discarded and another is formulated and tested. If a hypothesis continues to meet the tests of experiments over a long period of time, it is declared to be a *natural law*—the law of gravitation is an example.

Such is not generally true in geology. Rather than being black and white, the realm of geologists usually exists in shades of gray. The "experiment"—commonly on an immense scale of extreme complexity—has already been performed in nature, and all that can be seen is the result. Even the scope of observation is

Fig. 1-3. Sir Charles Lyell in 1863. *George Eastman House*

restricted because rocks at the earth's surface usually are not well exposed but are masked by soil and younger deposits. Yet, from careful observation tempered by a knowledge of the physical, chemical, and biologic processes that occur today, geologists are able to reach conclusions and to state hypotheses. Actually, they often formulate several equally plausible hypotheses to fit their observations—an approach called the method of *multiple working hypotheses*. Then the search continues for evidence that will negate one or another of the several hypotheses. As new evidence is turned up, all existing hypotheses may even be discarded and a new, seemingly more plausible one formulated. Geologic hypotheses, then, are much like present-day weather forecasts, in that both are based on an estimate of probability.

GEOLOGIC TIME—THE FOURTH DIMENSION

As is apparent by now, the study of the earth is involved with time. All of us have an awareness of time and fit our lives into a timetable (some people more stringently than others). It has only been through the study of the earth, however, that we have come to realize the true vastness of the fourth dimension, which extends back to the beginning of our world and beyond. In fact, as has been stated by A. Knopf, a well-known U.S. geologist, it is the concept of the immensity of geologic time that has been perhaps the single most significant contribution made by geology. Whereas in the historic world we deal with hours, days, and years, in geology we deal with thousands, millions, and billions of years.

To most of us, a few hundred years of history encompass a great deal of time. Fragmentary accounts and stories from the days of the early Greek, Roman, and Egyptian civilizations we class as "ancient history." The gulf between our own culture and those civilizations seems so great that we find it difficult to relate to them. So how do we stretch our imagination to conceive realistically of a length of time spanning many millions or billions of years? The truth is that we cannot. It is like trying to imagine the colossal distances that separate us from nearby stars or other galaxies. We can write down the six- and nine-digit figures, but we can never completely understand the immensity of time.

It is within the four-dimensional realm that the geologic processes have been and still are working. Given enough time it is possible for even the slowest working process to bring about large-scale changes. Time, then, is the essential ingredient to the uniformitarian view.

In order to estimate the passage of time it is necessary to use some device that records the passage of equally spaced events. We now use the ticks of a clock as monitors; in older societies people used such things as sundials and water clocks. The basis for all chronometers, or timepieces, is found in the motions of our planet in the solar system. In essence, the sun and encircling planets are like gigantic parts of a solar clock that is silently and relentlessly ticking in our corner of the Milky Way Galaxy. Yet even that natural clock, as we shall see, is not perfect and is prone to running down. The day is the obvious basic unit of time; its length is controlled by the speed of rotation of the earth on its axis and is defined as the time between two consecutive passages of the zenith of the sun (*solar day*) or of a distant star (*stellar day*) past a particular spot on the earth. The year, another basic unit, is designated as that time required for one passage of the earth around the sun. Inasmuch as the speed with which the earth travels around the sun varies throughout the year, the length of a day is variable, and we have been forced to establish the *mean solar day*—one that is always of the same length, regardless of the season. Now, it is well known that there are not an even number of days in a year; rather there are about 365¼. In order to retain parity between the calendar year and the planetary year we need to add one day every four years; that is, each Leap Year. But even that adjustment is inadequate, and additional minor resettings of the calendar are required about every 400 years.

Recently, technology has produced an improved timepiece based upon the electronic

Fig. 1-4. Coral about 375 million years old from the Sulphur Springs Range in Nevada; maximum width about 4 cm (1.5 in.). Measurements of growth lines on such corals have enabled determination of the number of days in an ancient year. *C.W. Merrian, USGS*

counting of atomic vibrations in a specified element. The *atomic clock* is capable of recording variations in the length of a day as small as billionths and trillionths of a second. With its use, scientists have become aware of an amazing fact—that the spinning earth is slowing down, a phenomenon recorded as a progressive but slight—very slight—lengthening of the day. Most scientists concede that the slowing of rotation is primarily the result of gravitational interaction between the earth and moon. The atomic clock has only been used for a few years; verification of the slowing over a considerably longer period of time comes from a study of the records of eclipses kept by the ancient Babylonians. When modern astronomers calculated the total elapsed time from the moment of a particular eclipse, they invariably came up with an "error" of about eight hours. The reason for the error becomes apparent when one multiplies the increase in a day's length as determined by the atomic clock by the number of days elapsed since the time of the Babylonians: the total increase amounts to about eight hours.

Their curiosity heightened, geologists immediately began to try to determine if the slowing of the earth had been going on throughout most of geologic time. The most fruitful approach has involved the use of a marine organism—coral (Fig. 1-4). Most marine animals living at the margin of the sea are under the influence of the tidal rhythms, which themselves are controlled by the passage of the moon. The tidal influence extends to many of the organisms' activities, including feeding habits. As coral grows, for example, it adds identifiable layers of shell material much in the way that a tree does. So, by painstakingly counting growth layers and recording their widths, it is possible to arrive at a figure for the number of days in a lunar month. When that method was applied to 300-million-year-old fossil corals, the astonishing conclusion was reached that 300 million years ago, a year would have been about 440 days long. Other kinds of marine fossils have been used, and the results are generally consistent. The slowing of the earth's spin apparently has been going on for as long as the moon has been its traveling companion.

Conventional clocks are fine for most purposes, but in recording geologic time unconventional methods are required. One rather unique attempt to determine the age of the earth was made in the mid-1600s by Archbishop Ussher (1581–1656) of the Irish Protestant Church. The Archbishop was a fundamentalist who firmly believed that the Bible contained a complete record of the world's events since its inception. Therefore, by adding up all of the genealogies (that is, all the *begets* and *begats*) recorded in the Old Testament, he was able to set the origin of the earth as the evening of October 22, 4004 B.C. Recently someone jested that he failed to tell us if it was standard or daylight saving time. Ussher's estimate was a refinement of a figure similarly determined in 1642 by John Lightfoot, a distinguished Greek scholar and Vice Chancellor of Cambridge University. Unfortunately, the date 4004 B.C. was referred to in the Great Edition of the English Bible published in 1701, and from then on it was incorporated into the dogma of the Christian Church. For a century thereafter, anyone who maintained that the earth's development took more than 6000 years was considered a heretic. In effect, a theological straitjacket was placed on all thinking in the Western

Fig. 1-5. Thick section of sedimentary rocks exposed in the Grand Canyon, Arizona. Hundreds of millions of years were required for the deposition of the section. Erosion rates commonly are measured in centimeters per thousand years, so you can readily estimate that millions of years were required to form a canyon of such grandiose proportions. *N.W. Carkhuff, USGS*

World concerning the antiquity of the earth. In order to explain the many geologic features and events it was necessary to invoke the occurrence of innumerable catastrophes. The grip was only broken by the patient efforts of Hutton and Lyell and the development of geology as a science.

With the emergence of geology, earth scientists became aware that a tremendous amount of time was needed to accomplish the changes they knew had taken place. How much time? It would seem that if we knew the rate at which sedimentary rocks, which generally occur in layers, or strata, were deposited, as well as their total thickness, we could estimate the length of time that erosion and deposition has operated on the surface of the earth (Fig. 1-5).

Unfortunately, there are a number of things wrong with such an assumption. No one knew then—or now, for that matter—the rate at which sediments were deposited. Obviously, the rate must have been different for a conglomerate made of boulders a meter across than it was for a limestone consisting of the remains of microscopic marine organisms. Furthermore, few if any sediments are deposited uninterruptedly. There may have been times of accumulation separated by intervals in which no deposition at all occurred, or during which erosion of the previously deposited sediments occurred. To add to the difficulties, no place was known on the land surface of the earth where strata representing the total age of the earth had been laid down in one continuous succession. Nonetheless, in 1883 the European geologist William Sollas (1849–1936) did use that method, estimating that 26 million years had been required to accumulate the total combined sequence of strata of which he was aware.

Another attempt to relate present-day processes to the age of the earth was made in 1899 by the Irish geophysicist John Joly (1857–1933). The quantity of salt dissolved in the oceans had been determined fairly accurately by that time, and Joly reasoned that if a fair estimate could be made of the amount of salt that rivers contribute each year to the oceans, he could establish a value for the age of the sea. His best estimate was 100 million years.

Joly did not see that there were many pitfalls in his method. Immense quantities of salt now found as salt beds have been removed from the oceans by evaporation of sea water in ages past. Furthermore, some of that salt is now being recycled—sent back to the sea—as the salt beds erode away. He also had no way of knowing that the amount of salt added to the sea each year probably has not remained constant throughout geologic time. The evidence now indicates that times of relatively high mountains and rapid erosion have alternated with times of low relief and correspondingly lower rates of erosion.

Relative time

While some earth scientists were trying to establish absolute ages for certain geologic features, most were involved in deciphering geologic events in their local areas of study. They placed all of the events recorded in the rocks into a *relative time* sequence—by determining whether an event occurred before or after another one. In our everyday lives, we commonly function in the relative time domain.

Only a few methods are available to use for establishing the relative time sequence. Some of them are so straightforward that they seem obvious; yet in the early days of geology, when little was known about the earth, such methods were unheard of. Each represents a discovery made by an individual only after long hours of observation. Nicolaus Steno (1638–86), a Danish physician and theologian who became curious about the earth, was just such a person. From his observations he arrived at the conclusion, in 1669, that in any sequence of layered rocks the age of one layer will be greater than the one above it and less than the layer below it. He reasoned that before the stratum above could be laid down, the layer below must have already been deposited. To that simple rule has been given the appellation *law of superposition*. It is used primarily in determining the relative ages of sedimentary rocks or lava flows.

Another rule established by Steno is that sediments were originally deposited almost horizontally. Therefore, if one finds a sequence of strata tipped up at an angle, then the event that produced the tilting must have occurred after the strata were laid down (Fig. 1-6). It has been found that exceptions to the rule, called

Fig. 1-6 A (above) Sand recently deposited on the floor of the San Gabriel Reservoir, California. Note that the layers are nearly horizontal, typical of many sedimentary environments. *John S. Shelton*   **B.** (opposite) Sedimentary rocks exposed near Bluff, Utah, looking north over the San Juan River. The rock layers were originally horizontal and were tilted to their present positions after they were deposited. *John S. Shelton*

original horizontality, do exist, but by and large the rule holds true.

James Hutton established another method of relative dating when he pointed out that if one body of rock cuts across another rock body, the latter must be older than the former. That is, one rock must already be in place for another, younger one to cut across it. For example, sedimentary rock layers not uncommonly contain younger tabular bodies, called dikes, that cut across the pre-existing strata. Similarly, faults (fractures along which movement has occurred) and most cracks that cut across a rock body must be younger than the rock in which they are found.

Finally, weathering and removal of material by erosion can occur only near the surface of the earth and can only affect rocks that are already there. The features used in relative time determination are shown in Figure 1-7, and by means of these and other simple criteria it is possible to establish a temporal sequence of geologic events for most areas on earth (Fig. 1-8).

Although it is comparatively easy to establish a relative time sequence in local areas, it is usually much more difficult to determine how rocks in one area correlate with rocks in another, even when the areas are close by. The best method devised thus far is to compare the physical features of the rocks themselves. If a sequence of strata made up of individually identifiable layers can be traced or mapped into another area, then the correlation is firm. Also, if each layer of a sequence of strata has diagnostic features that can be identified in the same sequence in a nearby area, then correlation can be made with some assurance.

Sometimes, however, strata change laterally in physical character, or the strata present in one area do not show up in another, nearby area. Correlation, of course, becomes much

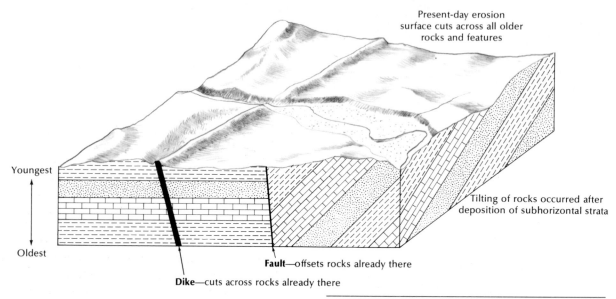

Present-day erosion
surface cuts across all older
rocks and features

Youngest

Oldest

Tilting of rocks occurred after
deposition of subhorizontal strata

Fault—offsets rocks already there

Dike—cuts across rocks already there

Fig. 1-7. Diagram of the features that enable relative age
determination.

Fig. 1-8. Cross section across the Grand Canyon in which
a number of geologic events can be relatively dated. The
oldest rocks—Vishnu Schist, granite, and pegmatite—were
eroded to form erosion surface A. Later, the sediments of
the Grand Canyon Series were deposited and then tilted;
the cutting of erosion surface B followed. Subsequently the
stack of flat-lying sediments was deposited and eventually
eroded into the present-day topography. After sketch portion of Fig. 28-1 (pp. 492–3) from THE EARTH SCIENCES, 2nd
Edition, by Arthur N. Strahler. Copyright © 1963, 1971 by Arthur N.
Strahler. Reprinted by permission of Harper & Row, Publishers, Inc.

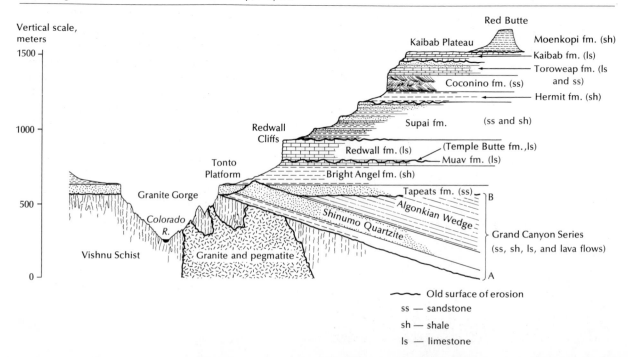

Vertical scale,
meters

Red Butte

Kaibab Plateau

Moenkopi fm. (sh)

Kaibab fm. (ls)

Toroweap fm. (ls
and ss)

Coconino fm. (ss)

Hermit fm. (sh)

Redwall
Cliffs

Supai fm. (ss and sh)

(Temple Butte fm.,ls)

Redwall fm. (ls)

Muav fm. (ls)

Tonto
Platform

Bright Angel fm. (sh)

Tapeats fm. (ss)

B

Granite Gorge

Algonkian Wedge

Shinumo Quartzite

Colorado
R.

Grand Canyon Series
(ss, sh, ls, and lava flows)

Vishnu Schist

Granite and pegmatite

A

Old surface of erosion

ss — sandstone

sh — shale

ls — limestone

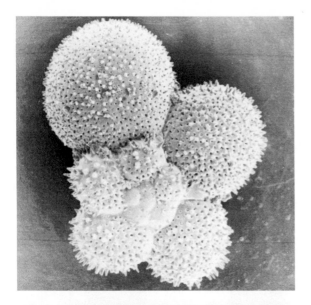

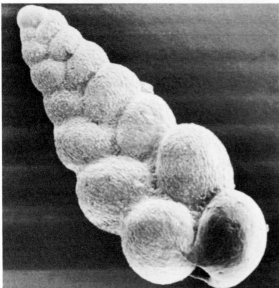

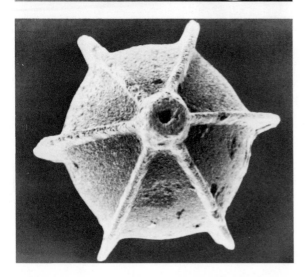

Fig. 1-9. Fossils of three species of one-celled animals (Foraminifera) from 80-million-year-old sedimentary rocks in Colorado. Each shell is about 0.5 mm in length. The photographs were taken on a scanning electron microscope. *Don Eicher*

more difficult. If the areas to be compared are quite far removed from one another, the problem becomes even more perplexing.

Fortunately, many sedimentary rocks contain *fossils*, the preserved remains or traces of prehistoric animals and plants (Figs. 1-9 through 1-11). In most cases it is the hard parts of an organism (for example, bone or exterior shell) that are fossilized. In some instances, however, soft parts can also be preserved. Tracks, trails, burrow structures, and feces may also be preserved and are also included under the broad definition of a fossil.

Fossils have proved to be of great value in correlating strata and in determining the relative age of the strata in which they are found. They are especially useful in making correlations between widely separated areas, even areas found on two different continents. Moreover, fossils form the basis for the relative geologic time scale, which represents one of the most unifying concepts in the science of geology. The working out of the evolutionary succession of fossil forms has been a major creative achievement of the human mind.

Since prehistoric times fossils have excited human interest. They are shown in ancient drawings, some have been employed in heraldry, and for centuries they have been cherished because of the interestingly decorative effect they give to polished stone. The word itself, a Latin one from *fossilus*, meaning something dug up, shows an awareness almost since the beginnings of language that a fossil was a thing of the earth.

Although such an awareness may have existed, it was accompanied by an abysmal ignorance of the true meaning of such testimonials to the existence of life in the remote past. To some, fossils were the creation of

Satan, placed in the world to confuse all mortals. To others, such as Avicenna (A.D. 980–1037), the Arab philosopher responsible for the revival of interest in the works of Aristotle, they were inorganic petrifactions that grew within the rocks as a result of the workings of a creative force, the *vis plastica*. Only by chance did they come to resemble bones or shells of living creatures. That ignorance extended to the discovery of the remains of prehistoric people, such as the Cro-Magnons and Neanderthalers, whose bones were sometimes carted to the village churchyard for a proper interment.

It remained for a singularly devoted, untutored, and eminently practical canal-builder, William Smith (1769–1839) to establish the existence of a relationship between stratified rocks and the fossils they contain. Fortunately for him, and for us, he lived and worked at the right time and place—during the period when canal building was active in England. The advent of the Industrial Revolution, coupled with the post-Napoleonic prosperity of Britain, placed a premium on the construction of such relatively cheap and modestly efficient trans-

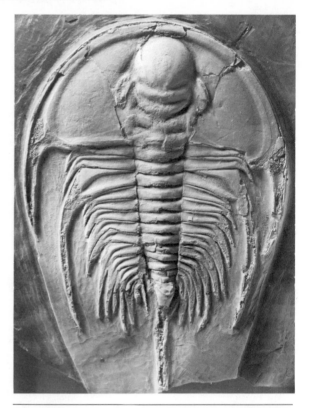

Fig. 1-10. *Olenellus fremonti,* a trilobite from the Marble Mountains, California. Its length is about 11.5 cm (4.5 in.). This animal, related to the hermit crab, lived in the oceans between about 600 and 225 million years ago; then disappeared from the face of the earth. *Takeo Susuki*

Fig. 1-11. Skeleton of *Ichthyosaurus,* a reptile that swam the seas about 175 million years ago. The modern porpoise resembles it somewhat in form. With a large mouth full of sharp teeth, an ichthyosaur no doubt presented a formidable appearance as it hunted in the ancient seas. *Smithsonian Institution*

portation as canals. In that boom, Smith was a willing participant.

In the construction of canals, the recognition of various rocks and an understanding of their physical properties are matters of prime importance. If the rocks are harder than expected, or if they slump and cave readily, or if they allow water to drain away, then with very little effort the contractor of a canal project can go broke. Not only is the structure of midland and eastern Britain relatively simple, but the rocks themselves are distinctive. Smith's approach was entirely empirical, and considering the rate at which canals were dug by hand, he was in no hurry. As a single example, he ran surveys on the Somersetshire Coal Canal for six years. All told, he tramped up and down the English countryside for 24 years making observations. When he was done, he had made the great fundamental discoveries that (1) the strata in southeastern England occurred in the same order—for example, chalk beds were always found above coal layers unless disturbed structurally, and never vice-versa, (2) different layers contained assemblages of distinctive fossils, and (3) the distribution of sedimentary rocks in southeastern England made a logical pattern that could be represented on a map.

The geologic map of England and Wales that Smith published in 1815 was the world's first geologic map. Geologic mapping subsequently has been found to be an effective means of depicting the geologic features of any given area and today represents one of the most important tools of the working geologist. The practical result was that, given a representative collection of fossils, he could determine from which layer of rock they came.

Smith had no awareness of the evolutionary significance of the fossils he collected so patiently over the years. To him they were little more than uniquely shaped objects. Yet, unknown to him, he had found the key by which strata could be correlated with one another, not only in a local region but whole continents apart.

Such a method of correlation is no simple matter, however, because we are dealing with the remains of organisms that lived in environments as varied as those of today. Furthermore, during the long span of geologic time some groups of animals and plants gradually evolved into more complex creatures, while others, having achieved perhaps an uneasy balance with their environment and their enemies, remained essentially unchanged. Still others, such as the dinosaurs, vanished completely.

This brings us to an important contribution that *paleontology* (the study of ancient life) makes to the whole realm of contemporary thought. It is the demonstration—from the fossils preserved in the thousands of feet of stratified rocks in many lands—that many creatures have evolved from relatively simple forms to complex hierarchies of plants and animals. In fact, the fossil record was one of the stronger bodies of evidence advanced by Darwin and his followers in support of the theory of evolution.

Geologic time scale

As earth scientists accumulated knowledge of strata in different parts of the world, as well as in a single region, they were able to compare rock layers with one another on the basis of fossil content. It was necessary first to determine the order of deposition in one region, and when that had been done, the succession of rocks there might be compared with a different succession of rocks—perhaps a whole continent away, and not necessarily the same kind of rocks at all, but ones containing fossils of the same geologic age.

The most valuable fossils for correlating rocks are those with comparatively short histories, geologically speaking—and yet which in their limited span on earth achieved a wide geographic range and underwent rapid evolutionary changes. Such ideal examples are called *index fossils*.

Through the application of the law of superposition, it was possible to ascertain, at least locally, which fossils were older than others. And eventually, as correlation extended to more and more stratigraphic sections, it be-

RELATIVE GEOLOGIC TIME			LIFE FORMS
ERA	PERIOD	EPOCH	
Cenozoic (Recent life)	Quaternary	Holocene	Rise of mammals and appearance of modern marine animals
		Pleistocene	
	Tertiary	Pliocene	
		Miocene	
		Oligocene	
		Eocene	
		Paleocene	
Mesozoic (Middle life)	Cretaceous	Late	Abundant reptiles (including dinosaurs); more advanced marine invertebrates
		Early	
	Jurassic	Late	
		Middle	
		Early	
	Triassic	Late	
		Middle	
		Early	
Paleozoic (Ancient life)	Permian	Late	
		Early	
	Carbon-iferous — Pennsylvanian	Late	First reptiles
		Middle	
		Early	
	Carbon-iferous — Mississippian	Late	
		Early	
	Devonian	Late	First land animals—amphibia
		Middle	
		Early	
	Silurian	Late	
		Middle	
		Early	
	Ordovician	Late	First fish
		Middle	
		Early	
	Cambrian	Late	Primitive invertebrate fossils
		Middle	
		Early	
Precambrian			Meager evidence of life

Fig. 1-12. Relative geologic time scale, based primarily on superposition of beds and the character of fossils in the strata.

came possible to establish a chronologic order on the basis of the fossils in the rock. That was the beginning of the relative geologic time scale (Fig. 1-12).

Geologic time is divided into units by using much the same philosophy employed in subdividing historic time. As a single example, in the Western World one of the longer time intervals is the Christian Era. True, there is a mea-

sure of unity within the nearly 2000-year chapter of history, but were we to find ourselves transported to ancient Palestine or Rome, very likely we should be more impressed by the differences than by the similarities.

It is a rare one-semester history course that, in attempting to cover such a broad panorama, succeeds in making it appear more lasting than the fleeting, multicolored patterns of a kaleidoscope. Fortunately, most history courses break that enormous block of time with all its crowded events into smaller units; sometimes they become very short indeed, and a semester may be devoted to a five- or ten-year period.

Carrying the analogy with human history a bit further, the most intensely studied parts of the record are more likely to be the later rather than the earlier parts. For example, 2000 years of Egyptian history may be sketched in with broad strokes, while a large fraction of a semester may be devoted to the decades following the Treaty of Versailles.

The same distortion colors our view of geologic time. Events closer to us leave more complete and decipherable records than those grown increasingly fragmentary over a long lapse of time and through the repeated deformation rocks undergo in such processes as mountain building. The result is that more is known about events in the later parts of earth history than in the earlier, and that knowledge is reflected in the larger number of subdivisions of the last part of the geologic time scale.

If we consider the names of the scale, the grand divisions, comparable in significance to such human episodes as the Stone Age, are the *eras*. In most cases they take their names from our concept of the dominant aspect of life during each era. The most ancient, the *Precambrian*, was a time interval of long duration characterized by a paucity of life forms. The eras that followed are called the *Paleozoic* (ancient life), *Mesozoic* (middle life), and *Cenozoic* (modern life). In broad terms, the Precambrian was a time of no life or scant primitive life. The Paleozoic was a time in which invertebrates (snails, clams, and the like) and simple backboned animals, such as fish, amphibians, and primitive reptiles were ascendant. The Mesozoic was high noon for the reptiles, and it is the chapter in earth history when the hegemony of the dinosaurs was complete. The Cenozoic, which is the present era, is a time when mammals became dominant.

The eras are divided into lesser units called

periods. From their names, it obviously is hard to find a common pattern. Two have a familiar aspect to Americans—the Pennsylvanian and Mississippian—and we can use them to illustrate the philosophy behind the naming of such time units. Most periods are named for regions where rocks containing fossils characteristic of their segment of geologic time were found. This means that strata deposited in the later Paleozoic in an inland sea covering much of the eastern United States were named for their occurrence in Pennsylvania, just as strata approximately 20 million years older were named for their occurrence in the Mississippi valley—not the state. Incidentally, neither of the latter terms is used in Europe; there the two periods are treated as a single unit, the Carboniferous, which acquired its name from the coal-bearing strata of England.

Another time term whose place of origin can be recognized readily is the Devonian, named for the rocks that crop out along the southwestern tip of Great Britain in Devonshire and Cornwall. A less familiar period is the Permian—named after the ancient Kingdom of Permia in Russia.

Only the small number of persons with knowledge of Latin may recognize the ancient name of Wales, Cambria, as the basis for the naming of the Cambrian Period. Two closely related periods, the Ordovician and Silurian, perpetuate the memory of Stone Age tribes whose homes had been in Wales, the Ordovices in the north and the Silures in the south.

Some of the other periods are named not for places, but for the physical characteristics of their rocks. The Cretaceous is a good example, since it is derived from the Latin word *creta*, chalk, and refers to the exposures of chalky rocks in the cliffed coast of southern England. In fact, most of the chalk of the world is of Cretaceous age. Many other kinds of rocks, however, were also formed during the Cretaceous. In the United States, Cretaceous rocks run the gamut from conglomerate through sandstone and shale to limestone and coal; or, as in the Sierra Nevada, may include the enormous bodies of granite intruded during that period. The Triassic takes its name from the fact that in Germany the rocks of the period are divided into three distinctive layers—a limestone in the middle, with reddish sandstones and shales above and below.

The subdivisions of the Cenozoic Era are called *epochs*, and their rather strange names are a special case. They were established for the most part by Charles Lyell according to the percentage of extinction of marine molluscan fossils. Thus in Eocene (from the Greek words for dawn and recent) strata about 1 to 5 per cent of the species found are still living. In the Miocene (Greek words for less and recent) 20 to 40 per cent of the fossil species are still alive. In the Pleistocene (from the Greek words for most and recent) 90 to 100 per cent of the fossil shells are those of species living in the seaways of the world today. Without perhaps fully realizing it, Lyell had divided that later segment of geologic time on a statistical basis—one of earliest applications of statistics to a natural science.

The Pleistocene epoch is regarded by many geologists as being coincident with the ice age, and, although the length of time represented by the multiple advances and retreats of the ice sheets is unknown, it probably is close to 2 million years in length.

The most accepted way of dividing the Cenozoic Era is into the Tertiary (meaning third)—which includes all but the Pleistocene and post-glacial time—and Quaternary (meaning fourth). The division represents a sort of cultural lag, since we no longer speak of the Primary and Secondary rocks as our forebears did. With the first and second of a series gone it seems strange to speak of a third and fourth, but you will see the terms employed consistently in most geologic writing today; such is the force of habit.

Absolute time

The relative geologic time-scale provides us with a framework within which to view the events that have shaped the earth, but we cannot tell from it when a specific event actually took place. We know, for example, that the dinosaurs lived after the plants found in the coal beds in Pennsylvania; but how long afterward and for how long did they exist? Or, at what specific time or over what time span was the Sierra Nevada range formed? Earth scientists, as we have already learned, contrived in various ways to answer such questions, but up to the end of the nineteenth century no method proved reliable or easily workable. In 1895, however, the door was opened wide by the discovery of new substances with unusual chemical properties—a discovery that led to techniques that have since enabled us to reach the ultimate goal of absolute age-dating.

Few people alive in 1895 were aware of the eventual significance to the world that a seemingly innocuous experiment by French physicist Antoine Henri Becquerel (1852–1908) was to have. He had been intrigued by Roentgen's discovery within the same year of the X-ray, and wondered whether the phenomenon of phosphorescence bore any relationship to the strange rays. Various substances, he reasoned, might be able to pick up energy from the sun, and accordingly, he exposed a number of them to the sun's rays—with no discernible effect. By chance (serendipity has played an integral part in many scientific discoveries) he observed that when some salts of the heavy element, uranium, were placed on photographic paper, the paper darkened as though subjected to some sort of radiation, and that was true whether or not the uranium salts had been placed in the sun's rays beforehand. Becquerel's discovery was the first demonstration of the phenomenon known as *radioactivity*.

Among those first experimenting with radioactive substances was the distinguished physicist Lord Rutherford (1871–1937), born Ernest Rutherford in New Zealand. In 1902, he discovered that when unstable radioactive atoms disintegrated, transforming into completely different atoms, radiation was given off. During the process heat was also liberated. The process of radioactivity, then, has some of the essence of medieval science of alchemy, which attempted to change base metals to precious

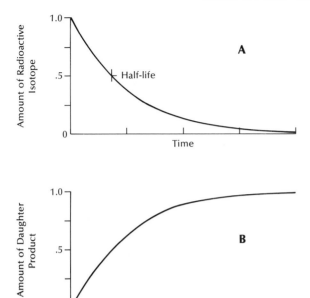

Fig. 1-13. Radioactive decay of a hypothetical isotope (A) and growth of the daughter product (B). The time required for one-half of the starting material to decay is called its *half-life* and is unique to each radioactive isotope.

ones. The radioactive substances, however, underwent natural spontaneous disintegration with the formation of different, and not at all precious, atomic species. Initial work by several people, including the French chemists Marie and Pierre Curie, indicated that both uranium and thorium were radioactive elements. During the early 1900s it was also found that several forms with the same elemental properties could exist, each differing only in their atomic weights. To those variants of the same element the radiochemist Frederick Soddy (1877–1956) gave the name *isotopes* (meaning same place). Eventually it was concluded that radioactive decay is a process that occurs within the nucleus of the atom and occurs in certain isotopes which, for fundamental reasons, are unstable. It was initially assumed, and later experimentally confirmed, that each radioactive substance breaks down at its own rate and that for many of them the rate of change is extremely slow, requiring, for some

substances, many millions or billions of years for completion of the process. Given a known quantity of a radioactive isotope (called the *parent*), the passage of time will bring about spontaneous disintegration of some of the parent, which will be replaced by an almost equivalent amount of decay product (called the *daughter*). The ratio of daughter to parent changes continually with time, and can, therefore, provide a measure of the total elapsed time. The time that it takes for a quantity of radioactive isotope to decay to one-half of its initial value is called its *half-life*. Each isotope has a characteristic half-life. The nature of radioactive breakdown is illustrated in Figure 1-13.

Rutherford recognized the importance of that result and in 1905, in a series of lectures delivered at Yale University, he proposed the use of radioactive decay as a means of absolutely dating rocks. There was not one clock but as many as there were usable radioactive substances. In 1906, he announced the first *radiometric date* ever obtained, from a mineral contained in a rock found in Connecticut—500 million years. The American chemist and physicist Bertram Boltwood (1870–1927) followed in 1907 with additional dates ranging from 410 to 2200 million years. A great many factors were not taken into account by the early age-daters, and many of the first dates were in error. The important thing is that a method had been discovered that held high promise of providing absolute ages of minerals and rocks, and that even the first dates determined by the method indicated that the earth was at least 2 billion years old.

The radioactive materials initially used in absolute dating were minerals containing three radioactive isotopes—uranium-238, uranium-235, and thorium-232. (The numbers indicate the atomic weight of the isotope.) Each of the three undergoes a long and involved process of nuclear disintegration, but all three reaction chains eventually produce an isotope of lead as the stable end-product. The gas helium is also produced during many of the intermediate breakdown steps. Many of the first dates were based on the ratio of radioactive uranium or thorium to helium; but later dates, which proved to be much more reliable, were based on the ratio of uranium or thorium to lead.

All told there are more than 1000 natural and artificial radioactive isotopes. Two naturally occurring isotopes, potassium-40 and rubidium-87, have also proved to be very useful in the absolute dating of rocks. Table 1-1 lists the radioactive isotopes commonly used in absolute dating and some of the pertinent breakdown data.

The business of extracting reliable dates from rocks and minerals is actually very complicated. It requires elaborate laboratory procedures and sophisticated equipment to make the incredibly precise measurements of the small amounts of parent and daughter isotopes. The work is carried out in relatively few laboratories, but they have now reached a level of performance such that little error is introduced into the dating from laboratory procedures. The principal source of error comes from the minerals and rocks themselves. If a mineral is to produce a reliable date it must have remained a closed system since its origin. What that means is that no parent or daughter materials have entered or left the mineral during its entire existence. It is not always easy to tell how closely the restriction has been met. One simple precaution that eliminates some of the uncertainty is to collect only fresh, unweathered rocks and to use no rock samples that appear chemically altered. Rocks dated by the potassium-argon method are particularly liable to loss of the daughter product, argon-40. It is a gas and as such is extremely difficult to keep trapped within the atomic latticework of the host mineral; it tends to diffuse slowly out of the mineral structure. And, as you might expect, if a rock is subjected to increased temperatures the likelihood of argon loss increases. Reduction in the amount of argon-40 leads to an underestimation of the age of the rock or mineral. If the temperature remains sufficiently high for a long enough period, all of the previously trapped argon will escape. In effect, the atomic clock is reset. Dating of such

Table 1-1		
Radioactive isotope	Daughter product	Half-life (years)
Uranium-235	Lead-207	710 million
Uranium-238	Lead-206	4.5 billion
Thorium-232	Lead-208	14.1 billion
Potassium-40	Argon-40	1.3 billion
Rubidium-87	Strontium-87	47.0 billion
Carbon-14	Nitrogen-14	5570

a rock will give the time of thermal resetting, which in many circumstances can be useful.

The best safeguard against hidden errors in the radiometric dates comes through the use of consistency checks. One obvious test is made by dating a mineral or rock by more than one radiometric method. For example, if both rubidium-strontium and potassium-argon methods are applied to the same rock and the results agree, the likelihood is high that the date is reliable. If the two dates are greatly different, then additional dating methods must be applied to determine which, if any, is the correct date. Another consistency test may be made through comparing the absolute and relative ages of rocks. For example, rocks that are known by physical correlation to be the same age should produce nearly identical dates.

Not all rocks can be dated by radiometric methods. First, the rock must have formed within a short time and have remained as a closed system thereafter. Datable rocks, then, are principally those which have formed as a result of igneous or metamorphic processes (see Chap. 4). The time of the deposition of sedimentary rocks, which commonly are made up of fragments of igneous and metamorphic rocks and other sedimentary rocks, cannot generally be radiometrically dated since any dating of the fragments would only reveal the age of the source rock—not the time when deposition occurred. It is possible, however, to date igneous rocks such as lavas or ash layers that are interlayered with sediments, or dikes that cut across sedimentary rocks. By using such methods, earth scientists can corre-

GEOLOGIC TIME			ABSOLUTE TIME*
ERA	**PERIOD**	**EPOCH**	
Cenozoic	Q Quaternary	Holocene	
		Pleistocene	
		Pliocene	—2—
		Miocene	—12—
	T Tertiary	Oligocene	—26—
		Eocene	—37–38—
		Paleocene	—53–54—
			—65—
Mesozoic	K Cretaceous	Late	
		Early	—136—
		Late	
	J Jurassic	Middle	
		Early	—190–195—
		Late	
	Ŗ Triassic	Middle	
		Early	—225—
Paleozoic	P Permian	Late	
		Early	—280—
	C Carbon-iferous ⟨ ₸P Pennsylvanian	Late	
		Middle	
		Early	
	M Mississippian	Late	
		Early	—345—
	D Devonian	Late	
		Middle	
		Early	—395—
	S Silurian	Late	
		Middle	
		Early	—430–440—
	O Ordovician	Late	
		Middle	
		Early	—500—
	Є Cambrian	Late	
		Middle	
		Early	—570—
Precambrian	PЄ	Late	
		Middle	
		Early	—3800+—

*Boundaries estimated radiometrically in millions of years before the present

Fig. 1-14. Geologic time scale, with boundaries between major divisions absolutely dated by radiometric methods.

late sedimentary rocks, particularly those containing index fossils, with the absolute time-scale.

Second, the rocks and minerals to be dated must initially have contained sufficient radioactive material to be measured. And last, enough time must have elapsed since the origin of the rock to enable measurable quantities of daughter products to accumulate.

Generally most rocks younger than about 250,000 to 500,000 years old cannot be dated.

Through the application of radiometric methods during the last fifty and particularly during the last twenty years, earth scientists have been able to tie down the geologic time scale, which was previously established on the basis of fossil data (Fig. 1-14). Not only that, but age-dating has enabled geologists to unravel the Precambrian rock record, in which fossil remains are scarce or nonexistent. It turns out that the Precambrian includes most of geologic time. Geologists are now able to determine accurately such things as rates of evolution and the time spans over which mountain building occurred.

The oldest dates obtained from the surface of the earth come from continental rocks and cluster around 3 to 3.2 billion years. Holding the distinction of being the very oldest—3.8 billion years—are rocks found in western Greenland. Some meteorites, however, have been dated at 4.5 billion years, and they are the most ancient materials ever found. Geologists consider the latter figure to be the best estimate of when our earth and solar system came into being—an estimate partially corroborated by dates as old as 4.2 billion years obtained from moon rocks gathered during the Apollo missions. As old as our solar system appears to be, its age falls far short of that of the Milky Way Galaxy (within which our solar system resides), which on the basis of astronomical considerations is estimated to be between 9 and 19 billion years old.

If we wish to reduce the immensity of geologic time to comprehensible dimensions, we can represent the age of the earth (4.5 billion years) by a single year. Then the record of abundant fossils extends back a scant 40 days or so; humans have been on the earth a matter of hours; and all of recorded history amounts to only a minute or so. Most of the year is taken up by the Precambrian Era. Interestingly enough, because the rock record is so poor and fossils are essentially nonexistent for that long span of time, we know the least about it. Without the advent of radiometric dating we would have known almost nothing.

Carbon-14 dating, dendrochronology, and lichenometry

One additional radioactive isotope used in age dating requires discussion. It has been left to last because its role in establishing the overall framework of the geologic time scale has been minimal. The isotope carbon 14 decays to nitrogen 14 with a half-life of only 5570 years. As such, it can be used effectively for obtaining dates only as far back as about 50,000 years.

That was a time, however, when human activities, which began slowly, progressed rapidly. By means of *carbon-14 dating* we are now able to place archeological traces of the earlier cultures in their correct historical perspective. Its use has also aided in the determination of details of geologic history during the later phases of the ice age.

The recognition of the existence of carbon-14 goes back to research in 1939 on cosmic rays in the upper atmosphere by the geophysicist Serge Korff. He noted that cosmic rays produce secondary neutrally charged particles (neutrons) during their initial collision with molecules of nitrogen gas in the higher levels of the atmosphere. Those neutrons, he predicted, when they collided with the abundant isotope, nitrogen-14, would react to free a positively charged particle (proton), forming the radioactive isotope carbon-14. Carbon-14 then would combine with oxygen to form carbon dioxide, which circulates in the atmosphere, is dissolved in the ocean waters, and is absorbed by plants and animals at the earth's surface. With time the carbon-14 decays back to nitrogen-14; however, new carbon-14 is continually being formed in the atmosphere so that an equilibrium level is maintained. As long as an organism lives it continually recharges its carbon-14 content; once it dies, however, the isotope that has accumulated in the body begins to decay, reaching an unmeasurable level in about 40,000 years. By measuring its radioactive level it is possible to determine how long an organism has been dead.

The basic assumption made by those who use the carbon-14 dating is that after an organism dies, it no longer assimilates carbon-14 from its surroundings. Yet some circumstances render that assumption invalid. For example, if the organic matter to be dated is buried, circulating ground water can provide a ready source of carbon-14, which then may enter into partial re-equilibration with the material. Or,

Fig. 1-15. Tree rings of a Douglas fir that began growing in A.D. 113 and was cut down in A.D. 1240. This specimen was used as a construction timber in a prehistoric structure south of Mesa Verde in southwestern Colorado. *Laboratory of Tree Ring Research, Univ. of Arizona*

grass roots, which can extend downward for two or more meters, may become so entangled with the organic matter that they cannot be completely separated from it. Because of such uncertainties and because of difficulties with the laboratory procedure any carbon-14 dates, but particularly those older than about 10,000 years, should be viewed with some reservation.

Carbon-14 dating (or *radiocarbon dating*, as the method is alternatively called), is also based on the assumption that the amount of carbon-14 produced in the atmosphere has remained relatively constant during the last 40,000 years. If variations in the production of the isotope did occur and lasted several hundreds or thousands of years, they would lead to systematic errors in the radiocarbon dates obtained from organisms that lived during those times.

In the early days of carbon-14 dating there was only a nagging suspicion that such might be the case, but there seemed no way to ob-

Fig. 1-16. Bristle-cone pine growing in the higher altitudes of California's White Mountains. Some bristle-cones in the area are more than 4000 years old.
V.C. LaMarche, Jr.

tain valid data to indicate the magnitude of possible errors. With the advent of tree-ring dating, or *dendrochronology,* a means for comparing carbon-14 dates with actual calendar dates became available. Each year that a tree lives, it normally adds one more growth layer to its circumference. In a cross section sawed from the tree the growth layers appear as rings, and they can be counted (Fig. 1-15). When the rings of very old trees were counted and measured, it became apparent that in some successions the rings were close together while in some they were wider apart. The spacing apparently reflects the times of slow and rapid growth in response to the climatic conditions prevailing at the time. The spacing of the tree rings provided a "signature," and all trees that grew during the same time span showed the same pattern of variation. Thus a key was provided for the matching of rings within a local

Fig. 1-17. Radiocarbon calendar age relationship as determined from carbon-14 dating of wood of a known calendar age.

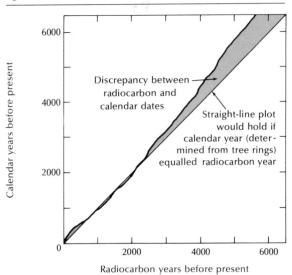

Discrepancy between radiocarbon and calendar dates

Straight-line plot would hold if calendar year (determined from tree rings) equalled radiocarbon year

Calendar years before present

Radiocarbon years before present

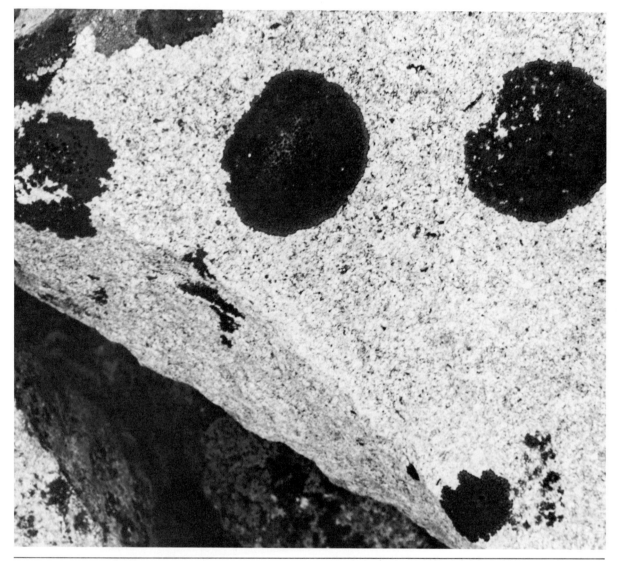

Fig. 1-18. Many of the numerous species of lichen are difficult to identify. These are *Lecidea atrobrunnea* growing on rocks in the mountains of southwestern Colorado. Their diameters (about 10 cm [4 in.]) suggest an age of between 900 and 1000 years. *Peter W. Birkeland*

geographic area of similar climate. By the counting and describing of patterns in successively older trees, a tree-ring master pattern could be compiled for a particular geographic area. Today, through the use of tree rings from the long-lived bristle-cone pine, one such tree-ring master template has been pieced together that extends back in time, year by year, from the present to about 7000 years ago (Fig. 1-16). Dendrochronology has provided an alternative method of dating archeological sites whenever logs and tree trunks can be found in association with cultural relics. In addition, by performing carbon-14 dating on the wood itself it is now possible to determine systematic variations in the radiocarbon dates (Fig. 1-17). The results indicate that from the present to about 500 years ago, and from 2100 to 7100 years ago, radiocarbon ages are younger than the calendar ages, whereas from 500 to 2100 years ago the two chronologies are essentially the same. For example, for dates near 7100 years ago, the error in the radiocarbon dates can amount to as much as 700 years. Most radiocarbon chronology labs now use that information to "correct" their carbon-14 dates.

Lichens are plants that can virtually coat a solid surface (Fig. 1-18), treasured by some because they enhance the beauty of rock used in fireplaces. They also can be used to date rock deposits, especially those lichens that are circular in form. The noted botanist Roland Beschel pioneered that unique age-dating technique when he noted that lichens on tombstones in Austria became progressively larger in diameter the older the tombstone. Other work has verified his finding (Fig. 1-19). In many environments for which the lichen growth-rate is known (Fig. 1-20), the plants can be used to date young surface deposits. Glacial, rockfall, or mudflow deposits in alpine settings commonly are dated in such a way, if they are less than several thousand years old.

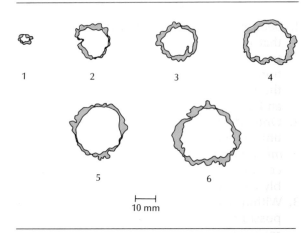

10 mm

Fig. 1-19. Tracings of a fast-growing lichen species (*Alectoria minuscula*) made over a four-year period (1963–67) in the eastern Canadian Arctic. Notice that the rounded shape is more or less maintained as size increases.

Fig. 1-20. Variation in growth rates with climate of the individuals of one lichen species.

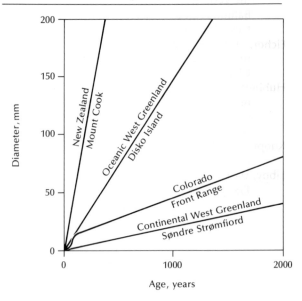

EPILOGUE

The pages of the earth's history are written in stone in a script that was unintelligible to our ancestors. The breaking of the code only became possible with the rise of geology as a science. As the record became clear it was apparent that the landscape we see today is only the last of a countless number that have formed and faded as forces and processes relentlessly worked on the earth. But only with the discovery of the atomic clocks, many of which have been ticking away since the earth's inception, have we been able to put order in the historical record and realize the immensity of geologic time.

SUMMARY

1. *Geology,* the study of the earth, is so broad that it is undertaken by scientists with extremely diverse backgrounds. The information that they supply is combined to provide the total picture of the earth, as it is today and as it was in the past.

2. One of the basic tenets that permits the unraveling of geologic history is *uniformitarianism,* which maintains that the processes we see in action today are probably the same as those of the past.

3. Within a local area of study it normally is possible to place the events recorded in the rocks into a *relative time* sequence on the basis of: *original horizontality* and *superposition* of beds, *cross-cutting relationships,* and determination of weathering and erosion surfaces.

4. Through the observation of *fossils* (the remains and traces of prehistoric life), which form an evolutionary succession, a *relative geologic time scale* was developed. From their fossil content it is possible, relatively, to date and correlate strata in widely separated areas, even on different continents.

5. The discovery of radioactivity in 1895 eventually led to methods that enabled *absolute dating* of minerals and rocks, and it was possible to tie the geologic time scale to an absolute time base. The oldest rocks yet found on earth are 3.8 billion years old, whereas meteorites date at about 4.5 billion years, which apparently marks the beginning of our solar system.

6. *Carbon-14 dating, tree-ring dating,* and *lichen dating* provide additional means of obtaining dates for very young geologic materials.

SELECTED REFERENCES

Adams, F. D., 1954, The birth and development of the geological sciences, Williams and Wilkins, Baltimore (Reprinted by Dover Publications, New York, 1954).

Eicher, D. L., 1976, Radiometric dating, *in* Geologic time, 2nd ed., Prentice-Hall, Englewood Cliffs, New Jersey, pp. 118–38.

Hubbert, M. King, 1967, Critique of the principle of uniformity, *in* Uniformity and simplicity, Geological Society of America Special Paper No. 89.

Knopf, A., 1949, Time and its mysteries; series 3, New York University Press, New York.

Libby, W. F., 1961, Radiocarbon dating, Science, vol. 133, pp. 621–29.

McIntyre, D. B., 1963, James Hutton and the philosophy of geology, *in* The fabric of geology, Claude C. Albritton, Jr., ed., Addison-Wesley, Reading, Massachusetts.

Roy, C. J., 1962, The sphere of the geological scientists, American Geological Institute, Washington, D.C.

Simpson, G. G., 1963, Historical science, *in* The fabric of geology, *see* McIntyre above.

Wendland, W. M., and Donley, D. C., 1971, Radiocarbon-calendar age relationships: earth and planetary science letters, vol. 11, pp. 135–39.

Fig. 2-1. The earth as seen from the Apollo 8 spacecraft, with the surface of the moon in the foreground. From that distance the earth appears generally blue in color. The streaky white swirls are large-scale cloud patterns. *NASA*

2

THE PLANET EARTH

The terrestrial planet, earth (Fig. 2-1), is one of nine solid bodies that circle endlessly in a plane around a relatively massive central star, our sun (Fig. 2-2). During the evening and early morning it is often possible to catch sight of nearby planetary neighbors as they travel through the lonely reaches of empty space. The existence of a solar system such as ours, astronomers tell us, most probably is not unique in our galaxy, the Milky Way. Of the nearly 10 billion stars in this single galaxy there are likely to be many thousands that possess planetary systems like ours. We can never hope to learn much about the nature of sister solar systems until we can venture into the galactic realm. The reality of intergalactic space travel lies far in the future, however, for even the closest star lies about four light years away (the distance traveled by light at a speed of 300,000 km/sec—186,000 mi/sec—during a four-year span).

The reason we cannot determine the existence of nearby solar systems telescopically is that planets are dark bodies—they do not emit light directly but only reflect that which falls upon them from the incredibly bright central stars. So the planets remain invisible even to our most powerful telescopes. Recently, how-

ever, the startling announcement was made that a nearby star, Barnards Star, has a planetary system consisting of at least two bodies, each about the size of our largest planet, Jupiter. Astronomers inferred the existence of the two large planets by observing oscillations in the position of the star itself. Yet, even if a planetary system does exist around Barnards Star, there is little chance we can learn more about it.

The only solar system about which we know a great deal is our own. Less than 15 years ago knowledge of our own system came mostly from direct observations of the earth, the study of meteorites that fell to earth, and telescopic observations of other planets. With the rise of space science, however, we have been able to augment that knowledge with direct observations on the moon (beginning in 1969) and data telemetered back to earth from unmanned space probes. Certainly the recent successful soft landings of two Viking mission spacecraft on Mars represent milestones in the accelerating effort to learn about the planets in our solar system. Data sent back from the spacecraft are helping us to answer some of the age-old questions, but they are also introducing new ones.

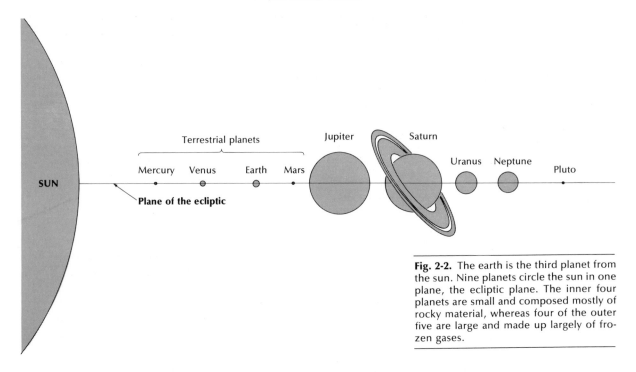

Fig. 2-2. The earth is the third planet from the sun. Nine planets circle the sun in one plane, the ecliptic plane. The inner four planets are small and composed mostly of rocky material, whereas four of the outer five are large and made up largely of frozen gases.

For some time to come, however, most of the information will have to come from the third planet.

The earth is in rough approximation a spherical body with a radius of about 6400 km (3977 mi) that spins once per day on its axis. The spin axis is inclined at about 23.5° with respect to its plane of orbit. Besides rotating, the earth circles once around the sun in about 365¼ days, at a mean distance of 149 million km (93 million mi). Because of the tilting of the rotational axis, the amount of radiation received from the sun generally is not divided equally north and south of the equator, and the result is the occurrence of seasonal variation.

Most of the features just mentioned are well known to us, and most of us accept them as true without a moment's hesitation. Yet human beings have not always known so much. Many centuries ago, when first they pondered the size and shape of the earth and its setting in the solar system, the "world" with which they were familiar amounted to only a small portion of the earth's surface. Adequate clocks

were not to be had, and even the simplest of surveying instruments were unheard of. Progress was slow at best; at times it came to a complete standstill. Commonly, following the wrong paths of investigation led to erroneous conclusions. Those who came after had to erase the errors and continue to push the frontiers of knowledge ahead. Those who made sizable contributions to our store of knowledge stand out like lighted lampposts along dark and lonely byways.

It is a good idea to recount some of the early and significant studies to see how scientists of those times, shackled by lack of equipment and workable techniques and confronted by the scientific and theologic dogma of their day, wrestled with the problems concerning the earth and its surroundings.

To most people living a few centuries before the birth of Christ, the earth's surface appeared flat. Yet a few observant men noticed that the shadow the earth cast on the moon during an eclipse appeared round; that when ships approached port the sails always came first into

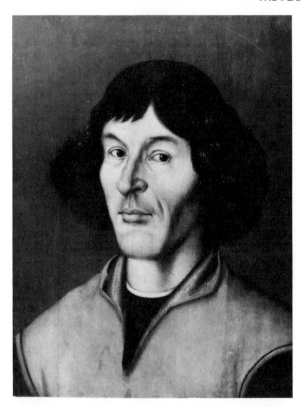

Fig. 2-3. Nicolaus Copernicus (1473–1543). Modern astronomy is based on the treatise Copernicus wrote about 1530 describing the sun as the center of a great planetary system. *Okregowe Museum, Toruń, Poland*

must be very large to appear the size that it does; so large, in fact, that it seemed more realistic for the earth to revolve around the sun than the other way around. Unfortunately, the views of Aristarchus were opposite to the earth-centered model of the solar system favored by Aristotle (384–322 B.C.), the dominant natural philosopher of his time and for some time thereafter.

Subsequently, the intellectual darkness of the Middle Ages spread across Europe and the Aristotelian views became ingrained. The scientific legacy of Eratosthenes, Aristarchus, and other scholars was all but forgotten.

By the sixteenth century a few people began to awaken from the intellectual stupor that had so long affected the Western World. Once again minds pondered questions about the earth and the solar system. Among the titans of the new wave was the Polish physician and astronomer, Nicolaus Copernicus (1473–1543) (Fig. 2-3), who also concluded that the simplest picture of the solar system was one with the sun at its center. Such a concept was completely contrary to the accepted dogma of the day, and it was not until shortly before his death that Copernicus summoned up enough courage to publish his results. The model he so imaginatively constructed has stood the test of time but was refined by other men of vision such as the Danish astronomer Tycho Brahe (1546–1601) and the German astronomer Johannes Kepler (1571–1630).

It was also in the late fifteenth and sixteenth century that Europeans developed a curiosity about the rest of the world and set sail across the oceans, some for commercial gain and others for glory of their countries. Perhaps the most heroic of the early adventures and the one that truly demonstrated the modern idea of the earth's size and shape was that of Ferdinand Magellan (1480?–1521) which began in 1519. After the reports of the seafaring men it was again accepted that the earth was round, and, with the development of new and improved surveying instruments, its dimensions eventually were measured.

Around 1650 the telescope, which had been

view; and that when one traveled relatively large distances north or south the stars in the night skies changed in position. The Greek astronomer Eratosthenes (c. 275–195 B.C.) not only concluded that the earth was round but on the basis of a very simple experiment made the first determination of its circumference—38,400 km (23,862 mi). Modern-day estimates are not greatly different from that figure. At about the same time another astronomer, Aristarchus, observed that the apparent position of the sun when viewed against the backdrop of the night sky did not change, no matter where the position of the observer. From that simple observation he concluded that the sun must be at a great distance from the earth and that it

invented in the days of Galileo (1564–1642), revealed that the two largest planets in the solar system (Jupiter and Saturn) were not exactly spherical but appeared to be flattened at the poles of rotation. Naturally it was asked if the same could be true of the earth. For a hundred years the question went unanswered while surveying groups from both England and France attempted to put the matter to rest. English scientists favored the concept of a flattened earth, basing their prejudice upon the theoretical consideration of the eminent Sir Isaac Newton (1642–1727)—one of the most important scientific figures who has ever lived (Fig. 2-4). He not only clearly stated the three fundamental laws of motion, originated calculus, and performed basic experiments on the nature of light, but was the first, in 1686, to provide a concise definition of the law of gravitation. That law states simply that for two bodies of mass m_1 and m_2, the force of attraction between them (F_g) is proportional to the product of the masses divided by the square of the distance between them at their centers of mass. Stated in mathematical shorthand:

$$F_g \propto \frac{m_1 m_2}{r^2}$$

In order to make the force uniquely determinable; that is, to substitute an equal sign for the proportionality sign, it is necessary to insert a constant—by convention G—in the equation. Then:

$$F_g = \frac{G m_1 m_2}{r^2}$$

What could be more simple? Here is a law that apparently applies to all particles in the cosmos, and yet it is tersely stated by means of only a few symbols. G in the equation is the gravitational constant; it is one of the fundamental properties of nature and is assumed to be the same everywhere in the universe. It is connected with the intrinsic properties of space itself in a way that is not yet understood. The equation is a description of the way gravity, as we know it, behaves; the law does not ex-

Fig. 2-4. Nineteenth-century engraving of the English physicist and philosopher Sir Isaac Newton (1642–1727), best known for his formulation of the law of gravitation. *New York Public Library*

plain the *cause* of gravitational attraction. At the moment, no one really knows why two masses are mutually attracted.

Gravitational forces are weak when compared to electromagnetic or electrostatic forces. When objects are as massive as the sun and earth, however, the resulting gravitational force is considerable. It is now quite clear, as it was to Newton, that the planets in the solar system are held in orbit by gravitational attraction between them and the sun. Similarly, the moons that circle around the planets are held in orbit by gravitational attraction. The earth has a sufficiently large mass that all objects are pulled toward its center and held to the surface by gravitational attraction. The nearly spherical shape of the earth is a result of the inward gravitational pull on all of the particles that make up the planetary body.

Newton noted that since the earth rotated on its axis it should behave like any other spinning body. If particles are to turn with the body and not fly off into space, an inward pull called a *centripetal* force (toward the center) must be exerted. Try swinging a rock at the end of a string—you will find that it is necessary to maintain an inward pull on the string or the rock will fly off in a straight-line trajectory. Because of the way forces behave, an outward, or *centrifugal*, force is also generated as a reaction to the inward pull. Newton reasoned that the centrifugal force generated in a spinning earth would be opposite to the inward, attractive force of gravity and that its effect would be therefore to lessen the pull of gravity. The effect should be greater near the equator since the distance to the spin axis is greater there than near the poles (Fig. 2-5). Because the earth's gravitational pull was most reduced near the equator the earth would tend to bulge outward in that region, resulting in a flattened shape at the poles.

By the mid-eighteenth century, field sur-veyors confirmed the idea that the earth was a flattened sphere. The shape corresponds to that which is generated by revolving (figuratively) an ellipse about its short axis and, accordingly, is termed an *ellipsoid of rotation*. The distance from the center to the poles was determined to be 6356.9 km (3950.2 mi) and that from the center to the equator, 6378.4 km (3963.5 mi). The ellipsoid of rotation of those proportions, defined as the *earth reference ellipsoid*, corresponds essentially to the surface of mean sea level. Since the reference ellipsoid has a regular definable shape, the earth's volume could be easily calculated and was found to be 1,024,000,000,000 km³ (246,000,000,000 mi³).

Although you may not realize it, the ellipsoidal shape of the earth provides some insight into the question of how strong the earth's rocks really are over long periods of time. Imagine for the moment that instead of solid rock material, the earth is composed of a liquid, such as water, which is capable of easily flowing and deforming. What shape would a "liquid earth," rotating on its axis once per day, take on? Surprisingly enough, it would be virtually the same as that of the actual earth—implying that over long periods of time the rocks inside the earth behave much as a liquid does, flowing and deforming in response to the forces that are applied.

In reality, the earth does not have a perfectly

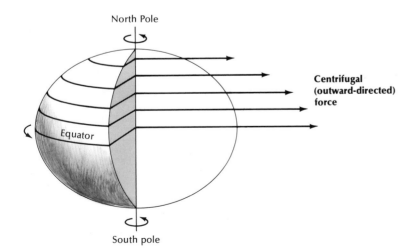

Fig. 2-5. Exaggerated ellipsoidal shape of the spinning earth. Centrifugal force is greatest at the equator, where distance from the spin axis is greatest, and decreases poleward.

North Pole

Centrifugal (outward-directed) force

Equator

South pole

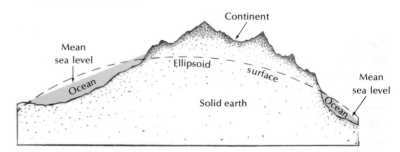

Fig. 2-6. Relationship of the earth's irregular shape to the surface of a reference ellipsoid, which provides a smoothed-out approximation of the earth's shape.

regular shape like the reference ellipsoid of rotation: there are mountains, continents, ocean basins, and such that spoil the regularity (Fig. 2-6). It is difficult to take all such irregularities into account, and geophysicists have finally agreed on a generalized shape in which the topographic highs and lows are represented but smoothed out and reduced. The form is called a *geoid* (meaning *earth shape*) and is unique to our planet. (In referring to Mars or the moon we might also speak of a "marsoid" or "moonoid.") The smoothed-out earth shape differs from the earth reference ellipsoid, but not by much, and the differences usually concern only mapmakers and geophysicists. A geoid may be defined as a surface upon which the attraction of gravity is everywhere the same, a surface upon which water cannot run downhill, for there are no inequalities of gravitational attraction. In the ocean basins the geoid corresponds to mean sea level. On land the surface can be imagined as that which would connect the mean water-surface levels in a set of narrow canals, crisscrossing the continents and connecting the seas (Fig. 2-7).

Since the dawn of the Space Age, countless satellites have been sent into orbit around the earth. Under the influence of the gravity field, which extends out into space, they circle the planet, and careful tracking of their paths has enabled scientists to determine the earth shape in even more precise detail. The results show that the earth is ever so slightly pear shaped; there are three bulges which protrude a maximum of 66 m (216 ft) above the reference ellipsoid. In 1500 Christopher Columbus stated that the earth had *la forma de una pera*. One can only speculate on the source of his idea, but to some degree it has proven true.

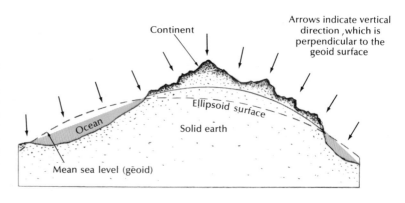

Fig. 2-7. Geoid surface, in relation to the reference ellipsoid and surface topography. In the oceans the geoid corresponds to mean sea level but under the continents, it rises above the ellipsoid surface and reflects the topography in a subdued way. The geoid is the best representation of a level surface which is everywhere perpendicular to the vertical direction.

MASS OF THE EARTH

Newton's law of gravitation ($F_g = Gm_1m_2/r^2$) can be used to describe the attractive force between any two masses. It can even be used to describe the gravitational attraction between a body of mass (m_1) at the earth's surface and the mass of the entire earth (m_E) itself. In everyday usage we term that attractive force the *weight* of the body. The mass of the earth acts as if it is concentrated at its center, at a distance of the earth's radius (R_E) away from the mass at the surface. The attractive force (F_g) between the earth and surface body can be given in mathematical shorthand as:

$$F_g = \frac{Gm_1m_E}{R_E{}^2}$$

In Newton's time, neither the mass of the earth nor the value of the universal gravity constant ("big G") was known. If in some way either one could be determined independently, then the other could be determined. The equation remained unsolved, however, until 1797 when another English scientist, Henry Cavendish (1737–1810), finally determined the gravitational constant in a simple, elegant experiment. (Unfortunately Newton did not live long enough to witness it.) Cavendish suspended from the ends of a balance arm two large lead spheres, each 30 cm (11.7 in.) in diameter. Next he fastened two smaller lead spheres, each 5 cm (1.95 in.) in diameter to the ends of a light, rigid rod which was hung from its center on a wire as thin as a cobweb. Placing the horizontally suspended rod between the large spheres, he waited until the system had come to rest and then moved the large spheres to a new position, closer to the smaller ones. Because the distance between the small and large spheres was shortened, the gravitational attraction between them increased, causing the rod assembly to rotate to a new rest position. Cavendish knew the elastic properties of the wire and was able to calculate the increase in gravitational force necessary to bring about the measured rotation of the rod assembly. Since he knew the force of attraction, the masses of

the spheres, and the distance between them, he could use Newton's law of gravitation to calculate "big G." In spite of its simplicity the apparatus was incredibly sensitive; the gravity force he measured was equivalent to the weight of about one ten-millionth of an ounce. The value of the gravitational constant he determined was not greatly different from that determined by modern methods. Once Cavendish knew "big G" he solved the Newtonian equation for the mass of the earth and found it to be about 6×10^{24} kg (the number 6 followed by 24 zeros), or 1.3×10^{25} lb. Since he knew the approximate volume of the earth he was able to calculate its density from the relation:

$$\text{Density} = \frac{\text{Mass}}{\text{Volume}}$$

The figure he came up with was 5.448 gm/cm³. As the methods for determining the mass and volume of the earth have improved the value has changed; today the accepted figure is 5.519 gm/cm³. In comparison the density of pure water is nearly one gm/cm³, so that the density of the earth as a whole is about 5.5 times the density of water.

The average density of rocks exposed at the surface of the continents is about 2.7 gm/cm³, or only about one-half that of the whole earth. Certainly some very dense material must exist inside the earth for the average density to be so high. Cavendish's experiment provided one of the first clues on the nature of the earth below the surface. Since his time quantitative information from other sources (see Chap. 17) indicates that some of the material near the earth's center is four to five times more dense than the average surface rock.

MEASUREMENTS OF GRAVITATIONAL ACCELERATION AND ISOSTASY

When an object is dropped, it starts its fall from the initial position and moves progressively more rapidly as it is pulled toward the center of the earth under the influence of gravity. The rate of change of the increasing velocity is called *gravitational acceleration* and is directly

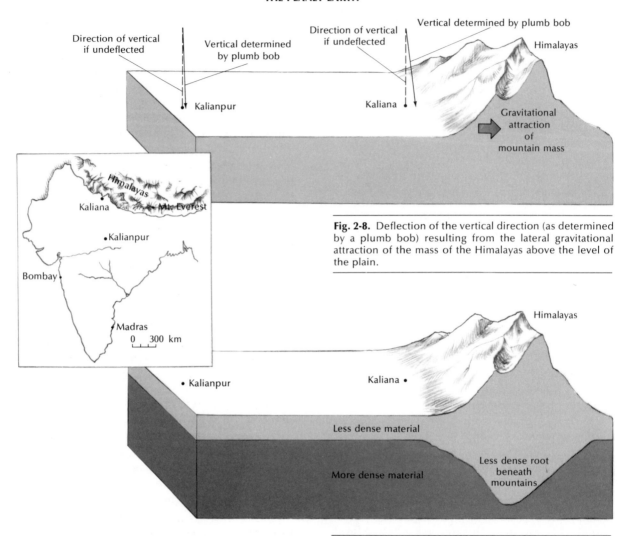

Fig. 2-8. Deflection of the vertical direction (as determined by a plumb bob) resulting from the lateral gravitational attraction of the mass of the Himalayas above the level of the plain.

Fig. 2-9. Airy's model, showing the root of the less-dense material which must exist beneath the Himalayas to account for the actual discrepancy of the vertical as measured at Kaliana and Kalianpur.

related to the force of gravity. In fact, the simplest way to measure the force of gravity at any spot on the earth's surface is to measure the gravitational acceleration, commonly denoted as *g*. And the simplest device for that purpose is a pendulum, which was first used by Galileo. The story is told that the idea of the pendulum came to Galileo one day while he was sitting in church. Rather than listen to the lengthy, turgid liturgy he watched the chandeliers sway to and

fro. The period of oscillation (the length of time required for one complete swing of the pendulum) depended upon two factors only: the acceleration of gravity at that location and the length of the pendulum. If either of the values could be measured, the remaining one could be calculated. In principle, then, by measuring the length of a pendulum and timing its period of oscillation, the force of gravity at that spot on the earth could be determined. But timing presented a problem, for in the late seventeenth century the most accurate clocks were based upon the pendulum, and the clock time varied with gravitational acceleration. Subsequently, as various kinds of precision chronometers came into use, the problem was alleviated, and the pendulum remained the principal means of determining acceleration of gravity until only recently. Today, even more accurate, simple-to-use instruments, called *gravity meters*, are used. The new instruments are so sensitive that they can detect differences in the acceleration between two stations less than 1 m (3 ft) apart in elevation. Gravity meters have been carted all over the earth and even to the moon. During the ascent of Mount Everest by a Japanese party several years ago, a geophysicist dragged (literally) a gravity meter nearly to the mountain top.

The results of the many measurements of g confirmed the idea that the shape of the earth was an ellipsoid of rotation. The acceleration of gravity is greater near the poles, where the distance to the center of the earth is less and where the centrifugal effect is nearly nil, than it is near the equator, where the earth's radius is greater and the centrifugal effect is more pronounced.

In addition to regular variations in g due to the rotation of an ellipsoidal earth there are variations related to irregularities in surface topography. A mountain peak like Mount Whitney in California is farther from the center of the earth than is the floor of a depression like Death Valley, the result being that readings of g on the peak are lower than those in the valley. It is possible to correct gravity-meter readings to take into account variations in altitude, and when such corrections are made, it becomes apparent that g at any one locality is also affected by the density of the rocks close below the surface. That density, of course, is closely related to the type of rock.

Earth scientists have found that in general the higher, continental parts of the earth are associated with rocks of lower average density whereas the lower oceanic portions are associated with rocks of higher average density.

You will recall from the discussion concerning the general shape of the earth that the rocks of our planet are not very strong over long periods of time. Rather, they flow and deform much as a liquid does. A continent of relatively low density or a mountain range rising high above its surroundings, therefore, could not be supported for any length of time at the earth's surface since the underlying rocks lack the strength to do so. How are they supported, then? The situation must resemble that of a low-density iceberg floating in relatively high-density water. If an iceberg juts far above the surface of the water it must also reach even farther below the surface. It is reasonable to expect that mountains and continents, too, extend far downward below the surface, like roots beneath a tree. And gravity data suggest that they do.

The first indication that mountains might have roots came from a precise mapping survey in India in the mid-nineteenth century, directed by Sir George Everest (1790–1866), for whom the highest mountain on earth was named. During the survey the distance between two stations, one at Kaliana and the other at Kalianpur, about 600 km (373 mi) apart along a north-south line, was measured with great precision (Fig. 2-8). Kaliana was situated close to the base of the Himalayas whereas Kalianpur lay out on the plains. Two different methods were used to determine the separation. One was the standard surveying technique known as *triangulation*, which is based on the construction of intersecting lines and involves the use of precision surveying instruments. The other was an astronomic method in which elevation angles measured from the horizontal to the polestar (North Star) were used to calculate earth distances. Although both methods were painstakingly carried out the results differed by about 153 m (502 ft). The surveys were rerun but the discrepancy still existed. John Henry Pratt (d. 1871), an English geophysicist, realized why the results were different. The problem as he saw it was related to the basic technique used in the astronomic method. In order to measure an elevation angle to a star a surveyor must first determine the horizontal plane. But if the horizontal planes at the two different stations are not the same, an error in the measured distance could be introduced. Only a very slight angular difference (5 sec of arc) was needed to produce the discrepancy in distance between Kaliana and Kalianpur. In his analysis of the problem, Pratt considered the vertical direction at each station using a plumb bob (a weight hung on a cord) rather than the horizontal (Fig. 2-8). If the horizontal planes between the two sites were askew, so must be the vertical directions. Pratt reasoned that the Himalayas could behave as a separate mass (like one of the large spheres in the Cavendish experiment), exerting a lateral gravitational force on the plumb bobs at the two stations and pulling the weights toward the mountains (Fig. 2-8). The amount of deflection, of course, would be greater at the station closest to the mountain. Pratt was quite sure he had solved the problem, and to demonstrate the correctness of his approach he calculated the gravitational effect that the Himalayas would have at each of the two stations. In order to do so he first estimated the mass of the mountain block, based on its size and density. Much to his chagrin, when he had finished his computations he found that instead of 5 sec of arc, the calculated discrepancy amounted to 16 sec of arc. Apparently he was on the right track but had failed to consider some other factor. Pratt concluded that the source of his error had to be in his estimate of the mass of the Himalayan block; it must contain less than he had originally thought.

While Pratt continued to cogitate, Sir George Airy (1801–92), a noted English astronomer, also became interested in the problem. He suggested that all blocks at the earth's surface are relatively low in density and that they "float" on top of denser rocks below. Lower-density mountain blocks that stand high must extend far downward into the denser material (Fig. 2-9). The part of the Himalayan block rising vertically above the surface of the plain is much denser than the surrounding air and

would exert a lateral gravitational attraction on a plumb bob at any one station, deflecting it *toward* the mountains. Also exerting a gravitational pull, however, is the root of the mountain block. Because the root is somewhat less dense than the surrounding subsurface rocks, the plumb bob would tend to be deflected *away* from the mountains. At Kaliana and Kalianpur the gravitational effect of the mountain mass was greater than the effect from the root below, so that the plumb bob was deflected toward the mountain but at a smaller angle than Pratt's simple model had allowed him to calculate. At about the time that Airy formulated his hypothesis, Pratt came to nearly the same conclusion.

Since the time of Pratt and Airy, the idea that roots exist beneath mountains and continents has been tested over and over again and has been shown to be valid. Conclusive evidence has come from the beaming of sound waves downward through the outer layers of the earth. Upon encountering the base of the continent or mountain root, the waves are reflected back to the surface. Geophysicists know how fast the waves travel in the outer earth, and by timing how long it takes for their return, they can determine the depth of the base of a continent or a mountain mass. Under the Sierra Nevada range in California, for example, the root extends downward for as much as 60 km (37 mi).

Pratt and Airy's theory—which stated that the blocks at the earth's surface are in floating gravitational equilibrium—was extremely significant to geology. To the concept has been given the name *isostasy* (from the Greek *isos*, equal and *stasis*, standing still). If the equilibrium is disturbed, the blocks will move up or down, and the denser material at depth will slowly deform and flow to enable readjustment of the blocks above. Isostatic readjustment can occur in several ways. Before looking at some of them, let us again consider the closely analogous system of ice in water. If ice is heaped on top of a floating ice cube, the entire mass will sink until it comes to a new rest position—the top will actually be higher

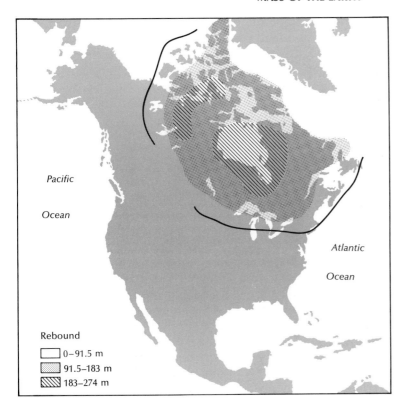

Fig. 2-10. North America, showing rebounding in the northeast after the rapid melting of the ice in the last glaciation. The regions of maximum uplift correspond to areas of maximum ice thickness.

Rebound
- ☐ 0–91.5 m
- ▨ 91.5–183 m
- ▧ 183–274 m

Pacific

Ocean

Atlantic

Ocean

than before and the root will extend deeper. If, on the other hand, ice is shaved from the top of the cube, it will rise to a new equilibrium position, but the top will not be quite so high as before, and the root will not project so far downward. The same is generally true of the earth. If material is loaded on a block at the earth's surface, that block will sink to some degree; if material is removed, the block will rise until a new position of isostatic equilibrium is reached.

During the ice ages of the recent past, parts of the continents were loaded with great masses of ice, up to about 3000 m (9840 ft) thick (see Chap. 12). The earth's surface was pushed downward in response to the loading. Today most of that ice has melted away, and most of the former ice-laden areas have risen and will continue to rise until isostatic equilibrium is re-established. As shown in Figure 2-10, the upward rebound in northeastern North America

has already amounted to as much as 275 m (902 ft). The rebound is still progressing, but its rate is tapering off. The last time the area was ice covered was about 18,000 years ago. Yet the large amount of rebound that has already taken place shows how quickly isostatic readjustment can occur in some areas.

High mountains are subject to high rates of erosion. As material is carved from a mountain top, the block undergoes continual isostatic uplift. Because of isostatic recovery, it takes a considerable length of time, perhaps several tens of millions of years, to wear down a high mountain range.

ATMOSPHERE, OCEANS, AND ROCKY SUBSTRATE

The features that stand out most noticeably when one looks at the earth from space are the oceans and the continents. Above the entire earth surface, water and land, circulates the tenuous invisible gaseous envelope we call the *atmosphere*. Yet the ocean waters and atmosphere together comprise only a small fraction of the total mass of the earth. The bulk is made up of higher-density material which lies beneath the continental land surface and the bottom of the ocean and extends all the way to the center of the earth—a zone we shall call the *rocky substrate*.

Atmosphere

Many of the beauties of nature result from the behavior of the atmosphere—skies of deep blue, sunsets blazing in lurid shades of red and gold, multicolored rainbows after summer showers, the skies in tempest. But it is the more mundane aspects of the gaseous envelope that concern us now.

The atmosphere is composed mostly of nitrogen (78 per cent by volume) and oxygen (21 per cent by volume). That leaves little room for the other gases, of which the most abundant are argon, carbon dioxide, and water. The lower boundary of the atmosphere is the sea and land surface, but there is no sharp upper boundary. The air close to the earth's surface, compressed under the weight of the air above it, is much more dense than that higher up. Nearly all lies within 96 km (60 mi) of the earth; above that height the amount is less than that in the best artificial vacuum we can make in the laboratory. Above a height of 960 km (597 mi) the atmosphere consists mostly of hydrogen and helium, and above 2400 km (1491 mi), only the small, fast molecules of hydrogen are found. As the earth moves through space, some of the light, active gases (mostly helium and hydrogen) are lost, left behind to become part of the interplanetary gas mixture.

Although the atmosphere close to the earth is composed largely of nitrogen, it is the presence of oxygen, carbon dioxide, and water vapor that most interests geologists. Most organisms cannot survive without oxygen, and carbon dioxide is vital to the metabolic func-

tion of plants. Water, which occurs in only relatively small amounts in the atmosphere, has several vital roles: it is essential to most plants and animals, it is the prime absorber of radiant heat in the atmosphere, and it provides the principal substance involved in processes of weathering and erosion.

The earth's atmosphere is unlike that of any other planet in our solar system. It appears to have evolved to its present state through geologic time through the slow action of many processes, among them the loss into space of many of the less dense gases, the addition of organic metabolic gases and gaseous emanations associated with volcanic activity, and the accumulation of gases such as argon and helium, which are produced as a result of the continued disintegration of commonly occurring radioactive isotopes.

The atmosphere is in continual agitation and flow, as is quite evident from global weather patterns. The driving force for the circulation of the air is in one way or another related to the energy received from the sun, 149 million km (93 million mi) away.

Fig. 2-11. The earth's heat balance in simplified form. The equatorial regions receive more heat than the polar regions, leading to thermal instabilities and the transport of heat by atmospheric flow.

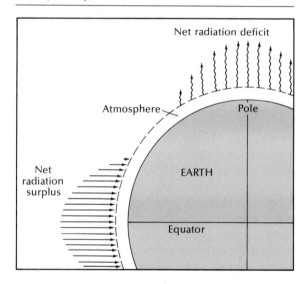

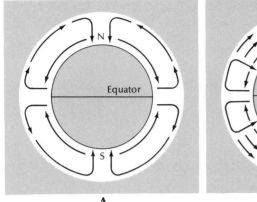 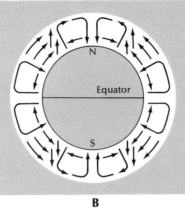 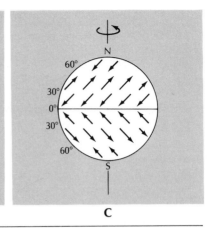

Fig. 2-12. Air flow patterns on a model earth. Cross section of flow on (**A**) a non-rotating earth and (**B**) a rotating earth. The surface wind pattern that would develop on a rotating earth is shown in (**C**).

Because of the spherical shape of the earth, the light that radiates from the sun is more concentrated in the equatorial parts of the world than in the higher latitudes and polar regions (Fig. 2-11). Thus, above the tropics the atmosphere is warmer than it is over the poles. Figure 2-12 shows patterns of air flow on a model earth. In response to inequalities in the atmospheric heat budget, hot air moves toward the polar regions at high altitudes while denser, colder air flows toward the equator at lower altitudes (Fig. 2-12A). Complicating that idealized pattern is the spinning of the earth on its axis: air begins to follow a curved path (deflected, apparently, by the Coriolis force) toward the right in the Northern Hemisphere and toward the left in the Southern Hemisphere. Circulation is short-circuited into the smaller air cells shown in cross section in Figure 2-12B. Figure 2-12C shows the surface wind pattern that would result.

Amazingly enough, the actual wind pattern on the earth closely conforms to the idealized pattern shown in Figure 2-12C. Differences between the two are the result of non-uniform distribution of the oceans and continents, topographic irregularities, and seasonal variations.

Oceans

The ocean, which covers about 71 per cent of the earth's surface, is a many-faceted realm. It has long provided a source of food and an avenue of transportation for peoples around the world. It possesses ever-changing aesthetic qualities that have captivated the imaginations of humans since time immemorial. Life, as far as we know, began in a simple fashion in the dilute but sustaining soup of the Precambrian seas. Today, after more than 2 billion years of evolution, life forms exist in all parts of the ocean basins in a bewildering array of life styles and complexity.

One of the most important functions of the oceans is to supply water to the atmosphere, through evaporation driven by the sun's rays. Water vapor is carried by prevailing wind systems far inland, where it precipitates to recharge the surface and ground waters and to provide nourishment for plants and animals. Chemical weathering of rocks would be minimal except for the moisture that lingers at the earth's surface after the rest has flowed away. And runoff of the excess water is without a doubt the prime agent involved in the erosion of the land.

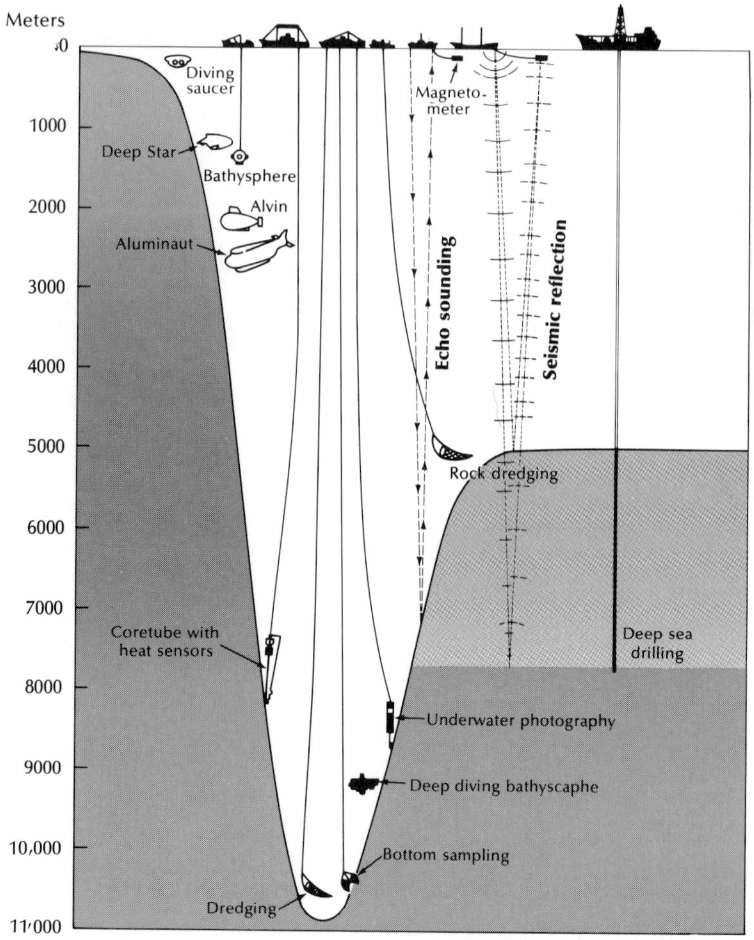

THE PLANET EARTH

Meters

0

1000 — Diving saucer

Deep Star
Bathysphere

2000 — Alvin

Aluminaut

3000

Echo sounding — Seismic reflection

4000

5000 — Rock dredging

6000

7000

Coretube with heat sensors

8000 — Underwater photography — Deep sea drilling

9000 — Deep diving bathyscaphe

Magnetometer

10,000 — Bottom sampling

Dredging

11,000

Fig. 2-13. Methods of sampling the ocean's waters and rock. Oceanographic exploration has become a highly technical field.

Scientifically, the water realm is several worlds wrapped into one. We can study, for example, the chemistry of water, the dynamics of current and wave action, the life forms, large and small, that exist there, or the geologic nature of the rocky basin that underlies the ocean. All are part of the field of *oceanography*, which began in earnest with the voyage of the British warship HMS *Challenger* in 1872. Before then, most of the lore of the oceans was collected from the tales of seafarers and consisted of myths and mysteries, salted with a dash of fact. During its three-and-a-half-year cruise the *Challenger* sailed more than 110,000 km (68,354 mi) in most of the major oceans and accumulated reams of data on their phys-

ical, chemical, biologic, and geologic aspects.

It was not until 1930 that centers for oceanography began to appear in the United States, and the rest of the world was just about as slow in the development of sea studies. Because of a lack of oceanographic vessels up until 10 or 15 years ago, most geological information came from the continents. We knew almost nothing about the geology over 70 per cent of the globe. Now a fair number of oceanographic institutes exist around the world, including within the United States such outstanding centers as Woods Hole Oceanographic Institution in Massachusetts, the Scripps Institute of Oceanography in California, and the Lamont-Doherty Geological Observatory in New York. Equipped with the most modern equipment, today's oceanographers are adding a great deal to our knowledge of the ocean world; knowledge which has led us to reconsider some of the earlier ideas concerning the earth's large-scale geologic features.

Because of their inaccessibility, the ocean depths remained virtually unexplored until modern technology provided the necessary sophisticated (and commonly expensive) equipment (Fig. 2-13). A single research vessel may take water samples from various depths, measure temperatures and current velocities, and collect samples from the ocean bottom by dredging or coring (Fig. 2-14). The sea floor can be examined visually by means of TV cameras, photographic cameras, or deep-diving submersibles (Fig. 2-15). One of the larger submersibles built in the United States has reached the incredible depth of about 10,900 m (35,752 ft). A number of the smaller submersibles, each carrying two to four people, can dive 600 to 2400 m (1968 to 7872 ft). Most contain windows for direct observation and are equipped with manipulator arms for sampling.

Fig. 2-14. Preparing to lower a piston corer from the research vessel *Atlantis*. Weights push the hollow coring tube as much as 30 m (98 ft) into the soft sediments of the ocean floor.
Woods Hole Oceanographic Institution

Fig. 2-15. A crane swings the research submersible *Deepstar* over the stern of the mother ship before a deep dive. The vessel carries three persons and operates at a maximum depth of 1220 m (4001 ft).
Naval Electronic Laboratory

Virtually all oceanographic ships carry an *echo sounder,* a sonic instrument that records ocean depths. The device operates by creating a loud noise or "ping" under water; the sound travels to the sea floor and is reflected back to the ship. Since the speed of sound in water is almost constant, the time it takes a sound impulse to journey to the sea floor and back is directly related to the distance to the bottom. On most deep-sea expeditions the echo sounder pings about every one-quarter second, resulting in a continuous visible recording of the bottom profile as the ship steams along. In 1969 Henry W. Menard, a well-known American oceanographer, described it as follows:

We cruised over hills and low mountains a mile and a half below and visible only on the echo-sounder. However, after a while the distance between the echo-sounder and the bottom is forgotten. As the marine geologist surveys a new range of undersea mountains, he senses them around him. I was once surveying with a captain new to the game. He remarked, as we headed again toward a peak on the map we were making, that he could not suppress a captain's feeling that we would hit it, even though he knew perfectly well it was a mile below the ship.

If the sonic energy of the "ping" is increased, some of the sound waves penetrate the ocean floor and are reflected back from the rocks and sediment layers beneath (Fig. 2-16). Marine geologists are able to "see" not only the shape of the bottom, but also the nature of the rocks below the surface. The process is called *sub-bottom profiling* and is quite enough to make a land-based geologist sigh with envy. With every track of the ship, another continuous cross section of the rocks below the ocean bottom is obtained. On land, geologists can see only the rocks on the surface and must imagine what they would look like in cross section. In contrast, marine geologists have cross sections provided for them by sub-bottom profiling but are called upon to imagine the map of the sea floor.

Until only recently, ships obtained information on rocks many tens of kilometers below the ocean bottom by detonating high explosives. Generally two ships, separated by several kilometers, were used in the operation, one to drop the depth charges and the other to "listen" to the sound waves as they bounce back to the surface. Today, because of environmental restrictions, the use of explosives in the sea has been curtailed severely. Most ships now produce an alternate signal source, by means of an air gun or a sparker system. Waves from such sources can penetrate a maximum of about 10 km (6 mi) beneath the ocean floor. Gravity meters and magnetometers, for measurement of the earth's gravity and magnetic fields, respectively, are also carried on many geophysical research ships.

One of the greatest advances in oceanography began in the early 1960s when a shipboard rig actually drilled into the ocean bottom to a depth of 180 m (590 ft). Many technical problems had to be surmounted before that first probe of the deep-sea floor. Imagine a ship rolling and pitching in rough seas while the end of a drill bites into rocks that lie perhaps 3000 m (9843 ft) below. The vessel must remain directly over the drill hole during the entire operation, a feat that is accomplished with a system of motors that push the ship this way and that in response to signals received from three sonar buoys attached, in triangular array, to the sea floor. No ship could possibly be anchored in waters thousands of meters deep. The method works so well that it is possible for a drilling crew to pull an entire string—which might be compared to a 3000 m (9840 ft) piece of limp spaghetti—out of the hole, change the drilling bit, re-enter the same hole, and resume drilling.

The premier drilling ship in service today is the *Glomar Challenger,* a U.S. research vessel that has voyaged to all parts of the world sampling rock material. It was designed to drill in water depths up to 6000 m (19,680 ft) and to obtain samples from 750 m (2460 ft) below the ocean bottom. It has been in nearly constant service since it first began operation in 1968. The core samples carrying a record of the ocean basins have been studied by all kinds of earth scientists, and the results of the studies have been geologically significant, sometimes startlingly so. For example, the record of the rocks beneath the ocean floor does not go farther back in age than the Jurassic Period (about 150 million years), and in most parts of the oceans, it does not go beyond the Cretaceous Period (about 135 million years). It appears, then, that the ocean basins are relatively young, whereas the continents are at least 2 billion years old. We will consider that unusual result in more detail in Chapter 19, "Continental Drift and Plate Tectonics."

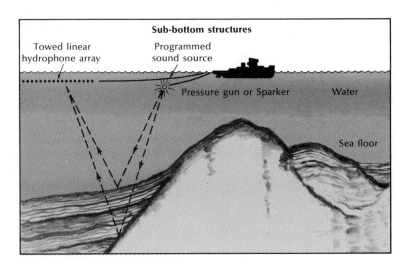

Sub-bottom structures

Towed linear hydrophone array

Programmed sound source

Pressure gun or Sparker

Water

Sea floor

Fig. 2-16. The method of sub-bottom profiling, with the actual profile superimposed. Strong sound signals penetrate rock layers beneath the ocean floor and are reflected back to the surface and the sound-wave receiver.

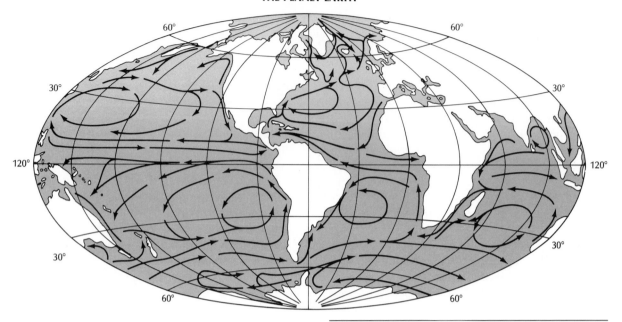

Fig. 2-17. Generalized world pattern of surface oceanic currents and gyres. The pattern in each part of an ocean basin is unique.

Sea water, as everyone knows, is salty. In fact, 100 lb (45 kg) of average sea water, when evaporated, leaves about 3.5 lb (1.6 kg) of salt. The word *salt* is a general chemical term and includes many compounds. The salt most familiar to us is table salt, *sodium chloride*. The base elements are sodium (a metal) and chlorine (a noxious gas). The salt compound, although comprised of the two elements, does not resemble either of them. During the formation of the salt the sodium atom loses a negatively charged particle (electron) and the chlorine atom gains one. The resultant atoms now carry a charge and are called *ions*. When table salt is dissolved in water the two ions split up and behave more or less independently.

The principal ions dissolved in sea water, making up more than 85 per cent of the dissolved material, are those of sodium and chlorine. The rest of the dissolved load is made up chiefly of the ions of sulfate, magnesium, calcium, potassium, bicarbonate, and bromine, in descending order of abundance. Traces of other salt ions, many of them necessary for the life processes of the marine organisms, are also present. Like the atmosphere, the chemical character of the oceans appears to have changed over long periods of geologic time. Much, if not most, of the water itself probably came from deep inside the earth. It was released as steam, along with molten rock, during volcanic eruptions that occurred down through the ages. For the most part, the complex array of salts in the ocean represents materials derived from the land through the process of chemical weathering. Some of the salts are used by organisms in constructing their shells, some are precipitated out of solution and fall to the ocean floor. At a few times in the geologic past, large-scale evaporation of sea water in restricted areas has resulted in the local precipitation of hundreds of meters of calcium sulfate, sodium chloride, and, to a lesser extent, potassium salts.

The waters of the ocean are constantly in

Fig. 2-18. Short-crested ripples on the ocean floor photographed at depths of 3895 m (12,776 ft) in the Drake passage (left) and 867 m (2844) ft) on the Blake Plateau (below). *National Science Foundation and Bruce C. Heezen*

motion. The currents, in some cases, are widespread and feeble; in others, they are concentrated and strong. The action of currents is not restricted to the surface layers. Flow patterns also exist at the deeper levels, patterns which bring about vertical as well as lateral transport of the ocean waters. Commonly, deeper currents are completely unlike those at the surface.

Surface flow is dominated by the prevailing winds. Blowing winds push at the surface of the water, eventually creating a slow-moving ocean drift. Like atmospheric currents, water currents are also affected by the earth's rotation; as they flow they, too, continue to curve until large rotating current cells—called *gyres*—are developed (Fig. 2-17).

Most of the deeper currents come about from differences in water density. Heavier water tends to sink whereas lighter water tends to rise. Basically, two things control the density of sea water: temperature and salinity. The colder or saltier the water, the greater the density. Water temperature is determined by climate, whereas salinity is determined by the relative effects of evaporation and precipitation. Cold, dense polar waters tend to sink and move at depth toward the equator; alternatively, warm, lighter waters in the equatorial regions move near the surface poleward.

It is usually assumed that deep currents are slow-moving ones. Recently, however, photographic evidence has indicated that currents moving at depths up to 7318 m (24,003 ft) have been, in some places, rapid enough to produce ripple marks in the bottom sediment (Fig. 2-18).

Rocky substrate

The earth beneath the ocean waters and land surface, all the way to the center of the globe, is composed of a higher-density, mostly solid material we shall call the rocky substrate. Our best information about the rocky substrate comes from the rocks at the surface. Drilling on the land and sea has enabled us to make direct observations of earth materials to depths of about 9150 m (30,012 ft), and analysis of the products of volcanism gives us some idea of the temperature gradient and rock material down to about 200–250 km (124–155 mi). For knowledge of the rocky substrate beyond those depths, we rely on geophysical studies—most of them from interpretations of the waves that emanate in all directions from an earthquake (see Chap. 17). Some useful data, however, have also been obtained from studies of the magnetic and gravitational fields at the earth's surface and of the regional loss of heat from the earth's interior.

All the data suggest that the earth is grossly layered in density, with rocks near the center being four to five times more dense than those at the outside (Fig. 2-19). The greatest variety of rocks, which average only 2.7 gm/cm^3 in density, are found on the continents and extend beneath them to maximum depths of around 70 km (44 mi). By contrast, the rocks just below the oceans average about 3.0 gm/cm^3 in density and project downward only 5–6 km (3–4 mi). Because of the greater thickness of lower-density rocks in the continents, they float higher isostatically than do the ocean basins. Geophysicists have concluded that the rocks down to about 2900 km (1802 mi) are rich in silicon and oxygen with lesser amounts of iron, magnesium, aluminum, calcium, sodium, and potassium. Below that depth is the earth's *core*, composed of, it is thought, nearly pure iron (possibly with smaller amounts of silicon, nickel, and sulfur)—some in a liquid state and some solid. The magnetic field of the earth appears to be generated in the liquid parts of the core (see Chap. 18).

Exactly how the earth came to be layered will probably forever remain a mystery. Some geophysicists consider the density stratification to have occurred quickly, as a result of the complete melting of the whole proto earth shortly after it came into being in the solar system. In the molten state the heavier materials could, under the influence of gravity, easily make their way toward the center of the earth, and the lighter ones would tend to move toward the outside. Some of the heat needed to produce melting could have come from decay of radio-

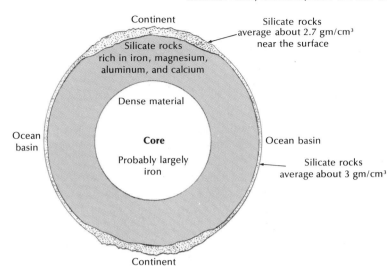

Fig. 2-19. Schematic cross section of a three-layered earth as determined by direct and indirect methods.

active isotopes which were more abundant initially, and some may have been released during gravitational compression of an initially larger, low-density proto earth into the smaller, more-dense earth that we know. Yet many earth scientists feel that such heat sources would have been insufficient to bring about complete melting of the earth. Alternatively, if temperatures became high enough to melt only the iron, then metal droplets could slowly make their way to the center of the earth, and the density-stratification process would be a slow-acting one. Whatever the process, evidence indicates that by about 3 billion years ago a molten core already existed that was large enough to produce a magnetic field.

The part of the rocky substrate most important to humans is that near the surface of the continents. From it we derive materials to construct our buildings and fossil fuels to keep civilization moving. Soils are formed from decayed rock material. The mountains and deep canyons we so cherish and in fact all of the landforms on earth are carved or modeled from rocks.

Many of the earth phenomena that we see (or feel), such as volcanism, mountain building, lateral movement of continental blocks, and earthquakes are the result of processes working far below the surface. Most of them appear to be particularly active at depths less than 250 km (155 mi), but some activity extends as deep as 700 km (435 mi). Many of the internal processes can be considered to be *constructional;* that is, the earth's surface is built as a result of such processes. The internally generated constructional process is discussed in a number of the succeeding chapters.

The earth's surface, however, is also subject to *destructive* activities, such as weathering, erosion, and deposition of sediments, that work at or close to the surface and tend to reduce the high spots and to fill in the low ones. Many of the subsequent chapters will discuss surface processes and how they go about the work of leveling the earth.

The face of the earth has changed continuously down through the geologic ages in response to the interaction between constructional and destructional processes. When land-leveling dominates for long periods, as for example during much of late Precambrian and early Paleozoic time, the continents are reduced to broad, low-relief features that barely rise above sea level. When, however, construction becomes dominant, as it has recently, the continental masses are built up until they stand, studded with lofty mountain peaks, high above the oceans.

The principal energy source that powers the

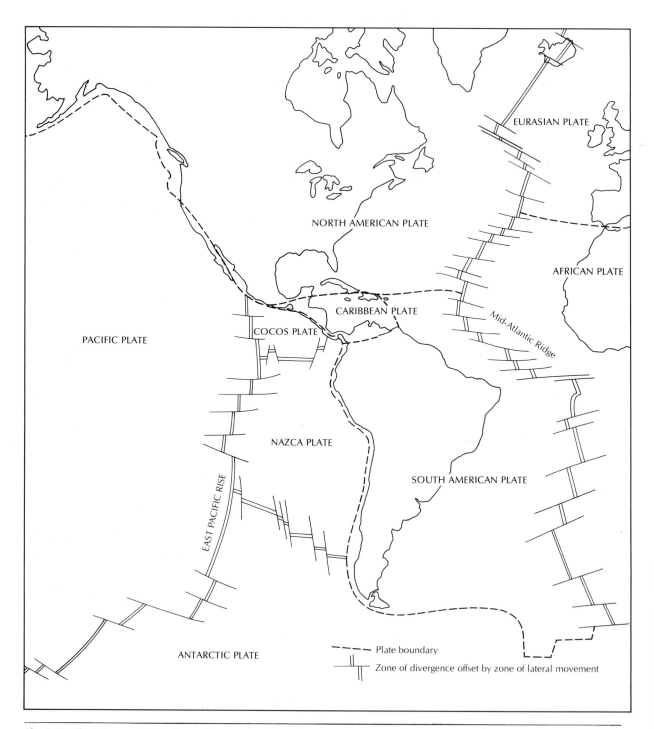

Fig. 2-20. The Western Hemisphere, showing the formally named rigid plates, bounded by zones of divergence, lateral movement, and convergence. According to the plate-tectonic model, essentially all of the internally generated phenomena are brought about by plate-margin interaction.

surface processes on earth, including life, is the sun. Our planet receives only a tiny fraction of the sun's radiant energy, and yet it has been enough to drive the surface destructive activities throughout geologic time. There seems little likelihood that the solar power source will fail for several billion years to come.

The source of energy that drives the internal constructive processes is not so easily defined, although most earth scientists consider it to be thermal. Some of the thermal source undoubtedly represents heat that was trapped far below the surface during the early stages of the earth's formation; and some represents heat that has been liberated continually by radioactive decay of unstable isotopes. How long the source of internal energy will continue to function is unknown. The moon and Mars both appear to be internally dead and subject primarily to land-leveling. Neither possesses a dense central core. So perhaps the source of the earth's internal energy is related to the development of its core. Whatever the source, activity within the earth appears to be as vigorous now as at any time in the 4.5-billion-year geologic history of the planet.

Earth scientists have long searched for a uni-

fied theory or model that would neatly tie together all of the internally generated processes. Today, many geologists believe that the geologic equivalent of the Holy Grail has been found. It is called by various terms, but is most recognized as the model of *plate tectonics* (from the Greek word *tektonikos,* meaning carpenter or builder). According to the model, the outer shell of the earth down to about 100 km (62 mi) is rigid, and is composed of a number of segments or plates—somewhat like the tiles on a bathroom floor—which can move with respect to each other (Fig. 2-20). At some plate boundaries the plates converge, at some they pull apart, and at some they slide laterally past one another. Supposedly, all of the internal processes and the attendant surface features and phenomena result from the plate motion. The plate boundaries therefore include most of the world's earthquake belts, volcanically active regions, deep-sea trenches, island arcs, and even some mountain chains.

An interesting aspect of the plate-tectonics model is that, although it provides a framework relating to internal phenomena, it does not provide an unequivocal answer to the cause of plate movement. Not everyone accepts all as-

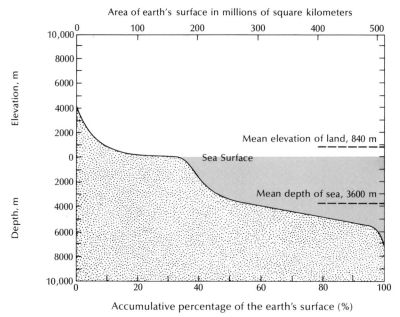

Area of earth's surface in millions of square kilometers

Fig. 2-21. Representation of the area of the earth's surface above any level by means of a single line. The most common average elevation above sea level (corresponding to the flat part of the curve at left) is 280 m (886 ft); the most common average depth in the oceans (the gently sloping part of the curve at right) is 4420 m (14,498 ft).

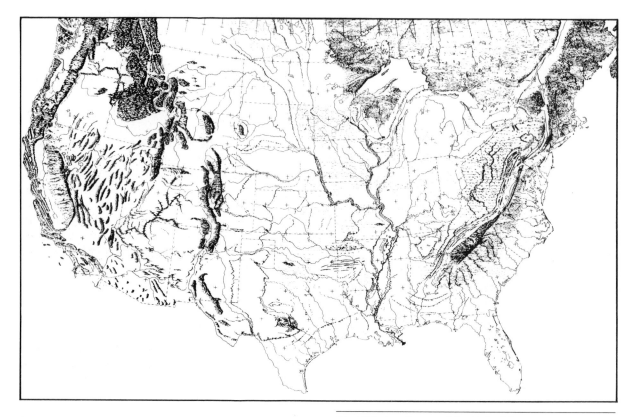

Fig. 2-22. Landforms of the United States.

pects of the model. Many believe that in our present stage of technology and knowledge we cannot adequately evaluate some parts of the theory. There is no doubt, however, that the dominant theory today concerning the internally generated processes of the earth is plate tectonics. Many scientists embrace it with an almost religious fervor. It remains to be seen, however, how well all parts of the theory meet the test of time.

Throughout the chapters in this book that deal with internal earth processes, we shall continually allude to plate tectonics, pointing out discrepancies as well as consistencies between the model and known geologic facts. In the last chapter, with all of the earlier chapters as background material, we shall examine, in detail, the tenets of the model, comparing it with others that have also been proposed to ex-

plain the "whys" and "wherefores" of activity beneath the surface of the earth.

Face of the rocky substrate

The first-order topographic features are the continents and ocean basins. Certain unique topographic features also exist at the junctions between the two features. As we have already mentioned, the contrast in topographic elevation between continents and ocean basins is a reflection of fundamental differences in the kinds of rocks that are found beneath each. The surface features on the continents are familiar to all of us and need little explanation. Those on the floor of the oceans, however, are rarely seen; and more space will be devoted to their description. A graphic representation of the distribution of elevations over the face of the earth is given in Figure 2-21.

Continents

Secondary features found on most continents are plains, plateaus, and mountains (Fig. 2-22). Smaller, more irregular landforms also abound but will not be considered in this summary.

Plains Plains are broad areas of low relief that occur generally at relatively low elevations. In the United States one could point to the Eastern Seaboard from New York southward and most of the drainage basin of the Mississippi River as prime examples.

Plateaus Plateaus resemble plains in being broad, relatively flat, or gently rolling regions, but they occur at relatively high elevations. Commonly streams and rivers will have dissected a plateau region to some degree (Fig. 2-23).

The austere but breathtakingly beautiful Colorado Plateau of the southwestern United States is a fine example of a plateau region. Incision by streams and rivers (principally the Colorado River and its tributaries) has produced color-banded, steep-walled canyons close to 2 km (1 mi) in depth.

Fig. 2-23. Two stages in the dissection of a plateau. In the upper diagram, the canyons are narrow and the upland surface is dominant; in the lower diagram, the canyons have deepened and widened to the point where the upland surface is nearly obliterated. (Adapted, by permission, from Fig. 32.13 in *Physical Geography* by Arthur N. Strahler. Copyright © 1969 by John Wiley & Sons, Inc.

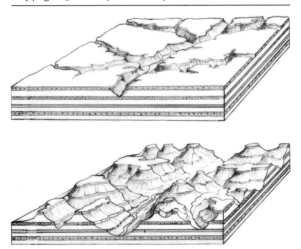

Mountains Mountains represent the highest parts of the continental masses. They can consist of single protuberances, but commonly they are found in linear groups of mountain blocks and peaks, called chains or ranges (Fig. 2-24). The highest peaks in the world are found in the Himalayas, where ranges rise in immense waves of sculptured stone. Swept by avalanches, and riven by glaciers they present supreme challenges to the best climbers in the world. In that mountain range, Mount Everest at 8848 m (29,021 ft) stands as the highest topographic feature in the world today. The highest mountain in the topographically disadvantaged contiguous United States, is Mount Whitney in the Sierra Nevada Range in California—a peak only about one-half as high (4418 m; 14,491 ft) as Everest.

Ocean basins

The common second-order topographic features in the ocean basins include abyssal plains and hills, submarine plateaus, oceanic ridges and rises, fracture zones and cracks, and volcanic cones.

Abyssal plains Throughout the world there exist relatively large, low–relief abyssal plains at depths of 4000–5000 m (13,120–16,400 ft) (Figs. 2-25 and 2-26). They are the flattest extensive surfaces on earth—a result of a covering of sedimentary rocks which blanket a somewhat irregular surface below. The veil of sediments has resulted from continued slow accumulation of sedimentary debris eroded from the continents and carried out to sea by oceanic currents.

In many of the deeper parts of the ocean basins there exist topographic protuberances, termed *abyssal hills*, which extend from 30 to 1000 m (98 to 3280 ft) above the sea floor. Although the origin of abyssal hills is not altogether certain, oceanographers generally consider them to be the result of volcanism.

Submarine plateaus In some parts of the oceans are found areas of relatively low relief

Fig. 2-24. Mount Tyree, Sentinel Range, Antarctica. Looking eastward across the range to the Edith Ronne Shelf. *USGS*

that markedly stand above the deeper parts of the ocean basin. One such topographic feature is the Blake Plateau, which lies in the Atlantic just east of Florida. The Melanesian Plateau to the east of Australia is one of the largest submarine plateaus on earth.

Oceanic ridges and rises One of the most significant discoveries in the world's seaways was the existence of a mountain chain that is found throughout the ocean basins of the world for a distance of 64,000 km (39,770 mi) (Fig. 2-27). The elongate suboceanic ranges are called *ridges* and *rises*. In general, a rise is a more rounded, less sharply defined topographic feature than is a ridge. In the Pacific the ridge and rise system (called the East Pacific Rise), runs nearly north-south, hugs the

eastern margin of the ocean basin off South and Central America, and appears to run aground and disappear just off the tip of Baja California; in the Atlantic, the Mid-Atlantic Ridge runs down the center of the ocean basin and extends from the Arctic nearly to Antarctica; in the Indian Ocean, the Mid-Indian Ridge begins near the Red Sea on the north and extends southward for about 5000 km (3107 mi), where it bifurcates to become the Southwest and Southeast Indian ridges. In a few places, such as Iceland, the ridge actually rises above the ocean's surface.

The cross section of the ridge is unlike that of continental mountains. In a traverse across a ridge the ocean bottom does not rise continuously to the axis but does so rather in a series of steps (Fig. 2-28). At the crest of some

Fig. 2-25. Sub-bottom profile of the abyssal plain off the coast of Nova Scotia. Notice how flat the sedimentary layers are beneath the plain. Abyssal plains are the flattest topographic surfaces known on the earth.

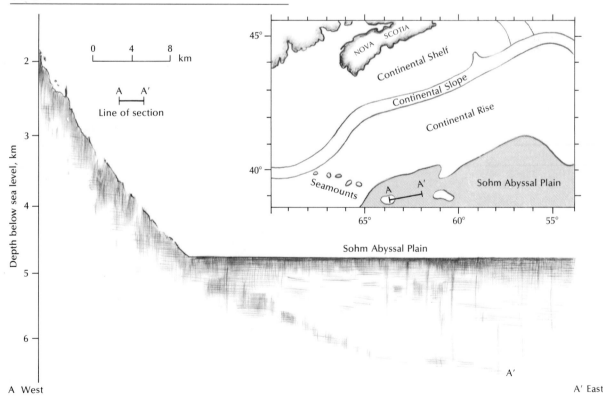

Fig. 2-26. Topography of the Atlantic Ocean floor off the east coast of North America, including the Blake submarine plateau and an abyssal plain.

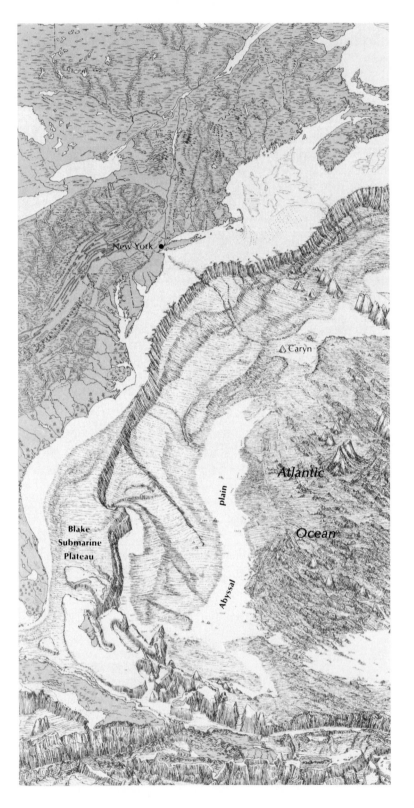

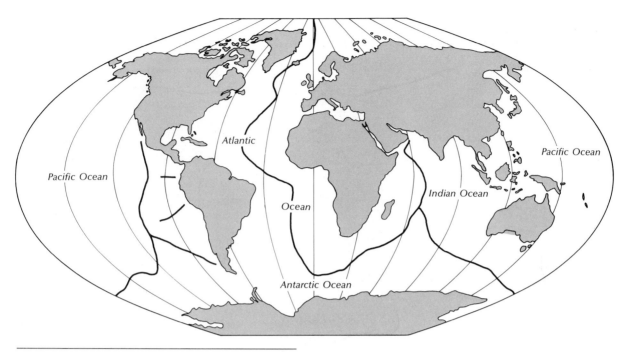

Fig. 2-27. General location (black line) of the oceanic ridge and rise system.

Fig. 2-28. Cross section of a typical oceanic ridge system showing a general increase in elevation toward the crest and a relatively depressed crestal region (the median valley). Step-like changes in topography appear to be related to movement along faults that parallel the ridge axis.

Ridge crest
(with median valley)

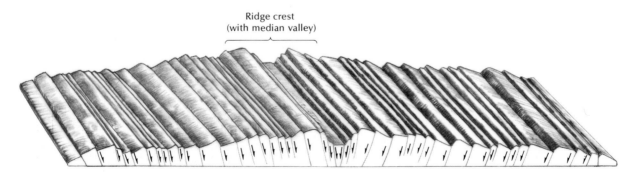

Fig. 2-29. The Mid-Atlantic Ridge in the North Atlantic, showing the fracture zones that transect and offset the ridge, and the elongate median valley.

ridges and rises is a keystone-like depression called the *median valley* (Fig. 2-29). It appears that the ridge system is an elongate welt along which the sea floor has been warped upward and in the process has been stretched and cracked parallel to the elongation.

Fracture zones and cracks   Cutting across the axis of the oceanic chain and extending laterally outward into the sea basins on either side are numerous fractures and cracks (Fig. 2-29). Often entire segments of the ocean floor have been uplifted or moved laterally along the fracture zones, slicing the range into a number of blocks sharply offset from one another.

Volcanic cones   Here and there, generally conical hills rise from the ocean floors either singly or in clusters; some are aligned along fracture zones (Fig. 2-30). From their shape, magnetic properties, and nature of samples dredged from some of them, they appear most certainly to be of volcanic origin. In a few cases volcanic eruptions have been so voluminous that cones have been built above sea level. The most magnificent set of such volcanic islands is found in the Pacific and includes the Hawaiian Islands. Once the islands rise above the sea, the erosive work of the pounding surf can begin. If volcanism cannot keep ahead of erosion, the island will be planed down to sea level.

Fig. 2-30. The northwest Pacific Ocean basin, showing a multitude of volcanic cones, many of which project above sea level; and the narrow, sinuous deep-sea trenches along the ocean margin.

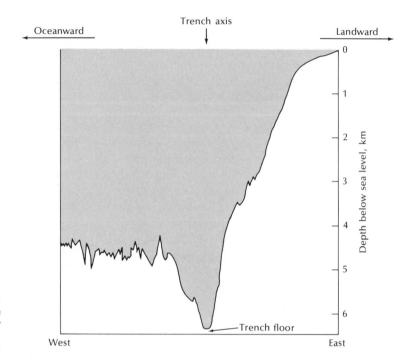

Fig. 2-31. Profile of the Peru-Chile Trench off the coast of Peru. The trench is nearly devoid of sediments.

Sea-margin features

The zones around the world where continents meet ocean basins are generally characterized by two different kinds of topographic features: island arcs and deep-sea trenches or continental shelves and slopes.

Island arcs and deep-sea trenches Some continental margins are marked by a volcanically active island archipelago and seaward of that, a deep-sea trench (Fig. 2-30). Typical examples of island arcs include the Aleutian and Japanese islands. The trenches are elongate (paralleling the trend of the island arcs), narrow, and deep (Fig. 2-31). In fact, the deepest places in the ocean—down to depths as great as 10,700 m (35,096 ft)—are found in the trenches. Island arc-and-trench systems are prone to a high level of earthquake activity. The deepest quakes in the earth—down to 700 km (435 mi) occur beneath the arcs.

Other continental margins, although volcanically active, deviate somewhat from that common pattern. For example, the west coast of South America, which is bordered by the Peru-Chile Trench, lacks an archipelago. The active volcanic mountains (the Andes) are located on the South American continent. The west coast of the United States is also unusual in that a volcanic chain—the Cascades—parallels the coast, but no ocean trench lies on the margin of that chain.

Continental shelves and slopes The other type of continent-ocean boundary is typified by the east coast of North America and is common all along the Atlantic and Indian ocean margins (Fig. 2-32). A relatively wide, shallowly submerged platform—the *continental shelf*—extends oceanward from the coastline. The continental shelf, for all practical purposes, can be considered part of the continent since it is composed of continental-type rocks and is commonly a seaward extension of the coastal plain. At depths between 150 and 500 m (492 and 1640 ft) the slope increases as it drops toward the depths of the ocean. The steep part of the gradient is termed the *continental slope*.

60

Fig. 2-32. Topography of southern South America and part of the South Atlantic basin. The nearly flat, gently sloping continental shelf extends along the entire eastern margin of South America, and is separated from the deep basin by a relatively abrupt scarp. If sea level were to drop only a small amount, South America would be appreciably wider.

PLANET EARTH—AN OVERVIEW

The overall density and composition of the earth were predetermined by circumstances that prevailed at the moment of its inception, 4.5 billion years ago or shortly thereafter. The nature of its internal structure, oceans, and atmosphere, however, are more than likely the product of internal processes that have been in action ever since it first came into being. The surface of the earth is, and has been throughout its history, a battleground between internally generated constructive processes that tend to rebuild surface features and externally driven destructive processes that tend to level the earth. The features of the earth today, so rich and varied in aspect, are the unique product of those two sets of opposing activities.

SUMMARY

1. The earth is one of nine planets that circle the sun, the central star of our solar system.
2. The earth is held in its orbit around the sun by *gravitational attraction*. The law of gravitation pertains to all particles in the universe and is given by:

$$F_g = \frac{Gm_1m_2}{r^2}$$

3. The slightly flattened ellipsoidal shape is the result of the inward pull of the earth on all its particles of matter and the rotation of the earth on its axis.
4. The average density of the earth is 5.519 gm/cm^3, about twice that of surface rocks. The obvious inference is that some very dense material must be present inside the earth.
5. Several lines of evidence indicate that continents are of relatively low density and float *isostatically* on the higher-density mantle rocks. When loading or unloading of the earth's surface occurs, the crustal blocks will undergo *isostatic readjustment* until they are again in floating equilibrium.
6. The atmosphere is composed principally of nitrogen (78 per cent) and oxygen (21 per cent). It also contains small amounts of other gases, many of which possess an importance far beyond their volumetric percentage. The latter include carbon dioxide and water vapor. It appears that the composition of the atmosphere during the early stages of the earth's history was different from what it is today. Circulation of the atmosphere is driven by energy received from the sun.
7. The world's oceans cover 71 per cent of its surface area. The waters of the oceans probably came from inside the earth and have been accumulating over the entire span of the earth's history. Most salts in sea water are derived from the land and brought to the sea by rivers. Surface currents are principally wind driven, whereas subsurface flow is the result of differences in density between water masses.
8. The earth is a layered body. The outer layer, composed of relatively light silicate rock, extends down to about 6 km (4 mi) beneath the ocean basins and to about 70 km (44 mi) under the continents. The next layer, composed largely of silicon and oxygen with lesser amounts of iron, magnesium, aluminum, calcium, sodium, and potassium in a solid state extends down to 2900 km (1802 mi). The core, the innermost zone, is composed mostly of iron and is in a partly liquid and partly solid state.
9. *Volcanism, mountain building, earthquake activity*, and *continental drift* result in rebuilding of the earth's surface. They are generated internally by processes that extend down to 700 km (435 mi).

    Other processes such as *weathering, erosion,* and *sedimentation* act to level the surface and are powered by the sun's energy. The face of the earth has changed continuously through time in response to the interaction of *constructional* and *destructional* processes.
10. *Plate tectonics* forms the modern basis for understanding and interrelating the internally generated phenomena. The outer shell of the earth appears to be composed

of a number of rigid plates that move with respect to each other: converging, diverging, or sliding laterally. Most of the constructional phenomena are developed at the plate margins, and little within the plates.

11. The first-order topographic features—the *continents* and *ocean basins*—are themselves made up of different kinds of secondary topographic features. On the continents there are *plains, plateaus, and mountains;* in the ocean basins there are *abyssal plains, submarine plateaus, oceanic ridges* and *rises, fracture zones* and cracks, and *volcanic cones.* At the sea margins are *island arcs* and *deep-sea trenches,* and *continental shelves* and *slopes.*

SELECTED REFERENCES

Barnett, L., and the Editorial Staff of Life, 1955, The world we live in, Time Inc., New York.

Bates, D. R., and others, 1957, The earth and its atmosphere, Basic Books, New York.

Beiser, A., and the Editors of Life, 1962, The earth, Time Inc., New York.

Buswell, A. M., and Rodebush, W. H., 1956, Water, Scientific American, April.

Cailleux, A., 1968, Anatomy of the earth, World University Library, McGraw-Hill Book Co., New York.

Carson, R., 1950, The sea around us, Oxford University Press, New York.

Davis, K. S., and Day, J. A., 1961, Water, The mirror of science, Doubleday and Co., New York.

Denn, W. L., 1972, The earth: Our physical environment, John Wiley and Sons, New York.

Editors of the Scientific American, 1950–1957, The planet earth, Simon and Schuster, New York.

Gamow, G., 1958, Earth, matter, and sky, Prentice-Hall, Englewood Cliffs, New Jersey.

———, 1960, Gravity, Anchor Books, Doubleday and Co., Garden City, New York.

Hamilton, H. C., 1854, The geography of Strabo, Henry G. Bohn, London.

Heezen, B. C., and Hollister, C. D., 1971, The face of the deep, Oxford University Press, New York.

Heiskanen, W. A., and Meinesz, V., 1958, The earth and its gravity field, McGraw-Hill Book Co., New York.

King-Hele, D., 1967, The shape of the earth, Scientific American, vol. 217, no. 4, pp. 67–76.

Krauskopf, K. B., and Beiser, A., 1973, The physical universe, 3rd ed., McGraw-Hill Book Co., New York.

Kuiper, G. P., and others, 1954, The solar system, Vol. II, The earth as a planet, University of Chicago Press, Chicago.

McAlester, A. L., 1975, The history of life, 2nd ed., Prentice-Hall, Englewood Cliffs, New Jersey.

Strahler, A. N., 1975, Physical geography, 4th ed., John Wiley and Sons, New York.

Sverdrup, H. U., Johnson, M. W., and Fleming, R. H., 1942, The oceans, Prentice-Hall, New York.

Urey, H. C., 1952, The origin of the earth, Scientific American, vol. 187, no. 4, pp. 53–60.

3

MATTER, MINERALS, AND ROCKS

Philosophers, alchemists, and chemists have considered the nature of matter down through the ages. To many of the ancient Greeks, matter was known simply to be the substance of which any object is composed. At that time curious persons had no means, other than their own senses, for investigating the physical world, and it was only natural that knowledge about that domain was either nonexistent or rudimentary. By about 400 B.C., however, the philosopher Democritus reached the conclusion, on the basis of logic, that matter must be made up of small indivisible bits which he called *atoms*, all of which are similar and eternal. About 50 years after that first pronouncement of the atomic theory, the highly regarded philosopher Aristotle (384–322 B.C.) proposed that the principal elements of terrestrial matter were fire, water, air, and earth, whereas the

only element of the heavens above was "quintessence." Earth was considered the essence that gave things the property of solidity. Matters of consequence concerning the world around us were simpler in those days, were they not?

As time progressed and the alchemists strived to attain the elusive goal of turning metals into gold, it became apparent that the physical world was not so simple as Aristotle supposed (Fig. 3-2). To be sure fire, water, and air were elemental, but earth came to be recognized as a mixture of many things.

Eventually the background work of chemists experimenting with gases as well as his own efforts enabled French chemist Antoine Lavoisier (1743–94), during the late eighteenth century, to clarify the subject. He correctly identified 23 elements which he defined as pure substances and indicated that other as yet undiscovered elements probably existed. He also stated that a chemical compound is a pure substance that consists of two or more elements in combination.

Today, there are 103 known elements, of which 92 are found naturally at the earth's

---

Fig. 3-1. Mineral group measuring $20 \times 25 \times 20$ cm from the Pala Chief Mine in San Diego County, California. The large pointed crystals terminated by six-sided pyramids are the common mineral quartz. The other minerals, less common, are tourmaline (elongate crystals), lepidolite mica (plates), and columbite (black). *Henry Janson*

Fig. 3-2. A sixteenth-century rendering of the mysterious powers of an alchemist. *Prints Division, New York Public Library*

surface. Analysis of the spectrum of light rays from the sun has shown that most of the naturally occurring elements exist on the sun, although not in the same proportions.

In the early 1800s the English chemist and physicist John Dalton (1766–1844) reactivated the atomic theory of the Greeks when he hypothesized that all matter is composed of tiny invisible atoms of different weights and chemical properties. An element, he concluded, is composed completely of the same kind of atom. He even went so far as to say that when two or more elements combine, their atoms form identical groups of atoms called *molecules*.

The next milestone in the quest for the fundamental building blocks of matter came mostly from experiments carried out in the late 1800s and early 1900s, particularly those with radioactive substances. Almost all atoms were found to be composed of three kinds of particles: negatively charged *electrons*, positively charged *protons*, and *neutrons*, which carry no charge whatever. In any atom there are as many negative particles as there are positive ones, and the electrical charge balance is preserved. In atoms of a particular element, the number of protons equals that of electrons and is always a fixed number (the *atomic number*). But the number of neutrons can vary, to a small degree, from atom to atom of that element. The atoms of an element that differ only in the number of neutrons (and therefore in *atomic weight*) are termed *isotopes* of that element—such as uranium-235 and uranium-238.

It was also found that electrons, protons, and neutrons are about equal in size—about 10^{-12} cm (a millionth of a millionth of a centimeter) in diameter. An electron has much less mass than the other two particles; it is only 1/1836 and 1/1837 as massive as a proton and a neutron, respectively. From its mass and neutral charge it is assumed that a neutron is essentially the combined form of a proton and electron.

An ingenious experiment by the English physicist Lord Rutherford in the early twentieth century demonstrated that in any single atom all protons and neutrons exist in a very dense central kernel, or nucleus, that contains more than 99.9 per cent of the mass but represents only about one-billionth of the volume of the atom. The nucleus is the center of an essentially spherical body with a diameter of about 10^{-8} cm (a hundredth of a millionth of a centimeter) in which the electrons move rapidly in continual agitation, now closer to the nucleus, now farther away, and resemble, to a degree, a swarm of gnats swirling around one's head. In heavier atoms, which contain many tens of electrons, certain electrons are constrained to a flight path close to the nucleus, whereas others tend to remain near the outer margin of the electron cloud. In chemical combinations it is the outer electrons that enter into the reaction. In a sense each atom is similar to a miniature solar system, with the nucleus as the sun and the electrons as planets (Fig. 3-3).

In their pursuit of the basis of matter, modern nuclear scientists have gone even farther. They have cracked the nucleus like a walnut, uncovering many strangely behaving particles with even stranger-sounding names—pion, mu meson, positron. We will merely note their existence and leave it at that.

Because atoms are so tiny it requires quite a large number of them in order to be visible to the eye. One-half pound of lead, for instance, contains the unbelievable number of about 1×10^{24} atoms (1 followed by 24 zeros). A telling demonstration of the large number of atoms (and molecules) that are present in small

volumes of matter can be shown by the following example: If you pour two quarts of water into the ocean and stir well, any cup of water subsequently dipped from the ocean will contain *one* molecule from the two quarts of water you poured in.

Atoms normally do not exist as solitary entities. Most are chemically active to some degree and when one atomic sphere comes in contact with another they join together, or *bond*. In some cases, the atoms that bond together are all of the same element; for example, the metal gold and the gas hydrogen are each composed of only one kind of atom. In most natural circumstances, however, the atoms of one element are combined with atoms of one or more different elements. When two or more elements are mixed in definite proportions, and the atoms combine through bonding, a *chemical compound* is formed. That compound has, in addition to a set chemical composition, characteristic chemical and physical properties. Generally it will look nothing like the initial elemental ingredients nor will it have any of their properties. A compound as innocuous as table salt (sodium chloride) is the result of the combination of sodium, a highly reactive metal, with chlorine, a very noxious gas (Fig. 3-4).

BONDING

The bonding of two atoms is accomplished through interactions of one or more electrons from the outer shells of those atoms. Several types of bonding are possible, and the particular type that the atoms of any one element can undergo is an intrinsic property of the element, determined by the number of electrons in the atom.

In one type of bonding, there is an actual transfer of one or more electrons from the electron cloud of one atom to that of another—one atom loses electrons and the other gains them. The resulting positively and negatively charged particles are called *ions* (Fig. 3-5). The ion carrying a positive charge is called a *cation*, and the one carrying a negative charge an

Fig. 3-3 A. The hydrogen atom contains one electron in its electron cloud and one proton in its nucleus. **B.** The helium atom contains two electrons in its electron cloud and two protons and two neutrons in its nucleus.

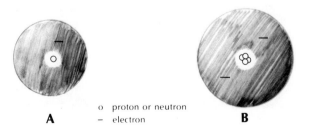

o proton or neutron
– electron

A   **B**

anion. Usually the positively charged cations form by loss of one or more electrons from atoms of the metallic elements, such as calcium, sodium, or potassium. *Ionic bonds* result from the electrostatic attraction between positively and negatively charged particles and are normally quite strong.

In another type of bonding, the electron, rather than transferring completely from the donor atom to the acceptor, remains halfway between the atoms and is shared by both. It is called *covalent bonding,* and it too is quite strong. In such a way, two atoms of hydrogen will join to one of oxygen to form a water molecule (Fig. 3-6). In most cases of covalent bonding, the electron does not remain equidistant from the two atoms. At times it will move into the electron cloud of one atom, and at times into the other; when it does so the bond will become, momentarily, ionic in type. The atoms in most solid geologic substances are held together by such alternating *ionic-covalent* bonding.

A third type of atomic glue, less common than the previous two, is the *metallic bond.* As you might expect from the name, such bonding is found principally among the metals in their uncombined state. In a metal, all of the atoms are exactly the same size, and they stack around each other like marbles. The tight packing of the equally sized spheres results in each atom being surrounded by twelve other atoms (Fig. 3-7). So the electron cloud of each atom has to react simultaneously with those of its 12 close neighbors. The bonding tends to be covalent at each point where the spheres touch each other. However, because more than enough electrons are available from all of the atoms to complete the bonding, those not in use at any moment tend to drift aimlessly through the metal. An electron forming a covalent bond may be at any moment replaced by an extra one. The replaced electron is then free to move on to another atom. Because of the musical chairs played by the atoms, metallic bonding has also been called *time-shared covalent bonding.* It is the abundance of moving,

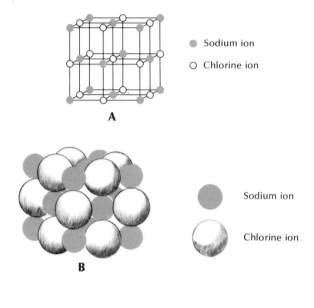

Sodium ion
Chlorine ion

A

Sodium ion

Chlorine ion

B

Fig. 3-4. The structural arrangements (**A** and **B**) of sodium and chlorine ions in the mineral halite. (**A**) is an exploded view showing the relative positions of the ions; (**B**) more closely represents the way in which the ions are actually stacked together. The external cube shape of the halite crystal (**C** below) consists of billions of sodium and chlorine ions and is a reflection of its internal ionic arrangement.

IONIC BONDING

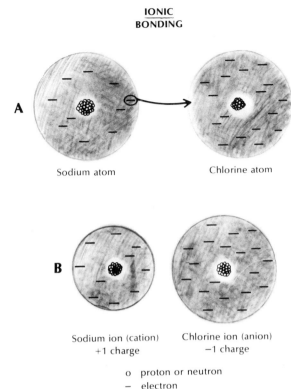

Sodium atom   Chlorine atom

Sodium ion (cation)   Chlorine ion (anion)
+1 charge   −1 charge

o   proton or neutron
−   electron

Fig. 3-5 A. A sodium atom has one loosely bound electron in its outer shell whereas a chlorine atom has many electrons, lacking only one from being completely full. **B.** When the two atoms come together, the sodium electron moves to the shell of the chlorine atom, producing a positively charged sodium ion (cation) and a negatively charged chlorine ion (anion). Electrostatic attraction holds the two ions together, and an ionic bond is formed.

loosely held electrons that gives the metals their characteristic abilities to conduct heat and electricity and to reflect light.

Molecules, that is integral groups of atoms, can also be bonded together. Such bonding does not rely on the sharing or exchange of electrons at all; instead it relies on relatively weak electrostatic forces between molecules. Accordingly, such bonds are also generally relatively weak. In many compounds, particularly those involving atoms of more than two different elements, several types of bonds, or levels of bonding, coexist.

STATES OF MATTER AND THE STRUCTURE OF MINERALS

Compounds exist in any of three different states—solid, liquid, or gaseous—depending on the strength of bond between the various atoms and on temperature and pressure as well. Water can exist at the earth's surface in all three states. As you know, when water is cooled sufficiently it freezes; if heated sufficiently it becomes a gas. It is apparent, then, that the principal difference between the three states is one of heat content, a measure of which is temperature. From experiments on gases in the 1800s it became clear that temperature was also a measure of the average translational speed of the gas molecules: the higher the temperature, the greater the speed. When you blow up a balloon, the molecules move in all directions, in a random, mixing fashion, hither and yon—order is totally lacking. At any one time many molecules are striking against the balloon wall, creating an outward pressure. If the balloon is heated the molecules become agitated, striking the wall harder and more frequently. The pressure thereby increases, resulting in an expansion of the balloon.

When a gas such as nitrogen is cooled, the atoms (or molecules) slow down until they can no longer remain completely independent of one another. Then, *attractive*, or bonding, forces become strong enough to cause the atoms or molecules to begin to stick to one another, and the liquid state is attained. Although relatively slow moving, each atom in a liquid still tends to move independently of every other one. There are moments, however, when two or more atoms jostle each other— and bonds are formed momentarily. When, in the next moment, those atoms resume their aimless wandering, the bonds are ripped apart. Around a particular molecule (or molecules) or within a small volume of liquid, therefore, there will be from time to time (when bonds exist) an ordered assemblage of molecules. *Short-range ordering*, as that state is called, is

Fig. 3-6. Covalent bonding of two hydrogen atoms with one oxygen atom (**A**) to form a water molecule (**B**). The electron of each hydrogen atom is shared with the oxygen atom to form a very strong bond.

COVALENT BONDING

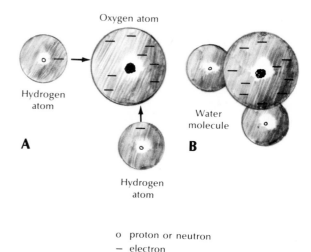

o proton or neutron
− electron

characteristic of all liquids. Because the atoms or molecules continually change partners, a liquid lacks rigidity; that is, it tends to flow. During flowage some portions move more rapidly than others, requiring the breaking of additional interatomic bonds. That results in a resistance to flow called *viscosity*.

Continued cooling of a liquid brings about a progressive slowing down of the atoms or molecules. Eventually it becomes possible for the bonds between them to remain firm and the change from liquid to solid will occur. In the formation of a solid the atoms fit together in the closest way possible for their size and shape and the type of bonds that exist. That is, a solid does not consist of a jumble of atoms or molecules; instead each occupies a particular, set position with regard to those surrounding it. An ordered latticework of atoms is created— a little like the repetitive pattern on a roll of wallpaper, but in three dimensions. *Long-range order* exists and the material is said to be crystalline. With few exceptions, only one particular atomic ordering is possible for each chemical compound.

The word *crystalline* comes from crystal, the Anglicized Greek word for ice. Crystalline chemical compounds and elements that occur naturally are called *minerals*. In fact, it was the study of minerals that eventually led to our knowledge of long-range ordering, or *crystallinity*, found in compounds in the solid state. Today *mineralogists* (those who study minerals) have described and named thousands of minerals. In the early days of history, however, only a few were known, generally those that were showy in appearance or that had been found useful to humans. For example, the common mineral quartz (Fig. 3-8), which often occurs in clusters of shiny, ice-like crystal spires, was well known in the Middle Ages, and

Fig. 3-7. Close packing arrangement of atoms of the same size; each sphere is surrounded by 12 others. The bond between any two atoms is covalent; since, however, any one atom can share electrons with any of its 12 neighbors, extra electrons are available for bonding. As a result, the electrons continually migrate, and bonding electrons come from different neighbors at different times. Such time-shared covalent bonding is typical of the metals and is called metallic bonding.

METALLIC BONDING

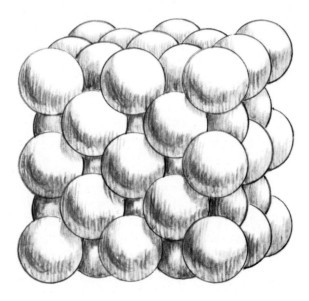

certain ore minerals which yielded iron, copper, lead, and silver have been recognized for several thousands of years. Progress in the identification and understanding of minerals was rapid throughout the eighteenth century, largely because chemists, who were trying to understand the basis of matter, commonly used naturally occurring minerals in their experiments. Eventually they began to understand the chemical properties of the materials they worked with, but neither chemists nor mineralogists were able to explain acceptably why many minerals were bounded by planar crystal faces.

By 1669 Nicolaus Steno, the Danish physician whom we met in Chapter 1, had demonstrated that the faces on a quartz crystal always meet at the same angle, regardless of the crystal size and shape (Fig. 3-9). As later *crystallographers* (those who study crystal forms) found, the angles are the same regardless of where in the world the quartz comes from. As the field of crystallography grew, the crystal forms of many kinds of minerals were studied. And, in spite

Fig. 3-8. Cluster of quartz crystals from Crystal Springs, Arkansas. These naturally occurring crystals are transparent and possess a characteristic external form composed of a number of relatively flat surfaces. *Smithsonian Institution*

of the large number studied, it became apparent that all of the many crystal forms of all of the minerals could be classified into just seven basic groups, or systems. That seemed to speak of an underlying order, but no one knew what sort of order.

As early as 1611 Johannes Kepler (1571–1630), the famous German astronomer, aware that snowflakes always possessed hexagonal symmetry (Fig. 3-10), conjectured that their regularity in form was probably due to the geometrical arrangement of the minute building blocks making them up. Later, in 1784 Réné Hauy (1743–1822) a renowned crystallographer in Paris, concluded that any mineral exhibiting an observable crystal form must be made up of small polyhedral units, each of which must have the same symmetry as the whole crystal.

By the early 1880s, following the announcement and acceptance of Dalton's atomic theory, mineralogists had concluded that minerals were chemical compounds with definite compositions, and that each compound must consist of a large number of atoms uniquely fitted together in a regular three-dimensional pattern. Crystallographers had finally come to realize that the planar faces

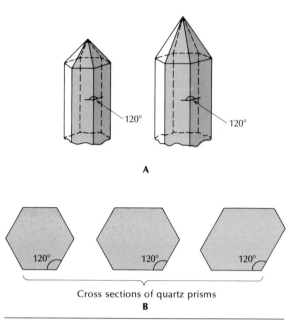

Fig. 3-9. Commonly occurring external forms on quartz crystals: the six-sided prism (straight-sided form) and pyramid (pointed form). The angles between two adjacent prism faces is always 120°, regardless of the difference in size between crystals (**A**) or apparent distortions in shape (**B**). The fact that crystals of a particular mineral have similar external forms led early mineralogists to suspect an internal order in atomic arrangement.

Fig. 3-10. Naturally occurring ice crystals (snowflakes). The crystals possess a six-sided symmetry—as do all ice crystals—suggesting a regular internal ordering. *Moody Institute of Science*

so characteristic of each particular mineral must be the outward reflection of the internal ordering—but they had no way of proving it for some time to come.

The proof, which eluded scientists until 1912, came about (as is so often the case) as the result of an extraordinarily fortunate and essentially intuitive experiment made by German physicist Max von Laue (1879–1960) and his associates. The mysterious X-ray had been discovered by William Roentgen in 1895, but little was known about it up to 1912. Von Laue and his companions conjectured that X-rays, like light rays, might be wave-like in character but of exceedingly short wavelengths. But in order to demonstrate those wave properties they needed a better analytical device than the diffraction grating available at the time. A standard *diffraction grating* consists of a glass plate engraved with fine, closely spaced parallel lines. When light passes through the plate the grooves interfere with, or diffract, the light waves in such a way that they are dispersed into a spectrum of colors, each color corresponding to light of a different wavelength. Finally, von Laue hit upon the idea that the tiny atoms in a crystalline substance might be systematically arranged in layers spaced closely enough to serve as a new kind of diffraction grating.

After the usual false starts and mishaps, he was eventually successful in sending an X-ray beam through a crystal of copper sulfate and onto a photographic plate placed behind the crystal (Fig. 3-11A). When the plate was developed there appeared upon it a pattern of dots von Laue interpreted as the evidence of X-rays, which had reflected from the electrons in the outer portion of the electron clouds of the crystal's regularly arranged atoms (Fig. 3-11B). Besides showing beyond a doubt that X-rays are wave-like in nature, the experiment revealed that the planes of atoms in a crystal are so regularly spaced that they could be used as a naturally occurring diffraction grating. X-rays had proved to be the magic key to the door of an unseen, ordered world in miniature, whose very existence previously could only be inferred from surface measurements of interfacial angles and the regular geometry of crystal faces.

Later X-ray experiments demonstrated that all minerals, whether they exhibit crystal faces or not, have crystalline long-range order. After von Laue's experimental breakthrough, many years and a great deal of effort were required before the "von Laue spots" for any particular mineral could be used to determine the detail

Fig. 3-11. Von Laue X-ray analysis. **A.** Technique in obtaining a Von Laue pattern. The X-ray beam is broken into a number of smaller pin-point beams by the internal crystal structure. **B.** Von Laue spots obtained by X-ray analysis of the mineral calcite. *Bernhardt J. Wuensch*

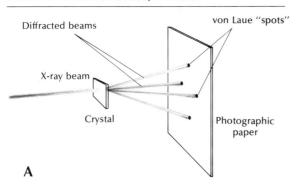

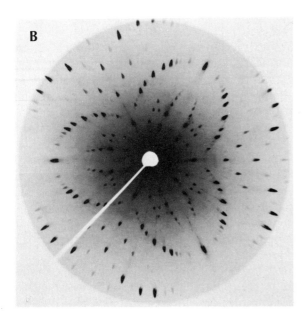

of its crystalline structure. Even the simplest of atomic lattices requires elaborate calculations and the application of abstruse theories of wave motion. Today, the use of automatic, high-precision X-ray equipment and high-speed computers has made accurate crystal determination almost routine. In modern biochemistry and molecular biology, X-ray crystallography is having perhaps its finest hour. The complex structure of DNA (deoxyribonucleic acid—a molecule in the nucleus of a cell which carries the information necessary for exact cell replication) has finally yielded to X-ray analysis, and, one by one, the structures of the various protein molecules are coming to be understood. The solution of the atomic structure of minerals is like child's play when compared to the structural analysis of a substance like hemoglobin, which contains nearly 10,000 atoms in one molecule.

PHYSICAL PROPERTIES OF MINERALS

With few exceptions the crystallographic structure of a particular mineral is always the same, no matter where the mineral comes from. Atomic structure is therefore a definitive property of a mineral and can be used to identify it. The crystallographic structure of a particular mineral is also a natural consequence of the way in which the atoms of the elements making up the compound bond together. But it is only one of several characteristic properties that each mineral displays.

Take, for example, carbon, which can exist as either of two minerals—diamond or graphite. Diamond is very hard, so hard in fact that it is used as an abrasive; it shines brilliantly, can take and hold a polish, and breaks smoothly along certain planes. Graphite, on the other hand, is dull and soft, and separates into small flakes that easily slide over one another.

The reason for such profound differences in minerals that are chemically the same substance lies in their completely unlike arrangement of atoms. In a diamond crystal the linkage between the carbon atoms is about as strong as it possibly can be. Each atom is in direct contact with four others, and covalently shares the four electrons in its outer shell with them. The result is a close-knit structure, geometrically arranged so that there exist four sets of parallel planes of weakness, which mark zones of weaker bonding. It is the existence of such planes that enables the diamond cutter to extract several marketable smaller stones from one large one.

The carbon atoms in graphite, on the other hand, occur in regularly spaced thin, parallel sheets. Because the bonds uniting the atoms within each sheet are many times stronger than the feeble ties between one atomic layer and its neighboring sheet above or below, graphite will split easily in one direction only.

Color

Color is the most obvious property that minerals possess, but it is not always the most definitive. When white light falls on a mineral, some of the wavelengths will be absorbed and some reflected. The color of the mineral corresponds to the wavelengths of the reflected light. Black minerals, for example, absorb essentially all of the light that falls on them. What causes a mineral to absorb certain wavelengths only is varied and complex. It can be a fundamental property, directly related to chemical composition, as in the blues and greens of some copper minerals. Or it may be unrelated to composition, depending instead on crystal structure and the type of bonding (as in diamonds and graphite). Commonly, it is caused by foreign atoms contained in the crystal structure. Minerals, being natural compounds, are not so pure as the materials a chemist uses in the lab. Small amounts of foreign atoms can be, and usually are, included in the lattice work as impurities. Pure quartz is colorless; the colored varieties of that mineral, so sought after by rock hounds, are due to small quantities of impurities. Therefore, we cannot use colors with the same confidence in identifying minerals as in naming birds or flowers. Considerable experi-

Fig. 3-12. Cleavage fragment of the common mineral calcite. Because of the characteristic regular arrangement of the atoms in the mineral, it always breaks, or cleaves, along smooth plane surfaces which meet at an angle of 74° 55'.

ence is needed to determine whether or not the color of a mineral is significant.

One way to avoid ambiguities is to grind the mineral into a fine powder and to observe its color. The intrinsic color will be seen, whereas colors due to impurities will be generally lost. A simple way to make the test is to rub the mineral on a piece of unglazed porcelain (a *streak plate*), leaving a thin film of powder, or *streak* of true color on the porcelain surface.

Cleavage

Cleavage is the ability of a mineral to split, or *cleave*, along closely spaced parallel planes (Fig. 3-12). Not all minerals have that property; many fracture along widely spaced surfaces that can be relatively smooth, but are often curved, or irregular and hackly. One notable (and familiar) type of fracture surface, commonly produced when glass is broken, resembles the markings on the outside of a conch shell—that is, a surface on which there are a number of concentric irregularities. It accordingly is called *conchoidal* fracturing (Fig. 3-13).

Fig. 3-13. Conchoidal fracture on a quartz fragment. Notice the nearly concentric arrangement of rounded grooves and ridges resembling the markings on a conch shell. *Jan Robertson*

Some minerals are characterized by only one set of cleavage planes; others by two, three, four, or (rarely) six. Because of crystal symmetry no minerals possess five, or more than six, sets of cleavage planes. Not only the number of cleavages but the angles between them as well are characteristic for any mineral. Some cleavages meet at right angles, but many do not. If splitting occurs easily and produces a nearly unblemished flat surface, the cleavage is said to be perfect. In some minerals the cleavage is less perfect and can be called *good, distinct,* or *indistinct,* depending on the roughness of the resultant surface. The geometrically repetitive nature of cleavage, its planar character, and the distinctive orientation of the planes are strong evidence that cleavage, like the crystal form of minerals, is a property determined by the regular geometric ordering of the atoms in the mineral.

Hardness

Mineral hardness, too, is related to atomic structure. Even a small scratch on the surface of a mineral requires the breaking of bonds and the separation of atoms, and the ease of separation will depend on the kinds of atoms and bonding inherent to the mineral. Hardness is an easy property to determine, and it can be definitive at times. A harder mineral will scratch a softer one. The scale we use today to judge hardness was devised more than a century ago by Austrian mineralogist Frederich Mohs (1773–1839), and it bears his name.

Mohs placed the hardest of all minerals, diamond, at the top of the scale and assigned to it an arbitrary value of 10. He ranked other softer minerals in a descending hierarchy, down to 1. The hardness numbers, although in sequential order, are not equally spaced. For example, the actual interval between diamond, to which he assigned a value of 10, and corundum, which he ranked as 9, is more than the rest of the scale combined. If absolute values were assigned to the various minerals used in the scale, diamond would be about 42. By means of the scale, the hardness of all min-

erals has been rated and those values listed, as in Table 3-1.

Luster

When a mineral is viewed in ordinary light, the amount of light reflected from its surface and the way it is reflected determines its *luster.* Essentially all lusters fall into two groups: *metallic* and *non-metallic.* The first term is applied to minerals that reflect light in about the same way that polished metals such as iron or copper do. In general a very high luster is characteristic of minerals possessing metallic bonding. Minerals with a metallic luster commonly are opaque (will not pass light), even along thin edges when held up against the light.

Non-metallic lusters are quite variable but certain common ones have been singled out. If a mineral reflects light to about the same degree as glass, it has a glassy, or *vitreous* luster. Other terms which are essentially self-explanatory are earthy, greasy, waxy, dull, resinous, pearly, silky, and fibrous.

Specific gravity

The specific gravity of a mineral may be defined as the weight of a specified volume of the mineral divided by the weight of an equal volume of water at 4°C (39°F)—the temperature at which water is most dense. It is a number

Table 3-1. Mohs hardness scale

Mineral	Hardness	Equivalent in hardness
Diamond	10	
Corundum	9	
Topaz	8	
Quartz	7	
Potassium feldspar	6	glass, knife blade
Apatite	5	
Fluorite	4	
Calcite	3	
Gypsum	2	fingernail
Talc	1	

that expresses how much more dense a mineral is than a common substance, water. For example, quartz has a specific gravity of 2.7, which means that its density is 2.7 times that of water. The specific gravity of any mineral is, within limits, characteristic of that mineral, and it is determined by the atomic weights of the elements that make up the compound and how closely packed the atoms are.

A rough estimate of the specific gravity of a mineral can be made (after some experience has been gained) simply by hefting the mineral by hand. Some minerals feel light while others, particularly those with metallic lusters, feel heavy.

Magnetic properties

A few minerals, particularly some of those made up of iron and oxygen or sulfur, are relatively strongly magnetic; most are not. A simple test of this property can be made by use of a small hand magnet.

MINERAL DESCRIPTIONS

A mineral, to briefly review, is a naturally occurring chemical compound (or element), crystalline in nature, and possessing a definite chemical composition or a restricted range of compositions. Nearly 2000 minerals have been recognized on the face of the earth. They do not, of course, occur in equal abundance; some are relatively common, others are rare. The abundance of a mineral more or less reflects the quantities of the component elements that are available for its formation at or near the earth's surface (Table 3-2).

Amazingly enough, only ten elements occur in sizable proportions, and they make up more than 99 per cent of the total mass of the minerals at the earth's surface. All the others, many of which are important to life and to the prosperity of humans, compose less than 1 per cent. It is strikingly evident that of the elements making up continental rocks, oxygen and silicon together make up nearly 75 per cent of the mass. The seven metals—aluminum, iron, cal-

Table 3-2. Chemical composition of the continental crust

Element	Weight per cent
Oxygen	46.6
Silicon	27.7
Aluminum	8.1
Iron	5.0
Calcium	3.6
Sodium	2.8
Potassium	2.6
Magnesium	2.1
Titanium	0.4
Hydrogen	0.14

cium, sodium, potassium, magnesium, and titanium, compose most of the rest. The average composition of the crust beneath the oceans is only slightly different from that of the continents.

Because of the overwhelming abundance of oxygen and silicon it is only natural that silicate minerals, which are composed of these two elements, are the most plentiful on earth. There are many *silicates*, most of which possess in addition to silicon and oxygen varying proportions of one or more of the seven relatively abundant metals and in some instances small amounts of hydrogen. Silicates are thought to predominate to the earth's core at a depth of 2900 km (1802 mi). The core itself appears to be largely iron.

We are all familiar with the organic compounds known as *hydrocarbons*, which are composed mostly of complex combinations of carbon, oxygen, and hydrogen atoms. The diversity of molecular combinations is primarily due to the chemical properties of carbon (and to a lesser extent oxygen and hydrogen) and its ability to bond in many ways. In the inorganic world, the silica tetrahedron—the basic building block of the diverse silicate group—functions similarly. Through various combinations of it and other, like units (or metal ions), the entire complex silicate group can be formed.

The silicon atom is relatively very small, and it bonds covalently to the four oxygen atoms

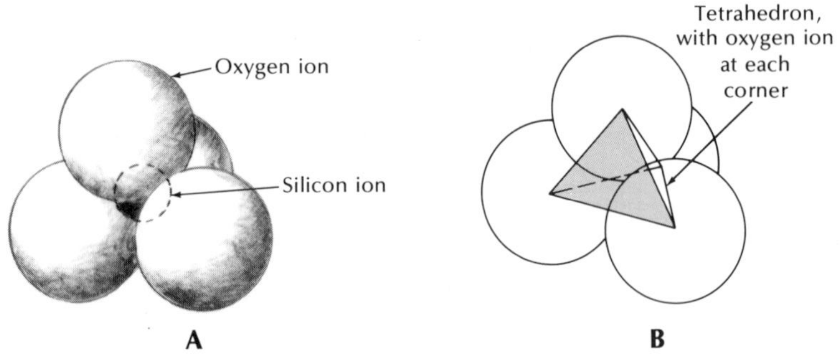

Fig. 3-14 A. The silica tetrahedron, consisting of one silicon ion surrounded by four oxygen ions. **B.** The same pattern with lines connecting the centers of the oxygen ions to emphasize their tetrahedral arrangement.

that can fit around it to produce a four-sided tent-shaped form called a *tetrahedron* (Fig. 3-14). The centers of the four oxygen atoms are at the points of the tetrahedron, each side of which is an equilateral triangle. The tetrahedra can exist singly, or they can be joined to other tetrahedra by sharing a common oxygen atom to produce a bewildering array of single- and double-chain structures, rings, or three-dimensional networks (Fig. 3-15). The common metal ions fit into available spaces in the crystalline lattices, helping to bond the tetrahedron units together.

Although silicates make up the bulk of the common minerals, a few *non-silicates*, particularly certain oxides and carbonates, are also locally abundant.

Fortunately for us, the minerals that make up most the solid earth are only about fifteen in number. Their names are given in Table 3-3—and in the following pages they are described.

The minerals naturally fall into two groups: silicates and non-silicates. In addition, a subgroup of four silicates has been designated the *ferromagnesian minerals*. All four contain cations of iron and magnesium and tend to be dark in color.

Before plunging ahead we should take a moment to explain the shorthand notation used by chemists to indicate the chemical makeup of a mineral compound. Rather than spelling out each element composing a mineral, they use one- or two-letter symbols to represent each element. Silicon is Si, oxygen is O, and SiO_2 is the formula for silicon dioxide,

the mineral known as quartz. The number 2 in the subscript indicates that quartz is made up of two parts oxygen to one part silicon, by atomic weights. By means of the chemical formula, one can quickly see what the constituent elements of a mineral are and in what proportion they occur. The symbols for the other common elements are: aluminum, Al; iron, Fe; calcium, Ca; sodium, Na; potassium, K; magnesium, Mg; titanium, Ti; and hydrogen, H.

Silicates

Quartz: SiO_2 Quartz has a vitreous luster, a hardness of 7, and when pure, is completely clear and colorless. In fact, the Greeks thought it was some kind of frozen water. It lacks cleavage, but it commonly fractures conchoidally (see Fig. 3-13). Should quartz grow free from interference it crystallizes customarily in a six-sided crystal form, which is terminated by a sharp-pointed pyramid at each end. If quartz grows into cavities, as it commonly does, it will possess only one pyramid on the end of the crystal that extends into the opening (see Fig. 3-8). Crystals that grow into openings may sometimes reach lengths of 0.3 m or more. Usually quartz occurs in association with other minerals as tiny grains two to three millimeters across that generally lack crystal faces. Where fresh and unweathered the disseminated grains often sparkle like tiny fragments of glass.

Feldspar The feldspars are the most abundant by far of the common minerals, probably making up at least 50 per cent of the rocks at the earth's surface. The two most common ones are *potassium feldspar,* which is rich in potassium, and *plagioclase,* which is rich in sodium and calcium. Silica tetrahedra in these minerals are joined in a strong three-dimensional network that possesses planes of weakness in two directions at or nearly at right angles to each other (Fig. 3-16). Cleavage along the planes of weakness and a hardness of 6 (on the Mohs scale) provide two of the most characteristic properties of the feldspars. In many rocks they occur in well-formed crystals with a tabular form, so that on weathered or broken surfaces they often resemble little gravestones or fence laths (Fig. 3-17).

Potassium feldspar: $KAlSi_3O_8$ There are several common varieties of potassium feldspar (among them orthoclase), which vary in the way their ions are arranged. For our purpose, little will be gained by differentiating between them. Potassium feldspar has a vitreous luster and may be colorless but is usually milky white or flesh pink. It sometimes resembles unglazed porcelain, like the dull surface exposed on a chipped dinner plate.

Plagioclase: $NaAlSi_3O_8 \cdot CaAl_2Si_2O_8$ Plagioclase, like potassium feldspar, has a vitreous luster. Its color is most likely to be white or

Table 3-3. Important rock-forming minerals

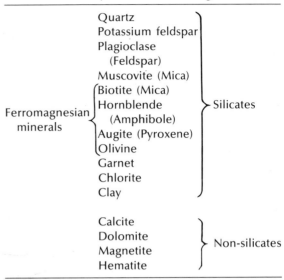

	Quartz	
	Potassium feldspar	
	Plagioclase (Feldspar)	
	Muscovite (Mica)	Silicates
Ferromagnesian minerals	Biotite (Mica)	
	Hornblende (Amphibole)	
	Augite (Pyroxene)	
	Olivine	
	Garnet	
	Chlorite	
	Clay	
	Calcite	
	Dolomite	Non-silicates
	Magnetite	
	Hematite	

pale gray, although some varieties show a beautiful iridescence, or play of colors, much like those of a peacock's feathers. In some cases it can be almost glass-clear, like quartz. Distinguishing plagioclase from potassium feldspar or quartz can probably best be done by examining the crystal or cleavage surfaces of the mineral for *striations*—a multitude of

---

Fig. 3-15. Examples of the diversity in the basic silica tetrahedron in silicate minerals.

SILICATE STRUCTURES

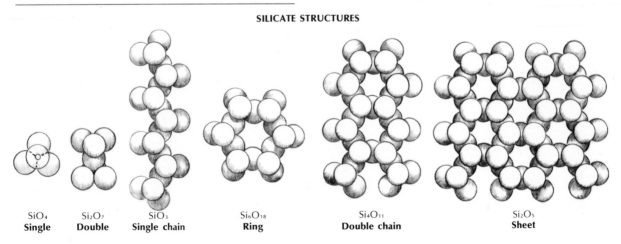

SiO_4	Si_2O_7	SiO_3	Si_6O_{18}	Si_4O_{11}	Si_2O_5
Single	**Double**	**Single chain**	**Ring**	**Double chain**	**Sheet**

very closely spaced parallel straight lines which look as if they had been engraved on the surfaces. Striations, which are well developed in plagioclase, are not found in potassium feldspar and quartz.

If we were to analyze the chemical composition of plagioclase taken from several different rocks, we would find that most likely the mineral contained calcium, sodium, aluminum, and silicon as the principal cations but that the proportions of each differed in each rock. The reason for the variability is easy to understand if we look first at the relative sizes of the four cations. Sodium and calcium are both relatively large and similar in dimension, whereas silicon and aluminum are both relatively small and not greatly dissimilar in size. Considering size alone, it is possible for the sodium and calcium or the aluminum and silicon to interchange, or substitute, for each other. Cations, however, are positively charged particles and during interchange, the neutrality of the electrical charge in the lattice must be maintained. The cations of sodium, calcium, aluminum, and silicon carry electrical charges of +1, +2, +3, and +4, respectively. Therefore substitution of a sodium ion (+1) for a calcium ion (+2) cannot

Fig. 3-16. Fragment of potassium feldspar typically displaying two cleavages at right angles. One cleavage plane parallels the front (illuminated) surface and the other parallels the upper surface. *Jan Robertson*

Fig. 3-17. Well-formed crystals of potassium feldspar, each about 3 cm (1.2 in.) long. On a broken rock surface they commonly have a lath-like outline. *Jan Robertson*

SILICA TETRAHEDRA IN MUSCOVITE SHEET

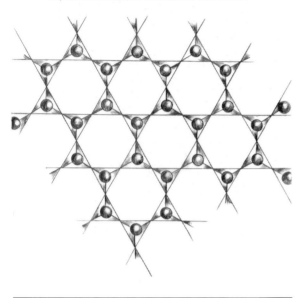

Fig. 3-18. Hexagonal arrangement of tetrahedra in a sheet of muscovite. The tetrahedra are strongly bonded through the sharing of a common oxygen atom at each point. A flake of mica is composed of many thousands of such sheets stacked one on top of the other (Fig. 3-19).

occur unless at the same time there is a substitution of a silicon ion (+4) for an aluminum ion (+3). The process of interchange of cations in minerals is called *solid solution*. In a case such as just described, where pairs of cations are involved, it is called *coupled solid solution*.

Substitution of one metal for another is very common in the mineral world, and is one of the major reasons that minerals are not pure chemical compounds. A simple way of informing the reader that two components have the same charge, are nearly identical in size, and can substitute in any proportion for one another is to enclose the symbols in parentheses; for example (Fe,Mg) indicates that iron and magnesium are freely interchangeable.

Mica The micas include a number of closely related minerals, all of which possess one perfect (or nearly so) cleavage and have a hardness ranging from 2 to 3. X-ray analysis shows that the crystal of mica consists of parallel sheets, like the pages in a book, which are made up of silica tetrahedra strongly bonded together at

MUSCOVITE STRUCTURE

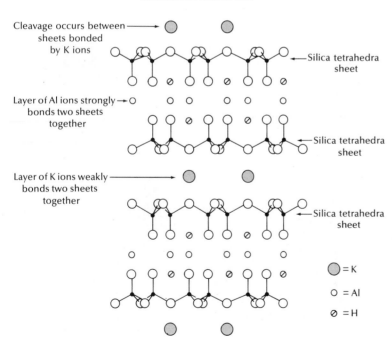

Cleavage occurs between sheets bonded by K ions

Silica tetrahedra sheet

Layer of Al ions strongly bonds two sheets together

Silica tetrahedra sheet

Layer of K ions weakly bonds two sheets together

Silica tetrahedra sheet

◯ = K

○ = Al

⊘ = H

Fig. 3-19. The double-sandwich structure of muscovite. The points of the silica tetrahedra in the upper two sheets face each other and are strongly bonded by aluminium (Al) ions. The lower sheets are similarly bonded. The double layers, however, are bonded by potassium (K) ions which form a relatively weak bond; the one perfect cleavage in mica occurs at and parallel to these planes of weakness.

their bases (Fig. 3-18). The sheets themselves are bonded to each other less strongly, so that they can split apart easily—thereby resulting in the perfect cleavage so characteristic of that mineral group (Fig. 3-19). The two most common rock-forming micas are *muscovite* (Fig. 3-20) and *biotite*.

Muscovite: $KAl_3Si_3O_{10}(OH)_2$ The common name for muscovite is *white mica;* and generally it is colorless, gray, or transparent, especially when split into thin sheets. Muscovite was used in the tiny windows of the houses of medieval Europe before the widespread use of glass brought more light to the gloomy interiors. It has a silky or pearly luster, and in sunlight the cleavage plates of the tiny mineral grains, which are common in many rocks, shimmer and shine—the German name of *glimmer* for white mica conveys an impression of that very property.

Biotite: $K(Mg,Fe)_3 \cdot AlSi_3O_{10}(OH)_2$ Biotite is commonly called *black mica*, and its chemical formula indicates that it includes iron and magnesium (in solid solution) in its composition, while muscovite does not. Thin cleavage sheets of it lack the transparency of muscovite. Black mica ranges in color from dark brown to black, and in many rocks it occurs as jet-black flakes that shine like satin in the sun.

Hornblende: $Ca_2Na(Mg,Fe)_4(Al,Fe,Ti)_3–Si_6O_{22}(O,OH)_2$ Hornblende, another ferromagnesian mineral, is the most common of a large and complex group with somewhat similar physical properties. *Amphiboles* is the name given to this group. Hornblende is a dark mineral, commonly dark green or jet black, shining as brightly as a lacquered surface in the sun. The crystals are as a rule long and narrow, and one of its most distinctive properties is its cleavage pattern (Fig. 3-21). The silica tetrahedra are joined in long, double chains that parallel the long axis of the crystal, and the two good cleavages, which intersect each other at angles of 56° and 124°, are parallel to planes of weakness (weaker bonds) that exist between groups of chains (Fig. 3-22).

Fig. 3-20. Model of the atomic arrangement in muscovite and an actual cleavage fragment (lower left). The large transparent spheres represent layers of silicon tetrahedra; the large dark spheres, potassium atoms, and the three layers of small black spheres, aluminum atoms. The one perfect cleavage in muscovite occurs at the layers of potassium atoms, where bonding is less strong than elsewhere. *Jan Robertson*

Fig. 3-21. Typical appearance of a hornblende crystal; a diamond-shaped (but six-sided) cross section, longer than it is wide. Two good cleavages intersect at 56° and 124°.

HORNBLENDE

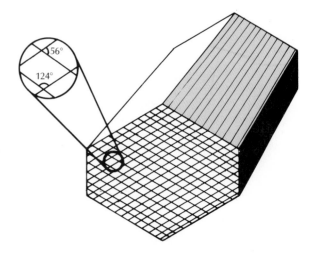

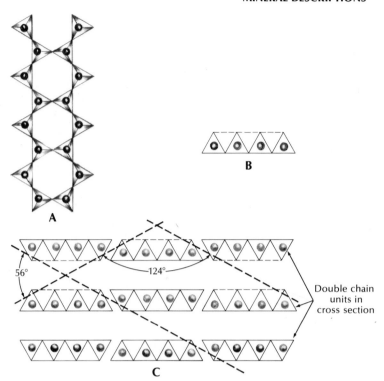

Fig. 3-22 A. In hornblende the silica tetrahedra join together to form strongly bonded elongate chains. B. Cross section of one such chain structure. C. The chains parallel the long axis of the crystal and are bonded to each other mostly by metal ions. The bonds are relatively weak and the two characteristic cleavages occur along the planes of weakness, as shown in the cross section.

Fig. 3-23. An augite crystal (pyroxene) is typically square in cross section and has two good cleavages that meet nearly at right angles; the crystal is generally as wide as it is long.

AUGITE

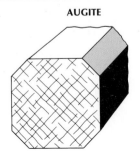

Augite: Ca(Mg,Fe,Al)(Si,Al)$_2$O$_6$ The ferromagnesian mineral augite looks superficially like hornblende in that it is dark and possesses two good cleavages and a vitreous luster. Augite crystals, however, generally are stubbier, to the point of being nearly equidimensional, and their two cleavage planes meet nearly at right angles (87° and 93°). Augite crystals seen in cross section are nearly square (Fig. 3-23). The cleavage planes in augite also parallel planes of weaker bonds between silicate chains but in this case the chains are single rather than double (Fig. 3-24).

Hornblende and augite are the more common of the darker rock-forming minerals. The principal distinctions between the two are: (1) hornblende crystals tend to be long and narrow, while augite crystals are short and stubby; (2) hornblende has two cleavages parallel to the long axis of the crystal that meet at 56° and 124°, whereas augite has two that

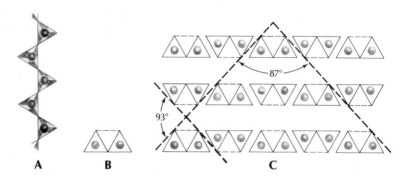

Fig. 3-24 A. In augite, the basic pattern is a single chain of silica tetrahedra. **B.** Cross section of one single-chain structure. **C.** Metal ions bond chains together but are relatively weak. Cleavage occurs at 87° and 93° along planes of weakness corresponding to planes of weaker bonds, as shown in the cross section.

intersect each other at approximately right angles; (3) hornblende crystals seen in cross section approach a rhombic pattern, while augite crystals are more nearly square.

Olivine: (Fe,Mg)$_2$SiO$_4$ Olivine, another ferro-magnesian mineral, usually occurs as rounded, granular glassy crystals, and from the name you are likely to guess that the color is green. When the crystals are large enough and free from blemishes they can be cut into attractive, though fragile, gemstones. *Peridot* is the name given to the gem variety of the mineral.

Olivine crystals in fresh, dark lavas often look like tiny bits of dark- to light-green bottle glass. As indicated by the chemical formula, olivine is a solid-solution mineral, which can contain variable proportions of iron and magnesium. Most varieties have an intermediate composition and most commonly are richer in magnesium than in iron.

Garnet: Complex silicate mostly containing Ca, Fe, and Al Garnet is the name given to a group of minerals that possess basically the same silicate structure. It has no cleavage and breaks with an uneven or conchoidal fracture. It possesses a resinous to a vitreous luster and a hardness of about 7. Colors of garnet are highly varied but most commonly they are red, brown, or yellow. Garnet almost always occurs in well-formed equidimensional crystals—perhaps its most distinctive property (Fig. 3-25).

Fig. 3-25. Garnet crystals about 2 cm across showing the characteristic equidimensional form. *Jan Robertson*

Chlorite: Complex group of hydrous silicates containing Mg, Al, and to a lesser degree Fe and other metal ions Chlorite usually occurs in scaly or thinly banded masses. Its luster is vitreous, it gives a greenish streak, and its grass-green to blackish-green color is one of its most characteristic properties. Other definitive properties are a hardness of only 1 to 2.5, and one perfect cleavage.

Clay: Hydrous aluminum silicate Clay is the common name of a group of minerals that result from the weathering of rocks. It has basically a white color but can be easily stained by impurities. Clay feels greasy to the touch, and often has a distinctive odor—somewhat like the air just after the start of a summer rain. Some varieties adhere to the tongue, or become plastic when moistened. Most commonly clay occurs in soft, compact earthy masses.

Non-silicates

Calcite: $CaCO_3$ and dolomite: $CaMg(CO_3)_2$ Normally, calcite is a light-colored (white or pale yellow) or colorless mineral, although—depending upon the amount and nature of the impurities—the color may range across a spectrum including yellow, orange, brown, and black. Calcite has a vitreous luster, a hardness of only 3, and occurs in a variety of crystal forms which commonly are six-sided. It possesses nearly perfect cleavage in three directions (see Fig. 3-12), and the intersections of the cleavage planes produce a rhombohedral pattern—that is, when the mineral breaks into fragments, each of the faces is a rhombus, or approximately diamond-shaped figure.

Dolomite looks very much like calcite superficially. A test to discriminate between the two is the placement of the unknown mineral in cold, dilute hydrochloric acid. Calcite reacts readily, with a vigorous release of bubbles, whereas dolomite reacts much more quietly and slowly. Dolomite also may be distinguished from calcite by its slightly greater hardness (3.5), its higher specific gravity, and by the fact that its crystal faces often are slightly curved, not sharp and well-defined planes.

Magnetite: Fe_3O_4 and hematite: Fe_2O_3 Magnetite, the black oxide of iron, occurs most commonly as disseminated grains in rocks rich in ferromagnesian minerals. It is opaque, has a metallic to submetallic luster, and lacks cleavage. Certainly its most distinctive property is its ability to be attracted to a strong magnet—hence, the name. Large pieces of magnetite are called *lodestone*.

Hematite looks like magnetite when in small grains, but it is less magnetic and produces a red to red-brown powder on a streak plate. At times it occurs as a very fine pigment that colors an entire rock red, sometimes so intensely that it looks as if the rock had been painted. Some varieties possess a metallic luster; when cut and polished they make fine semiprecious gemstones called black diamonds. Most hematite, however, has a dull, fibrous, or earthy luster. Some of the latter varieties are referred to as red ochre, a natural paint pigment.

ROCKS

Pick up any rock at the earth's surface and look at it closely (Fig. 3-26)—more than likely you will find that it is composed of a number of different mineral grains that differ in color, luster, grain size, grain shape, and so on. Because the mineral grains in most rocks are relatively tiny, many of the properties of the various minerals are hard to assess. For example, it is difficult to determine how many cleavages there are and at what angles they meet, when the crystal in question is no larger than a grain of rice. In time, however, if you are persistent you will be able to identify nearly all of the minerals in any common rock. And when that time comes, you will realize that all common rocks are almost entirely composed of the relatively few minerals just described. You will also become aware that rocks differ not only in the number and proportions of their constituent minerals but also in texture, which pertains to grain size, both absolute and relative, and how the grains fit together.

Geologists have found that most differences in mineralogy and texture are related to the way rocks were formed. By studying rocks and comparing their similarities and differences, we can obtain clues about the processes and events that formed them in the geologic past. The study of rocks has undoubtedly become the principal basis for our understanding of the earth.

As an aid to comparing similarities and differences between rocks, geologists have devised a *rock classification* system—imposing an order where none before existed. Biologists have done the same thing for the animal and plant kingdoms. To be sure, the scheme is arbitrarily defined, and there are the inevitable exceptions, borderline cases, and overlapping occurrences, but it works reasonably well for most rocks. Fundamentally the classification is a *genetic* one, based on the origin of the rocks. Each grouping represents rocks that have formed in much the same way, and each main group has been progressively subdivided. The smallest divisions are based on a rock's texture, mineralogy, and proportion of mineral constituents. Each division has been given a formal rock name; for example, *granite, shale, slate*. Once a rock has been classified, the name itself carries all the genetic, mineralogic, and textural information concerning that rock; which makes for ease of comparison of rocks of all ages from around the world, not to mention ease of communication among earth scientists.

In the following section we will describe only the major subdivisions of rock classification. We shall leave for later chapters the detailed description of the rocks within each group and the processes that form them.

All rocks can be placed into three main groups: *sedimentary, igneous,* and *metamorphic*.

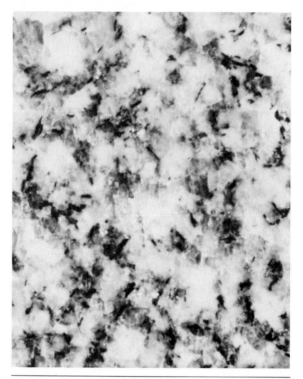

Fig. 3-26. A rock composed of a number of different minerals, some black, some gray, some white. The individual grains of each mineral generally have similar properties, which are different from those of other minerals. *Jan Robertson*

Sedimentary rocks

Of the three genetic rock families, sedimentary rocks are perhaps the most readily comprehended, because many of them bear a close resemblance to the materials from which they are made; also many of the processes responsible for their formation occur before our eyes or in reasonably accessible environments. About 75 per cent of the earth's surface is covered by sediments. Even so, they form only a thin, discontinuous veneer that is spread over the much more abundant igneous and metamorphic rocks, which are the true foundations of our continents and ocean basins.

Many sedimentary rocks can be thought of as secondary, or derived, rocks, in that they are

(Above) Mineral aggregate composed mostly of potassium feldspar (green-blue) and smoky quartz (black) formed as a result of igneous activity. *Floyd Getsinger* (Below) Rhodochrosite from Colorado. This mineral, here in a well-crystallized rhombohedron, is closely related to calcite. *William C. Bradley*

(Above) Biotite gneiss, viewed under the polarizing micro-scope. The colored minerals are mostly biotite; the light-to dark-gray minerals are predominantly quartz and feld-spar. *William C. Bradley* (Below) Basalt as viewed under the polarizing microscope; crossed polars. The blue and yellow-green crystals are augite, the gray are calcic-plagioclase, and the black mineral is magnetite. *Edwin E. Larson*

composed of bits of pieces of pre-existing rocks held firmly together by a suitable cement. Such a texture is called *clastic* (Greek, broken), and the fragments can be called *clasts*. Examples of clastic sedimentary rocks are (1) *sandstone,* which consists of sand grains cemented together; (2) *conglomerate,* which consists of larger, rounded fragments cemented together; and (3) *shale,* which consists of very small particles. Some sedimentary rocks may result from chemical precipitation in lake or sea water (for example, salt); others may result from the accumulation of a variety of organic remains.

Sedimentary rocks accumulate either on land or on the floors of lakes or seas. They are built up through the slow deposition of material, and so are typically formed layer upon layer. The layers are called *strata;* a single layer is a *stratum* (directly from the Latin—a blanket or pavement, derived from *stratus,* p.p. of *sternere,* to spread out). Individual layers may range from paper-thin sheets up to massive beds 30 m (98 ft) or more thick.

Igneous rocks

Igneous rocks have solidified from a molten silicate material to which the name *magma* is given. Most magmas are rich in silicon and oxygen; in most cases the two elements make up from 45 to 75 weight per cent of the molten material. The word *igneous* comes from the Latin *igneus,* having to do with fire. We use the same root in everyday language when we speak of the ignition system of a car.

The source for all igneous rocks is deep below the surface, where temperatures are relatively high. If the magma finds a path to the surface it erupts, cools, and solidifies to form *volcanic* rocks. The magma remaining at its source or magma which never reaches the surface ultimately cools and crystallizes in the deep, inaccessible subsurface domain to form *plutonic* rocks. The word *plutonic* comes from Pluto, the Greek god of the lower world.

In most igneous rocks the constituent mineral grains form an interlocking network of crystals, some perhaps with crystal faces, most without. The network is called the *crystalline* texture and results from the progressive growth of mineral grains during solidifaction, or crystallization, of the magma. The overall mineralogy, average grain size, and differences in grain size give clues to the conditions that existed during crystallization. For example, magma underground cools and crystallizes more slowly than magma at the surface, and for that reason the mineral grains in plutonic rock have a chance to grow much larger than those that crystallize quickly during the rapid cooling of a lava flow. In some cases, grains formed during the very rapid cooling of a lava flow are even too small to be seen without a microscope.

Metamorphic rocks

If metamorphic rocks were a puzzle to the first geologists we can appreciate their confusion. Such rocks do not form on the earth's surface but appear instead to be products of the action of internal heat, pressure, and chemical activity of fluids through long periods of time—long at least when judged by our time standards. These factors produce recrystallization, either partial or complete, of the minerals of the rock and the development of a crystalline texture. The name *metamorphic* means "change in form." New minerals appear, and they may develop a wholly new fabric or orientation with respect to each other. Instead of having a random orientation and heading every which way, as is true of many igneous rocks, the minerals undergoing metamorphic recrystallization may align themselves parallel to one another as in a deck of cards. Such layering is called *foliation,* and it can be weakly to strongly developed and on a coarse to fine scale. Some metamorphic rocks have such a strongly developed foliation that it superficially resembles the stratification of sedimentary rocks. The metamorphic layers, however, consist of interlocking crystals segregated into layers of dark minerals (ferromagnesian) and light-colored minerals. In some cases, elongate minerals like hornblende may become aligned with their long axes parallel, to produce a *lineation*.

In most types of metamorphism, the rock undergoes little or no change in chemical composition as its minerals recrystallize. The chemical elements already present simply regroup themselves under the conditions of higher temperatures and pressures to form new minerals which are stable in the new subsurface environment. In some cases, however, new minerals are formed because heated gases and fluids circulating within the earth very often associated with plutonic igneous activity have introduced new materials.

Metamorphic rocks are almost certain to be complex because they have no single mode of origin; in some cases temperature is the important factor, in others directed pressure, and in others the nature of the fluid phases. They can be made from all manner of rocks: igneous, sedimentary, or even from previously metamorphosed rocks. If they have any factor in common, it is crystallinity, and, like igneous rocks, they consist of a fabric of interlocking crystalline minerals.

SUMMARY

1. An *element* is composed of the same kind of atoms; a *compound* is composed of groups of different kinds of atoms.
2. Atoms consist of *electrons, protons,* and *neutrons.* In the atoms of a particular element, the number of electrons equals that of protons and is a fixed number (*atomic number*). But the number of neutrons can vary.
3. Atoms bond together through *ionic, covalent,* and *metallic bonding* processes that involve interaction of the outer electron shells. Molecules can also bond together through electrostatic forces.
4. Matter can exist in three different states: *gas, liquid,* and *solid.* The particles that make up a gas are *randomly ordered;* those

in a liquid possess *short-range ordering;* and those in a solid, *long-range ordering,* or *crystallinity.*

5. *Minerals* are naturally occurring crystalline compounds and elements of a definite chemical composition or a restricted range of chemical compositions. The crystalline nature of minerals has been established through *X-ray diffraction* studies.
6. Every mineral has a definite set of properties that can be used to identify it. Commonly used physical properties are: *color, cleavage* (or lack of it), *hardness, luster, specific gravity,* and *magnetic character.*
7. Oxygen and silicon make up almost 75 per cent of the minerals at the earth's surface. Most minerals are composed of those elements and are called *silicates.* The basic building block of silicates, *tetrahedra,* can exist singly or in an array of chains, rings, or three-dimensional networks.
8. Important rock-forming minerals are relatively few in number and include the following silicates: *quartz, potassium* and *plagioclase feldspars, muscovite* and *biotite micas, hornblende, augite, olivine, garnet, chlorite,* and *clay.* Four of them are relatively rich in iron and magnesium: biotite, hornblende, augite, and olivine. Four non-silicate minerals are also common: *calcite, dolomite, magnetite,* and *hematite.*
9. Rocks are composed of aggregates of one or more minerals, the grains of which fit together in various ways (*texture*). Most mineralogic and textural differences are related to the way in which different rocks formed.
10. The system of rock classification is genetic. The smallest units in the classification are based on the rock's texture and mineralogy and on the proportions of its mineral constituents. All rocks fall into one of three main categories: *sedimentary, igneous,* and *metamorphic.*
11. Most *sedimentary rocks* are made of fragments (*clasts*) of pre-existing rocks

cemented together. Some are the result of chemical precipitation or accumulation of organic remains. The source of *igneous rocks* is a molten material called *magma*. If magma solidifies below ground the rocks are called *plutonic;* if it solidifies on the surface, they are called *volcanic. Metamorphic rocks* form from the solid-state recrystallization of pre-existing rocks under the influence of heat, pressure, and the chemical activity of fluids.

SELECTED REFERENCES

Bragg, Sir Laurence, 1968, X-ray crystallography, Scientific American, vol. 219, no. 1, pp. 58–79.

Dana, J. D., 1962, The system of mineralogy; 7th ed., C. Palache, H. Berman, and C. Frondel, eds., 3 vols., John Wiley and Sons, New York.

Desautels, P. E., 1968, The mineral kingdom, Madison Square Press, Grosset & Dunlap, New York.

English, G. L., 1934, Getting acquainted with minerals, McGraw-Hill Book Co., New York.

Holden, A. and Singer, P. 1960, Crystals and crystal growing, Doubleday and Co., New York.

Mason, B., 1958, Principles of geochemistry, John Wiley and Sons, New York.

Pough, F. H. 1976, Field guide to rocks and minerals, 4th ed., Houghton Mifflin, Boston.

Sinkankas, J., 1966, Mineralogy: a first course, D. Van Nostrand Co., Princeton, New Jersey.

Turekian, K. K., 1972, Chemistry of the earth, Holt, Rinehart and Winston, New York.

Tutton, A. E. H., 1924, The natural history of crystals, E. P. Dutton and Co., New York.

Vanders, I., and Kerr, P. F., 1967, Mineral recognition, John Wiley and Sons, New York.

Fig. 4-1. Eruption of Cerro Negro, Nicaragua, which produced both the lava flow in the middle background, and subsequently, the steep-sided cone. Large rocks, explosively hurled from the vent, fall near the crater and cascade down slope leaving trailing white plumes of vapors (mostly water). *USGS*

4

IGNEOUS PROCESSES AND IGNEOUS ROCKS

Almost all igneous rocks owe their existence to the crystallization of minerals from the progressive cooling of a hot molten liquid called magma, or, alternatively, *melt*. The existence of magma and the relatively rapid formation of one group of igneous rocks—volcanic rocks—can be readily documented during and following the eruption of a volcano at the earth's surface (Fig. 4-1). Geologists have come to realize that the magma which feeds the fiery fountains of a volcano originates deep below the ground surface, where it resides in pools called *magma chambers*—pools surrounded mostly by solid rock. The subsequent cooling and crystallization of magma that remains in the chambers occurs slowly and leads to the formation of generally coarser-grained plutonic rocks. Later, erosion may strip off enough of the overlying rocks to lay bare the deep-seated plutonic rocks.

About 200 years ago, however, when the science of geology was in its formative stages, earth scientists had other ideas concerning the origin of igneous rocks. One such person was the influential geologist Abraham G. Werner (1750–1817), who gave inspiring lectures at the Freiberg Mining Academy in Saxony. In 1787 Werner published a general theory setting forth his ideas on how rocks of all kinds were formed. Volcanoes, he conjectured, resulted from the combustion of buried coal seams, which caused rocks near the earth's surface to melt. Since most coal is geologically of relatively recent origin, he considered volcanism to be a very young geologic phenomenon. At the time, Werner's hypothesis was completely reasonable. He knew for a fact that plutonic rocks occurred below the coals seams and therefore could in no way be related to volcanic activity. From the evidence at hand, he concluded that plutonic and metamorphic rocks were part of the original crust of the earth, which had been precipitated out of the sea waters that once covered the globe. Werner's dogma, called *neptunism*, survived for more than 50 years. The beginning of its demise was brought about, in 1795, by the efforts of the perspicacious Scot, James Hutton (whom we earlier encountered in connection with the principle of uniformitarianism). Hutton was duly impressed with the existence of subterranean heat, which could be demonstrated in mines, and inferred from hot springs and volcanoes. In his tramps through the countryside he found evidence suggesting that coal seams had themselves been seared and charred by previously existing molten material that came in contact with them—evidence in direct opposition with Werner's contention. In his wanderings Hutton also observed a body of rocks composed of the plutonic rock granite. After carefully inspecting its mineralogy he decided that sea water could not possibly have contained simultaneously in solution all of the different ions of which the granite was composed. Rather, it appeared to him that the granite resulted from crystallization of molten rock deep beneath the surface. That sagacious conclusion established the basis of modern thought concerning the origin of plutonic igneous rocks.

Unfortunately, when Hutton published his two-volume work, *Theory of the Earth with Proofs and Illustrations* (1795), he was ahead of his time. His theories directly opposed Werner's, whose influence was then at its zenith. Eventually, however, Hutton's views, which better fitted the observational data, came to be accepted.

Today we recognize many kinds of igneous rocks, each the result of the particular chemical and physical conditions that existed during their formation, and each with a different name. The naming of an igneous rock is based primarily on its mineral composition and texture, as we learned in the last chapter.

MINERAL COMPOSITION

The kinds and proportions of minerals found in any igneous rock depend on the chemical composition of the magma from which the rock crystallized; and magmas show a great deal of variation within limits. Essentially all are molten mixtures rich in, as you might expect, oxygen and silica with lesser amounts of the metal ions aluminum, iron, magnesium, calcium, sodium, potassium, and titanium. All contain a small percentage of volatile constituents such as water, carbon dioxide, and hy-

drogen chloride in solution. The temperatures of magma, as determined from direct measurements, experimental studies, and theoretical considerations, range from about 700 to 1250°C (1290 to 2282°F) depending on its chemical composition and the depth at which it was formed.

To some degree, igneous rocks resemble steel in the way in which they are formed. Steel is usually made in open-hearth furnaces, and the properties of any batch of the metal are determined largely by the addition of such elements as tungsten, molybdenum, or chromium to the original charge of pig iron and scrap. For example, stainless steel is made by adding 18 per cent chromium and 8 per cent nickel. So it is with igneous rocks. For example, if the silicon content of a magma is high, it is likely that quartz will be an abundant mineral in the resulting rock; if it is low, the rock will be more or less quartz-free. To a large extent, the initial composition of magma is determined by the chemical composition and percentage of the rocks that are melted to provide the magma.

During cooling, not all minerals crystallize from a magma simultaneously. For example, in

---

Fig. 4-2. Drawing of basalt as seen under the microscope. It shows well-formed grains of calcium-rich plagioclase (white) and slightly corroded olivine (dark rimmed), with grains of augite (stippled) that have crystallized out between the larger grains. A few magnetite grains (completely black) are also present.

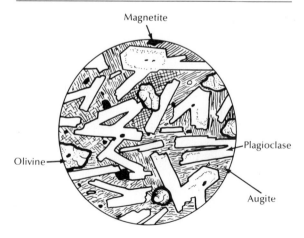

BOWEN'S REACTION SERIES

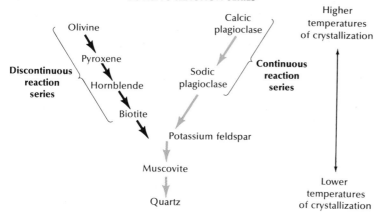

Fig. 4-3. Bowen's reaction series—a dual sequence of crystallization of minerals from a magma. The arrows indicate the direction of crystallization as cooling proceeds.

the dark lava called basalt, olivine and calcium-rich plagioclase occur as well-formed crystals—apparently they were the first minerals to crystallize. Augite, on the other hand, usually occurs in grains without crystal faces that apparently formed in spaces between the earlier-formed grains (Fig. 4-2). The various crystals form in sequence because the minerals contained by a magma do not have same melting points and temperature-dependent stability ranges. Those with higher melting points and higher temperature-dependent stability ranges will tend—if sufficient elemental material is present—to crystallize first. After the process begins, the first crystals are suspended in the magma. During subsequent cooling, and as minerals with lower melting points and lower temperature-stability ranges begin to crystallize, those earlier-formed crystals may react with the magma, forming new minerals that are more stable at lower temperatures.

Many years ago the petrologist Norman Levi Bowen (1887–1956) proposed that under favorable conditions minerals were likely to crystallize in the order shown in Figure 4-3. In a silicate magma rich in iron, magnesium, and calcium, but less rich in potassium and sodium, two minerals, olivine and calcium-rich plagioclase, are the first to crystallize. The plagioclase crystals, however, change continuously, becoming richer in sodium by reacting with the

remaining fluid magma as the temperature continues to fall and crystallization proceeds. The process is called a *continuous reaction series*.

Ferromagnesian minerals such as olivine, pyroxene, hornblende, and biotite form a *discontinuous reaction series* in which these earlier-formed minerals react within a particular temperature range to form completely new minerals. Olivine, if cooling is slow enough, reacts with the ions in the melt to form augite, and at a lower temperature augite reacts to form hornblende, and so on. Both continuous and discontinuous types of reaction go on only as long as the supply of melt holds out.

Toward the end of the cooling and crystallization process, when little melt remains, any silicon, oxygen, and aluminum ions remaining in the melt will begin to unite with potassium to form the low-temperature crystals of muscovite and potassium feldspar. If, at the end of the crystallization process, any silicon and oxygen remain, quartz will be formed.

What minerals are first crystallized and how far the earlier-formed crystals will react with the remaining melt depends largely on the initial chemical composition of the magma. If the melt is relatively rich in potassium and silicon, and poor in iron, magnesium, and calcium, the reaction path is still followed, but crystallization begins farther along toward the low-temperature end (Fig. 4-4).

By means of *Bowen's reaction series*, it is possible to explain some of the mineral variations in igneous rocks. For example, lavas spewed from a volcano will usually vary in composition; as an example, lavas rich in iron, magnesium, and calcium may be erupted first, followed by others less rich in those components. Earlier flows may contain little potassium feldspar, whereas later ones may contain considerable quantities.

When a magma rich in iron, magnesium, and calcium erupts and quickly cools it will form a dark lava composed mainly of olivine, calcium plagioclase, and augite. The remaining melt in the quietly cooling underground chamber may in time begin to form crystals of olivine and calcium-rich plagioclase. And those first-formed minerals may settle out and fall to the bottom, relatively depleting the magma of calcium, magnesium, and iron. Correspondingly the relative proportions of potassium, aluminum, and silicon will increase. Were the volcano to become active again, the lava erupted from the chamber would begin to crystallize somewhere in the middle part of the reaction series. The volcanic rock formed would be composed of pyroxene, amphibole, and intermediate plagioclase—an entirely different kind of rock than that resulting from the initial eruption.

If crystallization proceeded in the magma chamber and crystals continued to settle to the bottom, the remaining melt, which represents only a small part of the original volume, would be relatively rich in potassium, aluminum, and silicon. Subsequent eruption would produce lava flows that would begin crystallization quite close to the low-temperature end of Bowen's reaction series, and a rock containing abundant potassium feldspar and quartz would result.

The process just described, which involves the separation of earlier-formed minerals from the residual magma, is called *differentiation*. Introduction of new magma of the original composition into the magma chamber at any time during the middle and later stages of the process would cause an interruption of the differentiation process. As a result a different type of lava would be erupted, the exact composition of which would depend on the amount of new magma introduced and the composition and amount of residual magma and first-formed crystals that mixed with it.

Some geologists have pointed out that crystallizing magmas do not always follow, in detail, the sequence of the Bowen's reaction series. Even so, it provides a basis for understanding the crystallization sequence of most igneous rocks, and at least a partial explanation of the variability of their composition.

TEXTURE

The word *texture* is familiar to most people as having something to do with cloth; in fact it is derived from the Latin *textura*, a weaving. Today it means the arrangement and size of the threads in a woven cloth; for example, a burlap sack has a much coarser texture than silk. When we apply the term to such a thing as an igneous rock, we mean the size of the crystals as well as their mutual relationships.

Fig. 4-4. Lava formed from magma relatively rich in potassium and silicon, as seen under the microscope. Crystallization began at about the middle of the Bowen's reaction series and proceeded toward the lower-temperature end. The early-formed crystals of potassium feldspar, augite, and biotite are large and well formed. A mixture of fine-grained feldspar crystals and glass occurs between the larger crystals.

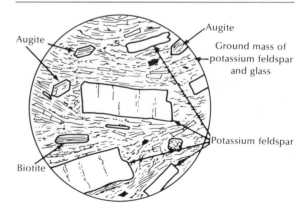

Fig. 4-5. Coarse-grained plutonic igneous rock with a crystalline texture, photographed under the microscope. Notice how the crystals have grown together to form a compact, interlocking network. *P.W. Lipman, USGS*

Crystalline texture

An igneous rock in which the minerals form an interlocking network is said to be crystalline (Fig. 4-5). If the crystals are visible to the unaided eye the rock is said to have a fine-, medium-, or coarse-grained texture, depending on the size of the crystals. An igneous rock whose texture is crystalline, as revealed by the microscope, but most of whose crystals are too small to be seen by the eye alone, has an *aphanitic* texture. The size of the crystals has nothing to do with their mineralogy; the mineral makeup of rocks with completely different grain sizes could be all the same. Why, then,

the differences in crystal size? The answer lies in the way in which the magma solidifies. If it cools slowly and under relatively undisturbed conditions, large crystals have a chance to grow around the various nuclei in the still-fluid magma. They may grow to fair size, up to 1.3 cm (0.5 in.) or more, in what is essentially a sort of crystal mush, with the last-forming minerals filling the interstices between the earlier-forming minerals when the whole mass finally congeals. The growth of a crystal occurs through the slow-acting diffusion, or movement, of the ions from the magma to the crystal surfaces, where the ions are added one by one. Should the magma cool rapidly, crystal growth goes on around the floating crystal nuclei as it did in the slow-cooling magma, but the whole process speeds up. Diffusion of ions over long distances in the magma is impossible, so many centers form and each nucleus continues to grow from diffusion taking place over very short distances. The result is a rock that, although crystalline, consists of a tightly knit fabric of much smaller crystals.

The texture of igneous rocks is determined in large part by what is known as their mode of occurrence—whether they solidify above ground or below it. To phrase the statement another way, are they volcanic or plutonic rocks? Volcanic rocks cool with relative rapidity, and therefore for the most part have fine-grained textures; intrusive, or plutonic, rocks cool more slowly, allowing the growth of larger crystals, and a coarse-grained texture is the result. In general, even the grain size of plutonic rocks is largely governed by the rapidity of cooling. If the magma chamber is small or close to the surface, it will cool more quickly than a larger one or one at depth, and correspondingly, the rocks will have a less-coarse grain size.

Two additional textures important among igneous rocks are *glassy* and *porphyritic* textures. They, too, provide information about the conditions that existed during cooling.

A glassy texture is typified by the volcanic rock *obsidian* (from the Latin *obsidianus*, after its describer Obsius), the most common type

Fig. 4-6. Obsidian, the most common type of volcanic glass, showing the characteristic conchoidal fracture. *Ward's Natural Science Establishment, Inc.*

of volcanic glass. It is wholly noncrystalline, because it passes quickly from the liquid to the solid state, leaving no time for the ions in the original magma to arrange themselves in ordered ranks as crystals. Obsidian commonly fractures conchoidally or irregularly into fragments with sharp edges (Fig. 4-6)—a property that made it an effective weapon when flaked or chipped into arrowheads, spear points, and knife blades by the people of ancient cultures.

Porphyritic texture takes its name from the Greek word *porphyra*, the word used for imperial purple—a highly prized dye extracted from an eastern Mediterranean shellfish. By extension the name was applied to a very specific kind of rock—*porphyry*, a dark, igneous rock from Egypt that contains small white feldspar crystals embedded in a purplish material composed of fine-grained crystals. Porphyry was greatly favored in the Roman world for carving busts of emperors, as well as their sycophants and the lesser dignitaries of the court. It made a striking contrast, especially when set off against an artfully draped white marble toga. Such a colorful arrangement

could scarcely fail to make even the least imposing Roman official regal.

Today, by a further extension of the original word, the term is applied to any igneous rock with a duality of grain size; that is, a rock containing crystals of two markedly different sizes (Fig. 4-7). Such a texture is interpreted to mean that the magma underwent two cycles of cooling, perhaps an earlier slow-cooling phase during which some crystals grew with well-formed faces to a relatively large size, followed by a later, more rapid phase when the smaller grains crystallized. In such a rock the larger crystals are called *phenocrysts* (from a Greek root, *phainein,* to show, combined with crystal), and the background material in which they are embedded is called the *groundmass*. Commonly, it is finely crystalline and even glassy (when the last stages of cooling are very fast).

A porphyritic texture is common in igneous rocks that solidified in small bodies at shallow depths. Initially the magma was in a deeper chamber, where some well-shaped crystals had begun to form. Movement of the magma to shallower depths brought about relatively rapid cooling, and it froze, in a sense, around the still-floating larger and earlier-formed crystals.

Because the larger crystals in a porphyry form while the magma is still largely a fluid, they grow without interference and often may achieve a nearly perfect crystal form. The minerals crystallizing later come out of solution more rapidly, and since the growth of each one is interfered with by its neighbors, very few have the opportunity to develop the external crystal form.

The quickness of cooling and variations in the cooling history probably are the principal reasons for variability in igneous rock texture, but not the only ones. Volcanic rocks rich in potassium, aluminum, and silicon tend to be more fine grained, even to the point of being glassy, than dark-colored volcanic rocks rich in iron, magnesium, and calcium. In fact, glass-rich varieties of ferromagnesium and calcium-rich rocks are virtually nonexistent. Quickness of cooling could not have been the principal factor in bringing about the difference in grain size, since both are volcanic rocks and must have cooled rapidly. The reason seems to be related to differences in the temperature of the lava when it was erupted and to the viscosity of the erupted molten material. As mentioned earlier, crystals grow through the process of diffusion. At higher temperatures the process is more efficient than at lower temperatures; also, ions can move about more easily in a magma that is less viscous. Lavas rich in iron, magnesium, and calcium originate at greater depths and are usually much hotter and therefore less viscous than those rich in potassium, aluminum, and silicon.

In plutonic rocks, textural differences are also related to the amount of volatile constituents (such as water, carbon dioxide, and hydrogen chloride) contained in the magma. It has been found that in any given magma, as the content of volatile materials is increased, initial crystallization takes place at a lower temperature and the efficiency of ionic diffusion is increased. The amount of dissolved gases in the magma is related to the amount contained in the rocks that melted to form the magmas as well as to pressure. At greater depths, where confining pressure is greater, magma can hold more volatile materials in solution than it can at lesser depths, where the pressure is less. Larger crystals should form from relatively volatile-rich magmas, other things being equal, than from those poor in volatile substances.

Clastic textures A clastic (fragmental) texture is usually associated with sedimentary rocks; yet volcanic rocks can also possess that texture.

Indeed, some clastic rocks have come about through the reworking of volcanic rocks by streams flowing on the flanks of a volcano. The deposits are composed mainly of rounded to subangular fragments, which can range from

Fig. 4-7. Porphyritic lava, seen on a flat surface. Large light-gray tabular crystals (potassium feldspar) and black tabular to irregular grains (biotite) are set in a medium-gray fine-grained groundmass composed mostly of potassium feldspar and quartz. In most cases a texture such as this is the result of two distinct phases of cooling, a slow one followed by a rapid one. *Jan Robertson*

boulders to clay-size grains. Such rocks are difficult to classify. But, since they are associated with volcanism, we shall consider them as volcanic with a clastic texture.

Clastic rocks can also be formed as a direct result of volcanism. The term applied to most such rocks is *pyroclastic* (Greek, fire and broken), which serves to differentiate them from normal sedimentary rocks and to emphasize their mode of origin. Pyroclastic rocks form from the volcanic material blown into the atmosphere during the explosive discharge of a vent and consist principally of fragments of volcanic crystalline rocks, volcanic glass, and phenocrysts.

CLASSIFICATION OF IGNEOUS ROCKS

All igneous rocks can be classified on the basis of what mineralogy and texture imply concerning the origin of rocks. More than several hundred names have been applied to igneous rocks, but we will discuss only a few major divisions here. The bulk of the named rock types are simply minor varieties of those listed below (Fig. 4-8).

Six of the major rock units, which comprise most of the igneous rocks, are shown. The names apply mostly to the crystalline igneous rocks. Clastic volcanic rocks are discussed separately.

Crystalline igneous rocks

The light-colored rocks, containing a large proportion of feldspar and silicon, are termed *felsic;* those containing abundant magnesium and iron (and calcium-rich plagioclase) are termed *mafic;* and those in between, *intermediate.* Even within a major subdivision, one rock can be described as being more "mafic" or more "felsic" than another.

Grossly, the three plutonic-volcanic divisions correspond to different starting magmas in the Bowen's reaction series. Gabbro and basalt form at high temperatures and from mafic melts; diorite and andesite at intermediate temperatures and from magmas of inter-

Crystalline igneous rocks			
	Felsic	Intermediate	Mafic
Extrusive fine grained (Volcanic)	Rhyolite	Andesite	Basalt
Intrusive coarse grained (Plutonic)	Granite	Diorite	Gabbro

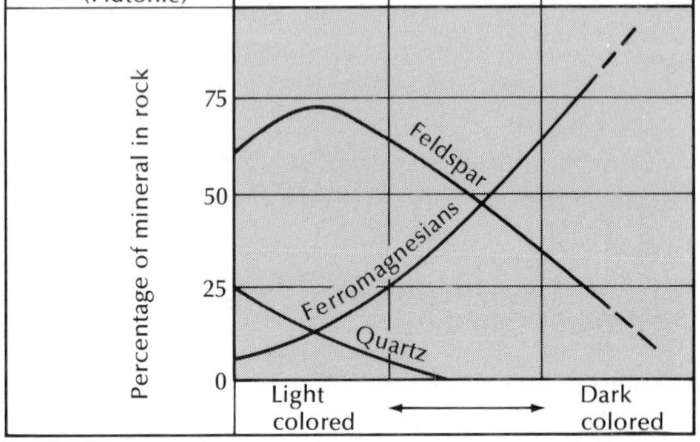

Fig. 4-8. The mineral composition and names of six common crystalline igneous rocks.

mediate composition; and granite and rhyolite at relatively low temperatures and from felsic melts.

Felsic rocks

Granite (the origin of the name is lost in antiquity, but it is believed by some to be derived from the Italian adjective *granita,* grained) is a relatively light-colored usually coarse- to fine-grained plutonic rock. Potassium feldspar and sodium-rich plagioclase together make up most of the rock (Fig. 4-9). Quartz is normally present in quantities up to 25 per cent; it is the last mineral to crystallize and occurs as gray rounded to irregular masses filling the spaces between earlier-formed minerals. Muscovite is also a common constituent in some granites. Black minerals, chiefly amphibole and biotite, are present in small quantities only.

Since granite is a widely used rock for such things as tombstones, monuments, and government and bank buildings (in fact, it has been used in the construction of stately

edifices for thousands of years), almost all of us readily recognize the characteristically speckled appearance of this familiar rock, with its white to gray background of feldspar and quartz flecked with the dark spangles of mica plates and needles of hornblende, as well as its coarse- to fine-grained texture. In the coarser varieties, some minerals (chiefly the feldspars) may reach dimensions of one cm (0.4 in.) or more across.

Rhyolite (from the Greek, *lava torrent* or *stream* + stone), a volcanic rock with the same compositional range as granite, has a wholly different texture. Commonly, it is fine-grained, glassy, or porphyritic; in the last case the groundmass may be so glassy or so fine grained that the minerals can be resolved only

Fig. 4-9. Granite from San Bernardino County, California, as seen under the microscope. The large crystal in center is potassium feldspar, the banded elongate crystals are mostly plagioclase, and light- to dark-gray equally spaced irregular grains are mostly quartz. *J. C. Olson, USGS*

Fig. 4-10. Rhyolite from Rio Grande County, Colorado, as seen under the microscope. Phenocrysts are mostly quartz and potassium feldspar with few plagioclase, biotite, and magnetite (black) crystals. The ground mass, originally glassy, has recrystallized to a network of extremely tiny grains of quartz and potassium feldspar. *P.W. Lipman, USGS*

with the aid of a microscope, if at all. Phenocrysts in the porphyritic varieties are usually small and composed of potassium feldspar, sodium-rich plagioclase, quartz, biotite, and, although rarely, other ferromagnesian minerals (Fig. 4-10). Quartz can be an early-crystallizing mineral during the formation of a rhyolite, and when it is, the crystals commonly appear completely formed and often terminated by pyramidal crystal faces at both ends. Rhyolite is ordinarily light colored, and may be white, light gray, or various shades of red. A characteristic textural feature is a streaked pattern known as *flow banding* (Fig. 4-11). As the name implies, the banding results from concentrations during

Fig. 4-11. Flow banding caused by viscous flow in a felsic lava that largely congealed as obsidian. The blocks are at the base of a volcanic dome south of Mono Lake, California. *John Haddaway*

flow of colored material or glass in layers just before the congealing, highly viscous lava solidifies. In many cases it appears that the lava must have flowed very slowly and with difficulty, much as heavy molasses does.

Rhyolite magma can cool so rapidly that crystallization of minerals is impossible. The resulting *volcanic glass* (already mentioned in discussing texture) is called obsidian. Volcanic glasses lack long-range internal crystal ordering and thus, from a geologic standpoint, are considered as supercooled or "frozen" liquids.

Window glass, another supercooled liquid, is colorless because it lacks metallic impurities. Obsidian, however, has such an abundance of impurities that the color is generally black, and less commonly red.

Volcanic glass tends to crystallize very slowly at the earth's surface. Over an extremely long period of time the ions in obsidian undergo solid-state diffusion (usually aided by the presence of water), which eventually results in total crystallization. Since diffusion takes place slowly, only tiny crystals can grow. The process is called *devitrification;* the resultant rock is called devitrified obsidian or devitrified glass. As a result, true obsidians are virtually unknown in the rock record farther back than the Cretaceous Period (135 million years).

Pumice (an ancient name, from a Greek word meaning worm-eaten and mentioned as long ago as 325 B.C. by the philosopher and naturalist Theophrastus) is a special variety of rhyolitic volcanic glass, rather like petrified froth, resembling the foam on the top of a beer (Fig. 4-12). Because of an abundance of gas cavities the rock is extremely porous and, accordingly, light in weight. Some pumice will even float on water; if pumice is blown out of coastal or oceanic volcanoes, it may drift for thousands of kilometers before becoming waterlogged and sinking to the bottom.

Intermediate rocks

Diorite (from the Greek word, to distinguish) is a coarse- to fine-grained plutonic rock whose mineral composition places it about midway between granite and gabbro—that is, it has an intermediate composition (Fig. 4-13). It is also a rock with an intermediate plagioclase feldspar as its chief constituent, with little or no quartz or potassium feldspar. Hornblende is its leading dark mineral, although biotite is often relatively abundant. The dark minerals can together be nearly as abundant as the feldspar (Fig. 4-13).

Because of a paucity of quartz and potassium feldspar and nearly equal proportions of plagioclase and ferromagnesian minerals, diorite tends to be a drab gray rock. It is not used so widely as granite for building stone because it is somewhat less abundant and possibly because its somber gray color is less pleasing.

Andesite (named for its occurrence in the Andean summit volcanoes of South America) is generally a gray to grayish-black, fine-grained volcanic rock. Commonly, its texture is aphanitic. In general, visible quartz is lacking,

Fig. 4-12. Pumice blocks and lapilli. Mammoth Lakes district, California.

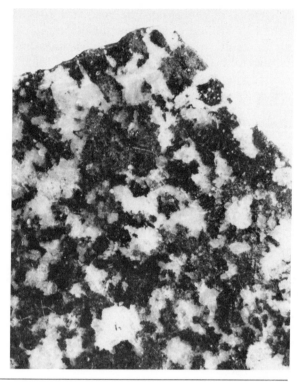

Fig. 4-13. Diorite from the Front Range, Colorado (left). As viewed under microscope. The large finely striped crystal in the center is intermediate plagioclase containing small crystals of quartz, hornblende, and biotite. The light-gray to white irregular grains are mostly potassium feldspar and quartz. *W. A. Braddock, USGS.* As seen in a hand sample (right). The light-gray patches are composed of intermediate plagioclase or, less commonly, of quartz; the black areas are made up of grains of hornblende, biotite, and magnetite. *Jan Robertson*

the chief feldspar is intermediate plagioclase, and the dark minerals are principally augite, hornblende, and biotite (Fig. 4-14). The same minerals also occur as phenocrysts in the porphyritic varieties. Andesite in general covers the same range of intermediate composition as its plutonic counterpart, diorite. Andesitic rocks are more abundant at the earth's surface than are rhyolites but are less widespread than basalts.

Mafic rocks

Gabbro (an old name for many of the dark rocks used in Renaissance palaces and churches in Italy) is a plutonic rock that consists typically of a coarse-grained intergrowth of crystals of pyroxene and calcium-rich plagioclase (Fig. 4-15). Many gabbros also contain olivine, and some possess minor amounts of hornblende. Unlike diorite and granite, gabbro contains larger amounts of ferromagnesian minerals than feldspars. There are exceptions, of course, and one gabbroic variety, *anorthosite,* consists almost entirely of calcium-rich plagioclase interlocked in a coarse-grained texture (Fig. 4-16). Another variety, much sought after as a decorative building stone for storefronts and banks, contains large dark-purplish calcium-rich plagioclase crystals that give a wonderfully impressive play of colors, like those of peacock feathers.

Basalt (one of the most ancient names in geology, since it apparently dates back to Egyp-

Fig. 4-14. Porphyritic andesite from Conejos Peak, Colorado, as viewed under the microscope. Most of the phenocrysts and small lath-shaped crystals are intermediate plagioclase. The ground mass is a mixture of glass and fine crystals of pyroxene. *P.W. Lipman, USGS*

Fig. 4-15. Gabbro from Victoria Land, Antarctica, as viewed under the microscope. The large dark-gray crystals are pyroxene, the light-gray crystals are mostly calcium-rich plagioclase, and the black crystals are iron oxides. *W.B. Hamilton, USGS*

Fig. 4-16. Anorthosite from Comanche County, Oklahoma, as seen under the microscope. The rock is composed almost entirely of large tabular crystals of calcium-rich plagioclase. *W.B. Hamilton, USGS*

tian or Ethiopic usage; one of the first references to it by name is by Pliny the Elder) is by far the most abundant of all volcanic rocks. Many regions of the world, such as the plateau bordering the Columbia River in the northwestern United States, were almost completely inundated by vast outpourings of basaltic lavas in the geologic past. The volume of basalt in the Columbia Plateau is estimated to be about 75,000 km³ (17,993 mi³). In addition, many oceanic islands such as Samoa, Hawaii, and Tahiti are volcanoes composed of basalt, and the rocks close beneath the ocean floors apparently are made up of that rock as well.

Basalt looks very commonplace for the dominant position it holds among volcanic rocks. When it is unweathered, it is ordinarily coal black to dark gray, and fine grained to aphanitic in texture. The two principal mineral constituents are pyroxene and calcium-rich plagioclase, and many varieties contain olivine as well (Fig. 4-17). Any of the three minerals may be present as phenocrysts in porphyritic varieties. Basalt is commonly frothy and cellular and filled with innumerable small holes called *vesicles*—gas bubbles trapped in the

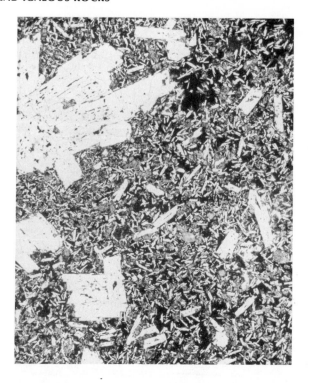

Fig. 4-17. (right) Basalt from the Columbia River Formation in Idaho, as seen under the microscope. The tabular phenocrysts and smaller lath-shaped light-gray grains are calcium-rich plagioclase; the dark-gray crystals are pyroxene; and the black grains are magnetite. *W.B. Hamilton, USGS*

Fig. 4-18. (right) Cellular or scoriaceous basalt, a fine-grained to glassy basalt filled with tiny bubble holes. *Hal Roth*

congealing lava. Such a structure, which is by no means confined to basalts, is denoted by the adjective *scoriaceous* (Fig. 4-18).

One common variety of basalt, *diabase,* has a texture in which the plates of calcium plagioclase occur in an interlocking network. The spaces between crystals are occupied by pyroxene, which crystallizes later (Fig. 4-19). Diabase of a coarser-than-normal grain size commonly results from the relatively slow cooling of shallowly buried, relatively thick bodies of magma. Basaltic melts, because of their high temperature (close to 1200°C; 2192°F) and chemical composition, have relatively low vis-

Fig. 4-19. Moderately coarse-grained diabase from Victoria Land, Antarctica, as seen under the microscope. Lath-shaped calcium-rich plagioclase crystals are present in an interlocking network of pyroxene crystals (dark-gray). Magnetite (black) also is present. *W.B. Hamilton, USGS*

cosities and flow more rapidly than felsic melts. Thus, when basaltic lava pools, or finally ceases to flow, it is still largely molten except for a thin outer skin. As the flow cools, and begins to crystallize from the outside inward, it contracts, and tensional stresses are developed in a plane parallel to the top and bottom cooling surfaces. With further crystallization, a number of vertical joints, or fractures, begin to form. In a fairly uniform sheet of lava such fractures tend to be relatively uniformly spaced and random in direction, resulting in polygonal (usually four to eight faces) vertical prisms, or columns, that look like giant fence posts, bunched together and stacked on end (Fig. 4-20). When the closely spaced columns are seen from above their pattern resembles that of hexagonal bathroom tiles (Fig. 4-21). Such fracturing is called *columnar jointing* and is most common in basaltic rocks although by no means confined to them. If conditions are right any sheet of magma, regardless of composition, can develop a columnar joint pattern upon crystallization (Fig. 4-22). Some of the finest examples of columnar jointing in the world can be seen at the Devil's Postpile in the Sierra Nevada Range in California and the Giant's Causeway in Northern Ireland.

Pyroclastic and volcanic-clastic rocks   During the eruption of a volcano, in addition to liquid lava, which flows down the flanks to subsequently congeal into typically crystalline rocks, there is also a great deal of lava hurled into the air above the vent (Fig. 4-23). Some falls back to earth near the erupting vent, some falls farther out on the volcano's flanks, and some is hurled high aloft to be carried long distances from the vent by the prevailing winds. As mentioned, all such explosively ejected material is termed pyroclastic. Some consists of large and small incandescent clots of liquid lava which cool and largely consolidate during their flight. Some are spindle shaped, but most occur in rounded, irregular forms. All such clots are called *volcanic bombs* (Fig. 4-24). Many have a crust a few centimeters thick that looks much like that of a loaf

Fig. 4-20. Columnar structure in basaltic lava flow; The Devil's Postpile, California. *Sierra Club*

of French bread. Some have a flattened snout, apparently resulting from impact with the earth's surface and indicating that the bomb did not completely solidify during its trajectory. Volcanic bombs cool rapidly and therefore are usually aphanitic to glassy in texture; they can be of any chemical composition, felsic to mafic.

Smaller volcanic rock particles are also commonly ejected from vents. Tiny fragments of chilled lava (about 2 cm or 0.8 in. in diameter) are termed *lapilli* (an Italian word for little stones), and rocks composed dominantly of that material are called *lapilli tuffs*. Even smaller pieces, down to the size of dust, are called *ash*.

Generally bombs and cinders are so heavy that they cannot be blown far from the vent; they fall mostly on the volcano flanks. Deposits

Fig. 4-21. Upper surface of the lava flow at Devil's Postpile, California, showing the ends or cross sections of the columns seen in Figure 4-20. Cooling of the lava causes it to fracture into columns of 4, 5, 6, or 7 sides. *Hal Roth*

Fig. 4-22. Columnar jointing developed in an obsidian flow, Obsidian Cliffs, Yellowstone National Park, Wyoming. *J.P. Iddings, USGS*

Fig. 4-23. Cerro Negro erupting at night. Luminous tongues of fire, composed mostly of pyroclastic material, are explosively discharged from the vent. White streaks represent trajectories of individual volcanic bombs. *Edwin E. Larson*

Fig. 4-24. A large elongate volcanic bomb amidst other smaller-sized pyroclastic material found on the northeast flank of North Sister volcano in the Cascade Range. *Edwin E. Larson*

composed of angular bombs, cinders, lapilli, and ash in an unsorted mixture but possessing crude layering are called volcanic *agglomerates* or sometimes volcanic *breccias*. If the ashy material dominates they are called *tuff breccias*. Lapilli and ash can be blown high over an erupting cone; subsequently some will fall back on the cone flanks and some, particularly ash, will be carried by prevailing winds and subsequently deposited in sizable amounts as far as 1000 km (621 mi) away. Ash that is carried into the air and later falls back to earth to form layered deposits is called *air-fall ash*, and the consolidated rock composed of that material is called an *ash-fall tuff*. The latter is fine grained in texture and is composed of pieces of pumice and jagged fragments of volcanic glass and phenocrysts—usually potassium feldspar, quartz, amphibole, and biotite—and small angular clasts of volcanic rocks ripped from the throat of the volcano during the explosion. Highly explosive volcanoes that pour forth great quantities of ash are usually those that erupt more viscous lava; normally they are felsic in composition.

One particular variety of pyroclastic rock

Fig. 4-25. Welded ash-flow tuff from near Gunnison, Colorado, as seen under the microscope. Partly broken phenocrysts of sodium-rich plagioclase (white) and biotite (dark gray) are in a matrix of fused, angular fragments of glassy ash (light gray). Note that the crude layering in rock is parallel to the long axis of the biotite grains.
J.C. Olson, USGS

composed largely of ash appears to have been extremely hot when it was deposited; so hot that the angular fragments of glass were able to weld themselves together after they came to rest. Under their own weight the deposits, which may be as much as 300 m (984 ft) thick and laterally extensive, settle and compact—even to the point that softened pumice fragments lose their porosity and completely collapse into dark-colored flattened blebs of volcanic glass. The resulting rock, called a *welded ash-flow tuff,* is dense and strong, with an easily visible banding parallel to the flow top (Fig. 4-25). Some of the most impressive layers of welded tuff are found in the vicinity of Yellowstone National Park in Wyoming. Single beds, resulting from catastrophic emissions of immense volumes of hot ash, are more than

160 km (99 mi) long, and up to 100 m (328 ft) thick.

Some volcanic deposits are bedded, composed of sub-angular to sub-rounded and rounded fragments ranging from boulders down to silt and clay size. Such deposits are the result of sedimentary processes acting on the slopes of the volcano. The upper slopes of a volcano commonly are piled high with relatively unstable pyroclastic debris and interlayered lava flows that are particularly susceptible to slumping and stream erosion. Often, when saturated with rainwater (which may fall as a direct result of the volcanic activity), the slopes become very unstable. Under the influence of gravity the rain-soaked material can move downslope within the confines of stream channels in torrents of muddy ash and coarser pyroclastic debris and out onto the more gentle slopes at the perimeter of the volcano and beyond (Fig. 4-26). Individual *mudflows*, as such slides are called, are found on the flanks of Mount Rainier in Washington, where they cover an area of 320 km² (124 mi²) and have a volume of 2 km³ (about 0.5 mi³).

Also, through the continued activity of streams on the sides of the volcano, material can be eroded from the steeper slopes of the volcanic vent and deposited in aprons surrounding its base (Fig. 4-27). Such deposits, which are composed of stream-worn clasts, are truly sedimentary rocks, like the mudflows, and are so named. Often stream-deposited material will be interlayered with strata composed of mudflows.

The term volcanic breccia, which we introduced earlier with respect to certain pyroclastic deposits, can also be applied to any layers composed of volcanic rock in which the fragments are angular (Fig. 4-28).

Fig. 4-26. Diagram of Mayon Volcano, Philippine Islands, showing tracks of mudflows generated during the 1968 eruption.

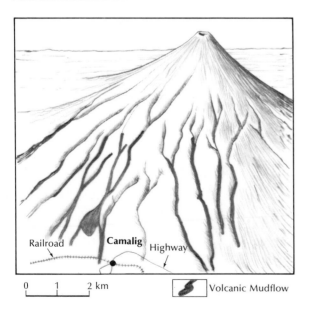

Railroad   **Camalig** Highway

0   1   2 km   Volcanic Mudflow

PLUTONIC BODIES AND IGNEOUS PROCESSES

It is now quite clear for a variety of reasons that molten material erupted from volcanoes does not owe its existence to the melting of rocks through the combustion of coal beds, as Abraham Werner so strongly believed. Rather, volcanic activity appears to be the surface manifestations—the frosting on the cake, so to speak—of igneous processes and activity that go on deep below the surface, even as far down as 250 km (about 155 mi). Although we can learn something about the deep-seated activity from the detailed study of the volcanoes themselves—where they occur and what the eruptive products are like—we never know exactly what is going on at depth. The situation is analogous in a sense to a lawn sprinkler system. When it is working we see the fountaining sprays of water but have little idea of the complexity of the subsurface plumbing system or where the source (the pump or water main) is located. The only way to find out is to dig up the pipes. Such a venture is not very feasible when one considers uncovering the "plumbing system" that carries magma to the volcanoes. Fortunately, however, natural processes have in part accomplished the task for us. Through erosion, acting over long periods of geologic time, large quantities of surface rocks have been successively stripped away in many parts of the world. In some instances erosion has cut down through several kilometers. Unfortunately, erosion has never cut deeply enough to expose the lower parts of the plutonic plumbing system, but at least we can examine the upper levels.

Remember, though, that when we observe crystallized plutonic bodies we are actually looking at a fossil, or petrified, system of conduits and pipes. The material composing the igneous body was molten long ago. Any eruptive vent that might have been present at the surface has been mostly stripped away. The study of igneous rocks, then, presents an insoluble perplexity: when a volcano is active we cannot see the conduit system below; when we

Fig. 4-27. These petrified tree trunks were once part of a forest that covered the lower slopes of volcanoes in Yellowstone National Park about 40 million years ago. They were buried by mudflows of volcanic material. The charac-ter of the material can be seen in the outcrop just above the geologist's head. Notice how angular the clasts are, the lack of good stratification, and the mixture of large and fine particles. *J.P. Iddings, USGS*

Fig. 4-28. Volcanic breccia, composed of angular fragments of andesite set in a matrix of fine-grained andesite fragments, ash, and mud. From Gunnison County, Colorado. *Jan Robertson*

can observe the subsurface channels the volcano superstructure has long since gone.

Intrusive rock bodies

Bodies of plutonic rock are found in a great variety of shapes and sizes. Many of the largest are elongate bodies nearly 1600 km (994 mi) in length. Some are characteristically tabular or cylindrical; some are irregularly ellipsoidal, circular, or even hot-dog shaped. All such bodies, regardless of size or shape, can be referred to by the general name *pluton*.

Most plutons did not result from the melting and later solidification of rocks in place. Usually, in fact, their chemical composition is completely different from that of the host rocks (commonly referred to as *country rocks*) that surround them. Rather, magma at depth, once formed, appears to be relatively mobile, at times amazingly so. It can push aside pre-existing rocks and force its way into cracks, some of which it may actually produce—the process is termed *forceful injection*. In some crack systems, magma has been able to move

distances of at least 50–100 km (31–62 mi). It can also move ahead by (1) melting away the surrounding country rocks or (2) wedging or prying out solid chunks of rock, which it then replaces. When magma makes its way into a pre-existing rock, it is said to have *intruded* into the host rock. Plutonic rocks, then, can be called *intrusive* rocks, and the surface of contact between the pluton and country rock is called the *intrusive contact*.

Magma can move and intrude both laterally and toward the surface. It is through vertical rise that it comes close enough to the surface to be uncovered by erosion. If magma can rise high enough or send out fingers or *apophyses* (Greek, meaning offshoot), that can do so it will eventually break through at the surface to produce a volcanic eruption. Some outbreaks of mafic magma have originated as far down as 250 km (155 mi).

The reasons that magma moves and intrudes where it does are varied and not always completely clear. One good reason for its tendency to rise, however, is that liquid magma is normally less dense than the solid country

Fig. 4-29. Mafic dikes cutting across Precambrian metamorphic rocks, Cross Lake, Manitoba. Basaltic magma moved into irregular fractures as it intruded laterally and upward. Note the sharp contact between the dike and the country rock. *Geological Survey of Canada*

rocks around it. A buoyant effect is created such that given an opportunity (for example, the presence of a fracture) it will move upward.

Once magma has stopped moving, it stagnates and begins to cool. At great depths it cools slowly and is often characterized by a coarsely crystalline texture, whereas material crystallizing near the surface, where cooling is more rapid, will form rocks with finer-grained textures. Magma that has risen very close to the surface may cool so quickly that the grain sizes of the resulting rocks are essentially equivalent to those of volcanic rocks.

Porphyritic textures are typical of intrusive rocks from shallow to intermediate depths be-

cause there may have been a considerable movement of magma from one environment to another in a complex conduit system—which results in different rates of crystallization.

Small Plutons Dikes and sills are tabular intrusions in which two of the dimensions are large as compared to the third—they have about the same geometry as that of a thin pad of note paper. The property that sets the two types of intrusions apart is the relation they have to the layering (commonly that of sedimentary rocks) of the surrounding rocks. Dikes are *discordant*, which means that they cut across the layers (Fig. 4-29), whereas sills are

Fig. 4-30. Diabase sill on Banks Island, Northwest Territories, showing an abrupt change in level of intrusion along a fracture. All of the rock layers now above the sill and those that since have been eroded away had to be uplifted during the emplacement of the sill. *Geological Survey of Canada*

concordant—they more or less parallel the layering (Fig. 4-30). Concordance does not mean that sills follow only a single stratum, because it is not at all uncommon for them to angle up or down from one stratum to another and to follow it for perhaps hundreds or thousands of meters before making another step-like change in level (Fig. 4-30).

Dikes and sills can be of more than geologic importance. Two such intrusions played a decisive role in Pennsylvania during three hot summer days a century ago in July of 1863. One is the thick basaltic sill that underlies Cemetery Ridge, a resistant, low-lying escarpment, and the other is the narrow but persistent dike that supports Seminary Ridge, both of which were the dominant elements of upland terrain at the Battle of Gettysburg. The Union Army held the former and the Confederate forces the latter. Both sides made the most of the so-called ironstone boulders which had weathered from basalt outcrops along the top of each ridge. Piled into fences the rocks made excellent defensive positions against the round shot and Minié balls of that day. Cemetery Ridge, along which the Union brigades were deployed in full strength on the fateful third

morning, makes such a continuous rampart, with a nearly unbroken forward slope up which Pickett's command had to charge, that the assault was virtually foredoomed to failure. That was doubly so when they were called upon to dislodge men sheltered behind a practically shot-proof basaltic barricade.

Dikes are seldom more than 30 m (98 ft) thick, but some are hundreds of kilometers long. An exceptionally long one, probably the largest in the world, is the so-called Great Dike of Rhodesia in southeast Africa. It is more than 480 km (298 mi) long but has an average width of only about 8 km (5 mi).

Dikes can occur singly, but commonly they occur in groups known as *dike swarms* (Fig. 4-31). Sometimes they are aligned on roughly parallel courses, or they radiate from centers, such as the host of basaltic dikes that lace the northern part of Great Britain, with some individual intrusions reaching lengths of 160 km (99 mi) or so. The focal point for a radial set of dikes usually is the throat, or conduit, of an extinct volcano or a cylindrical pipe-like intrusion.

Depending on their resistance to erosion relative to that of their host rocks, dikes may be

Fig. 4-31. Swarm of subparallel dikes of felsic rock cutting across a dark-colored gneiss, in the face of Painted Wall, Black Canyon of the Gunnison National Monument, Colorado. The Gunnison River (left) flows at about 686 m (2250 ft) below the top of the cliff. *W.R. Hansen*

Fig. 4-32. Aerial photograph of dikes northeast of Spanish Peaks, Colorado. Because of resistance to erosion, they stand above the surrounding terrain, like walls that run unbroken for many kilometers. Most of the dikes are part of a set radiating from an intrusive center that apparently fed a large volcano about 25 million years ago. *USGS*

distinctive features of the landscape. If they are more resistant they stand up somewhat as continuous walls (Fig. 4-32). Should they be weaker, they may be etched out.

Columnar jointing is also typical of dikes, but the columns generally lie horizontally if the dike itself is vertical, and look much like an immense stack of cordwood. The reason is that the columns grow inward horizontally from the dike's vertical side walls, which are the cooling surfaces. Dikes most commonly form when magma intrudes a fracture, the walls being pushed apart as it makes its way.

Sills may run for great distances across the country and have lengths comparable to those attained by dikes. An excellent example is the Great Whin Sill in Northumberland in northeastern England. It looms as a dark, north-fac-

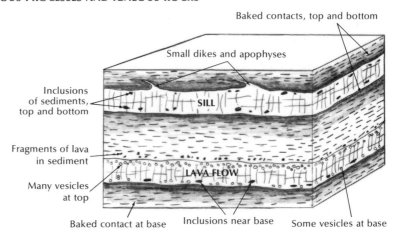

Fig. 4-33. Criteria used for distinguishing a sill from a buried lava flow.

ing ledge, about 30 m (98 ft) thick dominating the country for much of its course, with the result that the Romans, with their experienced eye for the military potential of the terrain, seized upon the natural barrier as a foundation for Hadrian's Wall, which was built to keep the Picts from ravaging northern Britain.

Certainly the most familiar of all sills in the United States is the abrupt cliff of the Palisades that follows the New Jersey shore of the Hudson River and is clearly visible to many thousands in Manhattan and to the multitudes of commuters who pass it every working day. The Palisades sill is a large one, as such things go; it is more than 300 m (984 ft) thick. It cooled slowly and without much internal disturbance, so that the crystals within it had a chance to develop a noticeable layering. The heavier minerals, such as olivine and pyroxene, sank to the lower levels of the large sill as they crystallized while the lighter plagioclase crystals tended to float and were concentrated near the top.

Because contraction also occurs in the solidifying magma of sills, columnar joints are characteristic of them, too; since most sills tend toward a horizontal position at the time of their origin, the columns in such an intrusion are likely to be vertical, having developed at right angles to the cooling surface.

Discriminating between buried lava flows and sills that have intruded into sedimentary rocks is not one of the simpler field problems in geology. Igneous rocks in both categories may have about the same texture, and both may have very well-developed columnar joints.

In many sills a "chilled" zone at the margins, where crystallization was the most rapid, is very common. The texture in that zone is fine grained and in some cases almost glassy. Additionally, there may be a "baked" zone in the invaded rocks immediately adjacent to their boundary with the sill, both above and below it. Such rocks are fired or hardened in much the same way that clay is fired in a kiln to make bricks or pottery. In a sill, fragments of the overlying host rock may be present in the igneous material, generally in the chilled zone. If the igneous body is an interbedded lava flow, however, the upper baked contact will be lacking altogether. Also, more likely than not, fragments of rock eroded from the flow will be found incorporated in the covering strata, which of course was deposited after the lava solidified. Figure 4-33 graphically displays the criteria used in distinguishing lava flows and sills.

Whereas dikes appear to continue to depths of several hundred kilometers, sills usually are found in the outermost part of the earth, particularly in the relatively thin sedimentary rock sequences that veneer the crystalline rocks beneath. Recently the American geologist M. R. Mudge studied the occurrence of nearly 100 concordant structures (mostly sills) in the western United States and found that none had been emplaced below a depth of about 2300 m (7544 ft). The reason that sills are confined to the outer layers is that when one of them intrudes, all the rocks above it must be lifted up by an amount equal to its own thickness. The force necessary to lift such immense weights comes from the fluid pressure in the magma body. Sill formation can occur only when the fluid pressure becomes equal to or greater than the overburden weight, a condition that can be met normally only when thin sections of overburden rock are involved; that is, in the outer skin of the earth.

Many if not most sills are offshoots from a dike. Magma intrudes upward until its fluid pressure exceeds the overburden weight; then sills branch off if the nature of the layered rock section is conducive to sill formation.

Volcanic necks are the channels or conduits

Fig. 4-34. Ship Rock, New Mexico, is a volcanic neck with two radiating dikes. The volcanic cone that at one time probably existed above the site has been eroded away. The neck is intruded into sedimentary rocks which are relatively more easily eroded than the neck. Throughout this part of the Colorado Plateau, volcanic necks such as Ship Rock form starkly picturesque features. *John S. Shelton*

that once served to connect a volcanic vent on the earth's surface with a magma reservoir at depth. In an erosional sense, volcanoes are vulnerable features; much of their interior consists of loosely consolidated ash and pyroclastic material, their steep slopes are easily gullied, and, if they are high enough they may intercept more snow and rain than the surrounding countryside. The result of augmented runoff working on an exposed structure is that, geologically speaking, the interiors of many volcanoes are bared to view relatively shortly after they lapse into dormancy.

Their internal skeleton of radial dikes and a central conduit often proves to be made of sterner stuff, with the result that dikes stand up as partitions, and the solidified magma of the conduit forms a central tower known as a *volcanic neck.* A well-known example is Ship Rock in New Mexico (Fig. 4-34).

More often than not, volcanic necks are found in clusters, rather than standing isolated. Prime examples in the United States are the buttes of the Navajo-Hopi country near the so-called Four Corners area, where more than 100 volcanic necks interrupt the surface of the plateau. Being composed dominantly of dark rock and projecting in jagged spires, they appear in sharp contrast to the flat-lying red and white sedimentary rocks into which they intruded. As the day draws to a close in that picturesque land, the volcanic centers stand silhouetted against the sky to produce an awesome scenery.

Another group of volcanic conduits that have achieved a measure of notoriety are the so-called pipes of *kimberlite,* a rock composed dominantly of ferromagnesian minerals, which are one of the sources of diamonds. In South Africa the pipes are deeply weathered near the surface into what is called "blue ground," a sticky clay from which the diamonds were separated by washing. The weathered pipe-like columns of dark rock were mined in the Kimberley pit to more than 1000 m (3280 ft) before those workings were abandoned. Since the pressures and temperatures necessary for diamonds to crystallize are not fulfilled short of

a depth hundreds of kilometers below ground, the diamonds must be phenocrysts carried from deep below the surface. Many kimberlites contain fragments of foreign rocks which apparently were ripped from the walls of the pipe as the magma moved toward the surface. From the presence of the diamonds and the nature of the foreign fragments, geologists have estimated that kimberlitic magma must come from depths of near 250 km (155 mi). In spite of this, most are only about 90 m (295 ft) across at the surface. Their general shape, then, is something like an extremely long drinking straw. Kimberlites containing diamonds are found in the United States in Arkansas and in an area near the Colorado-Wyoming border.

Laccoliths are concordant igneous bodies that are more or less circular in outline, with a flat base and a dome-shaped upper surface (Fig. 4-35). They are generally up to 3 km (2 mi) across at the base with maximum thicknesses near the middle of several thousand meters. Apparently laccoliths began development much like a sill. However, the magma injected between the rock layers was so viscous that lateral flow could not keep pace with the supply. More lava moved into the developing body than was transported marginally, and the body began to swell, or dome, upward, carrying the sediments above it into an arch. Because felsic magmas are generally more viscous than mafic ones, it would be expected that most laccoliths are composed of felsic rocks, and observation bears this out. Commonly several laccoliths will occur in a group, and all appear to have been fed from a single magma body. In the United States laccoliths are conspicuously developed in southwestern Colorado and eastern Utah. Some of them, for example, the LaSal and Henry mountains in Utah, stand in scenic beauty high above the surrounding countryside.

Figure 4-36 is a composite diagram depicting the characteristic mode of occurrence of dikes, pipes, sills, and laccoliths.

Large elongate or irregular plutons The largest bodies of plutonic rocks are composed

Fig. 4-35. Laccolith fed from a central pipe or dike. This type of occurrence is quite common.

of granite and diorite. They are found on all continents and range in age from Tertiary to Precambrian. In Labrador and northeastern Canada, where erosion has laid bare huge expanses of the earth's surface, plutonic rocks are found in bodies that may cover hundreds or even thousands of square kilometers. Large bodies composed mostly of granite with lesser amounts of diorite—we will lump them together and call them granitic rocks—are found also in the cores of many mountain ranges. The Coast Range pluton in British Columbia is an elongate body about 1600 km (994 mi) in length (Fig. 4-37).

Such huge bodies of rock are called *batholiths* (a word introduced in 1895 and derived from the Greek words for depth and stone). In modern usage a batholith is defined as a pluton whose exposure is in excess of 100 km² (39 mi²). If it is smaller than that, a pluton is called a *stock*. Many stocks, however, appear to be only the limited exposures of the top of a batholith. If erosion were to proceed further, it is likely that a batholith would be unroofed and exposed.

Much has been learned about the physical configuration of the upper parts of batholiths because erosion has cut down to different levels at different places within the same body. Deep mining operations have also provided information. We lack definitive data, however, on the lower regions of batholiths. The walls of many of them are very steep sided, which has led many geologists to hypothesize that such bodies extend downward to great depths. Gravity and earthquake-wave studies have given some insight, but the results are often inconclusive. For instance, it is suspected that a batholith, perhaps in part yet molten, is sitting just below the bubbling paint pots and geysers in Yellowstone National Park. In fact, it appears that the batholith has existed for at least 2 million years. Three voluminous eruptions from its chamber occurred at 2, 1.2, and 0.6 million years ago, respectively. Geophysical investigations indicate the existence of an anomalous rock body, perhaps partially molten, that extends downward for at least 60 km (37 mi).

A frozen batholith is also suspected of lying

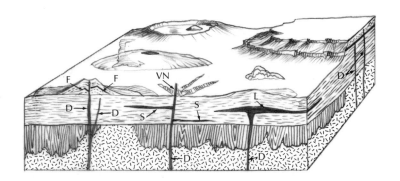

Fig. 4-36. The occurrence of dikes (D), sills (S), volcanic necks (VN), laccoliths (L), and flows (F).

under the pile of volcanic rocks that form most of the San Juan Mountains in southwestern Colorado (Fig. 4-38). From gravity measurements it is estimated to be 240 km (149 mi) long in an east to west direction by 160 km (99 mi) wide north to south. The top of the pluton is thought to be 2 to 7 km (1 to 4 mi) below ground and supposedly extends downward for about 19 km (12 mi).

In many cases batholiths are elongate (see Fig. 4-37). Many of the larger bodies have long dimensions of about 1600 km (994 mi) and are 32 to 240 km (20 to 149 mi) wide. Commonly, the long dimensions parallel the trends of mountain ranges in the cores of which they are found.

Batholiths are not simple bodies; any single large mass consists of smaller bodies of different kinds of plutonic rocks. One part may be a white granite almost entirely lacking in ferromagnesian minerals; another may be a standard granite; and yet another may consist of diorite or rocks intermediate between granite and diorite. The nature of the boundaries between the rock bodies as well as radiometric dating generally indicate that the smaller bodies were formed at different times and that the formation of the batholithic complex, as a whole, took many millions of years.

An indication of the heat and volatile content of a batholithic magma is the halo, or *aureole*, of metamorphosed rocks that surround most batholiths (see Chap. 7). The aureole represents the original country rock, recrystallized in place by the heat and chemically active fluids of the invading magma.

As voluminous as batholithic magmas are, there is ample evidence to suggest that once they formed they tended to move upward. In some instances, the magmas appear to have shouldered their way bodily into the country rock, displacing it. The boundary of the intrusion is knife sharp, and the granite is uncontaminated and homogeneous. The country rocks obviously are gone, but gone where? A study of salt domes has given geologists an indication of how country rocks could disappear without a trace.

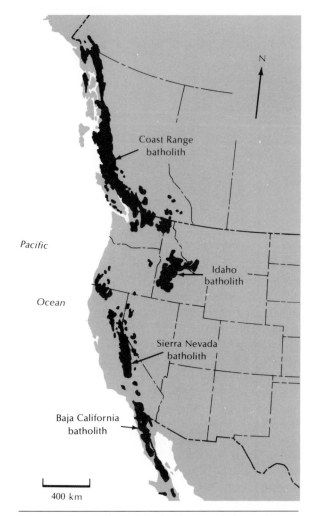

Fig. 4-37. Granitic batholiths (black) occur in large elongate bodies along the western margin of North America. Most of the plutonic rocks in these bodies intruded during the Jurassic and Cretaceous periods.

In some parts of the world, salt (mostly sodium chloride and calcium sulfate), which originated as precipitates from a restricted arm of the sea, lies in thick beds beneath thousands of meters of clastic sedimentary strata. Salt can deform easily and has a relatively low density; therefore, it tends to rise (Fig. 4-39). After a while, bumps of salt begin to protrude into the sediments above. With more time, the bumps rise higher and higher until an elongate,

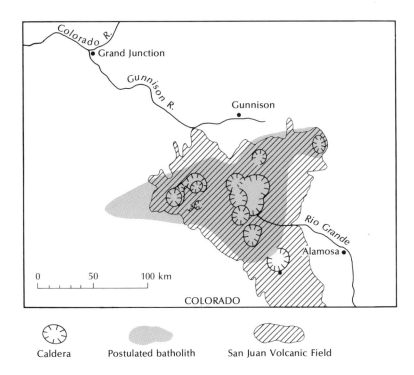

Fig. 4-38. Relation of a postulated batholith, calderas, and the San Juan volcanic field in southwestern Colorado. Many geologists think that the repeated explosive volcanism producing the calderas was the result of near-surface activity of the batholith.

Caldera   Postulated batholith   San Juan Volcanic Field

Fig. 4-39. Sequential development of a salt dome, beginning with undeformed sediments overlying the salt layer (A) and progressing to a tear-drop shaped salt intrusion (E).

DEVELOPMENT OF SALT DOME

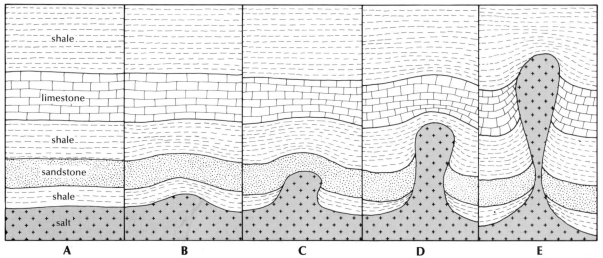

shale

limestone

shale

sandstone

shale

salt

A   B   C   D   E

Fig. 4-40. Light-colored granite in which are found a host of elongate, aligned dark bodies of rock that once were probably part of the country rock; north of Granite Pass in the Sierra Nevada, California. The tape measure is about one meter in length. Most likely, the dark blocks were pried loose (stoped) from the walls of the country rock by the intruding granitic magma. Subsequently, flowage caused subparallel alignment of the fragments.
G.K. Gilbert, USGS

cylindrical mass of salt projects above the original beds. Sometimes the cylinder completely separates from the salt layer and continues to move upward, much like a drop of oil moving toward the surface in a vessel of water. Eventually such low-density material will rise right to the surface. As the movement takes place the surrounding sedimentary rock remains in the solid state but slowly deforms to accommodate the rising plug of salt. Somehow, the salt intrusion is able to force aside the sediments. It is a slow process, to be sure, but effective. Since granitic magma is less dense than the surrounding rocks and is easily deformable, many geologists feel that, given enough time and in the right circumstances, the magma can displace the solid rock above, just as salt domes appear to do.

Is it also possible for batholiths to move by melting the rocks above? First, keep in mind that when a material changes from a solid to a liquid, heat is required, and that when it changes from a liquid to a solid, heat is given off. Many petrologists point out that apparently most melts contain little heat in excess of that required to keep the magma molten. Any melting that occurs in the country rock, therefore,

must be supplied by heat from a different source. Simultaneous crystallization (change of liquid into solid) in the magma could provide that source. Normally the first crystals to precipitate are more dense than the liquid magma and they would sink to the bottom of the chamber. Through continuous crystallization and melting of the wall rock, the magma chamber could move upward. Petrologists point out, however, that the process would only be effective under special conditions; most earth scientists, therefore, consider melting to be only of minor significance.

Some geologists have suggested, and there is some evidence, that magma could make a path for itself through the country rock by yet another process—*stoping* of wall rock. Commonly batholiths contain, especially near their margins, angular fragments of rock that are totally different from the granitic rocks. Many of them are crudely aligned as if by flow of the magma, and can be interpreted as small blocks that were pried loose, or *stoped*, from the roof or wall of the chamber and carried along during subsequent flow of the magma (Fig. 4-40). Apparently most of them contained minerals with higher melting points than the granitic

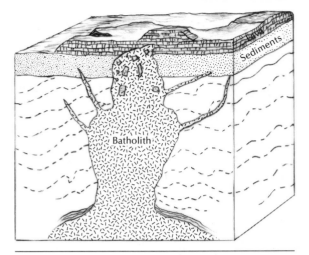

Fig. 4-41. Section through a hypothetical batholith depicting most of the ways that a batholith intrudes into the country rock. At the bottom on either side the granite is shown without boundaries, indicating that either it reacts with the other rocks, or it is produced when they melt. This would also be the zone of migmatite development. At the top, the batholith has penetrated sedimentary rocks and lavas, stoped blocks of which sink into the magma. Melting of the country rock by the batholith can occur at any level.

magma and therefore did not melt. A diagrammatic summary of the various ways that a batholith may make room for itself as it intrudes the country rock is given in Figure 4-41.

The basal portions of some batholiths, especially those that formed deep inside the earth, are surrounded with a wide marginal zone, and the transition between the invading and the invaded rocks is much less abrupt. The contact between the granite and its encasing shell may be blurred by a zone of *migmatite*, or mixed rocks—partially igneous and partially metamorphic. Migmatite has a typical *lit-par-lit*, or bed-by-bed, structure; which means that one layer of a rock may be granite, the next a metamorphic rock, the next granite, and so on (Fig. 4-42). Or knots and clusters of potassium feldspar, or other minerals typical of granite, may appear in the metamorphic envelope some distance out from the main body of granitic rock.

Such phenomena, taken together, have been interpreted by some geologists to mean that in some way the pregranitic rocks were digested, or replaced, by granite—in other words, they

Fig. 4-42. Migmatite from the Sawatch Range, Colorado showing lit-par-lit structure. Light-colored bands composed of quartz, potassium feldspar, and muscovite are plutonic; the gray and black bands are composed of calcium plagioclase and hornblende and are foliated metamorphic rocks. The light-colored bands in the center of photo are together about one centimeter thick. *Jan Robertson*

were converted in place from what they were originally into a brand-new rock that appears to be plutonic. The process is called *granitization* by its advocates—a vocal group—who strongly adhere to the belief that most granitic rocks are formed in place through the alteration of large volumes of sedimentary rock by chemically potent solutions ascending from greater depths in the earth's interior.

Some batholiths can, in fact, be seen to grade laterally from what appears to be plutonic rock in the center to plutonic metamorphic rock, then to metamorphic rock, and finally to essentially unaltered rock, even sediments. Ghosts of the sedimentary layering can be made out in the "plutonic" rock. However, because such occurrences are so rare, and because of experimental and theoretical considerations, most earth scientists do not endorse granitization as the major mechanism in the origin of granite.

Distribution of plutonic rocks

A glance at the occurrence of plutonic rocks around the world reveals a curious pattern. Mafic rocks are found almost everywhere, both on continents and oceans, but felsic rocks are restricted almost exclusively to the continents. There are no known occurrences of granitic rocks in the ocean basins. Why? One might think that erosion on the ocean floor has been only minimal, or that batholiths close to the surface have not been unroofed yet. However, consider the fact that granitic rocks have lower densities (about 2.7 gm/cm^3) than do most of the rocks of the ocean basin (about 3.0 gm/cm^3). Large bodies such as batholiths would be isostatically uplifted to protrude above the ocean floor as topographic highs—yet none do. Or gravity measurements would easily detect the existence of such large, low-density masses under the sea—yet none have ever been located by such means. Moreover, volcanic activity in all of the world's oceans has produced mafic rocks almost exclusively. Certainly the volcanoes are not being fed from a granitic magma. Studies of earthquake waves

that pass through the outer layers of the earth have also failed to indicate the existence of large bodies of granitic material beneath the ocean basins.

Granitic rocks, then, must owe their existence to processes that originate marginal to, within, or below a continent. Most geologists would contend that nearly all granitic magmas are produced in the lower to upper-middle regions of the continental crust. Some would place their points of origin a few tens of kilometers beneath the continental masses.

Some light can be shed on the issue from a study of the distribution of coarse-grained felsic plutons in the continents. Many occur (in association with metamorphic rocks) in linear trends, or belts, and many of the geologically more recent ones are found in the cores of mountain ranges—leading to the conclusion that most of the large plutons are related to mountain-building processes. And when we find long chains of old batholiths that extend across a non-mountainous terrain, as in large parts of Canada, we probably are looking at the deeply eroded roots of fossil mountains. If we knew what produced mountain ranges (see Chap. 16) we could speculate on the origin of batholiths.

The plate-tectonics model (explained in Chap. 19 in more detail) is relevant to both problems. Briefly, the theory proposes that where the large, rigid plates of the earth converge or come together, both mountains and igneous intrusions are produced. During the slow, convergent sliding of one plate against another, sufficient heat is produced to melt the rocks at depth, and magmas are formed all along the plate boundary. The visible effect of the activity at depth is the occurrence of linear chains of volcanoes—called *island arc systems*—such as the islands of Japan and the Aleutian Islands. The materials that are melted to become batholithic magmas are thought to be, in part, sedimentary debris eroded from the continents and trapped in deep, linear pockets along the boundary of collision, and in part the rocky material that makes up the edges of the plates themselves. Because the

126

sedimentary debris is relatively rich in potassium, aluminum, and silicon, the resulting magma is granitic in composition. The model seems to tie together many diverse phenomena, and most geologists have accepted the theory, at least in its general form.

Unfortunately, it is not possible to investigate the nature of the source magmas and their chambers until long after the volcanoes are dead and gone. It has been hypothesized that the Coast Range batholith in British Columbia and the elongate intrusions that run the length of the Sierra Nevada range and through Baja California probably represent plutons formed along a zone of convergence over a period of 90 million years, with the bulk of activity occurring between 100 and 80 million years ago.

Many zones of convergence today are marginal to continents. If that was true in the past and if granitic rocks were formed periodically along the margin, it would seem that, with time, as successively younger chains of batholithic material are added on, the continents should be growing outward. In fact, some continents do tend to show a "younging" toward their margins, which has led many geologists to suggest that during the early stages of earth history the continents were small or even nonexistent, and have been undergoing continental accretion ever since.

Mafic plutons are a different matter. They almost never occur in bodies of batholithic proportions—at least as far down as we have been able to investigate. Since the rocks are too mafic to have been generated by the melting of most of the continental rocks themselves they must come from below the continents. That view is supported by the observation that mafic rocks found on continents are not greatly different from those found in the ocean basins.

Studies of earthquake waves and artificially created explosions have provided much insight into the nature of subsurface rocks (see Chap. 17). The results strongly suggest that the earth below the continental masses and the ocean basins is made up of material that could, if melted, produce mafic magmas. Detailed study of earthquake rustlings related to magma movement deep beneath the island of Hawaii indicate that the magma feeding the fiery surface fountains must come from at least 50 km (31 mi) or so beneath the island. The geophysical investigations also have suggested that from about 100 to 250 km (62 to 155 mi) below the surface, the rock is in small part (perhaps up to 10 per cent) molten (see Chap. 17). Above and below that depth range the rocks are completely solid, raising the possibility that in some circumstances the molten material can pool together to form a magma which can then begin its journey to the surface.

The model of plate tectonics has also been applied to the occurrence of mafic plutonic (and volcanic) rocks. If magma is to rise, the most likely place for it to do so is where the earth's brittle layers are being stretched, cracked, and pulled apart, thereby providing easier access to the surface. The most obvious place for intrusion of mafic magmas, then, would be along the diverging boundaries (pull-apart zones) of the world's plate system. Today, in fact such zones are generally marked by higher-than-normal heat flow and mafic volcanism. Iceland, for example, sits full astride the Mid-Atlantic Ridge, and it is composed largely of mafic dikes, sills, and lava flows. Moreover, most of the dike swarms that fed the flows run parallel to the trend of the Mid-Atlantic Ridge.

Many of the larger basalt fields of the earth, where erosion has cut sufficiently deep, show evidence of being fed by swarms of sub-parallel dikes. Many geologists have said, therefore, that the magma that fed the flows also could have moved into fractures or rifts, that developed from cracking, brittle near-surface layers of the earth. But whether those fracture zones corresponded to extensive pull-apart zones such as the Mid-Atlantic Ridge or were isolated occurrences largely unrelated to plate boundaries is still an open question.

Those who do not subscribe to the theory of plate tectonics are faced with the problem of locating a source for the heat needed to melt large volumes of rock, particularly in the production of batholiths. It has been suggested that heat liberated from the decay of radioactive elements at depth is a possibility, but it seems improbable that there were enough radioactive elements to do the job.

Another theory is that during the first days of the earth's formation, heat was trapped at great depths and is slowly moving toward the surface. Or perhaps the heat is being generated in an unsuspected way.

Yet, if we accept the model of plate tectonics we must ultimately ask: What is the source of energy that drives the plates? At the moment nearly everyone in earth science is searching for clues that will explain the enigma.

SUMMARY

1. The kinds of minerals and their proportions in an igneous rock depend on the composition of the magma. During cooling, not all minerals crystallize simultaneously—a result of different melting points and temperature-dependent stability ranges. The general sequence of crystallization follows the *Bowen's reaction series*.

2. Earlier-formed crystals can become separated from the remaining magma. If that magma erupts, it will have a different composition from that of the starting magma, and a rock of a different mineralogic composition will crystallize. Such a process is called *differentiation*.

3. Most igneous rocks are *crystalline* in texture; that is, the minerals form an interlocking network. Grain size depends primarily on the cooling rate: plutonic rocks tend to be coarser-grained than volcanic rocks. If cooling occurs so quickly that the minerals cannot crystallize, the resulting rock is a *volcanic glass*. A *porphyritic* texture implies two cycles of cooling, one in which cooling was relatively slow, followed by one in which cooling was rapid.

4. Some volcanic rocks possess clastic textures, either as a result of reworking of volcanic rocks by streams or accumulation of volcanic material (*pyroclastics*) blown into the air during volcanic eruptions.

5. The three most common plutonic rocks are *granite, diorite,* and *gabbro;* the three most common volcanic rocks are *rhyolite, andesite,* and *basalt.*

6. Magma originates in small-to-large pools inside the earth and, once formed, moves laterally and toward the surface.

7. Relatively small *plutons* can be subdivided into the following classes: discordant tabular bodies (*dikes*), concordant tabular bodies (*sills*), *volcanic necks,* and domed concordant bodies (*laccoliths*).

8. The largest plutons are called *batholiths* and *stocks* and are composed almost exclusively of granitic rocks. They are found on all continents, in bodies that may cover thousands of square kilometers. Commonly they are present in the cores of mountain ranges, in association with metamorphic rocks.

9. Batholiths intrude into the surrounding country rock in a variety of ways: *forceful injection, stoping,* and *melting.* Given enough time, they may rise through the surrounding country rock much in the same manner that salt domes do through overlying sedimentary strata. Although there is some evidence to suggest that some batholiths originate through *granitization,* most appear to result from melting.

10. *Mafic* plutonic rocks are found both on continents and oceans, whereas *felsic* rocks are almost exclusively restricted to the continents. Mafic rocks apparently originate in the mantle, perhaps in the layer between 100 and 250 km (62 and 155 mi). Large-scale granitic plutons owe their existence to processes originating marginal to, within, or closely beneath a continent.

11. Batholiths and associated metamorphic rocks in many cases appear to be the result of mountain-making processes marginal to a continent. Today, it is believed that the environment for producing that association is found in convergence zones.

SELECTED REFERENCES

Bowen, N. L., 1928, The evolution of igneous rocks, Princeton University Press, Princeton, New Jersey.

Billings, M. D., 1972, Structural geology, Prentice-Hall, Englewood Cliffs, New Jersey.

Daly, R. A., 1933, Igneous rocks and the depths of the earth, McGraw-Hill Book Co., New York.

Hamilton, W. B., and Myers, W. B., 1967, The nature of batholiths, U.S. Geological Survey Professional Paper 554-C.

Mudge, M. R., 1968, Depth control of some concordant intrusions, Geological Society of America Bulletin, vol. 79, pp. 315–22.

Read, H. H., 1957, The granite controversy, Interscience Publishers, New York.

Simpson, B., 1966, Rocks and minerals, Pergamon Press, Oxford.

Turekian, K. K., 1972, Chemistry of the earth, Holt, Rinehart and Winston, New York.

Tuttle, O. F., 1955, The origin of granite, Scientific American, April.

Walton, M., 1960, Granite problems, Science, vol. 131, pp. 635–45.

5

VOLCANISM

Volcanic activity accounts for some of the most awesome displays of the forces of nature known to humans. Anyone who has ever witnessed a volcanic eruption, in person or on film, cannot help but be impressed by the immensity of the earthborn fireworks (Fig. 5-1). Of course, part of the allure of volcanoes is related to their potential for destructiveness. For example, the eruption of Krakatoa off the coast of Java and Sumatra in 1883 brought about the sudden demise of nearly 37,000 people and that of Mont Pelée in the West Indies in 1902 caused the nearly instantaneous destruction of the town of St. Pierre and the loss of all but two of 30,000 people within it.

Volcanism is worthy of study not only because of its spectacularity but because it provides us with our only direct source of information about the rocks and the conditions that prevail in the outer 250 km (155 mi) of the earth. Certainly the existence of liquid lava at the

Fig. 5-1. A volcano erupts from the sea; Myojin Reef, about 270 km (168 miles) south of Tokyo, Japan. The billowing white and gray clouds are mostly steam resulting from the interaction of hot lava with sea water. *U.S. Navy*

earth's surface is a positive proof that rocks can and do melt in the natural environment.

From evidence contained in the rock record it can be determined that the earth has undergone volcanism from its earliest days right up to the present—and there is no evidence to indicate that it is dying out. As you might expect, the activity has not been continuous through time; certain periods of earth history have been marked by more and others by less activity. Even a single, large volcano shows the same type of behavior; during its life history after lying dormant for long periods it may suddenly flare into lively display.

More than 516 active or recently active volcanoes dot the earth's surface, yet volcanism is a geologic phenomenon alien to most North Americans—in sharp contrast to the phenomena related to ice, wind, and water that are as common to us as the Sunday newspaper. There is at present no active volcano in the contiguous United States unless you wish to count Mount Baker in the Cascade Range, which has recently begun to pour out copious quantities of steam from fissures near its summit crater. Possibly in the near future that cone

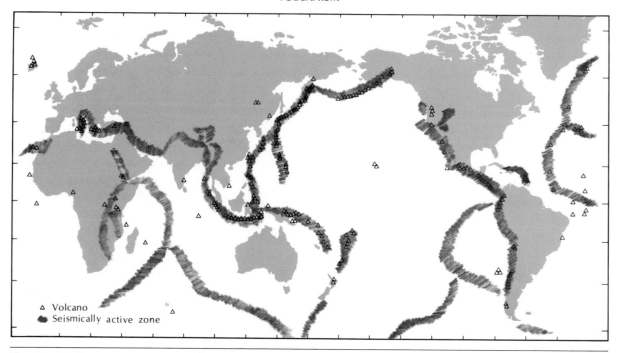

Fig. 5-2. Distribution of the active volcanoes of the world in relation to the active seismic zones. Each symbol represents one or more active or recently active volcanoes, of which there are about 516. Most volcanoes occur in belts marginal to the Pacific Ocean.

will progress to a higher level of eruptive activity. We can only wait and watch. In order to see volcanoes actually erupting in the United States it is necessary to travel to Alaska or Hawaii.

The word *volcano* comes from Vulcan, the ancient Roman god of fire and metalworking who manufactured the armor of the gods and the thunderbolts of Jupiter. One active volcano in the Mediterranean region carries the name Vulcano, and it was considered by the ancient Romans to be the forge at which Vulcan practiced his blacksmithing. Several other cultures, steeped in ancient lore and superstition, have associated volcanism with one or more deities. To the Polynesians, volcanoes were under the control of the demigoddess Pele; in the ancient world of the Greeks, Hephaestus, the ugly son of Hera, became the principal metalworker who set up his forge beneath one or another volcano in the Mediterranean region. A popular idea during the Middle Ages was that volcanoes were the gateways to hell or the prison of the damned. The cacophonous and often eerie noises that sometimes emanated from the volcanoes were thought to be the screams and moans of tormented souls.

DISTRIBUTION

Volcanoes occur in many parts of the world, but their distribution is far from uniform (Fig. 5-2). Many are concentrated within fairly well-defined zones or linear bands. Perhaps the most renowned is the "Ring of Fire" that girdles much of the Pacific Ocean. In that ring are found about four-fifths of the world's active volcanic centers. The rest of the active volcanoes are found mostly in the Caribbean and Mediterranean regions, in Asia Minor and the vicinity of the Red Sea, and in central Africa.

On the map of the world's volcanoes (Fig. 5-2) is also shown the distribution of generalized zones of earthquake activity. It is remarkable that the two zones coincide as closely as they do, a fact that has led most geologists to believe that there is a definite relationship between the two phenomena. This is not to say that earthquakes cause volcanic eruptions, or vice versa; almost never are the two phenomena in evidence at the same locality at the same time. (An exception to that rule occurred in Chile in 1960, when two days after a disastrous earthquake occurred, a small volcanic cone—Volcan Pujehue—went into mild eruption.) Rather, it seems that volcanism and earthquake activity are two different ways that the earth behaves in response to large-scale forces within it.

It can be pointed out that many of the linear belts of volcanoes more or less coincide (as do the earthquake belts) with the margins of the large and small plates that make up the outer rocky rind of the earth (see Chap. 19), and which fit together like the pieces of a gigantic three-dimensional jigsaw puzzle. Away from the edges of the plates, volcanism and earthquake activity are minimal. Where the plates converge, blocks push and slide past one another, and many volcanologists feel that the heat necessary for the melting of rocks and production of magma is created by the frictional energy released during such movement. Whatever causes the plate motion, it is thought, also accounts for the earthquakes in those zones. One dissimilarity between seismic and volcanic activity in zones of convergence is that earthquakes originate as far down as 700 km (435 mi), whereas the magma sources extend no deeper than about 125 km (78 mi) at most, and many may be much shallower. The Pacific Ring of Fire appears to owe its existence to the convergence of plates. The volcanoes associated with that convergence zone usually erupt andesitic to rhyolitic lava and pyroclastic rocks and do so, quite commonly, in a very violent manner. The volcanoes of the Mediterranean region, so well known in history and mythology, also occur at a zone of convergence and also pour forth mixtures of lava and pyroclastic rocks largely of andesitic to rhyolitic composition.

Volcanic activity is also notable at the zones where plates are pulling apart (diverging). The most conspicuous zones of divergence are found in the ocean basins and stand out as elongate rises on the ocean floor. If the water was drained from the seas, we would see the oceanic ridge- and rise-system (see Chap. 2), a nearly continuous range that snakes its way through all of the ocean basins and marks the zone of divergence. The evidence seems to favor the idea that as the earth's rocky skin pulls apart, magma rises from beneath to help fill the void, and volcanic activity ensues. Since most of the volcanic activity occurs beneath the sea, we are not exactly sure of the way it happens, or how often. But in a few places, such as the Azores and Iceland, eruptions have built volcanic edifices upon oceanic rises that project above the ocean's surface. At such places we have been able to study the nature of volcanic activity at a zone of divergence. Volcanoes in Africa's Rift Valley also appear to be associated with a diverging plate boundary. Unlike the Ring of Fire lavas, those associated with pull-apart zones are mostly basaltic in composition. They seem to be derived from a magma source at depths of 50–70 km (31–44 mi).

Not all active volcanoes can be easily fitted into a twofold classification based on the convergence and divergence of plates. Some, like those of the Hawaiian Islands, occur far out in the middle of the plates. No one yet has proposed a completely acceptable hypothesis for the origin of such eruptive vents. One thought, as discussed in Chapter 19, is that they are centered over "hot spots" in the outer part of the earth analogous to the top of a candle's flame, which have brought about the localized melting of rocks (see Fig. 19-24). Another view is that such volcanoes are at the leading edge of a large, active crack in the rocky skin of the earth that extends deep enough to encounter an earth layer containing some molten material. The crack would afford

Lava flow

 Pyroclastic layer

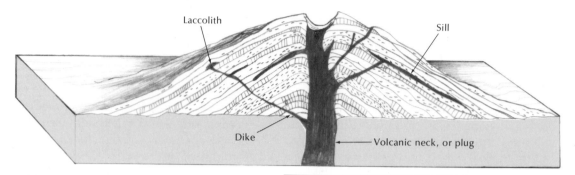

Laccolith   Sill

Dike   Volcanic neck, or plug

Fig. 5-3. Diagram of a composite volcano showing volcanic neck, dikes, sills, and laccolith.

Fig. 5-4. Plume of ash and vapor escaping from the summit of Shishaldin Volcano, Aleutian Islands, Alaska. Note the steep flanks of this stratovolcano. *USGS*

(Above) Volcanic eruption on Heimaey Island south of Iceland, January 1973. Fiery outbursts sent showers of pyroclastic material, a plume of volcanic gases, and ultimately a lava flow. (Below) Incandescent and chilled pyroclastic material bombard the ash- and gas-shrouded fishing town on the island. *Both photos, Icelandic Airways*

(Above) Pahoehoe (ropy lava), Hawaiian Islands. A convoluted glassy skin forms on the surface of still-molten basalt. *Glen Kaye, National Park Service* (Below) Bachelor Mt., Cascade Range, Oregon. This composite cone of flows, pyroclastic material, and volcanic mudflows was formed within the last million years. *Edwin E. Larson*

Fig. 5-5. Mount St. Helens, Washington, one of the most symmetrical volcanoes in the Cascade Range. The tracks of some of the more recent flows are visible in the right foreground. Mount St. Helens erupts about every 250 years; it last erupted about 150 years ago. Mount Rainier, visible in the distance, is now dormant but steam issues from the walls of the crater on top. *Ray Atkeson*

a channel for the molten portion of the layer to move to the surface. In fact, seismic data provide strong evidence to support the idea of a partially molten layer in the depth range between 100 and 250 km (62 and 155 mi). At present we cannot choose between the two explanations. Inasmuch as the roots and sources of all active volcanoes lie hidden many tens and even hundreds of kilometers beneath the earth's surface, we probably will not determine fully the reasons for their character and distribution for some time to come.

TYPES OF VOLCANOES AND VOLCANIC ACTIVITY

Volcanoes come in all sizes. Most mark localized centers of eruption, virtually at one spot at the earth's surface, which leads to the devel-

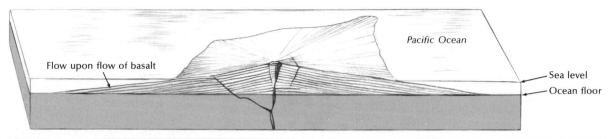

Fig. 5-6. Diagram of a Hawaiian shield volcano composed almost entirely of thin basalt flows. Because the magma is fluid, explosive activity occurs rarely, and pyroclastic mate-rial forms only a small part of the cone. As a result, the slopes are gentle and the volcano resembles a shield.

opment of a *volcanic cone.* The cone is in es-sence a heaping or piling up of the products of eruption around a volcanic vent, or orifice. Some volcanoes display a variety of eruptive styles, pouring forth flows of lava at one time and volumes of pyroclastic material at another (Fig. 5-3). Such composite cones or *strato-volcanoes,* as they are called, are rather steep sided (Fig. 5-4) and account for most of the scenically magnificent cone edifices in the world. Mount Fuji in Japan, Mount Hood in Oregon, and Mounts St. Helens and Rainier (Fig. 5-5) in Washington are prime examples of stratovolcanoes. The rocks erupted from such volcanoes are mostly andesitic to rhyolitic in composition.

In contrast, there are also cones built almost entirely of fluid lava flows, one flow upon the other. Cones built in such a manner generally have a basaltic composition. With gently rounded profiles and a nearly circular outline, they resemble turtles of colossal size or the shields once mounted along the gunwales of sea-roving Viking ships. As a consequence, they have been called *shield volcanoes* (Fig. 5-6). The volcanoes of the Hawaiian Islands and Kilimanjaro in Africa are some of the finest ex-amples of shield volcanoes in the world.

In some cases eruption is not restricted to a single, localized vent but occurs simulta-neously along a linear trend. The resultant con-structional feature then forms an elongate welt rather than a symmetric cone.

Most composite and shield volcanoes are built up over a considerable period of time, although most require less than a million years for their construction. Some vents, commonly on the flanks of larger cones, go through a complete eruptive life cycle in only a few weeks, months, or years. Because of the pau-city of eruptive products, they generally attain a height of only 100 m (328 ft) or so. A particular type of small cone, steep sided and composed almost entirely of reddened scoriaceous blocks, lapilli, and bombs is termed a *cinder cone* (Fig. 5-7). It may be found as a solitary isolated feature, but quite commonly tens of cinder cones will occur in close proximity, each marking a single, short-lived eruptive event. In some cases the growth of a cinder cone is associated with the eruption of small lava flows. The birth and growth of a larger-than-normal cinder cone was recorded in Mexico beginning in 1943. From its inauspicious beginning as a small crevice from whence emanated hot gases and gurgling sounds, the cone that was to be called Parícutin began to grow. The erup-tion lasted about nine years and built a cinder cone almost 369 m (1210 ft) high. During the early stages of its existence, activity was con-fined to the explosive expulsion of blocks, bombs, and lapilli; later, torrents of lava broke through the flanks of the cone and flowed out-ward, eventually reaching the nearby town of San Juan Parangaricutiro and covering all but the church steeple (Fig. 5-8). Its last days were punctuated by strong detonations during which fiery bombs and blocks up to 90 metric

Fig. 5-7. Classic form of a geologically recent cinder cone in the Cascade Range, northern California. The entire cone, which is an accumulation of cinders, lapilli, ash, and bombs, probably formed in a few weeks' time. In the background is Mount Lassen, which erupted last in 1914–17. *California Department of Water Resources*

Fig. 5-8. Parícutin Volcano, Michoacán, Mexico. In 1944, about a year after the cinder cone began to develop, lava issued from the vent and advanced toward the town of San Juan Parangaricutiro, eventually covering all but the church steeple. The photo, taken in 1950, shows the vent was still active. *F.O. Jones, USGS*

tons (100 short tons) in weight were hurled past the base of the cone.

Nearly all volcanic cones are characterized by a relatively small funnel-shaped depression—called a *crater*—marking the top of the conduit through which the eruptive products were channeled (Fig. 5-9).

There is considerable variety in the way volcanoes erupt. Even a single volcano may go through several phases of eruption, each one somewhat different than the one before. Nonetheless volcanoes in general, or any single eruption, can be categorized into broad types based on degree of explosivity. Some volcanoes erupt violently, with the explosive release of great clouds of pyroclastics, while others do so relatively quietly with only the outpouring of fluid lava. Even the quietest eruptions are, by our standards, spectacular to behold. Of course there are eruptions that fall gradationally at all levels between the two extremes. In general, volcanoes that erupt andesitic to rhyolitic material tend to be more explosive than those that erupt basalt. The most explosive volcanoes in the world are found in Indonesia and Central America; more-

over, the degree of explosivity is much greater in the circum-Pacific belt than in the other parts of the world.

Explosive eruptions result from the sudden release of large quantities of gases contained (dissolved) in the magma. All magmas contain gases. Our best estimates indicate a content of about 1–2 weight per cent in basaltic magmas and about 2–4 weight per cent in andesitic to rhyolitic magmas. By far the greatest proportion of the gases (commonly 50–80 per cent) is made up of water vapor, accounting for at least some of the vapor that swirls in dense clouds above an erupting volcano (Fig. 5-10). Carbon dioxide, nitrogen, and sulfur dioxide also contribute measurably to the gas content, but in varying proportions. Small amounts of carbon monoxide, hydrogen, sulfur, chlorine, fluorine, and hydrogen chloride are also detectable.

Some of the dissolved gases are undoubtedly derived from the country rocks which the magma contacts or engulfs as it moves upward in the crust; some, however, are indigenous to the magma and represent ancient volatile material trapped inside the earth during the early stages of its formation. In 1955 the geochemist W. W. Rubey reached the rather startling conclusion that essentially all waters on the face of the earth and in the atmosphere have probably been liberated from within the earth by volcanic activity operating throughout the history of the planet. Most geologists concur in that truly mind-boggling conclusion.

More gas is soluble in magma at higher confining pressures than at lower ones. Therefore, magma deep within the earth can contain more gas than can the same magma at a shallower depth. As the magma moves upward and nears the surface of the earth, and as crystallization of silicates that do not contain volatile-rich minerals begins to occur, the gas pressure increases until the contained gases begin to escape. If there is an open channel to the surface, and if the magma is fluid, the gases easily escape into the atmosphere. If the magma is less fluid, and if a plug of solidified material blocks the vent, escape of the gases is impeded. With time, pressure builds up. If the vent is suddenly cleared, either by release of the gases under high pressure or by eruption of magma, the pent-up gases burst out violently and explosively (Fig. 5-11). It is like popping the cork on a bottle of champagne. Expanding rapidly, the gases propel blocks of magma and solid material upward to great heights, as if shot from cannons. The nearly instantaneous release of the gases can cause the top of a magma chamber to froth, forming pumice or scoria, depending on the composition of the magma. In some cases, continued expansion of the gases within the innumerable vesicles in the chilled glassy froth can cause its fragmentation into fine ash and lapilli (Fig. 5-12). When large quantities of gas are catastrophically discharged, they can lead to nearly instantaneous emptying of the magma chamber.

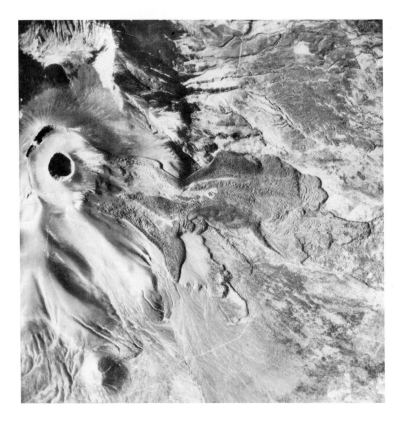

Fig. 5-9. Asama Volcano, Honshu, one of Japan's most active and dangerous volcanoes. Notice the sharp central crater marking the top of the active volcanic vent. Recent ash covers the flanks of the volcano. Andesite lava issued from the vent in 1783 to produce the molasses-like flow to the right. *USGS*

Fig. 5-10. Eruption of Hekla Volcano, Iceland. The billowing plume, carried windward, is composed of volcanic ash and volatile constituents, mostly water.
Thorsteinn Josepsson

Fig. 5-11. Eruption of a volcano. Viscous lava congeals in the throat, plugging the vent. Pressure builds up and eventually the pent-up gases are released explosively—completely or partially (as shown in the diagram) clearing the vent.

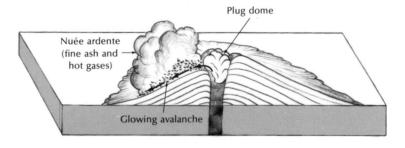

Plug dome

Nuée ardente
(fine ash and
hot gases)

Glowing avalanche

Andesitic to rhyolitic magmas, which tend to be less fluid than basaltic magmas, are more prone to explosive activity.

In the following pages we will document some of the possible variations in mode of eruption. Some of the information is taken from eyewitness accounts of historic eruptions, some from geologic studies of prehistoric volcanic centers.

Explosive eruptions

Krakatoa Three-quarters of a century ago captains of sailing ships, beating their way through the Sunda Straits which separate the great islands of Java and Sumatra in the East Indies, knew the island of Krakatoa well. Its conical, green-clad slopes rose uninterruptedly about 793 m (2601 ft) to the summit of the central peak. The straits were important since they

Fig. 5-12. Well-bedded ash and lapilli exposed in a stream cut about 1 km (0.6 mi) northeast of Cerro de Camiro, Michoacán, Mexico. The ash layers are the product of innumerable eruptions.
K. Segerstrom, USGS

were on the shortest sea road for the tea clippers en route from China to England. They were dangerous, restricted waters, haunted by sea-roving Dyaks who could give the crew of a becalmed vessel a bad time. In the same seaway many years later the U.S.S. *Houston,* harried by pursuing Japanese in the early years of World War II, blew up and sank with the loss of almost all hands.

Though the island of Krakatoa had been spasmodically active since May 1883, it seemed innocuous enough to the crew of the British ship *Charles Bal,* tacking under all plain sail through the hot tropical Sunday afternoon of August 26, 1883, until they arrived on one heading at a point about ten miles south of the island. Minutes later the mountain exploded. Seldom in the history of seafaring has the crew of any vessel been confronted by such a satanic outburst of energy. The entire mountain disappeared in clouds of black "smoke," and the air was charged with electricity—lightning flashed continuously over the volcano, as it very often does during eruptions—and the yards and rigging of the ship glowed with St. Elmo's fire. Immense quantities of heated ash fell on the deck or hissed through the surrounding darkness into the increasingly disturbed sea. As the vessel labored across broken seas through squalls of mud-laden rain, a thundering roar of explosions continued, much like a never-ending artillery barrage, accompanied by a ceaseless crackling sound which resembled the tearing of gigantic sheets of paper. The last effect was interpreted as the rubbing together of large rocks hurled skyward by the explosions. After an interminable night, the dawn, dim as it must have been, came as deliverance, and with the coast of Java in view and a gale rising rapidly, the *Charles Bal* was able to set all sail and leave the smoking mountain far astern.

It is well she did, for paroxysms of volcanic fury continued to shake the mountain until the final culmination of four prodigious explosions came on Monday, August 27, at 5:30, 6:44, 10:02, and 10:52 A.M. The greatest of them, the third, was one of the most titanic

explosions recorded in modern times—greater in intensity than some of our nuclear efforts. The sound was heard over tremendous distances: at Alice Springs in the heart of Australia, in Manila, in Ceylon, and on the remote island of Rodriguez in the southwest Indian Ocean, where it arrived four hours after the explosion had occurred 4800 km (2983 mi) away. It was truly the natural analogue of the "shot heard 'round the world."

The explosion seriously agitated the earth's atmosphere, and records of such a disturbance were picked up by barometers all over the world. They showed that a shock wave originating in the East Indies traveled at least seven times around the world before it became too faint to register on the instruments of that time.

Visibly, a more impressive phenomenon was the huge cloud of pumice and volcanic debris that blew skyward. The steam-impelled cloud of volcanic ash is estimated to have risen to a height of 80 km (50 miles) on August 27, and to have blanketed a surrounding area of 768,000 km² (296,525 mi²). The ash poured down as a pasty mud on the streets and buildings of Batavia—now Djakarta—133 km (83 mi) away. Pumice in far-reaching masses blanketed much of the Indian Ocean, and captains' comments recorded in logbooks of ships suddenly enmeshed in great floating rafts of pumice far offshore make interesting reading.

Volcanic ash hurled into the upper levels of the atmosphere was picked up by the jet stream—whose existence was not even suspected then—and carried with it as a dust cloud that encircled the earth in the equatorial regions in 13 days. The ash continued to spread across both hemispheres and produced a succession of spectacular and greatly admired sunsets over most of the world—even in areas as remote from Java as England and the northeastern United States—for the two years that it took the finer dust particles to settle through the atmosphere. There is ample evidence to suggest that the presence of such large quantities of ash in the atmosphere effectively screens the sun's rays and produces a world-wide cooling effect of about 1°C. That does not seem like much, but it has led some geologists to suggest that accelerated and explosive volcanism was the underlying cause of the ice ages in the recent geologic past.

The violent explosion of the morning of August 27 also set in motion one of the more destructive sea waves ever to be recorded. It spread out in ever-widening circles from Krakatoa much as though a gigantic rock had been hurled into the sea. About half an hour after the eruption, the wave reached the shores of Java and Sumatra, and on their low-lying coasts the water surged inland with a crest whose maximum height was about 37 m (121 ft). Since many of the people who inhabited the densely populated tropical coast lived in houses built on piers extending out over the water, about 37,000 people lost their lives.

The sea wave, after leaving the Sunda Strait with diminishing height, raced on across the open ocean. It was registered long after it was too faint to see, as a train of pulses on recording tide gauges along the coasts of India and Africa and on the coasts of Europe and the western United States.

After the explosions died down, returning observers were startled to find that where the 793-m (2601-ft) mountain had stood there was now a hole whose bottom was 274–305 m (899–1000 ft) below sea level, and that the sea now filled the bowl-shaped depression. All that remained of the island were three tiny islets on the rim. All told, although the estimates vary, the loss of material associated with the eruption amounted to just under 21 km³ (5 mi³) in volume. In popular accounts of the eruption the impression commonly given is that a volcanic mountain blew up and its fragments were strewn far and wide over the face of the earth. Were that the case, most of the debris covering the little islands that are the surviving remnants of Krakatoa would be pieces of the wrecked volcano, and the oversize crater, or *caldera*, now filled with sea water would be the product of a simple explosion.

Unfortunately for that seemingly plausible explanation, few pieces of the original volcanic mountain are to be found, and instead of such fragments the ground is covered with deposits of pumice up to 60 m (197 ft) thick. You may also recall the mention of great rafts of floating pumice in the Indian Ocean, which were a source of surprise to the mariners who encountered them drifting over much of the open sea. Pumice is original magmatic material, frothed up by gases contained in the magma, and has nothing to do with the internal composition of the vanished mountain. Thus, the abundance of pumice and the absence of pieces of the mountain lead logically to the conclusion that the volcanic cone foundered or collapsed on itself rather than having been blown to bits.

That explanation was advanced by a Dutch volcanologist, Reinout van Bemmelen, in 1929, and refined by Howel Williams, of the University of California, in 1941. The accompanying diagram (Fig. 5-13), adapted from Williams, shows the sequence of eruptive events that very likely were responsible for the disappearance of a nearly 800-m mountain and the appearance of a deep caldera in its place.

Crater lake Although Crater Lake, Oregon, stands at an altitude of 1846 m (6055 ft), rather than at sea level as Krakatoa does, their calderas have many similarities (Fig. 5-14).

The diameter of both circular depressions is disproportionately large compared to the dimension of the mountain of which each is a part, and in each case most of the mountain has disappeared in the formation of the caldera. Crater Lake is situated at the site where a volcanic mountain perhaps 3700 m (12,136 ft) high, and containing about 70 km³ (17 mi³) of material once stood; it has been called Mount Mazama. The formation of Crater Lake occurred about 6000 years ago at a time when only a sparse Indian population occupied the region. The event must have been witnessed by some Indians, for later descendants carried a legend in which the evil god of fire living within Mount Mazama waged a terrible battle with the good god of snow who lived atop nearby Mount Shasta. Good triumphed, but during the conflict the top of Mount Mazama was destroyed. The legend, although probably based on actual events, carries little information about the details of the formation of the Crater Lake caldera and the destruction of Mount Mazama. Yet from careful study of the character of the rocks exposed in the roots of the old volcano and of the distribution and composition of the vast amounts of pumice and ash that surround the site, the events before and during the catastrophic eruption can be pieced together. According to Howel Williams:

When the culminating eruptions were over, the summit of Mount Mazama had disappeared. In its place there was a caldera between 5 and 6 miles wide and 4000 feet deep. How was it formed? Cer-

tainly not by the explosive decapitation of the volcano. Of the 17 cubic miles of solid rock that vanished, only about a tenth can be found among the ejecta. The remainder of the ejecta came from the magma chamber. The volume of the pumice fall which preceded the pumice flows amounts to approximately 3.5 cubic miles. Only 4 per cent of this consists of old rock fragments . . . . Accordingly 11.75 cubic miles of ejecta were laid down during these short-lived eruptions. In part, it was the rapid evacuation of this material that withdrew support from beneath the summit of the volcano and thus led to profound engulfment. The collapse was probably as cataclysmic as that which produced the caldera of Krakatau in 1883.

Mont Pelée   In 1902, as today, Martinique in the French West Indies was one of the more

Fig. 5-13. Stages in the collapse of a volcano to form a caldera.

Stage I   The cycle starts with fairly mild explosions of pumice. The magma chamber is filled and magma stands high in the conduits. As the violence of the explosions increases, magma is drawn off more and more rapidly.

Stage II   The culminating explosions clear out the conduits and rapidly lower the magma level in the chamber. Pumice is blown high above the cone, or glowing pumice-laden clouds sweep down the flanks.

Stage III   With removal of support, the volcanic cone collapses into the magma chamber below, leaving a wide, bowl-shaped caldera.

Stage IV   After a period of quiescence new minor cones appear on the caldera floor.

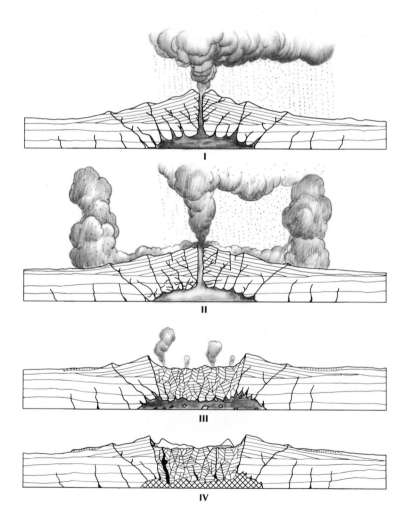

Fig. 5-14. Crater Lake, Oregon, looking southwest. The circular caldera is about 10 km (6 mi) across, the lake is nearly 610 m (2001 ft) deep, and the highest points on the rim are nearly 610 m above the lake. Wizard Island, a cinder cone that erupted on the floor of the caldera, is visible at the far right of the lake. Before the caldera collapsed the area was the site of Mount Mazama, a steep-sided stratovolcano. *Ray Atkeson*

picturesque links in the chain of islands reaching like green stepping stones across the Caribbean to join Cuba with the mainland of South America. Perhaps the most notable distinction of this small island is that it was the birthplace of Josephine, Empress of France, wife of Napoleon.

The island is quite mountainous; most of the interior is garlanded with a tropical forest; and the people then, as now, lived near the coast in villages, on plantations, and in the few large towns, of which St. Pierre (with a population of 28,000) was the most important and had a history of continuous settlement extending back to 1635. There was a moderately good open roadstead, and ships customarily moored in a

line standing off the beach and were kept headed seaward. St. Pierre was then the leading commercial town of the island and had a fair number of French and Americans in residence. Most of the people were Carib Indians or descendants of Africans imported to work in the plantations, sugar centrals, and rum distilleries.

Mont Pelée, about 8 km (5 mi) north of town, was known to be a volcano, but it had smoldered contentedly for several centuries in the mild climate of the trade winds. On April 23, 1902, it began to show signs of internal discontent, but they were little more than occasional rumblings, clouds of smoke, and spasmodic outbursts of ashes and cinders. The mildly petulant display rose to a more violent level of activity on May 4, when an outburst of hot mud, steam, and some lava broke through the crater wall, coursed down one of the radial stream canyons, buried a sugar central, and killed twenty-four persons.

By that time St. Pierre was thoroughly aroused, and not even the presence of the governor and his retinue, together with the issuance of the usual tranquilizing proclamation, served to quiet the multitude. In fact the city was kept in a continuous turmoil as country people and villagers, frightened into abandoning their homes, poured into town.

The tumult and confusion were stilled forever, with appalling suddenness, early in the morning of May 8, 1902, at 7:45 A.M. According to the few eyewitnesses, the top of the mountain vanished in a blinding flash, and almost immediately thereafter a rapidly moving, fire-hot cloud advancing at prodigious speed engulfed the city, whose population, swollen with refugees, probably numbered more than 30,000 (Fig. 5-15). All perished within a few moments, save two—one of them in the eyes of society perhaps the least deserving since he was incarcerated in an underground dungeon on a charge of murder. All others—men, women, and children—died in a blazing instant in a cloud whose temperature was high enough to melt glass (650–700°C; 1202–1292°F).

About the only credible account of the eruption came from some of the survivors on ships in the roadstead. Eighteen vessels were in port at the time, and of them only the *Roddam*, with more than half her crew dead, was able to up anchor and escape. The cable ship *Grappler*, directly in the path of the incandescent cloud, capsized and blew up. The purser of the *Roraima*, then approaching the harbor from the sea, left the most complete narrative of any observer. The *Roraima* was enveloped in the wall of flame that incinerated the town, was hurled over on her beam ends, the masts and stack were sheared off, her captain was blown overboard from the bridge and killed, and the ship herself burst into flames not only from the heat of the glowing cloud but, to add a bizarre touch to the holocaust, from the thousands of gallons of blazing rum that poured through the streets of St. Pierre and spread out over the waters of the harbor. Through heroic efforts the 25 injured survivors out of a crew of 68 were taken off by the French cruiser *Suchet* in mid-afternoon.

Within the town itself only two human beings lived. All were dead except for Auguste Cyparis, the occupant of the dungeon, who languished there deserted and in a state of shock for four days until his rescuers, who had despaired of finding any living thing in St. Pierre, peered through the barred window of his dungeon when his cries attracted their attention. The other survivor, Léon Compère-Léandre, was one of those unusually tough, resilient men who occasionally survive in disasters when all about them die. Although he was covered with burns, and no conceivable circumstance favored his survival over others, he made his way on foot through the burning city of the dead and lived to tell his tale.

The escape of entrapped high-temperature gases was the chief reason for the appearance of the catastrophically violent clouds that overwhelmed St. Pierre. Since there is no satisfactory English name for them, the French term *nuée ardente*, which might inadequately be translated as "glowing cloud," seems appropriate.

As was apparent to trained observers who flocked to Mont Pelée after the disaster, the nuée ardente that swept through St. Pierre, searing everything in its path, was but a minor aspect of the total eruption. The major feature of the eruption was an avalanche of incandescent rock and pumice fragments that shot out of the mountain top, cascaded down the canyon of the Rivière Blanche, and continued down to the ocean 3 km (2 mi) away. The nuée ardente was merely the hot dust and vapor cloud that rose above the *glowing avalanche*. Whereas the avalanche moved down the canyon, the fiery cloud jumped the ridge on the south side of the canyon and continued across the city. The destruction of St. Pierre was entirely the work of the dust and vapor cloud and the fires that followed it.

The 1902 eruption of Mont Pelée was impor-

Fig. 5-15. St. Pierre, Martinique, after the eruption of Mont Pelée in 1902. The charred, stark remains of what was once a thriving port and the capital city of 30,000 inhabitants. Only two people in the entire city survived the fiery breath of the nuée ardente. *Brown Brothers*

Fig. 5-16. Mono Craters, California, looking southwest into the Sierra Nevada. Mono Lake is in the foreground. At the lower right is Panum, a volcanic dome surrounded by a cone of pumice fragments. To the left is a larger dome of obsidian surrounded by a thick lava flow of obsidian bordered by a steep slope. Above and to the left of that is a two-pronged lava flow of obsidian that has almost buried the dome that was its source. Above that are three more domes and a large flow of obsidian. The chain of eruption centers continues to the base of Sierra Nevada, forming a line of volcanoes nearly ten miles long. *John S. Shelton*

tant geologically because it provided an unparalleled example of a kind not too well understood before. No lava appeared in the early, violent phase of the eruption, as was possibly true also at Krakatoa.

When the violently explosive phase ended, a viscous, stiff lava was extruded into the summit crater, ending with the construction of a dome-like protrusion of blocky lava, encrusted with lesser spires and pinnacles. By September 1903, the spire had attained a height of perhaps 305 m (1000 ft) and a diameter about twice as great. Estimates placed its volume at 100 million m³ (3531 million ft³).

Such *volcanic domes* are more common than many people think. Lassen Peak in northern California is an excellent example: a protrusion of blocky lava stands about 770 m (2526 ft) above its crater rim, with a volume of approximately 2.5 km³ (0.6 mi³). The mountain was last active in 1914–17, when steam explosions, after blasting a vent on the northern slope, melted the snow cap. The resulting mud and ash flows not only devastated the forest at the base but swept 18-ton boulders for distances of 8–10 km (5–6 mi).

Among other examples of volcanic domes are the Mono Craters in east-central California (Fig. 5-16). Some of the volcanoes of the Valley of Ten Thousand Smokes in Alaska (Fig. 5-17), as well as the Puys of the Auvergne region of France, are domes. Of the latter, the Puy de Dome is perhaps the best known.

Recognition of the deposits dropped by turbulently flowing incandescent avalanches and nuées ardentes was uncertain before the demonstration of their nature at Mont Pelée. Commonly their stratification may be chaotic. Large blocks are mixed with finer particles; the uniformly layered appearance characteristic of pyroclastic deposits where the ash had an opportunity to settle more gradually through the

Fig. 5-17. Novarupta Rhyolite Dome in the Valley of Ten Thousand Smokes, Katmai National Park, Alaska. *USGS*

atmosphere is lacking. Such deposits encircle Crater Lake (Fig. 5-18) for example; there, pumice flows swept down the Rogue River canyon for 56 km (35 mi). They may have attained velocities of 160 km (99 mi) per hour and were capable of carrying pumice blocks 2 m (6.5 ft) in diameter a distance of at least 32 km (20 mi).

The principal dissimilarity between the eruptions of Mont Pelée and those of Krakatoa and ancient Mount Mazama (Crater Lake) is one of scale only. Krakatoa and Mount Mazama each erupted such large quantities of pumice and ash that their cone superstructures became weakened and collapsed along ring-shaped fractures to produce calderas several kilometers across. On the other hand, eruptions from Mont Pelée were restricted to the explosive discharge of small tongues of incandescent material, followed by intrusion of the sticky vent plug. The magma chamber was never emptied to any great extent and collapse of the cone could not occur.

Yellowstone and the San Juan Mountains As spectacular as the catastrophic eruption must have been that produced the Crater Lake caldera, it pales by comparison with those that must have occurred not once but three separate times in the Yellowstone National Park region at the eastern edge of the Snake River Plain, Idaho, during the last 2 million years. Careful mapping in the land of grizzlies and geysers by members of the U.S. Geological Survey has uncovered three large calderas, the smallest of which measures 29 km (18 mi) wide by 37 km (23 mi) long and the largest 64 km (40 mi) wide by 80 km (50 mi) long. The latter is apparently the largest caldera yet recognized on the face of the earth. Calderas of such dimensions are commonly termed *volcano-tectonic depressions*. Yellowstone Lake itself occupies a portion of the youngest of these calderas. The formation of each caldera depression was associated with the cataclysmic eruption of hundreds of cubic kilometers of pumice-rich ash from the magma chamber. The temperature of the ash as it exited from the vent must have been close to

750–800°C (1382–1472°F). The ash traveled outward from the eruptive centers as far as 80 km (50 mi) so quickly that there was still sufficient heat left in the pyroclastic blanket after it came to rest—even at its distal ends—to bring about the softening and sticking together of the glassy ash fragments and collapse of the porous pumice fragments. The result was the formation of a dense, resistant rock called a welded ash-flow tuff. Fine-grained ash blown high into the atmosphere during each of the eruptions was carried eastward by the prevailing winds and has been recognized in deposits as far east as Kansas. Rivers, such as the Yellowstone, incised canyons into the three different ash-flow tuffs and have shown in most places that each deposit is more than 60 m (197 ft) thick (Fig. 5-19). An intriguing and somewhat sobering aspect of the Yellowstone eruptions is brought out by noting the sequence of the eruptions, as established by potassium-argon dating methods (see Chap. 1). The first took place 2 million years ago, the second 1.2 million years ago, and the third 0.6 million years ago. There was a lull of about 800,000 years between the first and second eruptions, but only 600,000 years between the second and third. Since the last eruption, 600,000 years have passed. Could the bubbling hot springs and fountaining geysers so common and so photographed in the Yellowstone National Park area be the harbingers of the next eruption? Any event of the immensity of even the smallest Yellowstone eruption in the past would bring about untold loss of life and property, and decline in economic prosperity. Because of that threat and because of the potential for geothermal power, the region around Yellowstone has been the subject of extensive geophysical investigations during the last several years. The

Fig. 5-18. Ash-flow tuff or consolidated volcanic ash, with lapilli and pumice blocks. These are the deposits of the "nuées ardentes" that preceded the collapse of Mount Mazama to form Crater Lake. Since their eruption, the relatively nonresistant deposits have been eroded into pointed spires. The view is of The Pinnacles, along Sand Creek in Crater Lake National Park. *Oregon State Highway Dept.*

Fig. 5-19. Canyon incised into welded ash-flow tuffs, Yellowstone National Park. The streaky layering visible in the tuff resulted in part from flowage during emplacement and in part from vertical compaction of the ash flow during welding after coming to rest. *J.P. Iddings, USGS*

results seem to indicate with some assuredness the presence of a rock body with anomalous physical properties just below the surface, 64 km (40 mi) in diameter and extending downward for 50 km (31 mi) or more. Most earth scientists have interpreted the body to be a magma chamber of batholithic proportions. The volcanic fireworks and caldera formations of the last 2 million years appear, in light of that finding, to be only the surface manifestations related to the development of the large, igneous body just below the surface. The question that remains unanswered is: how much of the body is still in a molten state?

In the San Juan Mountains of southwestern Colorado about 15 large calderas, each with

one or more associated sheets of welded ash-flow tuff formed about 30 million years ago (Figs. 5-20 and 5-21). Like those at Yellowstone, they appear to have been closely associated with the development of a large, granitic batholith that lay close to the surface beneath them.

Eruptions of intermediate explosivity

Vesuvius One of the world's most famous volcanoes is Vesuvius, the only one active on the European mainland today (Fig. 5-22). Its renown probably results from its well-publicized eruption of A.D. 79 with the accompanying destruction of the cities of Pompeii, Herculaneum, and Stabiae. Although the mountain

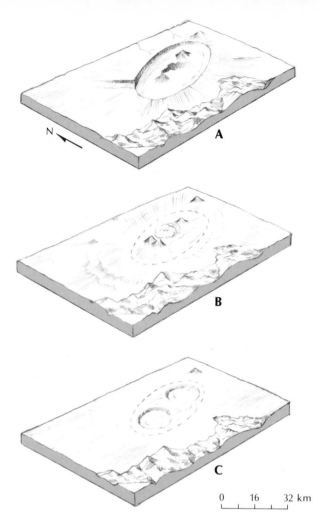

Fig. 5-20. The development of three calderas in the San Juan Mountains, Colorado. First formed was the San Juan caldera (**A**), which was subsequently nearly buried by volcanic products (**B**). Collapse within the central part of the San Juan caldera formed the Silverton and Lake City calderas (**C**).

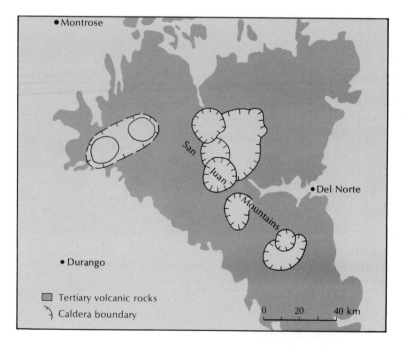

Fig. 5-21. Distribution of Tertiary volcanic rocks and calderas in the San Juan region, Colorado.

Fig. 5-22. Mount Vesuvius during the eruption of 1944. What appears to be a white-capped peak to the left of Vesuvius is a curved ridge known as Monte Somma, the margin of a caldera formed during the development of an older prehistoric volcano. *Brown Brothers*

had been active in prehistoric times, volcanic activity had been so long dormant that the Romans were unaware of its real character. Strabo, who visited the volcano about the beginning of the Christian era, surmised its volcanic origin from what he interpreted as burned and fused rocks near the summit, but he determined that the volcano was extinct. In fact the 1220-m (4002-ft) mountain we see today is superimposed in large part on the wreckage of the older and lower pre-A.D.-79 crater, to which the name Monte Somma is given. Its inactive, vine-covered bowl, encircled by steep cliffs, briefly made a stronghold for the gladiator Spartacus and his fellow slaves when they defied the power of Rome until they were slain by the legionaries of Marcus Licinius Crassus in 71 B.C. In A.D. 63 the volcano showed some stirrings of life when a succession of earthquakes commenced and caused some of the damage still to be seen around Pompeii. That, however, was but a prelude to the historic eruption of August 24, A.D. 79.

Fortunately, a relatively complete description of the destructive event has come down to us through two letters from the 17-year-old Younger Pliny to his friend Tacitus, the Roman historian. The letters were written primarily to describe the death of his uncle, Pliny the Elder, a leading philosopher of his day and also, rather surprisingly, an admiral of the Roman navy.

While Pliny the Younger turned to study of his books, the Elder Pliny, soon to achieve the distinction of being the world's first volcanologist, and a Roman of the old school, marched forth to his death on the mountain. Parts of Pliny the Younger's letters are cited here because they are such good examples of straightforward reporting, quite unlike the exaggeratedly impossible version in Bulwer-Lytton's novel, The Last Days of Pompeii, in which most of the populace dies while watching a gladiatorial combat in the arena.

Parts of Pliny the Younger's letters follow:

Gaius Plinius sends to his friend Tacitus greeting. You ask me to write you an account of my uncle's death, that posterity may possess an accurate version of the event in your history . . . .

He was at Misenum, and was in command of the fleet there. It was at one o'clock in the afternoon of the 24th of August that my mother called attention to a cloud of unusual proportion and size . . . A cloud was rising from one of the hills which took the likeness of a stone-pine very nearly. It imitated the lofty trunk and the spreading branches. . . . It changed color, sometimes looking white, and sometimes when it carried up earth or ashes, dirty or streaked. The thing seemed of importance, and worthy of nearer investigation to the philosopher. He ordered a light boat to be got ready, and asked me to accompany him if I wished; but I answered that I would rather work over my books. . . .

Ashes began to fall around his ships, thicker and hotter as they approached land. Cinders and pumice, and also black fragments of rock cracked by heat, fell around them. The sea suddenly shoaled, and the shores were obstructed by masses from the mountain. . .

My uncle, for whom the wind was most favorable, arrived, and did his best to remove their terrors. . . . To keep up their spirits by a show of unconcern, he had a bath; and afterwards dined with real, or what was perhaps as heroic, with assumed cheerfulness. But meanwhile there began to break out from Vesuvius, in many spots, high and wide-shooting flames, whose brilliancy was heightened by the darkness of approaching night. My uncle reassured them by asserting that these were burning farm-houses which had caught fire after being deserted by the peasants. Then he turned in to sleep. . . .

It was dawn elsewhere; but with them it was a blacker and denser night than they had ever seen, although torches and various lights made it less dreadful. They decided to take to the shore and see if the sea would allow them to embark; but it appeared as wild and appalling as ever. My uncle lay down on a rug. He asked twice for water and drank it. Then as a flame with a forerunning sulphurous vapor drove off the others, the servants roused him up. Leaning on two slaves, he rose to his feet, but immediately fell back, as I understand choked by the thick vapors. . . . When day came (I mean the third after the last he ever saw), they found his body perfect and uninjured, and covered just as he had been overtaken. . .

Pliny's letters clearly indicate that the destruction of Pompeii and Herculaneum resulted from the fall of hot ash, and in that shroud were buried 2000 of the 20,000 inhabitants who perished. Most of the dead were slaves, soldiers of the guard, or people who were too avaricious to leave their worldly goods. Most were suffocated by falling ash, by hot volcanic mud, or by volcanic gases, and the temperature of the ash was high enough that their bodies charred away. Centuries later when plaster of paris was poured into the cavities once occupied by their bodies, allowed to harden, and then excavated from the ash, their shapes as well as those of dogs and cats, loaves of bread, and all sorts of objects in similar cavities stood revealed. Hundreds of papyri were preserved in the library, along with murals on the walls of houses, and they give a most revealing insight into the interests and pursuits of those long-vanished Romans. The two cities of Pompeii and Herculaneum slept undisturbed for nearly 1700 years until the discovery of one of the outer walls in 1748 ushered in the period of modern archeology.

Vesuvius has continued its activity from A.D. 79 to the present; in A.D. 472, ashes drifted from its crater as far east as Constantinople. An especially violent eruption in 1631 is estimated to have killed 18,000 people, and came after a period of quiescence that lasted long enough for the volcano to be overgrown once again by vegetation. A large number of minor eruptions have been recorded, but major ones occurred in 1794, 1872, 1906, and 1944, in the midst of the Italian campaign of World War II. Lava then overwhelmed the village of San Sebastiano, but the most destructive effect, as far as the allied military effort was concerned, came from the introduction of glass-sharp volcanic ash into the moving parts of airplane engines.

The first lava is said to have appeared at Vesuvius in A.D. 1036, and its appearance has been a standard accompaniment of most eruptions ever since. Because eruptions in the current phase of the volcano's life history commonly include both the upwelling of large quantities of lava and violent explosions that blast great quantities of ash, cinders, bombs, and blocks skyward, the volcanic edifice that has built up since 79 A.D. is composed in part of solidified lava from flows and internally from dikes and conduits and in part of pyroclastic material blown out explosively. Most stratovolcanoes are similar to Vesuvius, in fact.

Quiet eruptions

Shield volcanoes, Mauna Loa and Kilauea The Hawaiian Islands, surely one of the most idyllic archipelagos in the world, owe their entire existence to volcanism. They are a chain of extinct, dormant, and active volcanoes built up from the depth of the sea and trending southeastward across the Pacific for 2560 km (1591 mi) from Midway on the north to the largest island, Hawaii, on the south in an arc bowed slightly to the northeast. The name Hawaii is believed to be from *Hawaiiki*, the ancestral home of the Polynesians. The eight largest islands are at the southeastern end, and the relative erosional age of their landscapes generally decreases southeastward. That means that Hawaii, the only island with active volcanoes, appeared above the sea more recently than Oahu, the island on which Honolulu stands. One way in which that conclusion is reached is from the more advanced state of stream erosion, valleys, cliffs, and canyons on Oahu as well as the deeper soil cover that has developed there. Even more substantial evidence of the progressive "younging" of the islands toward the southeast is provided by potassium-argon dating of the flows on the eight largest islands of the chain. Kauai, the northwesternmost of the eight, is about 5.3 million years old, whereas Hawaii at the southeast end has been built in the last 750,000 years.

There are five major volcanic centers on Hawaii. Two of them are immense volcanoes: Mauna Kea, 4205 m (13,792 ft) above sea level and Mauna Loa 4170 m (13,678 ft). The bases of those mountains sit on the ocean floor, about 4600 m (15,088 ft) below the surface, and they are as tall as Everest, but of enormously greater bulk since the circumference of Mauna Loa is about 320 km (199 mi) at the base. They are by far the largest volcanoes in the world.

Although some pyroclastic material is included in the mass of the huge volcanic piles, for the most part they consist of thousands of superimposed, relatively thin flows of basalt. Many of the flows were extremely fluid at the time they were erupted. The result is that the slopes of classic shield volcanoes are gentle because they are composed of thousands of overlapping, tongue-like sheets of once-fluid material, rather than loose piles of heaped-up volcanic fragments.

The fires of Mauna Kea are banked now, but Mauna Loa still maintains a high level of activity. Most of the historic lava flows have broken out on its flanks rather than simply overflowing from the summit caldera, known as Mokuaweoweo. In fact, activity in the caldera today is at a minimum. The caldera itself originated as the result of foundering through removal of support from below.

Mokuaweoweo is no circular crater; the very steep walls, which are about 180 m (590 ft) high, enclose a sink approximately 5.6 km (3.5 mi) long by 3.2 km (2 mi) wide. The long dimension trends northeast-southwest on the same line as the so-called Great Rift Zone, out of which so many of the historic flows have issued. The caldera itself has grown through the coalescence of several once-independent pit craters on the summit of the mountain, and it is Howel Williams's belief that the coalescence results from collapse brought about by the draining away of lava from the magma reservoir within the mountain through fissures on its flanks.

The theory is supported by the pattern followed by typical eruptions. They may commence with some volcanic dust being blown from the summit crater and a column of steam standing over it by day as well as a glow of light that illuminates the clouds at night. Somewhat later, lava may break out on the flanks, almost always on the Great Rift Zone, either to the northeast or to the southwest of the summit.

Lava flows on Mauna Loa seldom issue from a single vent, but almost always break out from great cracks, or *fissures*. The first phase of such an outbreak may be the appearance of a line of *fire fountains*, or geyser-like columns of lava that may spurt as much as several hundred meters into the air and line up as a nearly continuous curtain of fire along the fissure (Fig. 5-23). The basalt that streams from the fissures is at a high temperature, close to 1200°C (2192°F) and consequently is usually extremely fluid when it first pours out. It may flow down pre-existing stream courses with velocities approaching those of the rivers themselves; where there are sharp irregularities, the lava plunges over them like a waterfall. When the lava stream reaches the sea, a titanic conflict ensues between the forces of Neptune and of Vulcan, as it were. Immense clouds of steam boil upward, the sea seethes like a gigantic cauldron, and part of the lava is quenched so abruptly that it froths up as a tawny, cellular sort of volcanic glass (Fig. 5-24).

Not all the Hawaiian basalts flow in torrential streams; blocky, ponderously advancing flows are common, too. They march forward much like a tank, or caterpillar tractor, when the surface crusts over and is carried ahead by the still-molten interior. The advancing crust breaks up at the leading edge of the flow, and the blocks cascade over the front to make a carpet, or track, over which the flow can advance. The top and bottom of such a flow will constitute volcanic breccia when the whole flow has solidified, and for the most part the interior will be uniformly textured homogeneous basalt.

Hawaii has given us two Polynesian terms to describe the surface character of lava flows, and they have now won such general acceptance that they are used commonly in the literature of geology. Basalt with a rough, blocky appearance, much like furnace slag, is called by the remarkably brief name of *aa* (Fig. 5-25), while more fluid varieties with smooth, satiny, or even glassy surfaces that are commonly contorted and wrinkled are given the more euphonious name of *pahoehoe* (Fig. 5-26). It is not uncommon for a pahoehoe flow to turn into an aa flow as it moves downslope away from the vent. However, the opposite transformation never occurs.

Kilauea is like Mauna Loa in some respects and very different in others. For one thing it is

Fig. 5-23. Fire fountain at Kilauea, November 18, 1959. The lighter colored parts of the fountain are glowing fluid masses of basaltic lava. The lava cools in the air and turns dark at the top of the fountain. *G.A. MacDonald*

at a much lower altitude, about 1220 m (4002 ft); and it is no longer an independent mountain but is a partially buried satellite on the flank of the higher volcano. Perhaps within the next few millennia it will be inundated by flows from a flank eruption of Mauna Loa. Kilauea is the far better known of the two volcanoes since a paved road leads directly to its rim, and it has had a steady stream of visitors for more than a century. Kilauea has had more than 50 eruptions in historic time.

The caldera of Kilauea is always a surprise to the first-time visitor, for much the same reason

Fig. 5-24. (opposite top) Basaltic lava, pouring over a cliff and into the sea during an eruption on the island of Hawaii in 1955. As the extremely hot lava meets the water, copious amounts of steam are produced. *G. A. MacDonald*

Fig. 5-25. (opposite bottom) Blocky, or aa, lava flow, slowly advancing over a field during the 1955 eruption on Hawaii. The flames and smoke are from burning vegetation. Inside the blocky interior the lava is still liquid. Such flows move ahead much like a tank; the cooled angular blocks cascade down the steep flow front and are overrun by the advancing flow. *G. A. MacDonald*

as the Grand Canyon is. Both make an extreme contrast with their nearly level surroundings. The elliptical caldera of Kilauea, whose dimensions are approximately 5 km by 3 km (3 mi by 2 mi), is countersunk with almost vertical walls into the very gently sloping surface of the old volcano (Fig. 5-27). The bottom of the caldera is nearly level and is made up of only very recently solidified lava, which spreads like a tarry stream over the entire floor. Activity today is confined to only part of the caldera—the volcanic throat, or fire pit, of Halemaumau, which bears a relationship to the larger caldera much like that of a drainpipe in the bottom of a washbasin. Basaltic lava rises and falls inside the fire pit. At times it spills over Halemaumau's rim onto the caldera floor; at other times it sinks down more than several hundred meters below the surface. Then the floor of the pit is filled with long talus aprons of basalt blocks that have broken away from the vertical walls.

Commonly, lava swirls and seethes within Halemaumau without violent explosive activity, but occasionally there are impressive departures from that pattern. Such a one was the 1924 eruption, in which the sequence of events was as follows: (1) in January the lava lake was especially active and the level rose to within about 30 m (98 ft) of the rim; (2) in February it started to subside and by May had dropped to around 180 m (590 ft); (3) meanwhile the epicenters of a whole succession of minor earthquakes migrated steadily eastward along the line of the Puna Rift, accompanied by ground subsidence until almost certainly there was an eruption on the sea floor southeast of Hawaii; (4) immense quantities of lava blocks avalanched from the walls into the fire pit; (5) finally, those blocks and much of the debris that had accumulated on the floor of Halemaumau were hurled out in a series of violent explosions between May 11 and 27.

An interpretation of that sequence of events, which are a bit out of character for a quiet eruption, is that the lava column dropped because lava was being drained away through fissures from beneath Kilauea—whose entire

level dropped, incidentally, by about 4 m (13 ft)—southeastward along the Puna Rift. That permitted ground water to move into the area vacated by the sinking column of lava. When a sufficiently high pressure was built up, and the lava column subsided below sea level, the ground water was converted into steam under cover of the blocks of rock that had fallen into the fire pit. Then the pressure rose to a point high enough to shatter the rocks, hurling them out of the pit in what was a succession of steam explosions rather than ones produced by primary magmatic gases. To such a secondary eruption the name of *phreatic explosion* (derived from the Greek word for "water well") is given, and they are very characteristic of minor eruptions the world over; in Iceland, New Zealand, Japan, and possibly the 1914–17 eruptions of Mount Lassen in California.

When lava enters the sea over long periods of time or when eruptions occur, as they sometimes do, beneath the surface of the sea, it is also possible for *pillow lavas* to be formed. As the name suggests, pillow lavas are composed of lobes of lava material stacked one upon the other and resembling a pile of bed pillows (Fig. 5-28). They almost always form from pahoehoe-type basalt flows. The exact mechanism for their formation is not clear, but it appears that small extrusion toes, pushing out from cracks in the advancing pahoehoe flow, immediately chill when they come in contact with water to form an elastic, glassy rind. The rind continues to grow into a pillow form, which either remains attached to the flow front or becomes entirely detached from it. In the latter case the pillow can bounce and roll down to the base of the slope, where it piles up with other detached pillows. That unusual result of the meeting of water and lava is relatively common in the geologic record throughout the world.

The island of Hawaii has been the most intensely studied volcano in the world. The

Fig. 5-26. Surface of a pahoehoe lava flow; Mauna Loa, Hawaii. Pahoehoe is characterized by a glassy upper surface, which is contorted into ropy, undulating patterns. *Ansel Adams*

Hawaiian Volcano Observatory of the U.S. Geological Survey sits at the edge of the Halemaumau fire pit and is run by a staff who not only study the nature of the flows and gases that pour forth during each eruption, but also monitor almost undetectable changes in the surface shape of the volcano and earthquakes from deep beneath the island. Principally from the earthquake data, scientists have concluded that the magma originates about 50 km (31 mi) or more beneath the surface and moves upward periodically to collect in shallowly buried magma chambers within the volcanic cones themselves (Fig. 5-29). As magma fills the upper chamber, the cone just above it inflates—that is, balloons upward and outward—which causes a distinct and measurable tilting and stretching of the volcano's surface. The stretching continues to a point at which eruption occurs and the tumescence declines, in response to the draining of the upper magma chambers. Scientists are now able to predict when an eruption is likely to begin, how long it will last, and if the eruptive phase has definitely ended or merely reached a state of temporary quiescence.

Fissure eruptions

Several regions on the earth have been inundated by vast floods of lava that obviously did not come from a shield volcano or even from a chain of such volcanoes. Rather, the vast flows appear to be the result of eruption from innumerable cracks or fissures, commonly subparallel in their orientation, that extend over a considerable area. Prominent examples of *flood lavas*, which are universally of basaltic composition, are found in western India inland from Bombay, in South America near the Paraná River, in Antarctica, in South Africa, and in the area around Lake Michigan and in the Pacific Northwest in the United States (Fig. 5-30).

The last, one of the most extensive lava floods known, is designated the Columbia lava plateau. It covers an area of 51,200 km² (19,768 mi²). In places it is about 2 km (1 mi) thick, but the individual flows are much thinner, only a few being as much as 120 m (394 ft) thick (Fig. 5-31). Their composition is remarkably uniform, especially in view of the fact that the enormous volume of basalt was erupted over a time span of about 3 to 4 million years. The total volume of lava erupted, incredibly enough, amounts to more than 307,200 km³ (73,701 mi³), and the surface of the plateau covers a very broad area—extending from the Rocky Mountains on the east to the Cascade Range and Pacific border to the west. The hundreds, if not thousands, of fissures that the lava welled up through are seen today as subparallel basaltic dikes trending north-south to northwest-southeast. The country buried by the lava floods was one of moderate relief. As shown in the walls of canyons that cut across the basaltic plateau, the individual flows filled valleys and depressions, overtopped ridges, and ultimately coalesced to form a nearly uniform plain. They buried a wholly different sort of world beneath a sea of frozen stone. Such tremendous outpourings of relatively uniform basaltic composition require an extensive source, one of uniform composition. Most geologists feel that they must have come from the outer mantle of the earth (see Chap. 17), most likely from depths between 70 and 250 km (44 and 155 mi).

The nearest counterpart ever reported of a fissure eruption was a minor episode by comparison, impressive as it undoubtedly must have been to those who witnessed it. It was the eruption of the Icelandic volcano Skaptar Jokull that began on June 8, 1783. A stream of basalt amounting to about 3 km³ (0.72 mi³) in volume poured out from a fissure about 24 km (15 mi) long over a two-year time span.

Iceland has one of the most dramatic landscapes on earth, with more than 100 volcanic centers, of which at least 20 are active, a score of glaciers, and the all-encircling sea. In few other regions is the elemental conflict of fire, ice, and ocean more stark. In the long and remarkable cultural history of the island, extending back to A.D. 874, there have been many such encounters between outpourings of red-hot lava and streams of ice. The usual outcome

Fig. 5-27. Kilauea Caldera, looking westward to the summit of Mauna Kea, Hawaii. The depression is about 5 km (3 mi) long, by about 3 km (2 mi) wide. The smaller sharply defined countersunk oval within the main caldera is the fire pit of Halemaumau. Notice that Mauna Kea has the classic rounded form of a shield volcano. *USGS*

Fig. 5-28. Pillowed lava found at Gordon Lake, Northwest Territories, Canada. The black cores of each pillow are covered by white rinds of devitrified and altered material that was originally the glassy skin on the pillows. *Geological Survey of Canada*

is the melting of much of the ice, with the release of a sudden and devastating flood of water and mud.

That is exactly what happened in the disastrous eruption of 1783. With the huge fissure discharging basalt along its entire length, a broad tide of lava poured down the slope, filled the deep canyon of the Skapta to overflowing, and completely displaced a lake that lay in its path. The eruption continued for two years, and the two major lava torrents it produced had lengths of about 64 and 80 km (40 and 50

mi), respectively. Their average depth was 30 m (98 ft), but where canyons were filled to overflowing, they were as much as 185 m (607 ft) thick. Where the lava overtopped a stream valley and spread out across the plain it advanced along a front 19–24 km (12–15 mi) wide. The flow is estimated to have covered 90 km² (35 mi²).

The eruption produced one of the greatest disasters in the turbulent history of the island. The lava, blocking and diverting rivers and melting snow and ice, liberated huge floods,

Fig. 5-29. Probable substructure of two volcanoes on the island of Hawaii. Magma, which is derived from depths greater than 50 km (37 mi), moves up into the crust and pools in larger chambers within the cones.

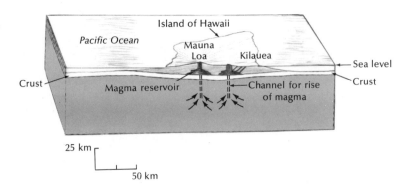

thus destroying much of the island's limited agricultural land. Twenty villages were overwhelmed by the lava, and many others were swept away in the floods. About 10,000 people, or 20 per cent of the population, died; 80 per cent of the sheep (190,000), 75 per cent of the horses (28,000), and more than 50 per cent of the cattle (11,500) perished.

Almost all the North Atlantic was obscured in the dust cloud, a phenomenon that greatly interested Benjamin Franklin. He wrote a brief description on the effect of the so-called dry fog in America (Griggs, 1922):

During several of the summer months of the year 1783, when the effects of the sun's rays to heat the earth in these northern regions should have been the greatest, there existed a constant fog over all of Europe, and a great part of North America. This fog was of a permanent nature; it was dry, and the rays of the sun seemed to have little effect toward dissipating it, as they easily do a moist fog rising from the water. They were indeed rendered so faint in passing through it that, when collected in the focus of a burning-glass, they would scarcely kindle brown paper. Of course, their summer effect in heating the earth was exceedingly diminished.

Hence the surface was early frozen.

Hence the first snows remained on it unmelted, and received continual additions.

Hence perhaps the winter of 1783–4 was more severe than any that happened for many years.

In Europe the ash cloud had a most deleterious effect on the weather, too, so much so that it was called the "year without a summer."

Crops failed in Scotland, 965 km (600 mi) away; fumes and ashes damaged crops in the Netherlands, and the ash cloud was reported from such widely scattered points as North Africa, Syria, eastern Russia, and Sweden. It rose to an altitude higher than the Alps, and the monks at the pass of St. Bernard correctly interpreted it as smoke and not haze. As might be expected, the event had a profound effect on people of the time, and among other things inspired the following passage of Cowper (Krakatoa Committee, 1888):

Fires from beneath, and meteors from above—Portentous, unexampled, unexplained. (*The Task*, BOOK II)

CLOSING THOUGHTS

We have explored in this chapter many of the more evident aspects of volcanism—the general distribution of active volcanoes, the character of their cones and eruptive products, and the nature of volcanic eruptions. Wherever possible, we have tried to point out reasons why volcanoes are located where they are and behave as they do. As is evident from the text there remain many unanswered or only partially answered questions. The basic reason for our uncertainty is clear; volcanism is only the surface manifestation of large and small-scale processes that occur hidden from view relatively deep within the earth.

The major unanswered question concerns

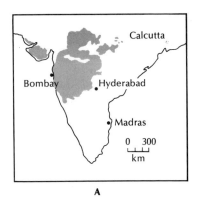

A

B

Fig. 5-30. The general extent of (**A**) the Deccan basalts in India, and (**B**) the Columbia River basalt and its equivalents, northwestern United States.

Fig. 5-31. Flow upon flow of the Columbia River basalt; east wall of Grand Coulee, downstream from Washington. Notice how the individual flows pinch and swell. Some are marked by crude columnar jointing. *John S. Shelton*

the ultimate cause of volcanic activity. The model of plate tectonics has provided us with a framework in which to view the phenomenon of volcanism, and there seems little doubt that the distribution and composition of many volcanoes are associated with activity at plate margins. But the nature of the subsurface activity and at what depth it occurs is imprecisely known. And of course the occurrence of many volcanic centers within the large plates themselves has required the formulation of axioms, not always well documented, that have had to be tacked on to the basic model. The process that leads to the generation of andesitic volcanism at converging plate boundaries is a point of considerable debate and controversy at the moment. Many believe it results from partial melting of rocks of basaltic composition at depths around 125 km (78 mi); others believe that additionally it requires melting of large amounts of rocks in the continental crust itself at depths of only a few tens of kilometers. Laboratory and theoretical studies have given some insights but have not yet pointed to any undebatable answers.

Another problem involves accounting for the colossal volumes of magma erupted as basaltic flood lavas of extremely uniform composition, such as those found in the Columbia lava plateau. Some geologists have suggested that the layer between about 100 and 250 km (62 and 155 mi), which appears to be partially molten, provides a ready source for such immense quantities of magma. If so, is it possible that basaltic magmas derived from different parts of the layer represent *parent magmas* from which all others are derived—by mixing of magmas, contamination to various degrees by rocks above the source layer as the magma makes its way to the surface, and by differentiation? Those three processes have some appeal in explaining the great diversity in composition of lavas that erupt at the surface, even from the same vent or from closely related conduits.

And further, we do not know the exact nature of a magma chamber or how molten material initially moves and coalesces at depth to form a body of magma.

These are but some of the problems related to volcanism. We can only hope that as scientists continue to learn more about the earth's interior and the processes that operate therein, we will be able to better resolve them.

SUMMARY

1. Today there are more than 516 active or recently active volcanoes on earth. Most occur in linear bands, principally around the Pacific Ocean but also in the Caribbean and Mediterranean regions, Asia Minor, central Africa, and near the Red Sea. Zones of volcanism generally parallel belts of earthquake activity, and those that occur at convergent plate margins are andesitic in composition, whereas those at divergent zones are basaltic. Some volcanoes (such as those of the Hawaiian Islands) occur away from the edges of a plate and cannot be explained readily according to the plate-tectonic model.
2. Most large volcanoes can be classified as *composite cones (stratovolcanoes)* or *shield volcanoes*. Composite cones are steep sided as a result of the eruption of lava and pyroclastic material, and most are felsic in composition. Shield volcanoes form from the accumulation of fluid lava and are gently rounded in profile.
3. In relatively viscous magma, gas discharge is inhibited. Commonly the pressure builds until it is released explosively. In some cases pressure release causes the entire upper part of a magma chamber to froth, forming pumice or scoria.
4. In 1883, the island of Krakatoa erupted explosively. Sea waves, generated by the eruption, inundated the nearby coastal regions and caused the demise of about 37,000 people. The eruption was related to *caldera* formation. Crater Lake also owes its origin to the foundering of ancient Mount Mazama during a catastrophic eruption about 6000 years ago. The eruption of Mont Pelée (1902) was the result of a relatively small explosion of gas-charged particles (*glowing avalanche* and *nuée ardente*). Gigantic explosive eruptions and extensive caldera formation have occurred in the Yellowstone National Park region and in the San Juan Mountains. In both areas, the volcanic activity was related to a large batholith that existed at shallow subsurface depths.
5. The Hawaiian Islands are shield volcanoes, the bases of which sit on the ocean bottom, 4600 m (15,088 ft) below sea level. When they erupt, almost no explosivity accompanies the release of dissolved gases from the fluid basaltic magma.
6. In some cases, basaltic flows are fed not by a single vent but by a *fissure* that might extend for several kilometers. One such fissure eruption was witnessed in Iceland in 1783. In the geologic past, fissure eruptions produced so many lava flows in short spans of time that in several extensive areas (such as the Columbia Plateau) the basalt is nearly 1.5 km (0.9 mi) thick in places.

SELECTED REFERENCES

Bullard, F. M., 1962, Volcanoes in history, in theory, in eruption, University of Texas Press, Austin.

Griggs, R. F., 1922, The valley of ten thousand smokes, National Geographic Society, Washington, D.C.

Jackson, Kern C., 1970, Textbook of lithology, McGraw-Hill Book Co., New York.

MacDonald, G. A., 1972, Volcanoes, Prentice-Hall, Englewood Cliffs, New Jersey.

Rittmann, A., and Vincent, E. A., 1962, Volcanoes and their activity, Interscience Publishers, John Wiley and Sons, New York.

Rubey, W. W., 1955, Development of the hydrosphere and atmosphere with special reference to the probable composition of the early atmosphere, *in* Crust of the earth, by A. Polderwaart, ed., Geological Society of America Special Paper 62, pp. 631–50.

Atlas of volcanic phenomena, 1972 U.S. Geological Survey, U.S. Government Printing Office, Washington, D.C.

Williams, H., 1941, Crater Lake: the story of its origin, University of California Publications, Bulletin, Dept. of Geological Sciences, vol. 25, pp. 239–346.

———, 1951, Volcanoes, Scientific American, vol. 185, no. 5, pp. 45–53.

6

SEDIMENTARY ROCKS

Widely spread over the surface of the earth is a relatively thin blanket of sediment which has been consolidated into rock through slow-acting processes that are relatively simple to understand. These sedimentary processes operate in land or sea environments at temperatures and pressures much more familiar to us than those prevailing in the crustal realm, where processes operate to produce metamorphic and igneous rocks.

Sedimentary rocks, for the most part, are secondary, or derived, rocks. One category of sediments consists of layers made up of clay, sand, or gravel particles derived from the disintegration or decomposition of pre-existing rocks. Layered rocks made of such fragmental material are called *clastic* sediments (Fig. 6-1).

Another large and economically important group of sedimentary rocks is chemically precipitated in water, such as evaporating lakes or shallow embayments of the sea. Perhaps the best-known example of that category is rock salt. Closely akin to it in origin are such well-known substances as gypsum and borax—both of which are chemically derived.

Organic sediments are a third category, and an important one to human beings. Coal, a vitally significant fossil fuel, is in the organic group, as are the oil shales now being studied intensely as a source of oil. Another familiar kind of organic sedimentary rock is limestone, and, of its many forms, several represent the slow accumulation over many centuries of deposits made by lime-secreting plants and animals.

In this chapter we will look at the main properties of sedimentary rocks, and comment on how those properties can be related to the environment in which sedimentary rocks were formed. Although such a relation may not seem particularly important, it can be of the utmost importance to a geologist hunting for oil, mineral resources, or even ground water. For example, a typical property, or set of properties, possessed by a sedimentary rock formation may indicate the presence of an important de-

Fig. 6-1. Sedimentary rocks commonly are characterized by horizontal layering, shown here in an aerial photograph. Layering probably shows up best in an arid region, where it closely simulates the contour lines of a topographic map. *William A. Garnett*

posit in that formation or in similar formations nearby.

ORIGIN OF SEDIMENTARY MATERIALS

Many of the materials that make up sedimentary rocks are derived from weathering reactions (see Chap. 8). A much simplified weathering equation is:

$$\text{any pre-existing rocks and minerals} + \text{weathering reactions}$$

$$\rightarrow \text{smaller-size solid particles} + \text{clay minerals} + \text{ions}$$

The smaller solid particles—gravel, sand, and silt—are derived directly from the source area and thus reflect the type of rock that crops out there. For example, a source area in which basalts predominate will produce sands high in olivine, calcium-plagioclase, and augite, whereas a granitic area will produce sands high in quartz, potassium-feldspar, and biotite. Weathering on the hillslopes in the source area before material is transported to its deposition site, however, can alter the mineralogy of the deposited sands. In climates in which intense chemical weathering is going on, more quartz is seen in granite-derived sediments than in the granite itself because quartz is more resistant to weathering than other minerals in granite.

Clay minerals form in soils and in other near-surface, low-temperature environments. They are fine-grained minerals with sheet-like atomic structures like the micas (see Chap. 8). Each of the many clay minerals recognized is sensitive to the environment in which it forms; however, if it is transferred to an appreciably different environment, it can change to another clay mineral. The interpretation of clay minerals in sedimentary rocks is complex, and is governed partly by the rocks and climate of the source area, and the chemical conditions underground where the process of transforming sediment into rock is slowly taking place.

Any weathering reaction produces ions that are dissolved in water. Some common ions are calcium (Ca^{2+}), sodium (Na^+), carbonate (CO_3^{2-}), and chloride (Cl^-); they and others are responsible for the salty taste of certain waters. Under the appropriate conditions, commonly in areas of fast evaporation, such ions precipitate out of solution. The minerals that form are complex, controlled in large part by the chemical composition of the parent waters. Thus precipitates divulge little information on the rocks or climate of the source area.

ENVIRONMENTS OF DEPOSITION

Sedimentary rocks accumulate in a wide variety of environments—about as many as there are different kinds of landscapes or climates. One thing most such environments have in common, however, is that they occupy the lower parts of the landscape—the parts to which material moves from higher areas. Two major realms of sedimentation commonly are recognized: (1) the sea (marine), and (2) the land (continental). That classification easily could be labeled arbitrary, for several kinds of sediments fit one category as well as the other; for example, the silts and clays that accumulate in the deltas of large rivers could be assigned readily to either province. Our purpose now is to describe briefly the main depositional environments; Chapters 9 through 14 will describe them in greater detail. It will be noted that each environment produces a more-or-less characteristic sediment, which later becomes a sedimentary rock.

A major concern of geologists is to examine rock outcrops, from which they attempt to piece together the depositional environment. The task is not an easy one—often a field geologist must examine miles of valley walls or mountainsides before clear evidence is unearthed. Further, when one considers how widespread the processes of erosion are, it is not unreasonable to suppose that some of the rocks holding the best evidence of depositional environments might have been eroded away. It is an excellent exercise in the method of the multiple working hypotheses to keep one's mind open, and to keep seeking more clues so that the nature of the depositional environment can be ascertained correctly.

Before continuing our discussion, we should describe briefly a property of sedimentary rocks that helps to distinguish one environment from another. That property is called *sorting*, and it refers to the degree of similarity in particle size in a sediment. In well-sorted sediment, for example, most particles are about the same size. In contrast, a poorly sorted sediment contains a wide assortment of particle sizes. Moderate sorting refers to an intermediate stage between the two.

Continental deposition

Many of us are aware of the ways in which sediments may be trapped on land. Although there are numerous examples, the following kinds of environments and deposits are the most typical (starting in mountainous terrain and progressing toward the seacoast):

Glacial deposits Glaciers, which today mostly are confined to higher mountains or to far distant Arctic and Antarctic shores, were once more widespread and their deposits—usually more disordered than those laid down by streams or in the sea—blanket much of North America and northern Europe. A good example of a typical glacial deposit is *boulder-clay*, which is literally that—rocks the size of boulders set in a clayey matrix with very little sorting of particles according to size.

Floodplains Although one main task of rivers is to deliver sediment to the ocean, they commonly store material along their margins, the floodplains (Fig. 6-2). Floodplains are those flat surfaces adjacent to streams, especially in lowlands, but in mountainous regions as well, over which streams spread in times of flood. During each flood a new layer of sediment is deposited. These depositional sites range all the way from extensive plains bordering the Nile, the Yangtze, and the Mississippi, down to narrow strips bordering small streams. Sediment size will vary with local conditions, but it

ranges from small clay particles to the largest of boulders. Sorting varies from good to moderate.

Alluvial fans These landforms are typical of arid and semiarid regions. When a stream comes rushing from mountains or hills carrying a great deal of rock debris and suddenly reaches a basin, its sediments are dumped in a spreading fan-like form (Fig. 6-3). Excellent examples are the fans bordering the mountains of our southwestern deserts. Some are so poorly sorted that they can be mistaken easily for glacial deposits. Such mistaken identity is a bit ironic, because the two deposits form in such different temperature regimes.

Wind Locally abundant sand dunes and virtual seas of sand testify to the effectiveness of wind in those parts of the world where factors such as an abundant supply of sand, little vegetation with which to stabilize it, and strong winds to move it about occur together (Fig. 6-4). Such combinations are likely to be encountered in deserts, along many of the world's coasts, and along the floodplains of large rivers, of which the Volga is an excellent example. Wind also sweeps along a fine dust—called loess—or silt that may pile up in vast windrows. A thick blanket of feebly consolidated dust is a dominating element of the tawny landscape of northern China near Peking. Unlike many other deposits, those moved and laid down by the wind are very well sorted, be they sand or silt.

Lakebeds Lakes are sure traps for sediment. Glaciers, rivers, wind action—all move sediment into lakes. Yet they are a relatively temporary feature on a geologic time-scale, being formed mainly by glaciers, landslides, volcanic activity, and faulting. Eventually they fill in with sediment; in turn, perhaps the sediments may find themselves in some other erosional and depositional environment. Lake sediments generally are well sorted.

Deltas Deltas form where sediment-laden

Fig. 6-2. The South Platte River, Nebraska, occupies only part of its floodplain. In times of flooding, however, it spreads out over much of the plain, depositing sediments. *USGS*

streams enter bodies of relatively still water, such as lakes or gulfs (Fig. 6-5). Although they form mostly below water level, their top surfaces can eventually rise above the water and so become floodplains. Deltas are complex environments indeed, because their origin stems from a combination of river-lake or river-ocean processes. In that much of the sediment is laid down in relatively quiet water, the sorting is good.

Marine deposition

At least two of the factors controlling the distribution of sediment in the sea are (1) distance from land and (2) depth of water. In order to simplify the story, we may say that sediments accumulate in four leading zones, sufficiently unlike one another to merit classifying them as separate units. In general, the sediments in the zones are well sorted.

Just seaward from the land is the *shore zone*—for all practical purposes the place where the surf breaks against the shore. On

Fig. 6-3. Alluvial fans flank the ranges in the Mojave Desert, California. The fan shape is best seen at the lower right; elsewhere fans have coalesced. To the left are the white saline deposits of temporary lakes. *USAF*

Fig. 6-4. Sand dunes in the Sahara. The forms range from relatively small ridges to star-shaped hills that may be as high as 100 m (328 ft) or so. *H.T.U. Smith*

many coasts with a large tidal range, a broad expanse of sea floor adjacent to the land may be laid bare at low water. Of the sediments carried to the marine environments, the coarsest—sands and gravels—usually are trapped in that zone, whereas the fine-grained sediments are carried offshore.

The *continental shelf* is much broader than the shore zone, extending seaward to an average depth of 133 m (436 ft). As a rule, the shelf is the region where most land-derived sediment comes to rest, although it makes up only 7.5 per cent of the area of the oceans. Most of

the sediments we see exposed today as marine rocks were deposited on the continental shelf; hence the importance of studying that environment. The sediments are usually silts and clays, with some sands.

In some shelf areas, carbonate-secreting organisms—such as coral—contribute greatly to the formation of sedimentary rocks. Inorganic precipitation of calcium carbonate ($CaCO_3$) occurs in such environments, but it is not too important. As a general rule, such areas are shallow with fairly warm water temperatures and a low supply of continent-derived sedi-

Fig. 6-5. Delta built by streams issuing from glaciers on the south side of Inugsuin Fiord, Baffin Island, Northwest Territories, Canada. *J.D. Ives*

Fig. 6-6. This monastery at Meteora, Greece, surmounts bluffs of layered, cemented gravel deposits (conglomerate) of Tertiary age. The area has undergone tilting since deposition, as shown by the fact that the layers are no longer horizontal. *Jean B. Thorpe*

ment. The remains of the organisms build up on the ocean floor, and, once hardened, they form limestone.

Seaward from the edge of the continental shelf, the topography steepens to form the continental slope which, in turn, continues on down to the deep floor of the sea, or the *abyss*—a region inaccessible by ordinary means of observation. It is now possible, however, to take photographs of the ocean bottom with underwater cameras, and, by means of deep-diving submersibles, to visit that dark, silent realm, almost as remote in its way as the world of space. There on the floor of the open sea accumulate the finest sediments, calcareous and siliceous oozes composed of the remains of free-floating and swimming microscopic plants and animals, and the extremely fine-grained clays that carpet the abyssal plains. Only under special circumstances can coarser-grained sediments be transported so far from the land and to such great depths.

FEATURES OF SEDIMENTARY ROCKS

Sedimentary rocks have several key features which distinguish them from the other major rock types. Some of those features have even survived metamorphism, and thus provide clues to the mysterious past history of meta-

morphic rocks. Sorting has already been mentioned; the other features follow.

Stratification

Most sedimentary rocks are made up of particles, ranging from very large to submicroscopic, that settled out through a medium such as air or water. In addition, most of them are layered (Fig. 6-6), and the layers too show a great range in dimensions—from those whose thickness is measurable in millimeters to those measured in meters.

Fig. 6-7. Cross section of an ancient river channel in Utah filled with conglomerate. *W.R. Hansen, USGS*

Fig. 6-8. Varved sediments from an ancient lake in Ontario. Each layer consists of a lower, light-colored silty unit, overlain by an upper dark-colored clayey unit. *Geological Survey of Canada*

Such depositional layers are called *strata;* an individual layer is a stratum. In everyday language the layers commonly are called beds if their thickness is greater than 1 cm (0.4 in.). If the layers are less thick, they are better called *laminae* (from the Latin, *lamina,* for thin plate, leaf, or layer). The term is used here in much the same sense that we speak of the laminations in plywood.

In "quiet water" depositional environments, such as the bottoms of lakes or the ocean floor, strata are laid down in a near-horizontal fashion and are uniformly thick. However, strata can vary in thickness, and in some layers ancient river channels can be seen (Fig. 6-7).

Layering can come about through variations in the energy of the transporting medium, or through variations in the size, amount, mineral composition, or color of the sediment being delivered to a particular depositional site. A high-energy flood, for example, could wash large particles out onto a lake bottom. Following the flood, silt and clay—the usual sediment carried to the lake—might be deposited on top of the flood debris. The result is two layers of contrasting particle size. Discontinuous deposition also can cause layering, because once deposition begins again, chances are slight that the newly deposited material will have exactly the same properties as the sediment that came before it.

Some sediments and sedimentary rocks are characterized by the rhythmic repetition of distinctive layers. Well-known examples are *varves,* which seem to form best in glacier-fed lakes. Each varve is made up of a couplet of two layers—a coarser silty layer overlain by a finer-grained layer of silt and clay (Fig. 6-8). Each couplet is thought to represent a seasonal event. The coarser fraction in the lower layer is thought to have been laid down during the spring and summer, the seasons of active snow and ice melt. The finer-grained upper layer, however, represents quiet sedimentation during the fall and winter, when melting stops and the surface of the lake is frozen over. Because each couplet represents a year of record, varves can be used with some degree of suc-

cess in determining the age of a particular lake. Thus, as a means of dating deposits or events, varves are not too different from tree rings.

Some sediments, ranging from coarse to fine grained, show a very different sort of stratification. The particles of the individual layers, instead of being distributed uniformly throughout, are graded—the larger particles concentrated at the bottom, the smaller at the top (Fig. 6-9). Such layers are said to have *graded bedding*—a feature that occurs when a mass of sediment is discharged suddenly into a relatively quiet body of water. The larger particles drop out quickly, followed by the medium-sized ones, and finally the finest particles are deposited. The change from one dominant particle size to the other is gradational, and the whole makes up the graded bed.

An excellent example of how graded beds are formed may be seen where the excessively muddy Colorado River flows into Lake Mead, which is backed up behind Hoover Dam on the boundary between Nevada and Arizona. The muddy river water seems to disappear as if by magic, and anyone who has seen the dark, blue-green water of Lake Mead cannot fail to be impressed by the contrast it makes with the turbid river. An explanation for the disappearance of the murky water is that, with its higher density, it sinks to the bottom of the lake where it moves as an underflow known as a *density current.* Every child has, at one time, formed a density current by poking at the edge of a mud puddle with a stick, causing the disturbed liquefied mud to flow out toward the middle of the puddle along its bottom.

Roundness of grains

An important property of clastic sediments is the roundness of the grains. How are they rounded off? One possible explanation is that the distance of transport affects the degree of rounding. Assume that a rock tumbles from a cliff to an adjacent river bottom. Initially the rock has very sharp corners, but as it slowly moves downstream it collides with and scrapes

Fig. 6-9. Sequence of graded beds in a playa in northern Sonora, Mexico. Each bed consists of sand or silt grading upward to fine clay and each probably records a flood. The bed at the center is about 2 cm (1 in.) thick. *W.B. Bull*

against neighboring particles and the corners become progressively more rounded. Some of the more rounded grains are found in beach and wind environments (Fig. 6-10), and good-to-moderate rounding characterizes river environments. In contrast, glaciers and mudflows do not allow much grain-to-grain contact; consequently rounding is generally poor.

Color

Igneous rocks, unaltered by exposure to the atmosphere, typically are shades of gray or black, because those hues prevail in the most abundant constituents of the rocks, feldspar and ferromagnesian minerals. Sedimentary rocks are often more colorful, however.

An important source of pigment in sediments is the very fine interstitial material that partially fills the space between individual grains. If the material should contain hematite (iron oxide, Fe_2O_3), the resulting rock is likely to be red in hue. Hematite is the source of most of the rich red coloring in the walls of the Grand Canyon. Other forms of iron may stain a rock brown, or even shades of pink and yellow. Iron may be responsible for the purple, green, or black shades of some sedimentary rocks, but often the origin of coloring matter simply is not known. Pigments may have been carried to the site of deposition along with the sediments or they might have been produced later by chemical weathering of the original sedimentary grains during eons of deep burial.

Fig. 6-10 A. (above) Well-rounded gravel clasts are characteristic of many beach environments. *William Estavillo* **B.** (left) Rounded river boulders of basalt stacked up to form a wall on a farm in southern Idaho. This amount of rounding occurred in 13 km (8 mi) of stream transport. *H.E. Malde, USGS*

Many of the dark sedimentary rocks owe their color to the organic material they contain. Coal, an excellent illustration, is entirely organic in composition, and its very name is a synonym for black. With varying amounts of organic material, sedimentary rocks may range from light gray to black. Some black muds, however, owe their color to finely divided iron sulfide dispersed through them rather than to carbonaceous matter.

The range in colors that sedimentary rocks may display is one of their more intriguing properties; in dry countries, where vegetation is lacking and the soil cover is sparse, the true colors of the rocks stand revealed in striking array, as in Grand, Zion, and Bryce canyons, and in Monument Valley, Canyon de Chelly, and the Painted Desert. It is the brilliant coloring of their sedimentary rocks as much as any other attribute that makes those places so renowned.

Fig. 6-11. Mud cracks formed by shrinkage accompanying drying of clay, Colorado River Delta, Mexico.
G.K. Gilbert, USGS

Mud cracks

When wet, clayey mud is exposed to the air it dries, shrinks, and upon shrinking, cracks—generally in a nearly uniform pattern of polygons (often four-sided) that resemble the tops of lava columns (Fig. 6-11). In lava, the reason is contraction upon cooling; in wet muds, it is contraction resulting from dehydration. As drying continues, the mud layers on the tops of the polygons may curl up at the edges, so much so that at times they form complete rolls, much like cardboard tubes or fancy pastries.

Mud cracks indicate that the sediment of which they were once a part was alternately wet and dry, and thus such cracks are very typical of mud-bottomed, shallow lakes that on occasion dry up. They are not so characteristic of muddy tidal flats because the time of exposure at low tide is too brief for much drying out.

Ripple marks

Nearly everyone has noticed the characteristic corrugated surface made by a stream or tidal current flowing across the sandy bottom of a lake or stream (Fig. 6-12), or has seen photographs of virtually the same pattern produced by the wind blowing across a desert sand dune. Such ripples, called *current ripples*, develop at right angles to the current and are likely to have steep slopes on the down current side but gentle ones upcurrent. The asymmetrical form results when an air or water current rolls sand grains upslope and gravity pulls them down the opposite side, or *slip face*. That slope stands at an inclination known as the *angle of repose* (the maximum slope at which sand grains will remain stationary without sliding down the slope). Current ripples can be used—once converted to solid rock—to establish the direction taken by long-vanished currents in the atmosphere or under water. In the past it was thought that such ripples in water-laid sediments indicated shallow depths, but in recent years underwater photography has shown ripple patterns on the sea floor at depths of several thousand meters.

Another type of ripple has symmetrical sides, sharper crests, and more gently rounded troughs than current ripples do. Such corrugations are called *oscillation ripples,* and presumably are the result of surface waves (called *waves of oscillation*) stirring up the sandy bottom of a shallow water body.

Cross-bedding

Earlier in the chapter the point was made that sedimentary rocks customarily are deposited in essentially parallel, horizontal layers. Yet there are several varieties of stratification in which the layers are inclined at steep angles to the horizontal plane. Such layering is known as *cross-bedding.*

One kind of cross-bedding forms in sand dunes (Fig. 6-13). Each layer inside the dune at some time past was part of the surface, and, since the dune's configuration was established largely through a balancing of wind transport

upslope and gravity-sliding downslope, most of the layers are sweeping curves which more often than not are concave upward. Because sand dunes are ephemeral landforms whose position and orientation change with the inconstant wind, it is not surprising that the sweeping, shingled layers may intersect one another in complex patterns such as those in the walls of Canyon de Chelly, Arizona, or in Zion National Park.

Another kind of cross-bedding forms in deltas. Streams carrying a fairly large load of moderately coarse debris are forced to deposit their sediment rapidly when they reach a body of water such as a lake or gulf. There, the sediment dropped by the stream constructs a leading edge out into the water, much as a highway-bridge fill is built out into a canyon by end-dumping from gravel trucks. The outer slope of the delta, like the slip face of a sand

Fig. 6-12. Ripple marks on a bedding plane in the Dakota Sandstone, Colorado. *J.R. Stacy, USGS*

Fig. 6-13. (opposite) Giant cross-bedding in ancient sand-dune deposits at Checkerboard Mesa, Zion National Park, Utah. *Ray Atkeson*

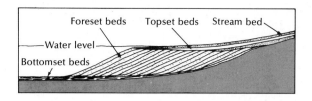

Fig. 6-14. Cross section through a delta illustrating the characteristic cross-bedding.

Fig. 6-15 A. (above) Small-scale cross beds in river-deposited sands, Idaho. *H.E. Malde, USGS* **B.** (below) Cross beds in sands and gravels, Washington. *F.O. Jones*

dune or a current ripple, also stands at the angle of repose. When such sediments are consolidated into rock, three distinctive layers may result. At the top and bottom of a deltaic deposit will be horizontal strata, known as *topset* and *bottomset* beds, respectively, and the steeply inclined layers in the middle that were once the delta front as it advanced out into the water are called *foreset* beds (Fig. 6-14).

River deposits also can be cross-bedded, but usually on a much smaller scale than those mentioned above (Fig. 6-15).

Fossils

No other property is so distinctively characteristic of sedimentary rocks as is the presence of fossils (Fig. 6-16). They are the remains of once-living things that on their death were buried in sand, silt, lime, or mud. Much of the organic matter that some of them originally contained gradually was replaced over the centuries by inorganic matter, until, to use petrified wood as an example, many of the woody fibers and cellulose were replaced by silica. Representatives of just about everything that crawls, walks, swims, or flies among the animals, or that simply stands in place, like plants, have been preserved as fossils. Even such improbable creatures as jellyfish, whose composition must be more than 95 per cent water, or such fragile things as the compound eyes of flies or the delicate tracery of dragonfly wings have been preserved. Those creatures

Fig. 6-16. Some examples of fossils preserved in sedimentary rocks. **A.** (above) Brachiopod fossils in the Trenton Limestone of Ordivician age, Watertown, New York. *Smithsonian Institution* **B.** (opposite top) Dinosaur bones being excavated around 1910 at Dinosaur National Monument, Colorado. *National Park Service* **C.** (opposite, bottom) Partial carcass of a juvenile mammoth recovered in 1948 from frozen silt near Fairbanks, Alaska. The head, trunk, and one front leg are shown. Radiocarbon dating indicates that the carcass was buried, frozen, for about 30,000 years. *American Museum of Natural History*

are the exception, however, because the organisms most commonly preserved as fossils are those that already have durable elements such as shells, bones, and teeth in their make-up. In fact, most fossils are the remains of shells or skeletons. In some instances an entire rock may consist of organic matter. A layer of coal is made up of plant fragments, and some limestones may be composed of the remains of coral or of calcareous algae, or may be a mass of seashells. In addition to the remains of or-ganisms, footprints, tracks, trails, and burrows also may be considered as fossils.

Concretions

Round, or almost round, solid bodies are sometimes found in sedimentary rocks. *Concretions*, as most are called, are sediments converted into rocks. Any small particle—a sand grain, a piece of shell, even a small insect—can act as a nucleus for a concretion.

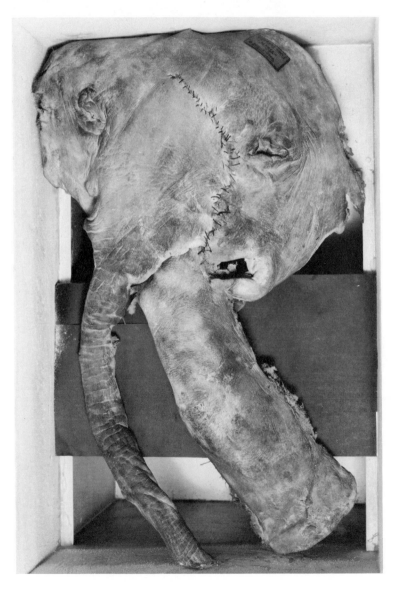

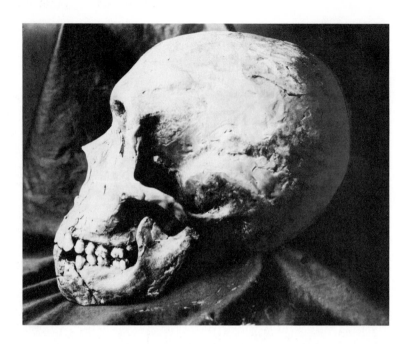

Fig. 6-16 (cont'd.) D. (above) Reconstructed skull of Neanderthal man, a race of *Homo sapiens* that lived in Europe and Asia about 100,000 years ago. *American Museum of Natural History* E. (below) This unique fossil is the jaw of a shark found in Idaho. Modern sharks shed their teeth as new ones grow in. The older teeth of this shark, however, were buried in a complex spiral in its jaw. *Jan Robertson, Geological Society of America* F. (opposite) Leaf impressions preserved in Tertiary volcanic tuffs near Corvallis, Oregon. *Jan Robertson*

A limy cement, which will eventually bind all the particles together, collects around the nucleus, and gradually enlarges the concretion, which may reach a meter (3 ft) or more in diameter. Most are much smaller, however. The nucleus is usually a fragment unlike the rest of the sediment—an alien particle, so to speak.

CONVERSION OF SEDIMENTS TO SEDIMENTARY ROCKS

Thus far, most of our discussion has had to do with sediments and the process of sedimentation, and very little has been said about the way in which sediments are converted into solid rock. What process, for example, converts loose sand, which at the beach can be idly sifted through the fingers, into a rock such as sandstone, which may be almost as unyielding as granite?

Is it pressure? The answer is an emphatic No. To apply enough pressure to force sand grains to adhere to one another would be to crush them into smaller and smaller particles. Pressure does play a role, however, in the process of *compaction*, which is the squeezing together of the particles in a sediment. If, for instance, enough pressure is applied to fine-grained muds, such as clay or silt, most of the interstitial water is squeezed out, the sediment shrinks markedly, and if clay is a dominant constituent, the particles tend to adhere to one another.

The closing up of the space between particles through compaction is an important precursor for *cementation*—the most significant process involved in the conversion of sediments into sedimentary rock (Fig. 6-17). Fundamentally the process involves the deposition from solution of a soluble substance such as calcite ($CaCO_3$) and its building up as a layer of film on the surface of sand grains, silt particles, or clay flakes, as the case may be, until all the pore space separating them is filled. Such a limy cement is precipitated in much the same way, although at a lower temperature, as the scale that forms inside a kettle.

Calcite is one of the most abundant natural cements. It is among the more soluble of the common substances that may be dissolved in ground water. Another important natural cement is silica (SiO_2), which is also soluble, although less readily than calcite. Iron oxide (Fe_2O_3), too, is a cementing agent.

WHICH WAY IS UP?

Many sedimentary rock layers have been so mildly deformed that it is no problem determining if the sequence is right-side up or up-

Fig. 6-17. A thin section of a sandstone from Utah as seen under a microscope. The rounded quartz grains are cemented together by fine-grained silica that fills the voids between the grains. *W.R. Hansen, USGS*

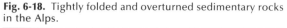

Fig. 6-18. Tightly folded and overturned sedimentary rocks in the Alps.

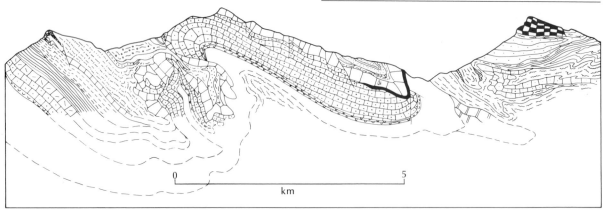

0   5
km

side down. Pity the poor geologist who tries to unravel the complex structural geology of, for example, some of our western mountain ranges, or the Alps (see Fig. 6-18). In places, the rocks have been so intensely folded that we now see them upside down (Fig. 6-18). Some sedimentary rock features can help determine which way is up (Fig. 6-19). Baked zones associated with volcanic rocks also are an indicator of the up direction. Interpretation of up versus down is slightly more difficult in metamorphic rocks, especially if intense metamorphism has taken place, and impossible in plutonic rocks.

TYPES OF SEDIMENTARY ROCKS

In the opening section of the chapter, the point was made that there are three major categories of sedimentary rocks: *clastic*, or fragmented; *chemical* precipitates; and *organic* deposits. Like many classifications of natural phenomena, those categories are more rigid than the actual state of affairs. Not only are there gradational types from one category to the other, but there are also varieties that might just as logically be placed in one pigeonhole as another, as well as a few that fit into none.

Fig. 6-19. Various ways to tell which way is up. For a true sense of direction tilt the book one way or the other. In each case the arrow points in the original "up" direction. **A.** Graded beds, with pebbles and sand shown in white; silt and clay in black. The sequence is upside down because each graded-bed layer is coarse at the bottom, gradually becoming finer at the top. **B.** These cross-bedded sediments are right-side up, because the base of each younger bed truncates the cross beds, which are concave upward. **C.** Basalt flow interbedded with sediments. Because the flow bakes only the sediments beneath it, the up direction is to the right. If the basalt was a sill rather than a flow, could you determine up from down? **D.** Beds of sand with oscillation ripple marks. Because such marks are concave upward when formed, the sequence is upside down.

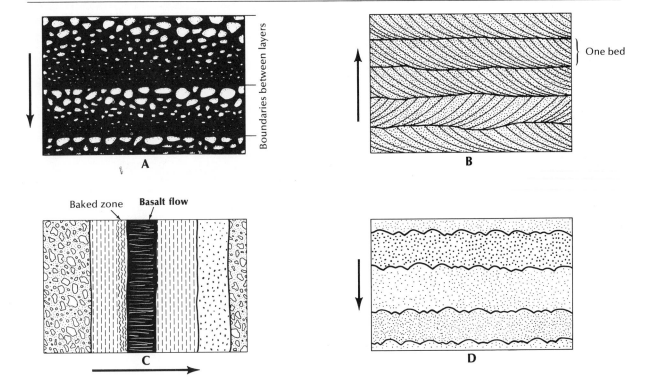

Clastic sedimentary rocks

The clastic rocks are truly secondary rocks because they consist of fragments of pre-existing rocks that range in size from blocks the size of boxcars down to colloids so fine that they remain in suspension almost indefinitely. Because clastic rocks consist of fragments of other rocks, they are very likely to show a wide range of composition. So much so, in fact, that in setting up a classification for them, the first property to be considered is the *size* of the particles that are cemented together rather than the material of which they are made.

Take sand, for example. To most people the word has a double connotation: (1) it is a size term—all of us are conscious of the fine grittiness of sand in a bathing suit or between our teeth; and (2) it has a compositional meaning—the beach sand that usually comes to mind ranges from white to a tawny yellow, and is likely to be thought of as consisting of quartz grains. Actually many beach sands contain mostly feldspar grains as well as a liberal sprinkling of other sand-size rock particles and mineral grains. Sand can consist of almost any material of sufficient durability. Along some of the rivers of the Atlantic states, sandbars are made of coal fragments. On some of the beaches of Hawaii the sands are coal black, too, but are composed of ground-up basalt. In the islands of the South Seas, the straw-colored sands of their fabled shores are made of fragmented coral heads, pieces of shells, and other organic debris. Think of how confusing it would be, therefore, if all sands of slightly different composition were given different names. The truth is that there are many names for sandstones, because geologists seldom pass up the opportunity to name something. They even have their own dictionary of geologic terms, which measures 4.5 cm (1.8 in.) thick! However, that multitude of names will be left to the hard-core geologists; in this book we will generally limit the vocabulary to the most important terms.

A classification that has won wide acceptance is a modified form of one originally proposed in 1922 by C. K. Wentworth (Table 6-1). It has the advantages that almost all the terms used are everyday words, and that the size ranges are close to the ones in common usage. The actual dimensions, however, are arranged so that they follow in geometric progression. The term *clay* as used in the classification refers only to particle size. It should not be confused with the term clay mineral, which refers to very small minerals with mica-like structures, some of which are described in Chapter 8, and most of which are of clay size. To make some order out of the confusion, most clay minerals are of clay size, but not all clay consists of clay minerals.

Environmentally, particle size can be related to the energy of the transporting medium. Large gravels, for example, are moved by glaciers, swift rivers, and debris-laden mudflows. Less energy is required to move sand, and so it is common to sand dunes and to some beaches and rivers. Because silts and clays settle so slowly, they commonly settle out only in quiet water; thus their association with lake and marine environments.

Conglomerates These rocks are cemented gravels, and the larger fragments may range in size from boulders with diameters of several

Table 6.1. Classification of clastic sediments and sedimentary rocks

Sediment	Particle term	Limiting grain size (in mm)	Rock
Gravel	Boulder Cobble Pebble Granule	256 64 4 2	Conglomerate
Sand	Very coarse sand Coarse sand Medium sand Fine sand Very fine sand	1 ½ ¼ ⅛ 1/16	Sandstone
Mud	Silt Clay	1/256	Shale or mudstone

Fig. 6-20. Conglomerate forms this hanging rock at the foot of Echo Canyon in Utah, photographed by Andrew Joseph Russell about 1868. *Beinecke Rare Book and Manuscript Library, Yale University*

meters down to particles the size of small peas (2 mm). More often than not, the interstices, or pore spaces, between the larger boulders, cobbles, or gravel are filled with sand or mud; the whole mass of sediment then is cemented together to form a single rock (Fig. 6-20).

Breccia is a variety of conglomerate with angular rather than rounded fragments. The same word was used for pyroclastic volcanic rocks in Chapter 5. And the same principle applies to conglomerates; if most of the large fragments in the rock are angular rather than rounded, the rock is a breccia—the adjective sedimentary or volcanic is usually added to indicate its origin.

Sandstones The sedimentary rocks called *sandstones* consist of cemented particles whose diameter ranges between 2 millimeters and 1/16 of a millimeter. Very commonly they include shale layers, or beds of sandstone may alternate regularly with beds of shale, or lenses of conglomerate.

Pure, well-sorted sandstone was often used as a building material before the advent of pre-stressed concrete or of lightweight aggregate. Quite a number of college campuses are adorned with examples of academic gothic—frequently, it seems, inhabited by the geology department—which were hewn from sand-stone blocks. The White House and the Capitol are both built of sandstone quarried a short distance down the Potomac from Washington, D.C. In the Victorian Era—especially the General Grant period—one of the favorite construction materials was the so-called brown-stone—a drab, red sandstone that regrettably will long outlast most of us. Many of Europe's celebrated landmarks are made of clastic sedimentary rocks—the castles at Heidelberg and Salzburg and most of the great ducal palaces of Florence are a few from among scores of famous examples. Sedimentary rocks were greatly preferred over granite by builders in the past because such stratified rocks split more readily along their bedding planes and could be worked far more easily with the primitive hand tools of the time.

The cement is what determines the degree of induration, or hardness, of sandstones, and therefore how well they perform as building stone. In some the cement is weak, and individual grains separate readily from their neighbors; in others the cement may actually be tougher than the material it holds together, and when the rock breaks, it breaks across the grains. Some cements dissolve readily, and the rock seems to crumble away, leaving a residue of sand grains behind.

Compositional differences affect the appearance of sandstone. Among the innumerable kinds of sandstones two leading varieties are *arkose* and *graywacke*. Arkoses are made up largely of quartz and feldspar grains and in color are commonly reddish to pale gray or buff. They result typically from the erosion of granitic rocks, and require relatively rapid transportation and deposition for their formation.

Graywackes (from the German *grauwacke*, gray stone) were originally named for distinctive sandstones in the Harz Mountains of Germany. They are darker than arkoses, and, although they commonly contain quartz and feldspar minerals, they have a much higher content of rock fragments—chiefly of the darker varieties of igneous and metamorphic rocks. The sand-size particles are set in a clayey or silty matrix which at the time of deposition was essentially a muddy or clayey paste. Characteristically, graywackes are dense, tough, well-indurated rocks whose colors are dark green, or gray, or black. Some of them appear to have been deposited in the sea, close to a steep mountain range. Muddy water carrying a large volume of sediment, including sand, was moved but a short distance from its source and deposited so rapidly that little weathering and rock decay occurred. This statement of the general characteristics of graywacke and its origin covers most of the points where a measure of agreement exists. Too much space would be required to review the points of disagreement.

Shale The original constituents of this fine-grained rock were clay-size grains and silt particles which matured into a typically laminated rock that splits readily into thin layers (Fig. 6-21). Shale is an ancient term in our language; it comes from the Old English word *scealu*, meaning scale or shell. When we use such an ancient word in geologic terminology it usually means we are dealing with a property so distinctive that it was recognized early enough to make its way into the root-stock of our native tongue.

Since shales are composed of clay grains and of individual mineral grains or rock particles less than 1/16 of a millimeter in diameter, few of the constituents can be distinguished by the unaided eye. Under the microscope they can be resolved; there we can see that most shales are made of minute grains of quartz, feldspar, and mica, and of larger rock fragments along

Fig. 6-21. Thin, even bedding is characteristic of this sandy shale outcrop in Utah. *C.B. Hunt, USGS*

means characteristic of all of them. Some varieties whose composition and grain size appear to be comparable are not fissile at all, but break in massive chunks or small compact blocks. They are best given the descriptive name of *mudstone*.

Precipitated sedimentary rocks

In addition to the clastic rocks consisting of fragments and mineral grains derived from pre-existing rocks, there is a second large clan of sedimentary rocks made of chemically precipitated materials. In the following pages chemically formed rocks are discussed according to their composition and their mode of origin. Such an approach, unfortunately, makes for confusion since some varieties of rocks—specifically, the carbonates—may have similar compositions but unlike origins, and thus of necessity the same term appears more than once in the classification.

Evaporites Rocks that result primarily from the evaporation of water containing dissolved solids are known as *evaporites*. As the water becomes concentrated the ions separate out from solution to form a crystalline residue.

Most familar of all such rocks is *halite*, or common table salt (sodium chloride, NaCl). Commonly it is formed when evaporation in an arm of the sea dominates the inflow of water from outside. Judging from the great thicknesses of the renowned salt deposits of the world, the process must have been repeated many times over. The evaporation of an inland water body, such as Great Salt Lake, can also produce the same result, as anyone knows who has seen the nearby Bonneville Salt Flats—widely known for the ideal surface the salts provide for speed trials.

Layers of salt deposited in the geologic past are sometimes interbedded with other sedimentary rocks, and where such layers are near the surface, salt licks may be found. From earliest times salt has been a highly prized commodity. Today we take it for granted, but in ancient times people gave their lives in battle to

with the ubiquitous clay. Shales constitute very nearly half of all sedimentary rocks.

Many shales are shades of dark gray or even black, especially if they contain organic matter. Others are dark red or green or parti-colored, depending upon their iron content or the presence of other kinds of coloring matter.

Although *fissility,* or the ability to split along well-developed and closely spaced planes, is a leading property of shales, it is by no

win control over salt deposits or to seize the trade routes over which it moved. Famous among historic deposits were those of northern India—the locus of a flourishing trade before the time of Alexander—as well as those of Palmyra in Syria from whence salt moved by caravan to the Persian Gulf. The salt mines of Austria are deservedly famous, and in the Salz-kammergut region around Salzburg they were in operation at least as early as 2000 B.C.

Gypsum ($CaSO_4 \cdot 2H_2O$) is closely related to halite in its origin, for it too is a product of the evaporation of sea water. Along with it is also found an anhydrous (water-lacking) calcium sulfate ($CaSO_4$) called *anhydrite*.

As a body of water evaporates, the specific minerals that precipitate out do so in a specific order. The order is determined by the solubility of the minerals in water—the less soluble come out first, and the more soluble stay in solution the longest and therefore precipitate last.

Gypsum is less soluble than halite and thus is precipitated first when sea water is evaporated. However, a lot of water must evaporate before either mineral forms. Both gypsum and anhydrite come out of solution after about 80 per cent of the sea water has evaporated, and halite appears after 90 per cent has gone. Following the precipitation of halite, the very soluble salts appear in such forms as sodium bromide (NaBr) and potash (KCl).

Some salt deposits have attained astonishing proportions—in places several hundred meters thick. Yet there is a vexing problem connected with such deposits: the evaporation of 300 m (984 ft) of sea water will produce a bed of salt about 4.5 m (15 ft) thick of which 0.1 m (0.3 ft) should be gypsum and anhydrite, 3.5 m (11.5 ft) halite, and the remaining meter (3 ft) potassium- and magnesium-bearing salts. Does several hundred meters of salt deposits represent the evaporation of thousands of meters of water without a fresh supply being added? Hardly so, for that would require the evaporation of our deeper oceans, an event that certainly has not taken place.

How then, do we explain the great thicknesses of evaporite deposits? Consider the fact that gypsum and anhydrite make up strata many hundreds of meters thick in west Texas and New Mexico. An immense quantity of sea water must have been evaporated there in the geologic past. But what took place was not simple evaporation, because extensive bodies of sodium chloride that should be associated with the gypsum beds are absent. An explanation advanced by the American geologist Philip B. King is that water in a shallow, sun-warmed lagoon reached the temperature and concentration at which calcium sulfate precipitated and settled on the floor of the lagoon. The salt-enriched residual water then flowed back to sea before the salt was able to come out of solution. Sea water kept flowing in and out of the lagoon, and the process was repeated. Were such a basin a subsiding one, an immense thickness of evaporites could accumulate without the water necessarily being deep.

There are many other kinds of evaporites, of minor significance in volume, but of major consequence economically. Among them are borax ($Na_2B_4O_7 \cdot 10H_2O$), and potash (KCl), both of which are found in lakes or ancient lake deposits of desert regions, such as the Mojave Desert in California. Years ago picturesque 20-mule teams hauled those salts across the incredibly rough desert floor (Fig. 6-22) in large wooden boxcars.

Carbonate rocks Rocks known as *carbonates* are made up of minerals consisting of calcium or magnesium compounded with carbonate, generally in the form of calcite ($CaCO_3$) or dolomite ($CaMg(CO_3)_2$). If calcite predominates, the rock is called *limestone;* if dolomite predominates the rock is called *dolomite.* The latter term is virtually the only one used to designate both a mineral and a rock. In order to avoid confusion, some label the rock *dolostone.* Limestones are for the most part organic in origin, and very few are true chemical deposits. Both limestone and dolomite look very much alike; the most practical field distinction between the two is the test described in Chapter 3.

Sea water is very nearly saturated with cal-

Fig. 6-22. Twenty-mule team on its way to the Lila-C mine in 1892. *U.S. Borax*

cium carbonate ($CaCO_3$). That means that very slight changes in the temperature of the water or in its chemical composition can bring about the inorganic precipitation of calcite. Usually, the initial mineral precipitated is *aragonite,* an unstable form of $CaCO_3$. There is a continuing argument as to how important inorganic processes are in forming the main limestone deposits of the world.

Travertine is a good example of a limy deposit that appears to have been deposited by spring waters saturated with calcium carbonate. It is of no great geologic significance, but plays a disproportionately large role in human affairs, because it is so greatly favored as an architectural material. It is soft and readily worked, has an interesting array of colors—generally pale yellow or cream colored if pure; brown and darker yellow if it contains impurities—and often shows pronounced banding in wonderfully complex, curving patterns. *Tufa,* or *calcareous tufa,* as it is sometimes called to distinguish it from volcanic tuff, also forms in springs and lime-saturated lakes, although to some degree its deposition seems to be fostered by the work of lime-secreting algae. Tufa and travertine, when cut and polished, make building stones much favored for the lobbies of banks, building and loan associations, and

the large railway terminals of a past era. Great quantities of tufa are imported from Italy and, as might be anticipated, much of monumental Rome is built of tufa, including Bernini's columns that nearly encircle the piazza in front of St. Peter's.

There seems to be little doubt about the inorganic origin of one curious type of limestone known as *oölite.* It is made of minute spherical grains of calcium carbonate the size of fish roe, from which it derives its name from the Greek words *oo* for egg and *lithos* for stone (Fig. 6-23). The grains form as layers of calcium carbonate build around a nucleus—perhaps in much the same way that layers of pearl shell are built up. A well formed oölite is the result of grains rotating again and again in a bottom current, and a logical place for such tumbling would be in a tidal area. Common environments of present-day formation are the coasts of Florida and the Bahamas.

The origin of dolomite has been debated long, and the end is not in sight. The argument centers around whether dolomite forms as a primary precipitate from sea water, or whether it forms through an alteration of calcite or aragonite crystals. In some places, dolomite masses cut across limestone layers or follow fracture patterns, cutting the limestone

 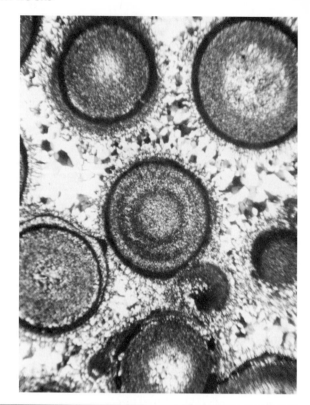

Fig. 6-23 A. (left) Oölite from the Bahama Banks as seen under the microscope. **B.** (right) Photograph of a thin section of oölites from Cambrian sediments in Pennsylvania, as seen under the microscope. Although silica has replaced the original carbonate, the original layering in the oölites is well preserved. *N.C. Nielsen*

in very much the same fashion that some igneous dikes do. It is rather widely held that such masses are the result of partial replacement of calcite in the limestone by magnesia-bearing solutions. In other places, however, dolomite occurs as widely spread layers, or beds, interlayered with limestone strata. The origin of such deposits is less clear-cut. Some geologists believe that the layers were precipitated directly on the sea floor. Others take the view that the dolomite layers represent selectively replaced layers of limestone—a view that calls for further debate: (1) was the original limy material replaced by magnesia very shortly after deposition, or (2) did chemical alteration occur long afterward, when the limestone was completely formed?

The pendulum has swung in favor of those who hold that dolomite does not form directly from water. In support of their theory, it should be noted that it is very rare in modern sediments. Also, studies of a desert lake have shown that calcite formed first, probably converting to dolomite in the decades or centuries it rested on the lake bottom. If a vote were to be taken, the "replacement" origin certainly would have it.

Organic sedimentary rocks

Rocks made of the remains of organisms, both animals and plants, are called *organic*. Coal is an excellent example because it consists of partially decomposed remains of land plants.

Fig. 6-24. Map of Australia's Great Barrier Reef.

Much coal contains finer plant remains, in spite of the popular view that it is the residue of a chaotic jumble of fallen trunks and tangled roots, once set in a miasmic marsh alive with monsters winging through a canopy of bizarre trees or slithering over the murky floor of the swamp.

Limestone The most abundant of the organic sedimentary rocks is limestone, and probably most examples of it are truly organic and not the result of chemical precipitation. By "truly organic," we mean that the deposits are either fossils of organisms in growth position, or calcareous shell material that has been reworked and moved about by currents, or a little of both. In a strict sense, limestones made up of shell debris that has been reworked by currents should be classified as clastic sedimentary rocks. Of the limestones made of large fossils

199

and fossil fragments, perhaps the most impressive are the reefs—and certainly the most impressive modern one is Australia's Great Barrier Reef (Fig. 6-24). That great carbonate barrier lies just off the coast, stretching a distance of 2000 km (1242 mi), and was the scourge of early sea captains, witness the experiences of Captain James Cook and crew:

June, 1770, found him cautiously sailing northwards in his tiny ship, the 70-foot *Endeavour Bark*. The hazardous journey ultimately led to parts close to those discovered by the Spanish explorer, Luis Vaes de Torres (1605). Going finally ashore on a sea-girt spot of land near the tip of Cape York, Cook named this Possession Island, and there took formal possession of the east coast of Australia for Britain.

Never before had a ship sailed so dangerous and unknown a sea as that skirting the Queensland mainland. Cook found the waters dotted with islands, shoals and coral banks. His way was through twisting passages and shallows into a strange world of mystery and beauty. He was for a long time unaware of a great barrier to seawards that was closing in upon his track. At a spot near the present site of Cooktown, the coral banks crowded in on his ship in baffling array. She finally ran aground and was all but lost on a treacherous patch. The stirring story of that mishap and the masterful saving of ship and crew is one of the highlights of Australia's early history.*

Modern reefs are made up largely of corals and carbonate-secreting algae (Fig. 6-25). In order to grow so profusely, they require an

*From K. Gillett and F. McNeill, 1959, The Great Barrier Reef and adjacent isles: Coral Press Pty. Ltd., Sydney, Australia.

Fig. 6-25. Examples of coral from the vicinity of the Florida Keys. **A.** (below) Star coral (*Montastrea*). **B.** (opposite, top) Brain coral (*Diploria*). **C.** (opposite, bottom) Staghorn coral (*Acropora cervicornis*). *D. D. Runnells*

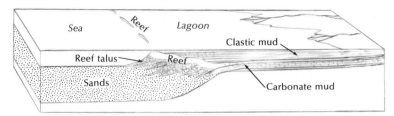

Sea — Reef — Lagoon

Clastic mud

Reef talus — Reef

Sands

Carbonate mud

Fig. 6-26 A. (above) Diagram of the zones commonly associated with an offshore reef. **B.** (below) Reef deposits about 350 million years old on northern Banks Island, Northwest Territories, Canada. The light-colored reef deposits, which are embedded in clastic deposits, are relatively resistant to erosion and so form prominent outcrops. *Geological Survey of Canada*

environment free of most land-derived sediment, and shallow, warm water that is both agitated by wave action and high in nutrients. Those conditions are met today off many coasts within about 30° latitude of the equator. The reefs most familiar to North Americans form the Florida Keys and the Bahama Islands, and part of the Hawaiian Islands.

Reefs are extremely complex in their structure, but three main zones are commonly recognized in cross section (Fig. 6-26). One is the reef itself, which is formed of organisms building upward from the shallow sea floor to sea level. Commonly such creatures are obstinate enough to grow in the face of the prevailing winds. Waves constantly crash against

202

Fig. 6-27. Well-bedded chert in the Cook Inlet region, Alaska. Originally horizontal, the chert has been subsequently folded, and a fault trends diagonally across the outcrop from the man's right foot. *G.K. Gilbert, USGS*

the reef, breaking off pieces of the delicate organisms, and forming an apron of steeply dipping organic debris—a sort of a submarine talus of reef material, on the windward side. Landward of the reef is a quiet-water lagoon in which carbonate and clastic muds accumulate.

Other limestone deposits consist of nothing more than the remains of microscopic limestone fossils heaped one upon the other. They commonly are made up of minute, free-floating, single-celled animals, the *foraminifera*. Some chalk deposits, such as those near Dover, England, are 100 million years old, and the truly remarkable thing about them is how little alteration or recrystallization they have undergone in all that time.

Siliceous rocks Such rocks are made largely of silica (SiO_2). In discussing their formation we will repeat the by-now familiar arguments of organic versus inorganic origin.

The most widely occurring siliceous rock is *chert*, a name used to cover a host of varieties of very dense, hard, nonclastic rocks made of microcrystalline silica. One familiar form is *flint*, which is dark-colored because it contains organic matter. Since flint is uniformly textured, has a conchoidal fracture much like obsidian, and is easy to chip while at the same time retaining a sharp edge, it proved to be the ideal strategic material for arrow- and spearpoints in the Stone Ages of Europe and in the United States. In what was perhaps a braver day than ours, flints were essential for survival on the frontier, not only to strike sparks from steel for fire but also to fire the flintlock gun of the eighteenth and nineteenth centuries. Red varieties of the same rock commonly are called *jasper*, the red color being derived from the iron oxide (Fe_2O_3) it contains.

On a volume basis, the most important bodies of chert are thick and well bedded (Fig. 6-27). Although some of them may have been precipitated directly from sea water, such is not the most common origin. The reason is that the amount of silica in sea water is too small for direct precipitation. If one could somehow enrich part of the sea with silica, as may happen in the vicinity of an underwater volcanic eruption, then chert might form as a direct precipitate.

A more likely origin for the thick cherts is organic. Microscopic marine animals, such as the *radiolaria* and *sponges*, and plants such as *diatoms* build shells by extracting silica from sea water (Fig. 6-28). In time, and in an environment in which both carbonate shells and

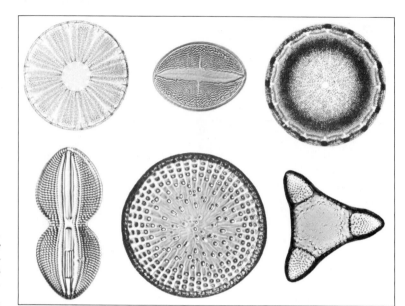

Fig. 6-28. A variety of diatoms, greatly magnified. The average size of these siliceous plant remains is about 50 microns (about $^2/_{1000}$ of an inch). *G. Dallas Hanna, California Academy of Sciences*

Fig. 6-29. Coal beds interbedded with shale and sand of Cenozoic age on Ellsemere Island, Northwest Territories, Canada. Plants in the rocks suggest a climate similar to that of present-day southern Ontario. *Geological Survey of Canada*

clastic material are excluded, a thick deposit of chert can build up.

Two other sources of siliceous deposits, although minor, should be mentioned. One is the direct precipitation of silica from hot springs; for example, pedestals at the bases of active geysers in Yellowstone National Park are found in such a way. Such deposits, called *sinter*, are quite spongy and porous. Another source has to be sought for the irregularly shaped, modular masses of chert contained in limestones. Because such deposits cut across the bedding planes, they clearly replaced part of the limestone at a later date. The process is similar to that in which silica replaces woody fibers to form petrified wood.

Coal and oil shale   Coal- and oil-bearing shales are perhaps the organic sediments most fa-

miliar to non-geologists (Fig. 6-29). No doubt such deposits will become more familiar as our reliance on them as a source of fuel increases.

Coal can be thought of as a somewhat metamorphosed sediment. The sediment itself consists of the plant remains that accumulate in a swampy area. A swamp environment is essential because its waters are low in oxygen content, and, as is well known, dead plants readily decompose in the open air or in an oxygenated environment. However, in a low-oxygen environment, vegetation accumulates to form peat, a brown, soft organic deposit in which plant remains are easily recognized. With burial under younger sediments, the pressure and heat send peat on its way to becoming coal. The first step is a brownish material called *lignite*. With further alteration, the carbon content of lignite is increased, and black *bitu-*

Fig. 6-30. Five cm (2 in.) of finely bedded oil shale in Colorado. The darker bands contain the most organic matter. *W.H. Bradley, USGS*

minous coal, then *anthracite* coal—the highest quality product—form.

Oil shale is in the news these days, mainly because it is being looked at as a source of oil when domestic sources are dwindling. In the United States, lake deposits laid down in parts of the Rocky Mountains some 50 million years ago are attracting the most attention (Fig. 6-30). Plant and animal life flourished in that environment. The deposits are well bedded, with the beds possibly being annual

features. Bottom conditions were such that a wide variety of organisms were preserved as fossils; perhaps the most famous of them are the fossil fishes (Fig. 6-31). The organic matter included in the bottom sediments is, in time, converted to *kerogen* which, on distillation and refining, yields oil—thus the name oil shale.

Problems remain in converting oil shales into usable fuel. One involves extracting the oil at a competitive cost. The other problems are en-

Fig. 6-31. Two armored herring (*Diplomystus dentatus*) from the Eocene Green River Formation in Wyoming. The fish were buried in this position by a shower of ash that fell on the lake. *Jan Robertson, Geological Society of America*

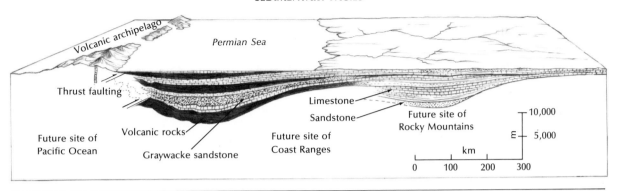

Fig. 6-32. Reconstruction of facies relationships of Paleozoic rocks in southeastern Alaska and British Columbia at the close of the Permian period. Notice that the thickness of the sediments varies. Also, the dominant volcanic rocks and graywacke sandstones to the west grade eastward to limestones and sandstones with distance from the volcanic archipelago.

vironmentally oriented and include (1) locating a sufficient water supply in that semi-arid region, (2) finding ways to avoid stream pollution from both the solid and dissolved wastes that are derived from mined materials, and (3) finding ways to quickly re-establish vegetation in mined areas and on waste piles. Engineers, geologists, and ecologists are joining together to attempt to solve those problems so that the oil can be extracted with the least damage to the environment.

SEDIMENTARY FACIES

Now that the major sedimentary rocks have been described, we should put forth a warning—the basic properties of any rock unit probably will change as the unit is traced laterally. For example, suppose we are examining a thick layer of conglomerate in a canyon wall. Over a distance of several kilometers, the layer might change gradually first to a sandstone, then to shale. The reasons for the change are fairly simple: in any depositional basin, conditions such as sediment supply, velocity of current, and biological production can vary laterally. The lateral differences are called *facies* changes; as conditions vary, so do the resulting sediments (Fig. 6-32).

In order to further understand such changes,

let us look at some present-day examples. A good example is a carbonate reef, which builds upward and seaward, but the deposit changes seaward to the reef talus, and landward to lagoonal deposits. Consider also going seaward from a gravelly beach. The gravels occur in the "high-energy" area, where the waves are hitting the shore, but farther offshore, where the deeper and quieter waters possess less energy to move particles, there are found sands and, eventually, silts and clays. Another example of facies changes is glaciated terrain, where glacial deposits grade downvalley to stream deposits, which in turn might grade laterally to marine deposits (Fig. 6-33).

Desert basins also exhibit lateral changes in environment and thus in the resulting sediments. Streams deposit gravels close to the mountain front, but with increasing distance from the front they deposit sand. In the center of the basin, silts and clays may be laid down, forming the floors of temporary desert lakes. The lowest parts of the basins sometimes contain salt deposits, formed when the temporary lakes evaporate. The examples are endless—just look around at the various sediments being formed in the wide variety of surface environments near you.

The concept of facies was exceedingly important to the early geologists; without it they

Moraine

Stream deposits

Fig. 6-33. Example of facies changes in recent sediments, Ekalugad Valley, Baffin Island, Northwest Territories, Canada. Glaciers occupy the uplands, and in places tongues of ice penetrate to the valley bottom. A ridge of glacial deposits (a moraine) has been left behind with the recent retreat of the ice. The ridge grades down-valley to sandy and gravelly stream deposits and these, in turn, grade into silts and clays deposited in the quiet waters at the head of the fiord. Although all of the sediments are the same age, their character is determined by their environment of deposition.
National Air Photo Library, Canada

could not have been aware that a limestone sequence in one place was the same age as a sandstone sequence 100 km (62 mi) or so away. It is a hard enough task to follow facies changes where outcrops are good and rocks are not deformed. Consider the plight of the geologist working with strongly deformed sediments in forested terrain with few rock outcrops! Only with hard work and a fertile imagination can the rock record be unscrambled.

SUMMARY

1. Sedimentary rocks result from the weathering, erosion, transporting, and deposition of any pre-existing rocks. Rock and mineral particles that form from weathering make up the *clastic* sedimentary rocks, and the ions from weathering recombine to form the various sedimentary rocks that are known as *chemical precipitates. Organic*

deposits form the third major subdivision of sedimentary rocks.

2. The source areas for sediments are usually mountains or hills, whereas the desert basins, river floodplains, lakes, and oceans are the common depositional sites.

3. The sediments have a set of field characteristics that can be used to identify the environment in which they formed. Such characteristics include stratification, sorting and roundness of grains, cross-bedding, fossils, mud cracks, and ripple marks. In most environments, the major stratification layers are horizontal.

4. Once sediments are deposited, *calcite* or *silica* may precipitate from the circulating ground water and cement the grains together to form sedimentary rock.

5. In classifying rocks, the primary criterion for clastic sediments is grain size, and for chemically precipitated rocks it is composition. *Organic sediments* are the products of the accumulation of plants or animals, and the common ones are limestone, coal, and oil shale.

6. The concept of sedimentary facies is essential to interpreting the origin of sedimentary rocks. In short, it means that the characteristics of sediments laid down in a specific time period will change laterally because the environments of deposition change laterally. Hence, stream deposits may grade laterally to clastic marine deposits and then to carbonate reef deposits.

SELECTED REFERENCES

Blatt, H., Middleton, G., and Murray, R., 1972, Origin of sedimentary rocks: Prentice-Hall, Englewood Cliffs, New Jersey.

Gillett, K., and McNeill, F., 1959, The Great Barrier Reef and adjacent isles: Coral Press Pty. Ltd., Sydney, Australia.

LaPorte, L. F., 1968, Ancient environments: Prentice-Hall, Englewood Cliffs, New Jersey.

Pettijohn, F. J., 1975, Sedimentary rocks, 3rd ed.: Harper and Row, New York

Fig. 7-1. Banded metamorphosed sedimentary rocks, near Convict Lake on the east side of the Sierra Nevada, California. *John Haddaway*

7

METAMORPHIC ROCKS

A geologist studying igneous and sedimentary rocks has an advantage over one investigating metamorphic rocks (Fig. 7-1), in that some of the two other kinds form on the earth's surface in environments where they can be observed.

One difficulty in understanding the origin of metamorphic rocks is that no one has ever seen a metamorphic rock being formed, and for that reason much of our thinking about them is pure conjecture. This is not to say, however, that it is as fanciful as the speculations of science fiction; there are strongly limiting physical and chemical boundaries within which any theory of metamorphism must operate.

Between a depth of about 20 km (12 mi), where under certain circumstances the temperature can be great enough to melt rocks, and the earth's surface, where weathering takes place, there is a wide range of temperature, pressure, and chemical environments. As a result of geologic processes such as mountain building, plate convergence, and sedimentation, some of the rocks in that zone are prone to change. Rocks that are mineralogically stable in one *physico-chemical environment* generally tend to become unstable when they are trans-

ported to another (Fig. 7-2). How the rock actually changes depends on the nature of the new conditions and the chemistry of the starting rock. We call the altered assemblage of minerals a *metamorphic* rock and the process by which it is transformed *metamorphism* (taken from the Greek, meaning to change in form). It should be emphasized that during metamorphism the minerals do not melt but remain largely in the solid state.

As in the case of plutonic rocks we can investigate the results of the metamorphic process only after the erosional removal of the overlying rocks and long after the actual metamorphic process has run its course. Geologists have to infer the nature of premetamorphic materials and the physicochemical environment during metamorphism: and field and laboratory studies of the altered rocks themselves, augmented by a knowledge of chemical reactions, help them to make their inferences.

It becomes apparent from studies of the rocks themselves that metamorphism is basically an *isochemical process*. That is, in most instances, the overall chemical composition of the rock is nearly the same before

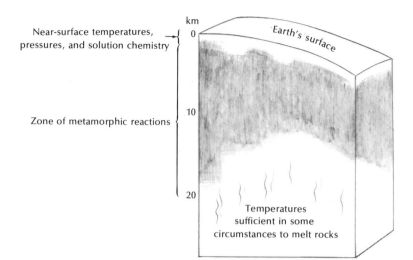

Near-surface temperatures, pressures, and solution chemistry

Zone of metamorphic reactions

Earth's surface

km

0

10

20

Temperatures sufficient in some circumstances to melt rocks

Fig. 7-2. Representation of the zone where metamorphism can occur, between the surface and about 20 km (12 mi), where temperatures can be high enough to melt rocks.

and after recrystallization. Little new material is gained or lost during the change. (Exceptions to the rule involve circulating fluids, commonly hot, which add other chemicals to the rock being transformed.)

Because of the isochemical nature of the recrystallization process, the chemistry of the original rocks largely determines what can be formed under any set of physicochemical conditions. For example, if limestone, composed solely of grains of calcite ($CaCO_3$) is metamorphosed, the resultant rock will be composed only of $CaCO_3$; whereas if a shale composed predominantly of clay (hydrous aluminum silicate) and quartz (SiO_2) particles is subjected to metamorphism, a number of silicate minerals will be formed, but calcite will be lacking.

FACTORS IN METAMORPHISM

As you might expect, there is a rich variety of metamorphic rocks. Any kind of rock—igneous, sedimentary, or metamorphic—can serve as the starting material, and the combinations of temperature, pressure, and chemical conditions that can bring about its alteration are limitless.

Heat

Thermal energy is probably the principal factor involved in the metamorphism of rocks. By increasing a rock's temperature, enough energy to activate recrystallization of minerals may be provided. As the temperature rises higher and higher, ions diffuse more quickly, leading to an increase in the speed and efficiency of the metamorphic transformation. Moreover, as a rock becomes hotter, minerals containing volatile components such as water and carbon dioxide become less stable and eventually break down, in the process releasing some of those components. The recrystallized minerals are, of course, proportionately less rich in volatile content.

It is also true that as the temperature of a rock is increased, it tends to decrease in strength; that is, it is more likely to deform or even to flow plastically in response to directed stresses. Many metamorphic rocks are highly contorted in appearance and look like they had been squeezed through a toothpaste tube.

If temperatures are high enough, rocks can actually melt. At first only certain minerals whose *melting points* are exceeded will melt, but eventually a nearly total fusion of the rock body and the formation of a magma reservoir will take place.

Pressure

Under increasing pressure the mineral grains in a rock are subjected to a squeezing action. Such pressure can cause recrystallization and the formation of minerals with more tightly packed atomic structures and greater density than the initial mineral grains.

Also, if a rock consisting of many separate mineral grains is compressed, much of the stress is concentrated upon irregularities along the boundaries of the individual grains. Such uneven pressure leads to local spot-melting at the points of high stress and, simultaneously, reprecipitation of material in the areas of low stress.

Much of the pressure to which a rock is subjected comes from the load of the rocks above it. The thicker the rock body above a metamorphosing rock, the greater the pressure. The situation in a swimming pool is similar: the deeper you go, the more the pressure increases, in response to the increasing load of the water above. At any point in the pool, the pressure is equal in all directions—otherwise the water would tend to flow. The all-sided pressure in water is called hydrostatic pressure, and the analogous pressure created at depth in the earth from the rock load above is called *lithostatic pressure,* or *confining pressure.*

Some of the pressure exerted on rocks results from the release of the volatile constituents (mostly water and carbon dioxide) liberated in response to thermal activation, as discussed earlier. Such pressure commonly is called *pore-fluid pressure,* and it too is an all-sided pressure.

In many instances *directed pressure* also plays a part in forming metamorphic rocks. It operates in a particular direction, and apparently is generated in response to mountain-building activity. No one seems to know quite how intense directed pressure may be, but all agree that it is important to the development of the

foliated and lineated rock fabrics that are so characteristic of many metamorphic rocks. It appears that in a rock recrystallizing under directed pressure, many of the newly forming mineral grains—especially of the platy species like mica and chlorite—are constrained to grow with their cleavage plates more-or-less parallel to each other and perpendicular to the direction of the pressure. Elongate minerals, such as hornblende, also commonly crystallize in bands (and, in some cases, with their long axes in parallel alignment) under the influence of the pressure.

Chemical activity

The chemical environment is also of great importance to the recrystallization of rocks, but its exact role in the process is hard to evaluate precisely. In most cases the chemical activity is related to the presence of small amounts of volatile constituents (mostly water and carbon dioxide) that permeate and fill pore spaces within and between grains in the metamorphosing rock. Some of the volatiles are derived, as we have already learned, from the breakdown of minerals as metamorphism progresses. Some of them, however, undoubtedly come from fluids already present in the cracks and pore spaces in the unmetamorphosed rock. Sediments, which are formed mostly in aqueous environments, particularly contain an abundance of initial pore fluids. In some way the permeating fluids increase the efficiency of the recrystallization process; that is, they *catalyze* the metamorphic reactions. In a progressively hotter environment, the fluids become correspondingly more active in their catalytic role. All earth scientists agree that without the presence of the pervasive, hot pore-space fluids, metamorphism would proceed at very slow, perhaps insignificant, rates. Laboratory studies have supported that view.

METAMORPHIC ROCKS

Metamorphic rocks are found throughout the continents and to a much lesser degree in the ocean basins of the world. A small portion occur in bodies of limited extent and include the *contact*, *hydrothermal*, and *cataclastic* metamorphic rocks. Most of them, however, occur in large, often elongate bodies that are exposed over thousands of square kilometers and include the *dynamothermal* metamorphic and mixed metamorphic-plutonic rocks. In the formation of each of the various kinds of metamorphic rock, different factors are important (Fig. 7-3) . Heat plays the dominant role in contact metamorphism; chemically active solutions and heat are necessary for formation of hydrothermally metamorphosed rocks; and localized pressure is paramount for the formation of cataclastic metamorphosed rocks. In the formation of dynamothermal and mixed metamorphic-plutonic rocks, heat, confining and directed pressure, and chemically active fluids all are necessary.

Metamorphic rocks of local extent

Contact metamorphic rocks   At the margins of smaller plutonic bodies the enclosing country rocks commonly are heated to such high temperatures that local metamorphism can occur in a relatively thin sheath, or aureole, as it is called (see Chap. 4). The recrystallized rocks are called *contact metamorphic* rocks, and the factor primarily responsible for their formation is the thermal energy derived from the hot magma within the plutonic body. Directed pressure plays an insignificant role in the recrystallization process and therefore such rocks generally lack foliation or banding.

The thin contact metamorphic zones generally are found at the edges of dikes and sills. Because of their relative smallness and the rapidity with which these igneous bodies cool, the contact zones are usually only a few centimeters wide at most. Where the magmas were injected into a shale, however, the sheath can be baked or hardened over distances of from several centimeters up to a meter or more. The clay minerals in the shale bake in much the same way that clay does when it is fired in a kiln to make bricks or pottery.

Larger intrusions, such as batholiths and stocks, which take many thousands to a million years to cool, make their influence felt over a wider area. There, the country rocks may be converted into a dense, hard, non-foliated rock called *hornfels* (Fig. 7-4).

Most hornfels rocks are generally aphanitic to fine-grained and are composed of nearly equidimensional recrystallized minerals. Some of the newly formed mineral grains are merely the result of recrystallization of previous grains on a one-to-one basis; others result from chemical reactions between two or more mineral species during recrystallization. If,

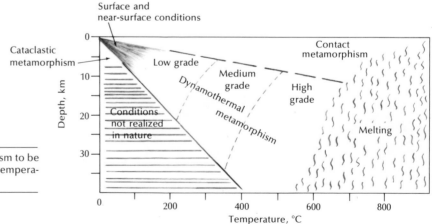

Fig. 7-3. The types of metamorphism to be expected under given depth and temperature conditions.

Fig. 7-4. Fine-grained hornfels, as viewed under the microscope. Quartz and feldspar (the larger grains) have been recrystallized on their margins only, whereas the original clayey matrix between grains has been totally recrystallized to muscovite (small grains in the background and upper left). This contact metamorphism occurred adjacent to a thick sill, Victoria Land, Antarctica. *W.B. Hamilton, USGS*

Hydrothermal metamorphic rocks Accompanying the intrusion of some batholiths is the release of a great deal of hot fluids and gases which travel long distances through enclosing host rocks, carrying in solution ions derived from the magma body. Such fluids and gases (called *hydrothermal solutions*) commonly are chemically potent and react readily with many of the minerals in the country rock with which they come in contact. The hot fluids not only act as catalysts in the recrystallization of the country rock but also bring in new chemicals which are incorporated with the newly recrystallized minerals. In some instances the hydrothermal solutions react with the host rock to remove chemical constituents, which are then carried away as the fluids pass through and beyond the rock. Not all solutions that enter into the recrystallization process come from the magma itself. Some is ground water that is heated and driven in convective circulation (like a boiling coffee pot) by the hotplate-like action of the nearby hot magma body.

Many minerals are particularly susceptible to alteration by hydrothermal fluids. For example, rocks rich in olivine ($(Mg,Fe)_2SiO_4$) commonly are altered to serpentine ($Mg_3Si_2O_5(OH)_4$) in that way. The reaction, called *serpentinization,* is principally one of water enrichment by the solutions. Many geologists now feel that the formation of serpentine is particularly prevalent in the vicinity of the oceanic ridge systems, where hot, chemically active solutions rise along the faults and fractures associated with the ridge.

During serpentinization of olivine-rich rocks there is an increase in volume and thus of internal stresses. Serpentine is a relatively low-strength rock, and when excessive stress is applied it starts to slide, flow, and fracture. Many of the multitudes of cracks that permeate a body of serpentine may be filled in later with white veins of dolomite, and they make a bold and striking contrast to the prevailing dark-green color of the rock. Because of its complex interweavings of white, green, and black, in a bygone age serpentine was greatly favored for

for example, a rock containing dolomite ($CaMg(CO_3)_2$) and quartz (SiO_2) is heated, the reaction between the two grains will result in the production of olivine (Mg_2SiO_4) and calcite ($CaCO_3$), with the coincident liberation of carbon dioxide (CO_2). The reaction can be represented in the shorthand of chemistry:

$$2CaMg(CO_3)_2 + SiO_2 \rightarrow 2CaCO_3 + Mg_2SiO_4 + 2CO_2$$

the walls of florist shops, funeral parlors, and the lobbies of small-town hotels—it bears the picturesque name of *verde antique*. The more common varieties of serpentine lack the white dolomite veinings and generally run to somber dark-green, black, or red colors. They nonetheless take high polishes and have been used as decorative facings in building lobbies and storefronts. In the United States they may be seen in the imposing columns of the rotunda of the National Gallery of Art in Washington, D.C., as well as in the United Nations building in New York.

Cataclastic metamorphic rocks In localized zones near the surface of the earth, stresses may build up to the point where rocks are sheared and crushed along large fracture planes known as faults. We shall see some of the effects of these forces when we discuss faulting (Chap. 16). In this chapter we are concerned primarily with the small-scale changes brought about by the crushing and granulation of rocks under the application of severe local stress (Fig. 7-5).

For rock to be granulated and crushed it must, in general, be brittle. Since low temperatures and confining pressures are conducive to brittle behavior, most shearing and crushing occurs near the outside of the earth. At greater depths, however, where rocks are hotter and under higher pressure, the application of the same stresses generally would result in plastic flowage rather than fracturing and granulation.

The principal change in rocks subjected to the lateral stress developed along a fault is the shearing, rolling out, and grinding up of mineral and rock fragments (Fig. 7-6). The resultant "cataclastic" (meaning dynamic and broken) rocks show markedly parallel, lens-shaped, and banded patterns. The process commonly is termed *cataclastic*, or *dynamic, metamorphism*.

Not all the minerals in rocks are equally resistant to shearing stress; they are much like people in that regard—some are stubborn and unyielding, others are pliant and readily molded. In a rock reshaped by dynamic meta-

Fig. 7-5. Early stages of cataclastic deformation of a granite from the Front Range, Colorado, as viewed under the microscope. The rock is cracked and broken, and in some places recrystallization, chiefly of quartz and potassium feldspar, has occurred. *W.A. Braddock, USGS*

morphism, susceptible, platy minerals such as mica may be drawn out in parallel streaky bands or layers, while the more obdurate ones, such as feldspar and quartz, are pulverized. Some particularly resistant minerals may stand out as rounded or even lenticular, eye-like clots—in fact they are called by the German word for eyes, *augen* (Fig. 7-6).

A common example of dynamic metamorphism is the rock *mylonite*. The name comes from the Greek word for mill, and it is rock that in a figurative sense was caught in the geologic mill and ground until it was reduced to powder. After the minerals in the original rock were crushed and pulverized, the tiny fragments were partially to completely

Fig. 7-6. Sheared, streaked-out appearance of a cataclastic metamorphic rock from the Bighorn Range, Wyoming. Some of the larger mineral grains are broken, rounded, and rotated. The large grain (augen) in the center is about 2 cm (1 in.) long. *Jan Robertson*

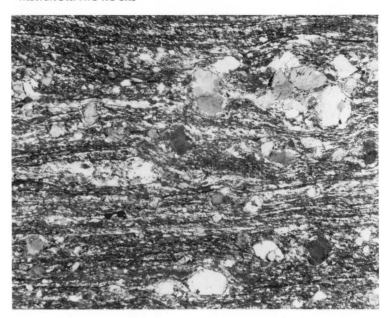

Fig. 7-7. Mylonite, as viewed under the microscope, from the Front Range, Colorado. The large broken grains (augen) are quartz and feldspar; most of the streaked-out matrix is composed of very small crystals of biotite and broken grains of quartz and feldspar. *W.A. Braddock, USGS*

recrystallized into an interlocking pattern, which left the rock as hard and durable as flint. In some cases, mylonite is transformed into a fine-grained glass-like material: with its dark, streaky appearance, it superficially resembles obsidian.

A microscope usually is needed to show that mylonite consists of angular, minutely brecciated mineral fragments which have been recrystallized to form a metamorphic rock (Fig. 7-7).

Metamorphic rocks of regional extent: dynamothermal metamorphic rocks

This term refers to metamorphic rocks that characteristically are exposed on the earth's surface over broad areas, sometimes many thousands of square kilometers. Commonly they are found in close association with large plutons (batholiths and stocks), and, like those igneous bodies, are often found in the axes of large mountain ranges.

On essentially all the continents of the world, leveling by erosion has produced large, flattened expanses of land floored almost ex-

clusively by plutonic and dynamothermal metamorphic rocks. Such wide expanses of crystalline rock—both igneous and metamorphic—were given the name *shields* many years ago, and they were regarded as the foundation of the continents. Two well-known examples are the Canadian Shield, that broad expanse of igneous and metamorphic rocks marginal to Hudson Bay and extending southward into Minnesota and Wisconsin and eastward across Labrador, and the Fenno-Scandian Shield, which includes most of Finland, Sweden, and Norway. Such rocks, with so wide a distribution, must have a more general cause than the heat generated by a single, igneous intrusion, or the reactions produced by chemically activated water, or the grinding effect of movement along a fault plane. Recrystallization is, in fact, brought about by heat and pressure (lithostatic and directed) working in unison with, to a lesser degree, chemically active solutions. In most large bodies of dynamothermal metamorphic rocks, it is apparent from the minerals they contain that temperatures were not uniform during recrystallization. Where temperatures were higher, high-temperature mineral

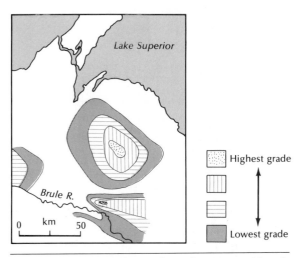

Fig. 7-8. Centers of dynamothermal metamorphism in northern Michigan. Relatively low-grade metamorphic rocks are found toward the outside and high-grade metamorphic rocks toward the inside of each center. Apparently the centers represent "hot spots" of thermal activity.

associations were formed. With distance away from such "hot spots," progressively lower-temperature mineral associations were formed, creating a zonation from *high-grade* metamorphic rocks to *low-grade* metamorphic rocks (Fig. 7-8).

Because of the role played by directed pressure most dynamothermal rocks are layered, or foliated (from the Latin *foliatus,* meaning leaved). Some, however, are relatively massive, or nonlayered, while others, which are relatively unusual, are composed of a closely layered mixture of high-grade metamorphic and plutonic rocks.

Foliated rocks All foliated rocks are characterized by the parallel orientation of their tabular minerals and varying degree of banding, or color layering (Fig. 7-9); some also can be split, or cleaved.

Low-grade metamorphic rocks. Slate, which takes its name from the Old French word *esclate,* or *slat,* is one of the most familiar foliated metamorphic rocks. It has been used for centuries for roofing material and blackboards, alike. Two properties are responsible

Fig. 7-9. Biotite gneiss from Uxbridge, Massachusetts, showing light and dark banding, or foliation. Such dynamothermally metamorphosed rocks characteristically show rock cleavage or foliation.
Ward's Natural Science Establishment, Inc.

Fig. 7-10. Rock cleavage in a slate. Sevier County, Tennessee. The original bedding is parallel to the 15-cm ruler, with the closely spaced cleavage planes developed nearly at right angles to the bedding. For a microscope view of the same slate, see Figure 7-11. *W.B. Hamilton, USGS*

for its desirable attributes: (1) it is dense and of uniformly fine texture and (2) it can be cleaved along smooth, closely spaced parallel surfaces. The latter property is called *rock cleavage* to distinguish it from mineral cleavage (Fig. 7-10).

Rock cleavage in slate results from the extremely fine foliation that develops during low-grade dynamothermal metamorphism. Microscopic investigation reveals that the cleavage planes of tiny recrystallized flakes of muscovite mica, which are everywhere present in the slate, are all aligned in near-perfect parallelism (Fig. 7-11). Splitting of a slate along one planar surface entails the sequential cleaving of numerous tiny mica flakes in that plane.

Slate is usually derived from finely laminated sedimentary rocks such as shale, but can be formed from other finely textured rocks, such as volcanic tuffs.

The original stratification planes in shale or tuff are usually still easily recognizable, and, in most cases, do not coincide with the rock cleavage planes (Figs. 7-10 and 7-11). Somehow, during metamorphism, the small planar clay grains in the shale (which were aligned parallel to the bedding planes) are recrystallized to mica platelets, all with their cleavage planes in parallel alignment but at an angle to the bedding. How the change comes about is not clear. Some believe that as the mica plates form, they are constrained to grow with their cleavages at right angles to the directed stress. Others feel that, once formed, the plates are rotated by mechanical movement (that is, deformation) in response to directed stress. And still others believe that the clay particles themselves are rotated during deformation and then are recrystallized.

The fact that original bedding can still be recognized, the fineness of the foliation, and the smallness of all of the individual mica grains are three bits of evidence indicating that slate is the result of very gentle dynamothermal metamorphism at relatively low temperatures. The metamorphic rock looks superficially very much like the original sediment from which it was derived.

In rocks that contain less clay minerals than shales and are accordingly less susceptible to alteration, low-grade metamorphism produces little outward evidence of having occurred except for the development of poorly defined rock cleavage and the growth of some grains of mica in subparallel alignment.

Intermediate-grade metamorphic rocks. Schist, from the Greek word *schistos,* cleft, or *schizein,* to split (we see the same word in schizophrenia, meaning a split personality, or in schism, meaning a division of opinion within a group), is probably the most widely occurring metamorphic rock. Slates grade into schists with increasing grain size. Whereas the platy minerals in slate are unrecognizable without the aid of a microscope, those in schists generally are visible to the unaided eye. All schists include tabular, flaky, or even fibrous minerals in their composition, and the extent to which those minerals developed in parallel orientation determines to a considerable degree whether *schistosity,* the characteristically wavy or undulatory rock cleavage, will develop. Many schists split readily into tabular blocks. They are the familiar flagstones so widely used throughout Europe in courtyards and for castle walls, and in North America for fireplaces, patios, and barbecues.

The individual folia laminations of schist are, for the most part, regularly spaced at distances up to about 0.5 cm (0.2 in.). The spacing is the principal attribute that separates schists from the more coarsely layered gneisses, to be described next. Sometimes adjacent folia bands possess different mineralogies: one band may contain mostly platy minerals (biotite, muscovite, or chlorite), and the adjacent one mostly quartz and feldspar. Generally the flaky minerals make up more than 50 per cent of the rock (Fig. 7-12).

Original bedding or other sedimentary structures are almost never identifiable in schists—indicating that their metamorphic grade is much higher than that of slaty rocks. From the mineral associations common to schists, and from the relatively large size and alignment of the grains, metamorphic petrologists (those who study the origin of metamorphic rocks) generally agree that most schists were formed over a range of intermediate temperatures under the influence of directed pressure (and to a lesser degree, chemically active fluids). Almost any rock can be transformed into a schist, but the likely candidates are shale, siltstone, and muddy sandstone.

Fig. 7-11. The fine structure of rock cleavage shown in Figure 7-10, observed under a microscope. The bedding, demarcated by dark, nearly horizontal bands, is cut at a high angle by the closely spaced cleavage planes. Note that the alignment of recrystallized flakes of mica is parallel to the cleavage planes. *W.B. Hamilton, USGS*

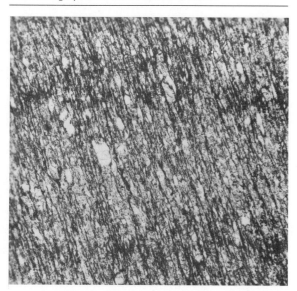

High-grade metamorphic rocks. Gneiss (pronounced as though it were actually spelled nice) is an old Saxon miners' term for a rock that is rotted or decomposed. It is a banded rock, usually with layers of light-colored minerals alternating with dark-colored bands (Fig. 7-13) and is much coarser-grained than schist or slate. In fact, the size of the quartz and feldspar crystals it contains is about the same as those in granite. Recrystallization of the rock under directed stress at relatively high temperatures has rearranged the minerals so that most of the light-colored ones are concentrated in one layer, while the dark ferromagnesian ones are grouped in another; the light- and dark-colored bands alternate rhythmically through the rock. The bands, unlike the cleavage planes of slate, do not make uniformly parallel planes that continue for long distances; more commonly they are strongly distorted (Fig. 7-13). They probably were deformed plastically; that is, although the rock was still in the solid phase it was able to flow, in about the same way that butter or sheet lead can, without becoming liquid at all. The common occurrence of plastic deformation in gneisses argues strongly for high temperatures during its formation—temperatures high enough to soften the rock and enable plastic flow.

Gneisses do not have the highly developed rock cleavage of slate or schist. In spite of the roughly uniform spacing of their bands, they break in about the same unpredictable fashion as a piece of granite does when struck a hammer blow.

Non-foliated rocks Non-foliated rocks are shaped by the same processes that bring about the foliated varieties, but as a consequence of their initial mineralogical composition they are not banded. Two leading examples are *marble* and *quartzite*.

Marble is the finely to coarsely crystalline equivalent of the sedimentary rock, limestone, and accordingly consists largely of the mineral calcite ($CaCO_3$). In the transformation of limestone to marble at relatively high temperatures

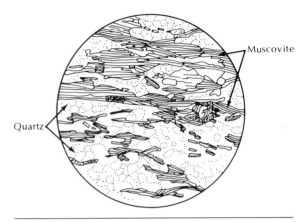

Fig. 7-12. Drawing of schist from Edge Hill, Pennsylvania, as seen under the microscope. It is composed almost entirely of grains of quartz and muscovite in sub-parallel alignment.

Fig. 7-13. Gneiss from the north rim of the Black Canyon of the Gunnison River, Colorado. The light-colored bands are composed mostly of quartz and feldspar, and the dark-colored layers mostly of quartz and biotite. Notice the crenulations and folds, indicative of plastic folding and flowage. *W.R. Hansen, USGS*

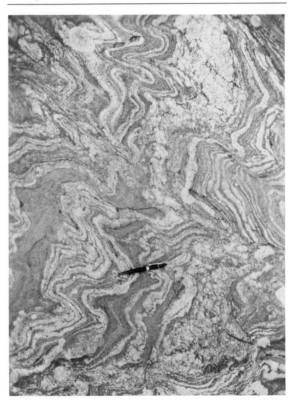

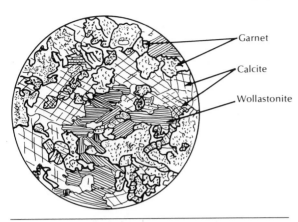

Fig. 7-14. Calc-silicate rock, Chihuahua, Mexico, as viewed under the microscope. Wollastonite and garnet grains are embedded in grains of coarsely recrystallized calcite. Calc-silicate rocks generally are the result of metamorphism of a sandy or silty limestone.

and pressures, the bedding and visible organic shell material is largely obliterated, and the result is a rather sugary textured rock composed of a fine- to coarse-grained aggregate of nearly equidimensional calcite grains. None of the grains are bounded by crystal faces. Because marble is composed of one mineral only, and the grains are essentially all of the same size and equidimensional, there is little possibility for any foliation or banding.

We should point out that not all marble is the result of dynamothermal metamorphism. Some is the product of recrystallization during contact metamorphism when, for example, a limestone bed is intruded by a plutonic body of at least moderate size. One of the finest white marbles comes from a quarry at Marble, in west-central Colorado. It owes its existence to the contact metamorphism that took place when a Tertiary laccolith intruded a Mississippian-age limestone.

Pure marble is snow white, and one of the most highly prized varieties from ancient times down to our day comes from the quarries of Carrara on the west coast of Italy. Carrara marble is a remarkably uniformly textured rock which is ideal for sculpture because it is free

from impurities, and has a hardness of no more than 3 on the Mohs scale.

Not all marble is snow white, which is immediately clear to anyone who has observed it in banks, building facades, lobbies, public lavatories, and on old-fashioned table tops and dressers. In general, the black and gray areas in marble are colored by carbonaceous matter, brown and red areas by iron oxide, and green zones by various iron- or magnesium-bearing silicate minerals.

Metamorphosed limestones are not always composed of only calcite. Commonly the sediments contain clayey or sandy material dispersed throughout the limey material. When such rocks are metamorphosed, reactions between the calcite and the other material result in the production of new minerals. For example, in the metamorphism of a limestone containing quartz sand, a new mineral, *wollastonite* ($CaSiO_3$), is formed, and carbon dioxide is liberated. The reaction in shorthand notation is:

$$CaCO_3 + SiO_2 \rightarrow CaSiO_3 + CO_2.$$

Wollastonite is a colorless mineral, commonly arranged in fan-shaped radiating needles that penetrate the granular marble host. It is representative of a wide variety of related minerals that develop with increasing clay and sand content in impure limestone to yield, after metamorphism, the so-called *calc-silicate rocks* (Fig. 7-14).

Quartzite is metamorphosed quartzose sandstone, and in it the pore spaces once separating the individual grains are filled with newly crystallized quartz. The ghost-like boundary separating the original quartz grain from the silica added to it may be barely discernible. The silica that fills the pore spaces of the original sandstone commonly proves stronger than the sand grains themselves, and when struck a hammer blow, the rock may break through the grains, rather than around them—as quartzose sandstones usually break. Most quartzites are the result of recrystallization at intermediate to high temperatures. Because they are composed dominantly of one mineral,

Fig. 7-15 Close view of mixed plutonic high-grade metamorphic rock found in the Sawatch Range, Colorado. The light layers are granitic; the dark layers, biotite gneiss. Rock flowage causes the crenulation of the layers. Notice the penknife for scale.
Edwin E. Larson

which occurs in equidimensional grains, they tend to show no foliation—regardless of the presence of directed pressure during their development.

Quartzites are nearly always light colored; light pinks or reds are very characteristic. Many are white or light gray, and with increasing amounts of impurities their colors darken to black. Very often they are interbedded with marble, calc-silicate rocks, and other rocks derived from sedimentary sources. Relic sedimentary structures, such as cross-bedding, are sometimes preserved; they are emphasized by slight color differences which superficially may resemble the banding in a gneiss.

Mixed high-grade metamorphic and plutonic rocks. At great depths temperatures can be so high—between 600° and 800°C (1112° and 1472°F) that melting of minerals can proceed. However, not all of the minerals melt simultaneously—those with lower melting points and temperature-stability fields melt first. Here, we reintroduce the Bowen's reaction series (see Chap. 4), but instead of considering

it in relation to the crystallization of minerals, we will view it in relation to their melting. As you might expect, the minerals that melt first are at the low-temperature end of the series and include quartz, potassium feldspar, and sodium-rich plagioclase. Only with a further temperature increase can the melting of the minerals toward the high-temperature end of the series—such as biotite, hornblende, and calcium-rich plagioclase—be started.

If a gneiss consisting of light-colored bands of quartz and potassium- and sodium-rich feldspars that alternate with dark-colored bands of ferromagnesian minerals is heated until melting begins, the bands of non-ferromagnesian minerals will be the first to melt. The layers rich in ferromagnesian minerals may become plastic and perhaps flow and fold, but they will remain solid. If the temperature were to stabilize at that point and then fall, the resulting rock would be made up of bands of dark metamorphic rock (composed largely of ferromagnesian minerals) alternating with light-colored granitic plutonic rock (Figs. 7-15 and 7-16).

If, in our example, the temperature had not stabilized but continued to rise, the entire rock

Fig. 7-16 Highly contorted layers of mixed plutonic high-grade rocks in the Coast Mountains, British Columbia. This type of rock is commonly associated with the lower parts of batholiths and appears to form deep inside the earth where the temperature falls just short of that necessary for total melting. *Geological Survey of Canada*

unit would have eventually melted to form a body of magma with a composition representative of the entire rock body—perhaps equivalent to a diorite.

Not all geologists agree with the explanation above, however. Some propose that the granitic layers are the result of intrusion of sill-like granitic bodies into a high-grade metamorphic rock. Others feel that the granitic bodies themselves are metamorphic rocks, formed by a process in which recrystallization is accompanied by large-scale diffusion of ionic material in and out of the layers, which finally become "granite-like."

Realm of dynamothermal metamorphism
Contact, hydrothermal, and cataclastic metamorphic rocks are of local significance only, and the reasons for their formation are fairly well understood, at least generally if not in detail.

On the other hand, earth scientists have yet to understand clearly why large belts of dynamothermal metamorphic rocks form where they do. We can note that they are commonly found in elongate bodies, are often closely associated with deep-seated plutonic rocks, like those in batholiths, and are found commonly at the axes of mountain chains. Most of them show evidence of subjection to moderate to high temperatures and confining pressures, and most also show indications of directed pressure during their recrystallization.

It has also been found, from radiometric-dating studies, that large metamorphic belts remain sporadically active for anywhere between 200 and 1000 m.y. (million years), with localized high spots of activity that run from 10 to 40 m.y. The times of increased activity are separated by quiescent intervals from 20 to 250 m.y. The active periods may or may not coincide with peak periods of pluton formation. In the Caledonide Mountains of Scotland, for example, peaks in metamorphic activity are recorded at 730 m.y., between 500 and 475 m.y., 440 m.y., and between 420 and 390 m.y. Plutonic activity, however, was restricted to

about 530 m.y., between 410 and 395 m.y., and at 365 m.y.

Most geologists believe that dynamothermal metamorphic rocks were formed at moderate to fairly great depths and in close association with the formation of plutonic rocks. Both processes are thought to be directly related to mountain-building.

Many geologists now feel that all three processes go on in zones of convergence where two plates are pushing together. The heat and directed pressure needed for the formation of dynamothermal metamorphic rocks are the result of the crushing action that occurs along the plate hinge line, and the rocks that represent the original pre-metamorphic materials are igneous and metamorphic rocks from the plates themselves as well as sediments that accumulate in elongate basins marginal to the continents, along the zone of convergence.

Just as in the case of the formation of batholiths, the plate-tectonic model looks promising. Unfortunately, we can only study metamorphic rocks at the surface long after the converging plates have ceased activity and after erosion has removed thousands and tens of thousands of meters of the overlying rocks. And in the zones of convergence of today we have no idea whatsoever of what metamorphic processes are at work deep beneath the surface.

SUMMARY

1. As a result of processes such as mountain building, plate convergence, and sedimentation, some rocks in the earth's outer 15 km (9 mi) are subjected to different *physicochemical environments. Recrystallization* of the minerals may therefore ensue. The process, which occurs while the rocks remain in the solid state, is called *metamorphism.*
2. Metamorphism is brought about by: *heat, lithostatic* and *directed pressures,* and the *chemical activity* of pore fluids.
3. Metamorphic rocks of limited area include *contact metamorphic* rocks, *hydrothermally metamorphosed* rocks, and *cataclastically*

metamorphosed rocks. Contact metamorphism is primarily a thermally induced process occurring at margins of plutonic bodies. The resultant rocks are called *hornfels.*

Hydrothermal metamorphism is the result of interaction of hot and chemically potent water-rich fluids with the rocks surrounding large plutons.

Cataclastic metamorphism represents recrystallization of materials that have been sheared and crushed along fault zones. A common cataclastic rock is *mylonite.*

4. Metamorphic rocks of wide extent are called *dynamothermal (regional) metamorphic* rocks. Recrystallization in such rocks results from heat, pressure, and contact with chemically active solutions. Because of directed pressure most dynamothermal rocks are *foliated.*

5. *Slate* is perhaps one of the best-known *low-grade metamorphic* foliated rocks. It possesses *rock cleavage,* resulting from the alignment of tiny recrystallized mica grains.

Foliated rocks of *intermediate grade* are called *schists.* Their schistosity, or foliation,

is determined by the parallel orientation of tabular, flaky, or fibrous minerals. Shale, siltstone, and muddy sandstone are the rocks most often changed to schists.

High-grade metamorphism produces the foliated rock, *gneiss.* It is coarser grained than schist, with a dark-and-light banding in which ferromagnesian grains are concentrated in the dark layers, and light-colored grains (quartz, feldspar) in the light layers.

6. *Marble* (recrystallized limestone), and *quartzite* (recrystallized quartz sandstone) are two common non-foliated rocks.

7. At great depths, temperatures are so high that melting of the lower-melting-point minerals takes place, forming bands of light-colored plutonic rock, which alternate with dark-colored bands of metamorphic rock. Such mixed rock represents the highest grade of metamorphism.

8. Dynamothermal (regional) metamorphic rocks are commonly found near batholithic rocks and occur in the cores of mountain ranges. Many geologists feel that such rocks are formed in zones of plate convergence.

SELECTED REFERENCES

Barth, T. F. W., 1962, Theoretical petrology, John Wiley and Sons, New York.

Ernst, W. G., 1969, Earth materials, Prentice-Hall, Englewood Cliffs, New Jersey.

Harker, Alfred, 1930, Metamorphism, Methuen and Co., London. (Repr. 1976, Halsted Press, New York.)

Hyndman, D. W., 1972, Petrology of igneous and metamorphic rocks, McGraw-Hill Book Co., New York.

Mason, B., 1966, "Metamorphism and metamorphic rocks," Chap. 10 *in* Principles of geochemistry, 3rd ed., John Wiley and Sons, New York.

Miyashiro, A., 1973, Metamorphism and metamorphic belts, John Wiley and Sons, New York.

Ramberg, Hans, 1952, The origin of metamorphic and metasomatic rocks, University of Chicago Press, Chicago.

————, 1960, "Metamorphism," *in* Encyclopaedia Britannica, vol. 15, pp. 321–26.

Simpson, B., 1966, "Metamorphism," Chap. 22 *in* Rocks and minerals, Pergamon Press, Oxford.

Spry, A., 1969, Metamorphic textures, Pergamon Press, Oxford.

Turner, F. J., and Verhoogen, J., 1960, Igneous and metamorphic petrology, 2nd ed., McGraw-Hill Book Co., New York.

Tyrrell, G. W., 1929, The principles of petrology, E. P. Dutton and Co., New York.

Williams, H., Turner, F. J., and Gilbert, C. M., 1954, Petrography, W. H. Freeman and Co., San Francisco.

Fig. 8-1. Arches National Park, Utah. Different rates of weathering have sculptured these narrow bedrock ridges into forms that rival works of art.
Jesse Kumin © 1976

8

WEATHERING AND SOILS

In the days of the Pharaohs a cherished status symbol was the obelisk. Those hieroglyph-bedecked stone columns early became collector's items for a procession of conquerors of the Nile, beginning with the Caesars and ending with Napoleon. Even the United States collected an obelisk in 1879. After prodigies of effort, involving among other things cutting a loading port in the bow of one of the primitive steamships of the time (whereupon it nearly foundered), the obelisk was finally set up in New York's Central Park (Fig. 8-2) to take its place among similar far-wandering artifacts in cities such as Paris, London, and Rome—in which alone there are twelve of them.

New York's climate is considerably more humid than that of Egypt. A mixture of cold winters with many freeze-thaw cycles, and hot, steamy summers was bound to have an effect on the obelisk. No wonder that, in about 70 years, many of the hieroglyphs spalled off and the whole surface of the obelisk started to disintegrate, while its counterparts still standing in Egypt have survived nearly unscathed beneath the desert sun for almost 4000 years (Fig. 8-2).

That brief story makes the point that climate is one of the leading factors in determining the rate and manner in which rocks disintegrate or decompose, or, as we say, *weather* —using the word in about the same way we do when we speak of a weather-beaten face. Another critical factor in determining the effectiveness of weathering is the kind of rock that is exposed to atmospheric attack. Evidence for that can be found in New England graveyards, where slate headstones carrying the salty epitaphs beloved by some of our forebears survive from the 1700s, while the words carved on limestone or marble markers of much more recent vintage may be partly or wholly obliterated (Fig. 8-3). We can use data derived from tombstone studies to help us in selecting the most durable rock for building construction in various areas. For the most part, carefully selected rocks will last for the life-span of the building. Yet the influence of air pollution, especially in industrial areas, is greatly accelerating the natural rate of rock decay. In parts of Europe, for example, centuries of weathering only slightly modified some fine stone sculptures, whereas the past 100 years, in which

Fig. 8-2 A. (left) The surface of this obelisk, still standing in the desert at Karnak, Egypt, is scarcely marred by exposure to weathering processes over four millennia. *Jean B. Thorpe* **B.** (right) The obelisk of Thothmes III, from the temple of Heliopolis, Egypt, now in Central Park, New York. The lower part of the granitic column shows a loss of detail due to weathering. The monument was brought to New York in 1879, where most of the weathering took place in the next few years. *Metropolitan Museum of Art*

industrialization has been quite intense, have almost destroyed the same works of art (Fig. 8-4). Yet weathering can transform rocks into works of art (Fig. 8-5), or add stark beauty to the landscape (Fig. 8-6).

Most observant travelers have probably noticed what a difference there is in soils in various parts of the world. In humid areas, where all traces of the original rock structure have been obliterated, the soil may be dark colored at the

(Above) Weathering and erosion of sedimentary rocks has etched out these picturesque forms in Bryce Canyon National Park, Utah. The red color is the result of tiny crystals of hematite throughout the sediment after deposition. *National Park Service* (Below) Remarkable Rocks, Kangaroo Island, Australia. Weathering of a uniform early Paleozoic granite has produced extreme pitting. *William C. Bradley*

(Above) Weathering and erosion over millions of years carved these formations in Monument Valley, Arizona, the vestiges of once-continuous sandstone beds. *Edwin E. Larson* (Below) Strongly developed soil formed in a pre-Wisconsin stream deposit in western Nevada. The profile consists of a dark A Horizon, a reddish B Horizon, and a slightly oxidized C Horizon. *Peter W. Birkeland*

surface, grading to reddish hues at greater depth. In contrast, soils in dry regions are quite light colored and thin, and the rock structure is readily visible. Climate and vegetation have long been known to explain regional variations in soils, but rock type and the length of time of formation are major influences as well.

Soil is perhaps the most valuable mineral resource on earth. Without it life such as we know it would be impossible, and needless to say, it is a resource rapidly gaining in importance in the face of an expanding world population. Yet in many rapidly urbanizing areas, housing tracts and industrial communities are built on the low-lying flat ground that also harbors our most productive agricultural lands. Common sense would dictate that such lands should remain agricultural and that any construction should take place on adjacent land with less productive soils. Although some land-use planning is directed toward that goal, it is too slow in coming, and all too often it is simply ignored. While we could survive without a number of other substances such as gold or diamonds—which are admittedly more attractive than plain, ordinary dirt—we would never make it without the latter.

In this chapter we shall examine the various

Fig. 8-3 A. (left) Weathered marble tombstone in a cemetery in Boulder, Colorado. A little more than 70 years of weathering have slightly roughened the original polished surface and rounded the corners of the stone. Lichens growing on the rock no doubt speed up the weathering process. *Jan Robertson* **B.** (right) Weathered limestone grave marker in a cemetery in New Braunfels, Texas. Ninety years of weathering have nearly obliterated the inscriptions. *Pauline Baker*

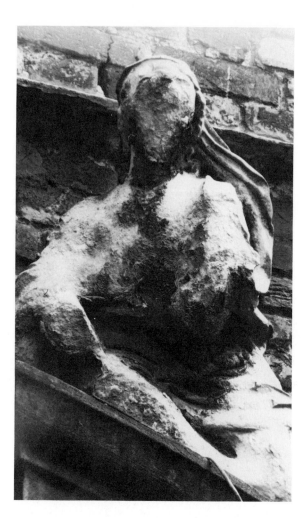

Fig. 8-4. Decay of a seventeenth-century sandstone sculpture from the Rhein-Ruhr region, Germany. **A.** (left) The sculpture as it appeared in 1908. **B.** (right) The same sculpture in 1969. **C.** (below) A plot of the amount of decay with time. *Landesdenkmalamt Westfalen-Lippe, Münich*

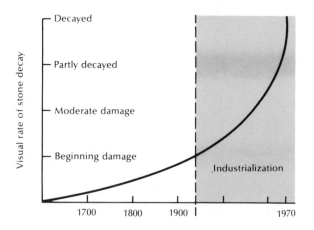

processes that cause rock to weather and to form soils. We shall also look at the overall influence of the environment on the kind of soil that forms and then review some of the practical reasons that cause geologists and soil scientists to study soils so intently.

WEATHERING

Most rocks found in the top several meters of the earth's crust are exposed to physical, chemical, and biological conditions much different from those prevailing at the time the rocks were formed. Because of the interaction of those conditions, the rock gradually changes into soil-like material. Collectively the changes are called *weathering*, and two main types are recognized: chemical and mechanical. In humid, warm regions where vegetation flourishes and organic acids are abundant, *chemical weathering* is dominant and rocks are prone to decompose, or to decay. In harsher climates where frost action may dominate, rocks break up mechanically, or disintegrate,

Fig. 8-5. Chemical weathering of a limestone breccia has produced this artistic rock displayed in the Imperial Palace, Peking, China. The sculpture with its carved base dates from the Ming Dynasty (1368–1244). *H.E. Malde, USGS*

Fig. 8-6. Different rates of weathering were responsible for carving Balanced Rock in the Garden of the Gods, Colorado. *Photographed about 1880 by W.H. Jackson*

without undergoing much chemical alteration; and the process is known as *mechanical weathering*. When rocks decompose, they are changed into substances with quite different chemical compositions and physical properties than those of the original rock. In contrast, if rocks disintegrate mechanically, they break up into smaller fragments, much as if they had been struck a hammer blow. There are few areas where only chemical weathering or only mechanical weathering operates to the exclu-

sion of the other process, but there are many areas where one or the other rules.

Mechanical weathering

Some aspects of mechanical weathering are irritatingly familiar to all of us, such as the wedging apart of sidewalks, foundations, and walls by the roots of grass, trees, and shrubs. The same thing goes on in the mountains, where a common sight high on the slopes is an isolated pine clinging to a sheer granite ledge. With no soil in which to take hold, the tree's roots succeed in forcing their way into crevices, and root-volume expansion with growth tends to force the rocks still farther apart. The process is much like the one known millennia ago to Egyptian slaves, who pried out granite blocks for obelisks using water-soaked wooden wedges.

Almost all rocks are cut by cracks, large and small—sometimes as closely spaced as a fraction of a centimeter; at other times many meters apart, as they are in the stupendous cliff of El Capitan in California's Yosemite Valley. Such cracks are called *joints,* and they provide an ideal path for roots and organic acid-bearing waters to penetrate far into the rocks and cause weathering (Fig. 8-7).

Fig. 8-7. Vertical and horizontal joints in granite in the Sierra Nevada, California. The joints provide avenues for moisture (which may aid in mechanical disintegration when it turns to ice) and also enable plants to send down their roots, which wedge blocks apart. *Cedric Wright Collection, The Sierra Club*

Fig. 8-8 A. Angular blocks produced by frost wedging of well-jointed granitic rock in the Sierra Nevada. *Cedric Wright Collection, The Sierra Club*

Freezing and thawing Water is a highly unusual substance. The property of the greatest significance to us is the expansion water undergoes when it crystallizes to ice. The expansion amounts to about 9 per cent by volume. Should water freeze in a confined space, it is capable of delivering an enormous outward pressure against its containing walls—as anyone who has glumly contemplated a cracked engine block or ruptured radiator knows all too well. Few people realize, though, how great the force actually is. Under confined conditions at

$-22°C$ $(-7.6°F)$, pressures can reach 2100 kg/cm² (4.3 million lb/ft²). No wonder the need for repairs in the wake of water frozen in the plumbing can devastate a household budget!

For such great pressure to build up in nature there must be a completely enclosed system with no air—a condition seldom realized in an environment such as a crack in a rock. Consequently, the pressure created by the freezing of water in rocks is less than the above figure but high enough to sunder most rocks exposed in high mountains. Water in an open

Fig. 8-8 (cont'd.). **B.** The summit of Mount Whitney in California is composed of a large field of loosened granite boulders that rest on closely jointed bedrock. Most gently sloping mountain crests above the timberline are blanketed with shattered blocks resulting from frost wedging. *Tom Ross*

crevice freezes from the top down and thus is sealed in with a cover of ice. Then, if confined in the lower part of the crevice, it can act as a wedge to spring rocks apart along planes of weakness such as joints or bedding planes. The process is called *frost wedging* and is thought to reach its peak effectiveness where temperatures rise above the freezing point by day and drop below it by night. It is thought to be a rather prominent weathering process near and above timber line on high mountains. As a result, summit uplands may be carpeted with frost-shattered angular joint blocks (Fig. 8-8). In contrast, if temperatures remain far below freezing for much of the year, as in some high latitude polar areas, frost wedging proceeds at a much slower rate.

Not all of the block fields at high altitudes or in the high latitudes have formed under present-day frost wedging. Indeed, many fields seem to be inactive, as shown by the weathered character of the blocks, the lichen that grows on them, and the presence of a well-developed cover of vegetation between them. Such fields were formed some time in the past and probably are a testimony to more rigorous mountain or polar climates in ice-free areas during former ice ages.

Salt crystal growth A mechanical weathering process somewhat similar to frost wedging is associated with the growth of salt crystals in rock. In arid regions, ground and soil water, as well as water within rock pores and cracks, commonly contains dissolved salts in ionic form. As the water evaporates, the salt ions are left in the remaining water and their concentration increases. In time, the concentration reaches a point at which salt minerals crystallize from the solution. Pressures accompanying the crystallization can be quite great—certainly great enough to dislodge individual minerals or flakes of rock or to shatter objects (Fig. 8-9).

Another salt-related mechanical weathering process is *hydration,* or the volume expansion of salt minerals when water is added to their crystal structure. If the salts are part of the rock, the pressures generated can break up the rock. An excellent example is the weathering of the obelisk in New York (see Fig. 8-2). Before it was brought to the United States the monolith rested on its side in Egypt for about 500 years. During that time, salt-laden ground water penetrated the column, and salt minerals crystallized out as the ground water evaporated into the hot desert air. But in the humid climate of New York, the same minerals absorbed the water from the atmosphere and in so doing expanded in volume. The expansion from within caused considerable mechanical weathering in the form of flaking of surface fragments over a short period of time. Freezing and thawing also contributed to the weathering of the obelisk.

Surface unloading In many parts of the world the rock close to the surface is cut by joints that more or less parallel the surface, giving it the appearance of an onion skin or a giant's staircase (Fig. 8-10). Several processes, collectively called *exfoliation* from the Latin *exfoliatus*—stripped of leaves—or sheeting are

Fig. 8-9. Telephone pole damaged by the crystallization of salts, Bonneville Salt Flats, Utah. Saline ground water is drawn upward in the pole, and, as the water evaporates, the salts crystallize, shattering the wood. *W.C. Bradley*

Fig. 8-10 A. (above) Exfoliation sheets show well in Independence Rock, Wyoming, a noted landmark for early travelers in the area. *W. H. Jackson, USGS* **B.** (left) Exfoliation in sandstone, Arches National Park, Utah. The width of the rock shown is about 0.5 m (1.6 ft). *Jesse Kumin* © *1976*

Fig. 8-11. Granite domes are well developed in Yosemite National Park, California. Perhaps the most famous of them is Half Dome, the dome with the vertical cliff on one side (lower center). *USGS*

called upon to explain the onion-layered appearance of many rocks. A common origin is the upward expansion of rock as an overlying or confining rock burden is removed. At depth, the rock is under high confining pressures equivalent to the weight of the overlying mass. As erosion removes the overlying rock, the remaining rock can expand—usually upward or toward the valley walls. The release of pressure results in joints oriented at right angles to the direction of the release, hence they usually parallel the land surface. Many of the rock domes of the world are thought to have been formed in that way (Fig. 8-11). Workers in mines and quarries know the expansive properties of rock all too well. Not uncommonly, as new rock faces are exposed the pressure release is instantaneous, and the resulting rock bursts send dangerous missiles flying through the air, often threatening the safety of the miners.

Temperature changes A generation ago textbooks made much of rocks disintegrating as a result of alternate expansion and contraction induced by severe temperature changes. The favorite locale for such performances was the desert. There, according to most versions of the story, rocks expanded drastically under the noonday sun and contracted sharply with the falling temperature at night. Presumably the dimensional changes were greatest on the surface of a rock and least in its interior, because rocks are such notoriously poor conductors of heat. The result could be exfoliation on a small scale.

There is no doubt about the existence of exfoliated rocks; their number truly is legion, but there is uncertainty about the way in which they are formed. The peeling off of concentric rings of heated surface layers from a cooler interior by differential expansion is an appealing solution, but how can it be explained then, for example, that, in the Sahara and the Arabian desert many stone monuments and buildings have survived for 4000 years with scarcely any blurring of their inscriptions?

Perhaps the most conclusive evidence that temperature changes alone are incapable of disrupting rocks comes from an experiment made by the American geologist D. T. Griggs. He alternately heated and cooled the surface of a highly polished block of granite—for five minutes the heat was on and then for ten minutes a fan cooled the block. The process subjected the surface to a temperature range of 110° and was repeated 89,400 times, or the equivalent of 244 years of weathering, should each of the 15-minute cycles be considered a day. But even in the fiercest desert the diurnal range is less than 100°; if we were to consider the experiment in more realistic terms, 1000 years of actual weathering may have been more closely approximated.

What happened as a result of the punishment the rock received? Nothing. The surface remained unblemished, retaining its original bright polish throughout the entire ordeal.

Griggs then produced a little rain, as it were, by introducing a fine spray of water during the cooling cycle. Water was used for only ten days (the equivalent of two and a half years of weathering), and in that brief time a number of notable changes occurred. The granite lost its polish, the surface of feldspar crystals clouded up, and exfoliation cracks started to appear. All the changes occurred in the equivalent of two and a half years, as compared with the absence of visible results after the laboratory equivalent of some ten centuries of total aridity and extreme temperature ranges.

That isolated experiment supports observations made in Egypt by an American geologist, Barton. He noticed that almost no discernible change was visible on granite inscriptions that faced the sun, while those that were in the shade, and thus remained relatively damp, showed much more spalling of rock surfaces and hieroglyphs. An alternative hypothesis, however, states that in the hot deserts the amount of time required for weathering due to temperature changes is more than the amount represented by the age of the monuments Barton observed in the field or more than that

Fig. 8-12. Exfoliation of granite that took place during a 1976 forest fire near Boulder, Colorado. High temperatures associated with the fire caused the thin white shell of rock to expand and peel off. *Scott Burns*

represented by the duration of Griggs's laboratory experiment.

The conclusion of many geologists is that temperature changes by themselves are incapable of causing rocks to exfoliate, and that water plays an important role in the process. A plausible explanation is that in many deserts, no matter how arid they may appear to be, a little water may be available from sporadic showers or from nocturnal dew. When the water combines chemically with the more susceptible minerals in a rock, they swell. It is the increase in volume that may give the needed shove to lift off the outer layers of a rock in concentric shells. To summarize, exfoliation appears to be essentially a mechanical or disintegrative process, accomplished, however, by chemical alteration.

High temperatures associated with fires also seem capable of bringing about exfoliation, as witness the exfoliated rocks around campfire sites. In studies throughout the western mountains of North America, surface rocks within forested areas commonly show signs of exfoliation, probably the result of occasional forest fires, whereas rocks above timberline do not (Fig. 8-12).

Chemical weathering

Chemical weathering takes place in all environments but is dominant in hot and humid lands where temperatures are high, a large amount of water is available, and vegetation flourishes. Organic acids, which are potent agents of rock decay, are readily generated and rocks that can stand up to their onslaught are rare. Carbonic acid is a common organic acid, and it results from a union of water and carbon dioxide:

$$H_2O + CO_2 = H_2CO_3$$
(water) (carbon dioxide) (carbonic acid)

Generally water is readily available from atmospheric sources, and the carbon dioxide is derived partly from atmospheric sources and partly from root respiration and the decay of organic matter. Although relatively weak as acids go, carbonic acid is common in most natural environments.

Of the manifold processes involved in chemical weathering, three of the most important are *solution, oxidation,* and *hydrolysis.*

Solution Solution is perhaps the easiest process of chemical weathering to visualize because a rock may literally dissolve away, much like a sugar cube in coffee, but at a slower rate (Fig. 8-13). Limestone is especially susceptible to attack by solution, as shown by the following equation:

$$CaCO_3 + H_2CO_3 = Ca^{2+} + 2HCO_3^-$$
(calcite) (carbonic acid) (calcium ion) (bicarbonate ion)

What the equation says is that a solution containing carbonic acid reacts with calcite, the chief mineral in limestone, to form two ions. Those ions are removed from the site of weathering by percolating water. Thus, where a layer of limestone may once have been, there may well remain nothing, because the calcite has been completely dissolved. The process ex-

plains the profusion of caverns, underground channels, and disappearing rivers in limestone regions. Indeed, the growth of such underground voids sometimes leads to the sudden collapse of the overlying rock into a cavern (see Chap. 15).

Oxidation Rusting is a process familiar to most of us. In anything but the most severe climates, such as the central Antarctic ice sheet, all unprotected objects made of iron will rust away within a lifetime. In a rainy tropical climate the struggle to maintain steel bridges, ships, rails, and automobiles is a relentless one.

Most rocks contain some iron-bearing minerals. When they are exposed to atmospheric attack, like an old Volkswagen frame in an auto graveyard, they rust. The rocks, originally gray, are stained a wide variety of colors, such as red, yellow, orange, or red-brown when weathering goes on in an environment with ample oxygen. The equation describing the process is:

$$4FeO + O_2 = 2Fe_2O_3$$
ferrous iron oxide oxygen ferric iron oxide
(gray-green)   (rust-colored)

Indeed, discoloration of rock to yellowish-brown and red colors is one of the first visible signs of chemical weathering. It is thought that the exact color is determined by the kind of iron oxide that forms.

Hydrolysis The common rock-forming minerals weather by a process called *hydrolysis.*

Granite   Limestone   Gypsum

Fig. 8-13. In an experiment to demonstrate chemical weathering, acid water was dripped on these rocks for six months. Gypsum (right) weathers the most rapidly, and limestone (center) shows substantial weathering. In contrast, granite (left) is more resistant to weathering, and the effects of weathering are barely discernible. *Jan Robertson*

The by-product of hydrolysis is a substance much different from the weathering minerals. The weathering of feldspar (an aluminum-silicate mineral) is a common example of hydrolysis and can be shown by the following equation:

$$2KAlSi_3O_8 + 2H^+ + 9H_2O$$

feldspar   acid   water

$$= H_4Al_2Si_2O_9 + 4H_4SiO_4 + 2K^+$$

clay mineral   silicic   potassium
(kaolinite)   acid   ion

Feldspar eventually breaks down completely in the presence of carbonic acid, and the aluminum and some of the silicon combine with hydrogen from the acid to form a clay mineral. The silicic acid and potassium ions remain in solution, and may be carried away, leaving only the clay mineral behind. Some of the potassium, however, might be absorbed by plants or enter into combination with specific clay minerals.

Clay minerals—of which there are a great variety—are so small that they can be identified only by X-ray methods. The kind of clay that forms is influenced by the environment. Kaolinite, the clay mineral formed by the reaction shown in the above equation, commonly forms in humid climates where large quantities of water move through the soil, leaching from it the ions released by weathering. In dry regions, where there is little leaching, the most common clay mineral is montmorillonite (Fig. 8-14). Clays vary in their physical and chemical properties. Some react well to heat and are used in pottery and ceramics, some are especially useful as muds used in drilling wells, and others enhance the fertility of soils. In-

Fig. 8-14 A. (left) Montmorillonite grains as seen under the scanning electron microscope (SEM), formed from the alteration of silicate minerals at some depth in ancient sediments that once filled a basin in Baja California, Mexico. The magnification is 2200 times. **B.** (right) SEM photograph of montmorillonite grains that formed from the alteration of the hornblende grain on which they rest. Note that the surface of the hornblende grain is pitted, a result of chemical weathering. Magnification is 1100 times. *T. R. Walker*

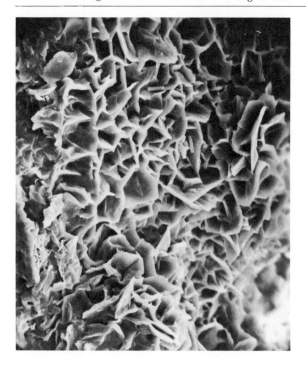
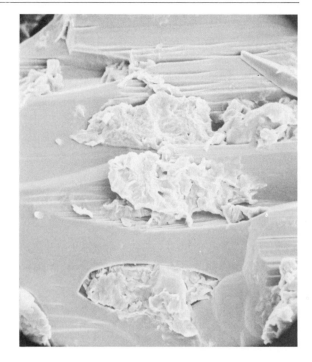

deed, some even are sold for facial mudpacks, as seen in a recent ad extolling the virtues of a montmorillonite skin treatment!

Granite weathering—an example of the hazy boundary between mechanical and chemical weathering

As mentioned earlier, the boundary between mechanical and chemical weathering is not always clear-cut. The weathering of granite illustrates the point. In many deep road-cuts, small fragments of loose and broken granite (mainly the size of the individual mineral grains) can be seen. Because evidence of chemical weathering—notably the presence of clay minerals and iron oxide colors—may be missing, such weathering has been commonly ascribed to a mechanical process. Recently, however, careful X-ray studies of the minerals in granite indicate that the most common alteration is a subtle change of the biotites to a very similar mineral of greater volume. Apparently the little water that does penetrate granite first reacts with biotite, changing it ever so slightly. The accompanying expansion of all the biotites produces enough internal pressure to break up the rock. Thus, volume expansion due to a subtle form of chemical weathering mechanically shatters the rock.

Research in the western United States suggests that about 50,000 years of weathering are needed to bring about such a change—an example of the snail's pace at which weathering takes place. Reactions that proceed so slowly are extremely difficult to duplicate in the laboratory. Some processes can be speeded up (as in Griggs's experiments), but it is difficult to guess how closely they approximate what actually goes on in nature.

Relative weathering rates of minerals and rocks

Not all minerals and rocks weather at the same rate, as shown by the study of tombstones. Examples of *differential weathering* are numerous (Fig. 8-15). Rocks that weather more rapidly erode more rapidly, so it is not uncommon to find that the less resistant rocks form slopes and low areas, whereas the more resistant rocks form prominent ledges or cliffs (Fig. 8-16). Many of the common rock-forming minerals are relatively susceptible to chemical weathering. Calcite, for example, weathers the most rapidly because it dissolves so readily in water. Silicate and aluminosilicate minerals, however, consist of more tightly bonded atoms and thus are better able to resist weathering. Yet even they weather at different rates (Fig. 8-17). Quartz and feldspar appear to be the most resistant to weathering, and olivine and calcium-plagioclase the least resistant. The sturdiness of the former helps to explain the predominance of quartz grains in many sandstones. With repeated cycles of weathering and transportation, the durable quartz grains persist whereas the less durable grains weather away, often forming clay minerals and soluble ions.

Rocks weather chemically according to the rate at which their constituent minerals weather. Hence granite, because it contains an assortment of resistant minerals, chemically weathers much more slowly than gabbro (Fig. 8-17). Grain size also comes into play since fine-grained rocks weather more slowly than coarse-grained rocks of the same mineral composition.

SOIL

The product of weathering can be rearranged into layered materials that are quite different physically, chemically, and biologically from the parent rock. That material is soil. Because soil is so fundamental to life, it has been studied intensively for more than a century. Leaders in the study have been the Russians V. V. Dokuchaev (1846–1903) and K. D. Glinka (1867–1927), whereas in the United States E. W. Hilgard (1833–1916), C. F. Marbut (1863–1935), and Hans Jenny have carried out the basic research. In the United States, the emphasis of study is on soil genesis, classification, and the relationship between soils and environment, as well as crop production. Most recently there

Fig. 8-15 A. Differential weathering produces rows of alcoves in sandstone layers in these cliffs in Arches National Park, Utah.

has been a trend to use soil data to help solve a multitude of environmental problems.

Many soils consist of three basic layers, or *horizons*, the sum total of which comprises the *soil profile* (Fig. 8-18). At the surface is the *A horizon*, a dark-colored layer rich in decomposed organic matter, or humus, derived from the decomposition of surface vegetative debris and roots. Beneath the A horizon is the *B horizon*, which characteristically displays the greatest amount of weathering in the profile. The B horizon is enriched in both clay and iron oxides derived from the weathering of minerals contained within the layer itself and from the A horizon above. Because of the iron-oxide enrichment, B horizons form the brownest or reddest layer within each profile. Beneath the B horizon is the *C horizon;* then comes the parent material, which is thought to have been present in the position of the soil profile before soil formation. The C horizon is only slightly altered parent material.

Factors of soil formation

From the early work of Dokuchaev and Hilgard, it became apparent that soils differ from place to place because environmental conditions vary from place to place. But it was the work of Hans Jenny in the 1930s and 1940s that made possible a clear understanding of the en-

vironmental impact on soils. Five factors are generally considered in describing the soil environment: (1) climate, (2) vegetation, (3) time, (4) parent material, and (5) topographic position. Because the influence of topography is mainly of local importance, we will not discuss its effect on soil formation.

Climate and vegetation Climate has the greatest impact on soil properties from place to place, and, because the effects of vegetation

Fig. 8-15 (cont'd.). **B.** Indians took advantage of differential weathering in sandstone cliffs by building their dwellings in places that offered protection from the elements and hostile neighbors. These structures in Mesa Verde National Park, Colorado, were inhabited from about A.D. 1200 to 1300 and then abandoned for unknown reasons. *Jesse Kumin* © *1976*

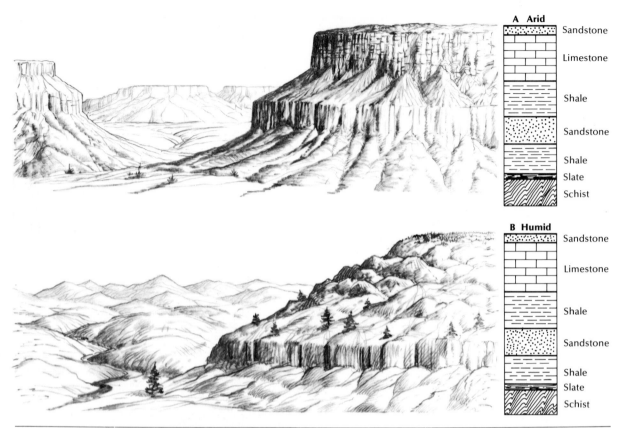

A Arid

- Sandstone
- Limestone
- Shale
- Sandstone
- Shale
- Slate
- Schist

B Humid

- Sandstone
- Limestone
- Shale
- Sandstone
- Shale
- Slate
- Schist

Fig. 8-16. Rocks vary in their resistance to weathering and subsequent erosion. Whether a rock forms a steep cliff or a gentle slope is partly dependent upon climate, however. In an arid climate (**A**), limestone and sandstone are cliff-formers and shale is a slope-former, often covered by talus. In a humid climate (**B**), sandstone also is a cliff-former, but limestone weathers by solution to form irregular slopes. Again, shale is a slope-former, often covered by a thick soil.

Fig. 8-17. Relative rates of weathering of various common minerals and rocks.

Relative rate of mineral weathering		Relative rate of rock weathering	
		Coarse Grained	Fine Grained
Quartz		Granite	Rhyolite
Feldspar	Na-plagioclase		
Biotite		Diorite	Andesite
Hornblende			
Pyroxene	Ca-plagioclase	Gabbro	Basalt
Olivine			

Most resistant

Least resistant

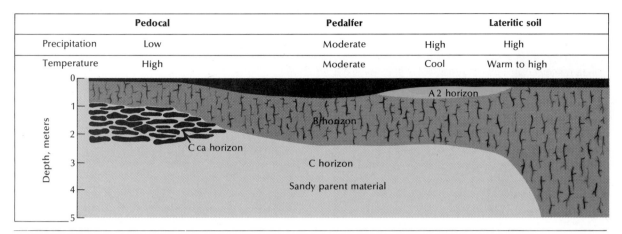

	Pedocal		Pedalfer		Lateritic soil
Precipitation	Low		Moderate	High	High
Temperature	High		Moderate	Cool	Warm to high

Fig. 8-18. Transect from a dry to humid environment showing soil profiles that are characteristic of each environment on fairly old landscapes. The desert profile could be found in the valleys of the western United States, the humid region profiles in the Great Lakes region and the Northeast, and the warm, humid region profile in the Southeast.

are not easily separated from those of climate, both factors will be covered here.

Pedalfers are the common soils in humid temperate regions under either grass or forest vegetation. The word is derived from the Greek *pedon* for "ground" and the symbols Al and Fe for the aluminum and iron such soils contain. Pedalfers consist of relatively thick A, B, and C horizons with a fairly high content of organic matter in the A horizon (Figs. 8-18 and 8-19). Mainly in the cooler climates toward the northern limit of trees but also in some warm humid climates a special kind of pedalfer known as a *podzol* forms (Figs. 8-18 and 8-19B). Podzols are characterized by a whitish layer called the *A2 horizon* that lies between the A and B horizons, and from which most of the iron oxides have been removed by downward-percolating waters.

On old landscapes in humid warm climates, especially tropical climates, we see the end product of extreme weathering—the *lateritic soils* (Figs. 8-18 and 8-20). For the term laterite we are indebted to Buchanan Hamilton, an observant Scotsman who, while traveling in India in 1807, was greatly impressed by the ease with which the red-brown tropical clay could be

transformed into a building material. Hindu laborers simply excavated the clay and shaped it into bricks that needed only case-hardening in the sun before they could be used. The bricks that Hamilton described were much like the sun-dried blocks, or *adobe*, of the arid Southwest, even though both are very different in origin.

Because the tropical clay could be used readily as a construction material, Hamilton called it *laterite* (from the Latin word *latere*, or brick). Its origin was a source of wonderment to him. As a construction material, laterite has served to build enduring monuments; among others, much of the long-forgotten city of Angkor Wat in Cambodia is built of laterite (Fig. 8-21). Despite a humid tropical climate the buildings of that city not damaged by warfare in recent years, are well preserved, alone an indication that laterite consists of a virtually insoluble material.

Clays that harden into actual laterites are few. But lateritic soils, deep, highly weathered, and reddish, are widespread in tropical climates. A lateritic soil can be regarded as a "soil skeleton" since the soluble elements, such as calcium, sodium, and potassium, have been

A horizon Dark brown to black due to high amount of organic matter, and well aggregated so that rainfall readily enters soil with little surface runoff.

B horizon Brown to reddish brown due to iron oxides, and high clay content as shown by vertical shrinkage cracks that develop on drying. Grades downward to slightly altered C horizon material.

Fig. 8-19. Two different kinds of pedalfers: **A.** (above) Grassland soil formed from loess in Iowa. The section shown is about a meter deep. **B.** (opposite) Forested soil (podzol) formed from glacial till in Québec; about half a meter thick. *Roy W. Simonson*

leached out, and even such a relatively insoluble substance as silica (SiO_2) has been removed (Fig. 8-22). What remain are mainly iron oxide (Fe_2O_3), derived from the weathering of iron-bearing minerals, and quartz, if the parent rock contained any. It is the crystallization of the iron oxides that form the bricks, or true laterite. Should the rocks from which lateritic soils are derived have a high content of aluminum, then bauxite ($Al_2O_3 \cdot 2H_2O$), a chief ore of aluminum, can form.

In some parts of the tropics, soils may be high enough in iron content to be mined as ore. Cuba's iron mines and those on the Surigao Peninsula on Mindanao are examples. Bauxite shares a nearly common origin, except that it is formed from the weathering of rocks

richer in aluminum than in iron. The bauxite ores of Little Rock, Arkansas were formed under climatic conditions that probably were very much like those of the tropical savanna today. Most of the aluminum ore now processed in North America no longer comes from Arkansas, but from the high-alumina clays of Surinam, Jamaica, and other South American and Caribbean lands.

A much different soil forms in semi-arid to arid regions. It is named a *pedocal* because it contains an accumulation of calcium carbonate at some depth (Figs. 8-18 and 8-22). Because rainfall is slight and vegetation scanty, the A horizon is thin and not too rich in organic matter, and the common clay mineral in the B horizon is montmorillonite. Beneath the B

A horizon Dark brown to black due to high amount of organic matter.

A2 horizon Light colored because both iron and humus have been removed from the layer by downward moving water.

B horizon Dark brown to reddish because the iron and humus, removed from the A2 horizon, accumulate here. Horizon has less humus with depth, and red color decreases in intensity.

horizon, at about the depth to which the annual rainfall penetrates, is a white accumulation of calcium carbonate ($CaCO_3$) known as a *Cca horizon*. On some old landscapes the amount of calcium carbonate is so high that it forms a highly indurated layer known as the *K horizon*, or *caliche*. The carbonate (CO_3) in calcium carbonate is derived from the reaction of carbon dioxide (CO_2) and water, whereas the calcium ions are derived from the weathering of calcium-bearing minerals, or from atmospheric or ground-water sources.

The three main types of soil—pedalfers, lateritic soils, and pedocals are all found in the United States. The 100th Meridian is the approximate boundary between pedalfers to the east and pedocals to the west. In the West,

however, pedocals occur only in the dry basins; the adjacent mountains, in contrast, receive enough rainfall to harbor pedalfers. In the East, podzols occur in the Northeast and in the Great Lakes area, and lateritic soils are widespread in the Southeast, south of the glacial boundary.

Time All of the key properties of soil take considerable time to form (Fig. 8-23). The A horizon forms most rapidly—several centuries in humid environments to several thousand years in arctic and alpine environments. B horizons take much longer to form because minerals weather so slowly. Discoloration due to weathering of iron-bearing minerals shows up in about 1000 years, and the reddest colors

A horizon

B horizon

Fig. 8-20 A. (above) Roadcut about 7 m (23 ft) deep near Rio de Janeiro, Brazil, exposes highly weathered lateritic soil formed from gneiss. The rock is so thoroughly decomposed that it can be readily cut with a knife or a shovel, yet quartz veins (diagonal white lines) are essentially unaltered. **B.** (left) Lateritic soil formed from gneiss near Rio de Janeiro. The A horizon is dark colored and enriched in organic matter. Below that for several meters is a red, highly decomposed B horizon. *Roy W. Simonson*

Fig. 8-21. The building blocks of laterite that compose this temple at Angkor Wat, Cambodia, are highly resistant to weathering because they are composed chiefly of residual iron oxide, all the other original materials having been removed by chemical weathering. In contrast, the sandstone columns and statues exhibit considerable weathering. *Leonard Palmer*

may require 100,000 or more years to develop. Clay-rich B horizons, if formed from sandy parent materials, require at least 10,000 years for initial detection and more than 100,000 years for maximum development. In arid regions, the strongly cemented K horizons may take 100,000 to 500,000 years to form. Deep lateritic soils take the longest to form, and our best guess of how long is in the order of one or more million years. No wonder, then, that we should protect our soils from erosion! Once they are lost they are probably lost forever.

A horizon Thin and light colored due to small amount of organic matter. Most of the A horizon has been eroded.

B horizon Reddish brown due to the kind of iron oxides that form at high temperatures. The clay content is high, as shown by shrinkage cracks.

K horizon (caliche) Contains 50 per cent or more $CaCO_3$, and can reach the hardness of concrete, especially between 0.30 and 1 m. Grades downward to the parent material at about 2.2 m.

Fig. 8-22. Desert soil (pedocal) formed from alluvium in New Mexico. *L.H. Gile*

Fig. 8-23. Time necessary in the western United States to develop diagnostic soil horizons from a sandy parent material.

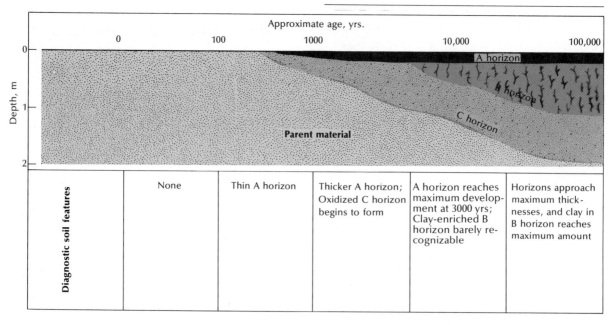

| | None | Thin A horizon | Thicker A horizon; Oxidized C horizon begins to form | A horizon reaches maximum development at 3000 yrs; Clay-enriched B horizon barely recognizable | Horizons approach maximum thicknesses, and clay in B horizon reaches maximum amount |

Fig. 8-24 A. (above) Dust storm in Prowers County, Colorado, photographed in the 1930s. **B.** (below) Abandoned farmland near Stillwater, Oklahoma, resulting from accelerated erosion. *Soil Conservation Service*

Parent material The influence of parent material depends mainly upon weathering and the rate at which clay is produced. Granite, for example, produces a fairly deep sandy soil because, although biotite weathering breaks the rock down to sand-size particles, the minerals chemically alter very slowly to clay minerals. In contrast, soil on basalt in the same area would be fairly thin because fine-grained rocks weather fairly slowly. What soil there is, however, could be richer in clays because the minerals in basalt are more susceptible to chemical weathering.

Benefits of soil research

We can benefit greatly from the study and mapping of soils. Perhaps some of the benefits seem academic to us, but others are of immediate concern.

The study of soil development on various deposits has aided earth scientists in estimating the age of those deposits. In areas of glacial deposition, for example, soil studies not only helped to develop the idea of multiple glaciations in the United States, but also helped to determine the frequency with which ice ages occur. Armed with such knowledge we might attempt to predict future climatic changes.

We can also estimate the frequency of geologic hazards, such as landslides, rockfalls, and major flooding from the study of soils. For example, areas in which such events took place in the recent past would have poorly developed soils, whereas better developed soils would be associated with areas in which hazards occurred long ago. Soil dating can also provide key data for building sites; and those where nuclear power plants are to be built are probably the most crucial. The installations must be located in areas free of the ground breakage (faulting) associated with earthquakes, and soil studies can aid in detecting recently active faults near a plant site.

Soils are also valuable aids in determining the nature of past climates. Buried pedocals in a humid region tell us that at one time a drier climate must have existed there. In the more arid parts of Australia, pedocals cover young deposits, but on ancient landscapes lateritic soils are found. What a change—from tropical landscape to parched desert in a few million years—a relatively short time in the earth's history. Why did the change occur? We are still developing theories for that and other climatic reversals in various parts of the world. The solution to the problem may help us to predict the future march of climatic change.

Soils vary in productivity and manageability for crop growth. Maps produced by the U.S. Department of Agriculture help us to locate the best—and worst—soils for the production of crops. People involved in land-use planning should consult such maps when laying out subdivisions, attempting to keep the most productive soils for agriculture. Less productive soils should be utilized to the maximum benefit for all. Once a good soil is paved over or covered by a house or factory, it is a resource lost. And when crops are moved to less productive land it only serves to drive the cost of foodstuffs higher and higher. Citizens and politicians should take heed of those arguments.

Although lateritic soils are no longer considered a major problem in tropical agriculture (they make up but 7 per cent of the tropical land area), we still need a better understanding of how they behave under cropping. Such soils are extensively weathered and thus are low in the nutrients necessary to plant growth. Cropping only reduces the nutrients further, and fertilizers are required to restore the supply. Some lateritic soils harden if dried out; once hardened they can no longer be tilled. Intensive research and fitting crops to soils rather than soils to crops will be needed to help solve the problems of food production in tropical areas.

An adequate soil is necessary to protect land from accelerated erosion, as we learned so dramatically in the Dust Bowl in the 1930s (Fig. 8-24). The organic matter, clay particles, and ions in soil have the ability to bind or aggregate the soil particles together to form

Fig. 8-25. Gully developed on a slope in the Coast Range north of San Francisco. Excessive grazing caused increased runoff and gullying. The sharp topographic angle between the gully wall and the original land surface means that gullying was still active when the photograph was taken in 1906. If erosional deepening subsides, the gully walls will assume more gentle slopes. *G. K. Gilbert, USGS*

Fig. 8-26. Manipulation of the land is not the only reason for accelerated erosion. Shown here are gullies formed in only a few months on a hill of ash deposited during the eruption of Parícutin volcano, Mexico. It takes time for vegetation to become established in such material, and thus protect the land from erosion. *K. Segerstrom, USGS*

small clumps or blocks of various sizes (see Fig. 8-19A). Water will penetrate such a sponge-like structure more readily than it will run off the surface. And the more water that sinks into the ground, the less the runoff and hence the less the surface erosion. The properties of the A horizon are especially critical to the ability of the soil to take up water. If that horizon is removed or compacted, runoff and erosion can be accelerated (Fig. 8-25). And once erosion starts it is very hard to curb (Fig. 8-26).

SUMMARY

1. No rock or sediment is stable at the earth's surface. With time they will alter or weather to products more stable in the surface environment.
2. Two main kinds of weathering are recognized. *Mechanical weathering* is the disintegration of particles to smaller size without appreciable change in either the chemistry or mineralogy of the original material. In

contrast, through *chemical weathering* the chemistry and mineralogy of the original material are changed. Clay minerals found in soils and sediments originate from chemical weathering.

3. Soils are layered bodies consisting of an *A horizon* rich in organic matter overlying a clay-enriched *B horizon* which, in turn, rests on the only slightly altered *C horizon*.

4. Soils vary from site to site for a variety of reasons. The main factors controlling the variation are parent material, climate and vegetation, topographic position, and time. Most well-developed soil profiles have taken thousands or tens of thousands of years to form, and the very deep ones of the warm humid areas may have required a million years to form.

5. Soils can provide much information on the younger geologic history of the earth, and their properties and distribution should be taken into account in all land-use decisions.

SELECTED REFERENCES

Bartelli, L. J., and others, eds., 1966, Soil surveys and land use planning, Soil Science Society of America and American Society of Agronomy, Madison, Wisconsin.

Birkeland, P. W., 1974, Pedology, weathering, and geomorphological research, Oxford University Press, New York.

Bridges, E. M., 1970, World soils, Cambridge University Press, London.

Buckman, H. O., and Brady, N. C., 1969, The nature and properties of soils, The Macmillan Co., London.

Buol, S. W., Hole, F. D., and McCracken, R. J., 1973, Soil genesis and classification, The Iowa State University Press, Ames.

Carter, V. G., and Dale, T., 1974, Topsoil and civilization, University of Oklahoma Press, Norman.

Hunt, C. B., 1972, Geology of soils, W. H. Freeman and Co., San Francisco.

Jenny, Hans, 1941, Factors of soil formation, McGraw-Hill Book Co., New York.

Sanchez, P. A., and Buol, S. W., 1975, Soils of the tropics and the world food crisis, Science, vol. 188, pp. 598–603.

Winkler, E. M., 1973, Stone: Properties, durability in man's environment, Springer-Verlag, New York.

Fig. 9-1. This rockfall from the Flimerstein, Switzerland, occurred on April 10, 1939. It buried not only forests and arable land but also a building and 11 persons, all in a moment of time. *Swissair-Photo*

9

GRAVITY MOVEMENTS AND RELATED GEOLOGIC HAZARDS

On the night of October 9, 1963, a torrent of water, mud, and rocks plunged down a narrow gorge in Italy, shot out across the wide bed of the Piave River and up the mountain slope on the opposite side, completely demolishing the town of Longarone and 2600 inhabitants in it and in adjoining towns (Fig. 9-2). It has been called history's greatest dam disaster, but when it was over, the Vaiont Dam in the narrow gorge was still intact. What could have caused the water in that reservoir, which was not even half-filled, to rise up over the dam and proceed on its destructive course? One clue is that one shoulder of the dam was supported by Monte Toc, nicknamed *la montagna che cammina*—"the mountain that walks"— by the local inhabitants. Despite assurances by engineers regarding the safety of the dam and the expensive efforts that had been made to stabilize its slopes, Monte Toc not only walked that night in October, it galloped. About 240 million m³ (314 million yd³) of mountainside slid instantaneously into the lake behind the dam. The water rose 240 m (787 ft) above its previous level and one great wave rose 91 m (299 ft) above the dam and

dropped into the gorge below. There, constricted by the narrowness of the gorge, the water increased in speed tremendously, and snatching up tons of mud and rocks, raced on its destructive path. It was all over in seven minutes. What went wrong at Vaiont? Surely the dam was strong enough, for it withstood the onslaught of tremendous forces, estimated at 4 million metric tons, and is still standing today. In fact, at that time the dam was the world's second highest (266 m; 873 ft).

California has long been known as earthquake country, but it also is renowned for its landslides (Fig. 9-3). A major one occurred in 1956 and for the following three years in the Palos Verdes Hills near Los Angeles, where a development had been built directly over an old landslide that started to move again. Many homes were totally demolished and the value of property destroyed ran over $10 million. Eventually damage suits of several million dollars were collected against the county. The upshot of that and similar events was that planners and developers began to consider more seriously local geologic settings before attempting to build houses or roads on certain

261

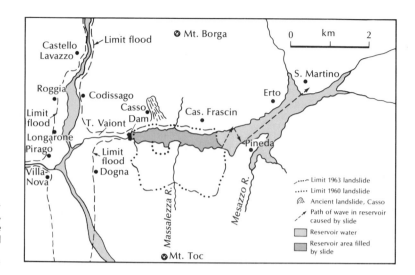

Fig. 9-2. Map of Vaiont Reservoir area, Italy, showing limits of the landslide, the area of the reservoir filled by the slide, and the extent of downstream flooding.

Limit flood
Mt. Borga
0   km   2
Castello Lavazzo
S. Martino
Roggia
Codissago
Erto
Casso
Cas. Frascin
Limit flood
T. Vaiont
Casso Dam
Longarone
Pirago
Pineda
Limit flood
Dogna
Villa-Nova
Massalezza R.
Mesazzo R.
Mt. Toc

····· Limit 1963 landslide
····· Limit 1960 landslide
🌀 Ancient landslide, Casso
⟋ Path of wave in reservoir caused by slide
▨ Reservoir water
▨ Reservoir area filled by slide

Fig. 9-3. Landslide in the Palos Verdes Hills, California. *George Cleveland*

kinds of terrain. The cost of such tragedies always convinces the skeptics; for example, in one two-year period, landslides accounted for $21 million worth of damage in eight San Francisco Bay counties. Much of the work of geologists, especially those working for governmental agencies, involves the recognition of potential landslides as well as other geologic hazards.

Material at the earth's surface, be it weathered or fresh rock, can fail under special circumstances and move downslope under the ubiquitous pull of gravity. Such movements can take place very slowly and with little immediate evidence or happen so quickly and involve so much material that they boggle the mind. Our goal in this chapter is to discuss the major kinds of gravity movements, to assess their causes, and to point out some remedies.

Mass movement is responsible for the downslope transfer of material to rivers, which then act as continuously moving conveyor belts to carry it away. The way in which the walls of the Grand Canyon flare outward from the Colorado River results largely from the gravity transfer of rock fragments and mineral grains downslope to places where the river and its intricate network of tributaries can carry the material out of the Colorado Plateau.

How much or how little material will be shifted downhill by gravity and how rapidly or

Fig. 9-4. Bent tree trunks are good evidence of active creep. The trees are rooted at some depth in material that is not moving or is only slowly moving downslope. Closer to the surface, however, the soil moves more rapidly, pushing against the trunks and bowing them downslope. The trees respond by maintaining vertical growth in their upper parts—hence the curved trunks. In this example, however, G. K. Gilbert, the geologist who took the picture, thought the bending was caused by the pressure of snow.
G.K. Gilbert, USGS

Fig. 9-5. A common example of creep as shown by bent strata in sedimentary rocks. The effects are more pronounced toward the surface because creep is mainly a near-surface phenomenon. *W.C. Bradley*

how slowly it will move are a consequence of many factors, such as (1) climate, (2) nature of the bedrock, (3) vegetation, (4) amount of weathering, (5) steepness of slopes, (6) local relief, (7) the presence or absence of earthquakes. Few landscapes consist solely of bare rock, and most hillslopes show some degree of rounding. Soil cover serves to alleviate the starkness of a rock-dominated landscape, and soil-blanketed and smoothed slopes are an indication that surface-weathered material is not stationary but has a motion of its own.

The transition from gradual, virtually imperceptible movement at one end of the scale to a free-falling mass of rock avalanching down a mountain face and filling valleys with thunderous echoes at the other, is blurred. In fact, such distinctions are nonexistent, since the series is entirely gradational. A practical solu-

Fig. 9-6. The mechanism of creep due to expansion and contraction. Before an expansion-contraction cycle, particles on the original surface rest at (A). On expansion, the surface is elevated, and all particles move upward at right angles to the original surface (A → B). On contraction, the surface more or less assumes its original position as all particles move vertically downward (B → C).

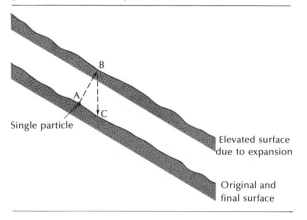

Single particle

Elevated surface due to expansion

Original and final surface

tion to the problem of classification is to divide mass movement into two obvious categories: slow and rapid. The division is highly subjective, however, because people are bound to disagree with the definition of those words. We are constantly confronted, for example, with differences of opinion in controversies involving interpretations of traffic speeds with local authorities.

SLOW MOVEMENT

Creep

This descriptive word is used for the slow, glacier-like movement of shallow soil material downslope. We are likely to be oblivious of such movement, although we may observe with dismay building foundations thrown out of line, power and telephone poles tilted, and sidewalks and retaining walls cracked. Trees even react to creep, as shown by bent trunks (Fig. 9-4). Cuts made in hillsides generally will reveal active creep, commonly shown by bent-rock layering—look carefully for such layers should you ever think of building a hillside house (Fig. 9-5).

Several mechanisms account for creep. One is subtle expansion and contraction of the ground surface, the result of which is an imperceptible movement of material downslope (Fig. 9-6). Expansion and contraction can result from a wet-dry cycle in clayey material, or from freezing and thawing. Another important mechanism is the saturation of the ground with water, such as during storms. Water adds weight to the slope and causes it to lose some of its cohesiveness. Gravity then takes over.

Earthflows

Earthflows are transitional between slow and rapid earth movement. They are more visible than creep, yet slower than mudflows or some landslides. And, although they are usually minor features, some may cover large areas (Fig. 9-7). Earthflows usually have a spoon-shaped sliding surface with a crescent-shaped cliff at the upper end and a tongue-shaped

bulge at the lower end. Thus they differ from creep deposits, which have little or no form and no sharp boundary with material at depth that is not moving. They also differ from true landslides in that the flowing material breaks up internally during movement. Earthflows are most likely to occur when the ground is saturated. Water not only increases the weight of the material but drastically reduces its stability by lowering its resistance to deform and flow.

Some earthflows move quite rapidly, but special circumstances are necessary. They involve special clays, aptly called *quick clays*. Quick clay is composed primarily of flakes of clay minerals, and it has a water content which can often exceed 50 per cent by weight. It is commonly part of the debris deposited on the sea floor adjacent to glaciers, so it is not surprising that in Norway, Sweden, and parts of eastern Canada several such earthflows take place every year (Fig. 9-8). Quick clay has a most amazing and treacherous quality: ordinarily it is a solid capable of supporting 1 kg/cm² (14 lb/in.²) of surface, but the slightest jarring motion immediately turns it into a flowing liquid. When quick-clay layers were originally deposited, generally in salt water, they contained sodium ions that somehow kept the fine particles together. But when the clays are exposed to weathering, the sodium ions are leached out by rainwater and their cohesive effect is lost. Any sudden shock can produce liquefaction. In one slide in Sweden the trigger is believed to have been the hammering of a pile driver. The result was that 32.3 million m³ (42 million yd³) of soil and gravel slid down into the nearby river, carrying with it 31 houses as well as a paved highway and a railroad it picked up along the way. One person was killed, 50 injured, and 300 homes were destroyed in less than three minutes.

The most damaging quick-clay earthflow on record occurred in 1893 in Norway where a 9 km² (3.5 mi²) area was wrecked, killing 120 persons. And we are all familiar with the extensive damage that resulted from the 1964 Alaska earthquake. The city of Anchorage and other coastal towns in Alaska were especially hard-

Source area

Younger, active earthflow

Older, inactive earthflow

Lake San Cristobal

hit, and much of the damage was due to lique-faction of a quick clay. In that case, however, the liquefying shock was provided by the earth-quake.

RAPID MOVEMENT

Landslides

A multiplicity of downslope movement is in-cluded in the broad term "landslide," and no useful purpose is served here by reviewing the many schemes for classification proposed by geologists, engineers, and other specialists. That so many people are concerned indicates in itself the menace that landslides are. The

Fig. 9-7. Slumgullion earthflow in southwestern Colorado. Derived from highly altered volcanic rocks at 3500 m (11,480 ft), the flow has descended to 2500 m (8200 ft), where it has dammed the valley and formed a lake. A younger, active earthflow may be seen advancing over the older, stable one. The rate of movement varies from 6 m (20 ft) per year at midpoint on the flow to less than 1 m (3 ft) per year at its lower end. *C.W. Cross, USGS*

Fig. 9-8. Quick-clay landslide along the South Nation River, Ontario, Canada. The landslide occurred during a severe thunderstorm in May 1971. About 2 to 6 m (6.5 to 20 ft) of silty sand overlies the sensitive clay responsible for the slide. *National Research Council of Canada*

problems they create—and they are consider-able as well as expensive—are largely of our own doing. Without our disturbance of natural slopes, landslides would occur chiefly in re-mote mountainous terrain or on hill slopes underlain by notably unstable rocks. Today, landslides are a problem of increasing mag-nitude as urban areas spread out and the de-mand for high-capacity expressways mounts. Both trends require larger excavations for building foundations and deeper cuts and higher fills for highways. Oversteepening of slopes is a likely cause of ground movement.

A landslide may involve the bedrock alone, or it may be limited to the overlying soil mantle, especially if the latter is deep and water saturated. Usually, however, it involves both soil and rock, and differs from an earth-flow in that the rock remains more or less intact—in earthflows, the moving mass is rather thoroughly broken up internally. The geologist David Varnes recognized two major categories of landslides: (1) *glides* and (2) *slumps*. In a glide, the slippage is dominantly planar; that is, a large mass of rock may become separated from its fellows and glide outward and down-ward along the surface of an inclined bedding plane (Fig. 9-9). The motion of a slump is rotational—usually along a concave-upward slip-plane—so that the upper part of the land-slide is dropped down below the normal ground level and the lower part is bulged above it (Fig. 9-10).

Slumps have a very characteristic form, and since by far the greater number of landslides are variants of that form, some of its details are worth noting (Fig. 9-11). Most such slides start abruptly with a crescent-shaped *scarp*, or cliff, at their head, sometimes known as a *breakaway* scarp (Fig. 9-12). Lower down there may be a number of lesser scarps, which on a plan view almost always appear concave downslope. Between the individual scarps the surface of the slide customarily is tilted or rotated back-ward against the original slope of the ground. The backward-rotated wedges cause more in-stability since they create collecting basins in which small lakes or ponds can form. Another factor that greatly increases instability is the seepage of water along the margins of the slide. The concave-upward slip-plane (surface of rupture) down which the jumbled mass of soil and rocks moves may approximate a cross section of a cylinder whose axis parallels the contour lines on the ground surface, if the slide is sufficiently broad. Otherwise, the surface of rupture is likely to be spoon shaped. The slide may advance downslope from the point where the surface of rupture intersects the ground as a glacier-like lobe of jumbled debris whose surface typically is a chaotic pattern of hummocks and undrained depressions. If the slide moves down a forested slope it often

Fig. 9-9. Glide-type landslide at Point Fermin, California. Although there are minor slump features at the rear of the slide, most of the movement was along sedimentary strata that dips gently seaward. The glide surface is located just below sea level. The maximum average movement was 3 cm (1.2 in.) per week. *John S. Shelton*

Horizontal beds

Fig. 9-10. (above) Cross section of an ancient slump exposed by a recent landslide on the shore of Franklin D. Roosevelt Lake, Washington. Although little surface form is visible, the surface of rupture (A) and rotation of the beds (B) are clearly visible. *F. O. Jones, USGS*

Fig. 9-11. (left) Slump-type landslide blocking Highway 24 near Oakland, California. *USGS*

Backward rotated surface

Main slump scarp

Minor slump scarps

Fig. 9-12. Crescent-shaped scarp at the head of a slump, Madison County, Montana. At the front of the scarp the original surface rotated backward during movement. Tilted trees also are a sure sign of landsliding. *J.R. Stacy, USGS*

creates a desolate scene of broken trunks and trees. The inexorable thrust of the foot of a slump against a building or other structure almost invariably leads to its collapse, and it is the foot that commonly is responsible for blocking canals, highways, railroads, and engulfing other types of excavations.

Among such slumps, the immense ones that closed the Panama Canal at Culebra Cut shortly after it was opened in 1914 and that kept it closed more or less continuously until 1920 are impressive examples. Of the 128 million m³ (167 million yd³) excavated in the Gaillard Cut, landslides made necessary the removal of at least 55.5 million m³ (73 million yd³). Great masses of loose unstable volcanic ash, shale, and sandstone slid on a gently inclined rupture surface toward the canal excavation. One unexpected result was that the bottom of the canal was heaved upward—once as much as 9 m (30 ft)—until what had been the canal bottom appeared as an island in mid-channel.

Conditions favoring landslides All landslides are the result of the forces of gravity acting upon earth materials in an unstable condition or position. Although the movement itself may be extremely rapid (or imperceptibly slow), a landslide does not suddenly spring into being, but rather develops gradually, step by step, with one element triggering another.

Instability does occur naturally but human beings often act as the trigger in an unstable area when they disrupt precarious balances. The natural process of erosion can create an unstable condition by oversteepening a slope, as, for example, when a cliff is undercut. The material making up the upper part of the slope is held in place by the weight of the material at the base of the slope, and when the anchoring material is removed, the upper part is rendered unstable. If erosion wears away the toe of an ancient landslide it could start on the move again.

Highway engineers are particularly sensitive to the problems of oversteepened slopes. In making highway cuts, they are called upon to create slopes steeper than the original ones—a rather perilous business, for the slope must be not only steep but stable as well. Shallow or artificial cuts are avoided because more material must be removed, and that is considerably more costly. Cuts in hillslopes for house foundations can also create stability problems on slopes.

The inclination of rock stratification relative to the ground surface is another common cause of landslides. If layers of rock slant down toward a road-cut or railroad-cut, the rocks may slide along the bedding planes between the layers. Or, in the absence of an artificial cut, a normally stable area may become unstable when a heavy load, such as a building or excavated material, is placed upon it. If layers of rock slant into the hillside, of course, the chance of sliding is usually minimal.

Water is the hidden devil in the ground, for it is the cause of many landslides (Fig. 9-13). When natural drainage conditions are changed, either by natural process or by human activity, the ground-water situation can become potentially dangerous. An unusually high content of water adds considerable weight to the soil or rocks, making it difficult for those containing clay to stick together. Water also acts as a lubricant along potential sliding planes.

Excess water can be added to slopes in a variety of ways—some obvious and some not so obvious. We can start by saying that many natural slopes that are not patently sliding are relatively stable. Human activity can easily tip the balance, however, and usually unbeknown to those involved. In some cases the mere watering of a sloping lawn can trigger a slide by increasing the water content to the critical point, at which the slope becomes unstable. Clear-cutting forests on slopes can have a similar effect. Trees might be thought of as water pumps, returning the water to the atmosphere by transpiration. Remove the pumps, however, and the slopes probably will retain more moisture—an invitation to instability. An artificial reservoir can also be the cause of a landslide; if enough rising water soaks into the banks, they lose much of their strength. Of course, a natural cause such as a climatic

Fig. 9-13. Freshly formed landslide in the Berkeley Hills, California. Although some small ground cracks are thought to have opened during the 1906 earthquake, most of the movement took place during the next rainy season, when the ground was soaked. *G.K. Gilbert, USGS*

change involving greater precipitation can also set landslides in motion. We should be aware of both the natural and artificial causes of such events, because if personal property is damaged the case might easily find its way to court. Putting the finger on the real culprit, however, is no easy task.

What caused the Vaiont Reservoir disaster (see Fig. 9-2), for example? In the first place, the valley is steep-sided, due to relatively recent river downcutting (Fig. 9-14). Engineers commonly seek such narrow gorges for dam sites, however, because less concrete is needed to plug the valley. But at Vaiont, the bedrock is made up of limestone and layers of slippery clay, and the bedding planes are in the worst possible configuration—headed right toward the valley axis. Further, the limestone

itself is rife with underground solution caverns that serve as collection basins for water that helps to saturate the ground. The valley has been the site of other landslides as well; witness the ancient one near Casso and the 1960 slide into the reservoir along its south side. In 1963 the trigger was excess water introduced into an already unstable system. Some seeped in laterally from the reservoir and some was provided by the downpours mentioned in the description at the beginning of the chapter. Creep in the slide area had increased from about 1 cm/wk (0.4 in./wk), the average rate after the reservoir was built, to 80 cm/day (31 in./day) on the day of failure. Even animals grazing on the slopes sensed the potential danger and moved away in time. What could have been done to prevent the disaster? The

Fig. 9-14. Cross section of the valley in which the Vaiont Reservoir is situated, showing the geologic setting and the pre- and post-landslide topography.

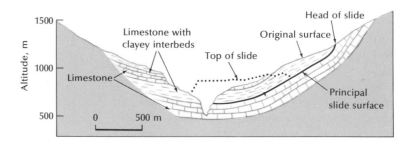

only thing that comes to mind is that the geologic setting of the area could have been studied more carefully. Certainly, sites that in a geologic sense have so many things wrong with them should be avoided. No amount of engineering could have saved the reservoir, at least within acceptable costs.

A side effect of the Vaiont slide was that towns near the head of the reservoir were also damaged. The destruction there, however, was from slide-generated water waves that ricocheted from one valley wall to the other, wreaking havoc in their path (see wave path in Fig. 9-2).

Stabilizing landslides Civil engineers have devised many ingenious methods of controlling unstable slopes, although not enough people are aware that it is easier, cheaper, and safer to apply the methods before, and not after, gravity movements take place. Consolidation of unstable materials is one method of control, and an unusual example of its use was the freezing of surficial materials in a dormant landslide that threatened to start moving again during the construction of Grand Coulee Dam in Washington.

Another common way of stabilizing landslides is to regrade the slide area to a low angle, one that is stable under the newly imposed conditions, be they natural or artificial. At the same time, the amount of water in the slope should be reduced drastically, either by surface or subsurface drainage or by covering the surface of the slide with impermeable material. An example of slope treatment and drainage is provided by the history of the large slides—some covering as much as 64 hectares (158 acres)—that interrupt the hilly terrain of the Ventura Avenue oil field in southern California. In the rainy winter of 1940–41 one 24-hectare block (59 acres) slid as a single unit for a distance of approximately 30 m (98 ft). Because the slides sheared oil wells as they moved (in the 1940–41 episode 23 wells were cut off at depths as much as 30 m (98 ft) below the ground surface), unusually extensive and expensive efforts to curb the movement were made. Partial success was achieved by covering the surface with tar and by drilling horizontal drainage holes into the slides—64 km (40 mi) of them. Vertical wells were also drilled through the slides to a porous layer of sandstone, which served as a conduit to carry water away from the slides and into adjacent solid ground. There the excess flow could be pumped out.

Large corporations can afford such expensive remedial measures, but most homeowners cannot. Often the owner's only recourse is to cover the landslide with sheets of plastic—a tactic that works but one that hardly enhances the beauty of a homesite.

Another common stabilizing method is retention of the toe of small slides. Steel or concrete walls commonly are put up to hold back the force of the slide, or, in places where bedrock is easily available, the toe is weighted with large blocks of rock in an attempt to thwart future movement.

Mudflows

Mudflows are an intermediate type of gravity movement—with increasing water and velocity and less load the movement becomes an ordinary stream flow, whereas with decreasing velocity it grades into an earthflow. A typical mudflow is a streaming mass of mud and water moving down the floor of a stream channel, such as a desert arroyo. Such a viscous mass, with a specific gravity much higher than clear water, very often carries along a tumbling mass of boulders and rocks, some of which may be as large as automobiles. The huge rocks are often found on the floor of desert basins far beyond the base of a bordering mountain range. There they linger, long after the enclosing mud that once rafted them out beyond the mountains has been eroded away.

Mudflows are an impressive feature of many of the world's deserts. In arid lands normally empty stream courses may fill almost at once with a racing torrent of chocolate-colored mud, following a cloudburst. Where arroyos are shallow, the flow may exceed the channel's

273

Fig. 9-15. Cabin buried to the eaves by a mudflow in 1941 at Wrightwood, California. The flow resulted from rapid snowmelt in the headwaters of the drainage. Cement-like mud surged down the valley some 24 km (15 mi) at velocities that averaged close to 3 m/sec (10 ft/sec) in the more fluid parts of the flow. *R.P. Sharp*

capacity and spill out over the desert surface.

Mudflows not only are capable of transporting large natural objects, such as house-size boulders, but may trap and sweep along trucks, buses, or even locomotives. Houses inundated by mud streams have been buried all the way up to the eaves (Fig. 9-15).

Arid or semi-arid lands are by no means the only regions where mudflows may be seen. They are characteristic of alpine regions, too, and are likely to be exceptionally destructive where a combination of steep slopes, a large volume of water freed by melting snow, and a great mass of loose debris prevail. Geologist R. R. Curry describes such a flow in the Colorado mountains in 1961:

Direct observations of the mudflows were hampered by very intense rain and by the fact that the author was about 900 m from the cirque headwall, at the rain gauge, at the time the flows began. At about 4 p.m. on August 18 a loud roar became clearly audible above the thunder. A series of what appeared to rockfall avalanches were noted in four different localities around the cirque headwall. These appeared confined to areas previously covered with talus cones and, even though the talus had been soaked by 48 hours of intense rain, large

rock-dust or water-vapor clouds accompanied the disturbances.

. . . Individual flows occurred as a series of lobate pulsations which, in the case of the largest unit, lasted for 1 hour. This unit was made up of ten more or less distinct flow pulses, each traveling with a maximum surface velocity of 915–980 m/minute near the center of the flow at an elevation of 3750 m. The velocity of the flow pulses dropped to 1 m/minute or less where the flow went out onto the valley floor beyond the base of the talus slope or where a relatively small pulse breached the side of the 0.6–0.8 m high natural levees and slowly flowed out over porous talus. Intervals of 4–15 minutes of relative quiescence elapsed between the flow pulses. Velocity measurements were made by timing the travel of a given flow front between two reference points slightly less than 300 m apart and by analysis of 8 mm motion pictures.

Curry was interested in obtaining samples of the moving material and he did so

by forcing wide-necked glass liter-sized jars into the slowly moving side of the flow. The jars were inserted 30–45 cm into the flowing debris about 1 m above the base of the flow. Some bottles were broken and it was generally difficult to force the bottles through the armoring surficial boulders moving up to 150 m/minute, but once beneath the outermost boulders the flow was liquid enough to fill most of the jars. A total of almost 3 liters of matrix of a median diameter of 5 cm or less (a limit imposed by the size of the jars) was collected.

Curry's observations were important to the understanding of these processes. He was also fortunate in having been at the right place at the right time. How many of us, though, would have rushed *to* the mudflow rather than *away* from it on that rainswept August afternoon?

Mudflows are a common occurrence on the steep slopes of andesitic volcanoes because loose material is abundant, steam can alter the volcanic rocks to a clayey goo, and rainfall or rapid snowmelt can set huge masses in motion. Volcanoes in the Pacific Northwest are no exception. One very real hazard on any of those volcanoes is the rapid melting of snow or ice that could be caused by an outburst of steam or lava. Vast amounts of water would be unleashed, which could pick up surface debris and turn into a mudflow downvalley.

More than 55 mudflows have originated at Mount Rainier in the last 10,000 years (Fig. 9-16), and future occurrences are a major threat. Because mudflows can travel so far and with such high speed, it has been recommended that valley floors within 40 km (25 mi) of Rainier should be evacuated in the event of an eruption. Residents of those areas would be hard-pressed to get out if a mudflow were already in motion. For example, one in Japan traveled at a velocity of about 90 km/hr (56 mph). In Indonesia, mudflows are called *lahars*, a term now being accepted and used in many parts of the world.

Another associated hazard is that mudflow debris might quickly fill in reservoirs in valleys flanking the mountain, displacing the water and producing downstream floods. An ingenious way of controlling smaller mudflows is to empty the reservoir, trapping the mudflow in it. There is no such hope of containing the large flows, however. At the writing of this book, Mount Baker, another glacier-shrouded

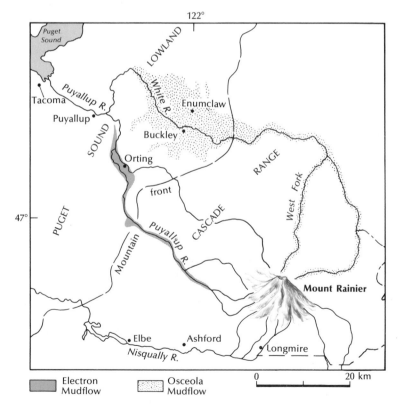

Fig. 9-16. Extent of two recent mudflows that originated on the flanks of Mount Rainier, Washington. The Osceola Mudflow, which is 5800 years old, extended far from the mountain and buried the area now occupied by Enumclaw under 21 m (69 ft) of debris. 500 years ago the Electron Mudflow moved downvalley and deposited 5 m (16 ft) of debris near the present site of Orting.

Fig. 9-17. Rockfall on the headwall of a cirque in western Colorado. The rock is well jointed and the frequent rockfalls in the summer may be due to loosening of the blocks by a freeze-thaw process.
Peter W. Birkeland

volcano in the Pacific Northwest, is steaming near the summit crater, and people living nearby are being cautioned on the possibility of mudflows.

Rockfalls

When rock material drops at nearly the velocity of free fall the event is called a *rockfall*. It may range from the plummeting of an indi-

divual block to an avalanche weighing hundreds of thousands of tons (Fig. 9-17). After the furor of an avalanche individual blocks commonly come to rest in a loose pile of angular rocks, or a *talus*, at the base of a cliff (Fig. 9-18). Should large blocks drop into a standing body of water, such as a lake or fiord, immensely destructive waves may be set in motion with no warning at all. Such waves are particularly feared in Norway, where small deltas often provide the only flat land at sea level. Should a rockfall-induced wave burst through a village, destruction is likely to be as complete as it is sudden, since the waves often range as much as 6–9 m (20–30 ft) in height.

A spectacular wave was set off by a rockfall at the head of Lituya Bay, Alaska, on July 9, 1958. The rockfall was set in motion by an earthquake and began at an altitude of about 900 m (2952 ft), and involved some 31 million m³ (41 million yd³) of rock. Water surged to a maximum height of 530 m (1738 ft) above the level of the bay at its head, and moved to the mouth of the bay at about 160 km/hr (99 mph). Don Miller of the U.S. Geological Survey has described the experience of a couple who were anchored at locality B in Figure 9-19:

Mr. and Mrs. Swanson on the *Badger* entered Lituya Bay about 9:00 p.m., first going in as far as

Fig. 9-18. Cones of rocky debris at the foot of steep slopes built up by successive rockfalls in the Sierra Nevada, California. Such debris is commonly called *talus,* a term borrowed from medieval military engineering for the slope at the base of a fortification wall. *Tom Ross*

Fig. 9-19. Lituya Bay, after the giant wave of July 1958 washed over it. The path of forest destruction is readily seen. (R) marks the rockslide, (D) the maximum altitude (524 m or 1720 ft) of forest destruction, and (B) the location of the *Badger* before it was carried over the spit. *D.J. Miller, USGS*

Cenotaph Island and then returning to Anchorage Cove on the north shore near the entrance, to anchor in about 4 fathoms of water near the *Sunmore*. Mr. Swanson was wakened by violent vibration of the boat, and noted the time on the clock in the pilot house. A little more than a minute after the shaking was first felt, but probably before the end of the earthquake, Swanson looked toward the head of the bay, past the north end of Cenotaph Island and saw what he thought to be the Lituya Glacier, which had "risen in the air and moved forward so it was in sight. . . . It seemed to be solid, but was jumping and shaking . . . Big cakes of ice were falling off the face of it and down into the water." After a little while "the glacier dropped back out of sight and there was a big wall of water going over the point" (the spur southwest of Gilbert Inlet). Swanson next noticed the wave climb up on the south shore near Mudslide Creek. As the wave passed Cenotaph Island it seemed to be about 50 feet high near the center of the bay and to slope up toward the sides. It passed the island about 2½ minutes after it was first sighted, and reached the *Badger* about 1½ minutes later. No lowering or other disturbance of the water around the boat was noticed before the wave arrived.

The *Badger*, still at anchor, was lifted up by the wave and carried across La Chaussee Spit, riding stern first just below the crest of the wave, like a surfboard. Swanson looked down on the trees growing on the spit, and believes that he was about 2 boat lengths (more than 80 feet) above their tops. The wave crest broke just outside the spit and the boat hit bottom and foundered some distance from the shore. Looking back 3 to 4 minutes after the boat hit bottom Swanson saw water pouring over the spit,

Fig. 9-20. The Blackhawk Slide, on the north slope of the San Bernardino Mountains, California. The surface form of such deposits and those of glacial moraines are somewhat similar and both deposits are poorly sorted. How would you go about telling the two landforms apart? *John S. Shelton*

carrying logs and other debris. He does not know whether this was a continuation of the wave that carried the boat over the spit or a second wave. Mr. and Mrs. Swanson abandoned their boat in a small skiff, and were picked up by another fishing boat about 2 hours later.*

Some rockfalls starting high on a mountain face may plunge down the slopes and sweep completely across a valley with a velocity as great as 160 km/hr (99 mph) or more. They are called *rock avalanches*. An excellent example is the prehistoric Blackhawk Slide in California (Fig. 9-20), which started as a rockfall in the source area. Its initial velocity probably was between 200 and 270 km/hr (124 to 168 mph), and, after leaving the canyon walls, it spread out across the valley floor at velocities of no less than 120 km/hr (75 mph), overtopping a 60-m (197-ft) hill. The front of the slide now rests 9 km (6 mi) beyond and 1100 m (3608 ft) below the source.

The most celebrated historic example of a rockfall in North America is one that occurred at Frank, Alberta, in 1903. At the crest of Turtle Mountain a mass of strongly jointed limestone blocks, possibly undermined by coal mining carried on below the thrust fault at the foot of the mountain, broke loose and plunged down the steep escarpment. About 37 million m³ (48 million yd³) thundered through the little coal-mining town of Frank—killing 70 people on the way—and swept to a high point 122 m (400 ft) above the valley floor on the slope facing the source.

The great rockfall-rockslide at Gohna, India, in 1893 remains as one of the most impressive examples of modern times. There an enormous mass of rock, loosened by the driving monsoon rains, dropped 1200 m (3936 ft) into a narrow Himalayan gorge. A huge natural dam was formed by the mass of detritus—perhaps 275 m (902 ft) high, 915 m (3001 ft) across the gorge at the crest, and extending for 3350 m (10,988 ft) upstream and downstream. That pile of

* From pp. 58–59, U.S. Geological Survey Professional Paper 354-C.

broken rock, about 3.6 billion m³ (4.7 billion yd³) impounded a lake 240 m (787 ft) deep in damming the waters of the river.

The British engineers, then in India, proved to be a remarkably foresighted lot. They predicted the dam's failure within ten days of the time that it actually occurred, over the two years of its span of life. While the dam still existed, all bridges were removed downstream, the river channel was cleared of obstacles, a telegraphic warning network was set up, and everything was prepared for the imminent flood. When it came it set a record. Around 280 million m³ (366 thousand yd³) of water were discharged in four hours, causing a flood whose crest was 75 m (246 ft) high. Interestingly enough, after the flood was over the river channel close to the dam, instead of being deepened, was raised 70 m (230 ft) by the sand and gravel deposited after the flood crest had passed and the river flow had returned to normal.

The western United States has been the site of several recent rock avalanches. In Montana one occurred in the canyon of the Madison River where in 1959 an earthquake jarred loose enough rock from the canyon walls to dam the valley and form a lake. As at Gohna, there was flood danger to downstream areas from rapid spillover across the rock dam. Bulldozers were pressed into service to cut a spillway and to control the discharge from the lake.

In 1963, rockfalls cascaded down the slopes of Little Tahoma, a subsidiary peak on the flank of Mount Rainier, quickly moving some 6 km (4 mi) downvalley on a lower slope. The debris descended some 2000 m (6560 ft), and the velocity was estimated at 130 to 145 km/hr (81 to 90 mph). Those falls may have been triggered by volcanic steam explosions.

The most tragic rock avalanche recorded, however, was related to the 1970 earthquake in Peru. About 15 km (9 mi) east of Yungay, a town of 20,000 inhabitants, rise the lofty Peruvian Andes and the snow-covered peak of Nevados Huascarán, which soars to about 6665 m (21,861 ft). Ground motion from the quake shook the mountain and severed a huge block

Fig. 9-21. View eastward over Yungay across the Rio Santa toward Nevados Huascarán. During the 1970 quake a large slab of ice and rock broke from the north face (dark scar visible on the left side of the mountain), forming a mudflow that brought death and destruction to those living downslope from the peak. *USGS*

of ice and rock from the top of the precipitous north face (Fig. 9-21). The mass fell in free flight and impacted 1000 m (3280 ft) below, where it knocked loose a large volume of rock and burst into thousands of smaller fragments. All of the debris continued to cascade down the north face. Frictional heating of the ice during impact caused some melting. The entire mass was transformed into a rapidly moving mudflow. It raced along previously established avalanche and stream channels, and by the time it had moved away from the mountains, contained an estimated 80 million m³ (2824 million ft³) of water, mud, and rocks. Onlookers estimated its speed at about 160 km/hr (99 mph). Just a little upstream from Yungay the channel in which the mudflow was racing made a sharp bend. Most of the surging mass of mud and boulders made the turn. However, some slopped over the embankment, and within seconds the entire city was buried. All 20,000 residents were interred almost instantly (Fig.

9-22. The mudflow continued downslope where it buried another town—Ranrahirca (1800 people)—and caused extensive damage to road and rail routes, power and communication lines, and the Hualanca hydroelectric plant (Fig. 9-23). Eventually it reached the bottom of the valley and the Río Santa, where its momentum carried it across the river as far as 60 m (197 ft) up the opposite bank. There it finally came to rest.

How do we account for such high velocities and such long distances of transport over relatively flat surfaces, as evidenced by the events just described? Some believe that the great velocities are due to air entrapped and compressed beneath the falling mass of debris. The material is temporarily buoyed up in much the same manner that air temporarily buoys up a sheet of plywood dropped onto a flat surface. When the falling debris is pitched into the air a good, compressed air cushion forms beneath it, and some geologists have noted the occurrence of such geological "ski-jumps" in the field. The eventual loss of the air cushion brings an end to the truly remarkable transport mechanism. Another theory of transport is that countless semi-elastic collisions of blocks with each other may keep the debris acting as a fluid, causing it to move great distances beyond the cliffs from which it was derived.

EDUCATING THE PUBLIC

Because the conditions favorable to gravity mass movements build up slowly, it should be possible to determine ahead of time those areas that could become troublesome. As population continues to increase and construction follows the growth, such prediction becomes ever more important for the prevention of human anguish. Obviously, the more we know about the geology of a region, the easier it will be to point out the regions where earth movements are likely to occur. Such information is gathered by geologists and recorded on geologic maps showing the distribution of different types of rocks on the surface, and in

Fig. 9-22. The statue of Christ at Cemetery Hill was all that remained of Yungay. *USGS*

Fig. 9-23. Block weighing about 630 metric tons that was transported to the former town of Ranrahirca. The largest block carried by the avalanche is estimated to weigh 12,800 metric tons. *USGS*

cross sections which are informed guesses of the distribution of rocks below the surface. When such work is related to geologic hazards—landslides, earthquakes, earth subsidence, and the like—it is called "urban geology" or "environmental geology," and it is becoming increasingly important as a direct application of science and technology to the humanizing of our surroundings.

Not only is geologic knowledge of hazardous areas vital, but the knowledge must be available to other scientists and to those persons in authority who can and will use it properly. Cooperation among geologists, engineers, city planners and managers, and the public can prevent many gravity mass movements and predict the possibility of others, thus ensuring proper land usage as land itself becomes more and more a rare and valuable commodity. M. R. Hill says: "We must focus our efforts on converting the unforeseen to the predictable, transforming the predictable into the preventable, preventing the preventable, and restraining the foolish."

PERMAFROST

Mass wasting seems to be particularly active in high northern and southern latitudes, and at high altitudes. The reason is that close beneath the surface the ground is frozen solid. In the following section we shall examine frozen ground, some of the landscape features associated with it, and a few of the engineering problems that result from building on such a precarious surface.

Ground which remains frozen from one year to the next was called *permafrost* during World War II, and all efforts to substitute the term with one that sounds a little less like a trade name for a refrigerator have been successfully resisted. Permafrost is defined only on the basis of temperature, it being ground that remains below 0°C (32°F). Hence, the content of ice in permafrost can vary from very small to very large concentrations. A knowledge of how

much ice and where it is located is extremely important, as we shall see, to engineers and builders working on such terrain.

Permafrost, which is more widely distributed than many people realize, underlies almost 20 per cent of the land surface of the earth, including about 85 per cent of the land area of Canada and the USSR (Fig. 9-24A). Its maximum thickness is reached around the margins of the Arctic Ocean in Alaska, Canada, and the Soviet Arctic. Siberia holds the record for maximum thickness: 1400–1450 m (4592–4756 ft); whereas sites of maximum thickness in Alaska and Canada are about half that. In a general way the thickness decreases southward until finally it thins to zero at about the southern boundary shown on the map. Permafrost also formed in ice-free areas during the Pleistocene, and that which now occupies former glaciated areas could have formed after the ice melted.

Two major kinds of permafrost distribution are recognized (Fig. 9-24B). To the north the distribution is continuous, whereas to the south it is discontinuous, occurring only in patches. It is hard to predict where those patches are located, and as a result land-use planning can be more difficult. Permafrost also occurs in high mountains south of the border shown on the map. It has been reported, for example, on the summit of Mount Washington in New Hampshire, in Colorado's Rocky Mountains, and even on Mauna Kea, Hawaii.

Overlying the permafrost is a thin layer of soil in which *ground ice* (subsurface ice) thaws in the spring and freezes in the fall to remain frozen throughout the winter. That soil is the active layer, and it varies from less than 1 to about 3 m (3 to 9 ft) in thickness. The upper surface of permafrost is called the *permafrost table*, and below that the pore spaces are filled with ice. Hence water in the active layer cannot sink underground—one reason why much of the Arctic tundra is so boggy and water soaked. Precipitation over large segments of the Arctic is very slight, although many lakes and muskegs seem to belie that fact. Water remains at the surface because there is no chance for it

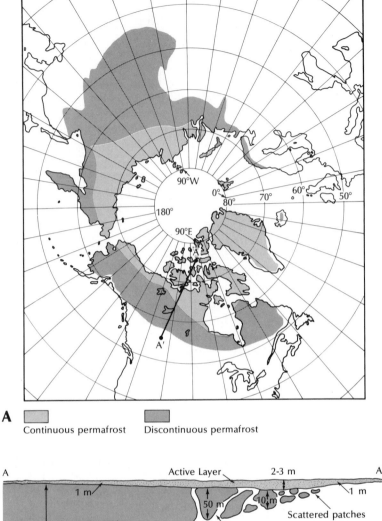

A

Continuous permafrost   Discontinuous permafrost

Fig. 9-24 A. Extent of permafrost in the Northern Hemisphere. **B.** Cross section of permafrost along line A-A'.

to sink underground. Further, the evaporation rate is relatively low.

Patterned ground and solifluction

A bizarre but common manifestation of ice-churned ground in permafrost areas is a curiously regular patterned surface (Fig. 9-25). Sometimes from the air the ground looks like a gigantic tiled floor. Some geometrically shaped polygonal areas are thought to be the result of *frost heave*, which acts much more effectively in fine-grained soils than in coarse.

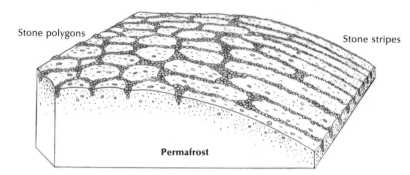

Stone polygons

Stone stripes

Permafrost

Fig. 9-25. Patterned ground comes in many shapes and sizes. On level terrain, stone polygons form, whereas on sloping terrain where creep is active the polygons are stretched out, forming stone stripes.

When frost heaving takes place year after year in a soil of mixed composition, the coarse materials, such as boulders and gravel, are gradually shoved radially outward from the central area, and the finer materials lag behind and become concentrated. Other patterned ground results from a network of vertical ice wedges (Fig. 9-26). In some places, patterned features were formed in the past and are not actively forming today. By studying them, we can estimate former Pleistocene climates (Figs. 9-27 and 9-28).

Solifluction is an extreme sort of creep that reaches a maximum development in cold climates (Fig. 9-29). Hilly terrain underlain by permafrost exemplifies it best, for while the surface layers freeze and thaw, the permafrost table remains constant. As we have noted, surface water cannot sink into the permafrost, so that water which would normally percolate far down into the ground is concentrated in the active layer. The active layer, then, is far more susceptible to creep than similar terrain would be in a more temperate climate because (1) the opposing forces of ice crystallization and melting of ground ice are more active, (2) it holds

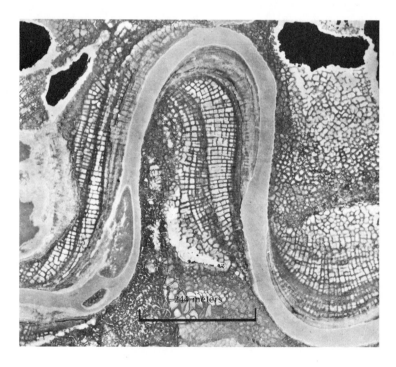

244 meters

Fig. 9-26. Patterned ground along a river southeast of Barrow, Alaska. The pattern is formed by interconnecting ice wedges, in extremely cold climates when ice-cemented permafrost cracks. The wedges are widest at the top and taper downward. Evidence for former ice wedges may be seen in many countries just south of the Pleistocene glacial boundary. *O.J. Ferrians, Jr., USGS*

Fig. 9-27. Foliated ice wedge in late Quaternary silt along the Aldan River, central Yakutia, U.S.S.R. *T.L. Péwé*

more water than it would under a similar precipitation regime without permafrost at depth, and (3) the water-saturated, unstable ground rests on a frozen base over whose surface it can readily slide.

Active solifluction produces a landscape that bears some resemblance to the wrinkled hide of an aged elephant. Different parts of the water-saturated surface layer creep downslope at different rates, so that hillsides where solifluction is active are festooned with soil lobes, or tongues, some of which advance rapidly and some slowly (Fig. 9-30). Maximum rates of lobe movement approach 40 m/1000 yr (131 ft/1000

Fig. 9-28 A. Mounds near Fairbanks, Alaska, that formed when a mosaic of ice wedges melted, causing the ground beneath them to subside. The intervening areas were left as mounds. The melting began when the trees were cleared for agriculture. *R.F. Black and T.L. Péwé, USGS*

Fig. 9-28 B. Low mounds in the southern part of the Great Valley of California. Similar forms occur in places far removed from former glaciers, leading to speculation of widespread cold climates and permafrost in the past. Many alternate theories of origin have been put forth to explain the mounds, with construction by gophers gaining in popularity.
F. E. Matthes, USGS

Fig. 9-29. Solifluction distorts railroad tracks along a tributary of the Yukon River, Alaska. *W.W. Atwood, USGS*

yr). A curious aspect of solifluction is that it tends to produce a rounded, smooth terrain that stands in strong contrast to the very rugged terrain in glaciated valleys.

Engineering problems in permafrost terrain

Before 1942 in the United States, little attention was given to permafrost and its manifold problems, although the Russians had studied the phenomenon intensively for more than half a century. Visitors to the interior of Alaska, especially to the area in and around Fairbanks, are invariably surprised by the tilted houses and the "drunken forests" with their tipsy-looking trees thrown out of line by melting of the permafrost or by its upward growth into the active zone.

From the scores of problems that permafrost can produce, only a few need be mentioned. Heated buildings in the Arctic are likely to thaw out the ground beneath them and to melt their way down into the soggy, unstable mush under the foundation (Fig. 9-31). Usually the structures sink unevenly, so that floors sag, walls tip, and doors stick. It appears that the most practical solution to the problem is to build houses and barracks on stilts, so that cold air could circulate under the structures. Disturbance of the permafrost would then be minimal.

On the other hand, unheated structures, especially large ones such as hangars or warehouses, *insulate* the ground surface below them so that the active layer does not thaw in the summer. The result is that the permafrost table rises, possibly blocking the flow of ground water through the active layer. If the ground water is forced to the surface, where it freezes, a phenomenon known as *icing* occurs. Icing may be spectacular indeed, converting the interior of an unheated building into a huge block of ice, with ice cascading from the doors and windows.

Vegetation is so critical to the thermal regime of permafrost that the slightest alteration of

Fig. 9-30. Solifluction lobes on a 14° slope in the Ruby Mountains, Yukon Territory. *Larry W. Price*

Fig. 9-31. A geologically famous road-house built in 1951 on the Richardson Highway, Alaska. By 1962 the right side of the house, which was heated, had settled considerably. An unheated porch (left), however, subsided much less. Damage was so extensive by 1965 that the building was razed. *T.L. Péwé, USGS*

delicate plant life can have disastrous effects. Even driving a vehicle across the tundra can result in a scar that will not heal for generations (Fig. 9-32). Indeed, road building without proper care for the permafrost terrain could result in the world's longest and narrowest lake!

In order to diminish the problems inherent to permafrost country, special construction procedures are necessary in the building of roads, railroads, and airfields. The simplest method is to insulate the paving surface with a layer of gravel so that melting does not take place. Quite commonly, however, such insulation is so effective that the permafrost rises beneath the gravel layer and may even penetrate its base. Depending upon conditions in the active layer and the flow of ground water on the permafrost table, large-scale icing could result on the paved area. However, with proper foresight and engineering practice such embarrassment can be avoided.

The list of problems permafrost can create seems endless, but the degree to which Arctic pioneers have solved them is a testimonial to their ingenuity and perseverance. Even such a simple thing as developing a water supply in a permafrost area can become a major frustration. Ground water in the active layer is available only during the summer, and is usually at

so shallow a depth that it is readily contaminated by surface wastes. Although there may be ground water below the permafrost, it is deep, and well sections drilled through the frozen ground are almost certain to freeze. Delivery of water also poses a vexing problem. If water pipes are buried underground, they freeze; if placed above ground, they freeze, too, and are likely to be thrown out of line as the ground under them either heaves when it freezes or sinks when it thaws. Expensive insulation procedures are the only solution.

One solution to water supply in permafrost areas is to build dams and collect the summer snowmelt for year-round use. Yet climatic conditions can create difficulties for dams and reservoirs. The Russians, for example, ran into trouble when a far northern dam started to leak shortly after the reservoir behind it was filled. The dam was built on volcanic rock whose tiny cracks were permanently filled with veinlets of ice. Ordinarily, ice-filled rock below the permafrost level can be treated as solid rock. In the Russian case, however, the filled reservoir with its insulating layer of ice on the surface acted as a heat trap. The ice veinlets melted, and the bottom of the dam became virtually a sieve. Newer dams built in similar cold areas are refrigerated by pumping cold air into them to prevent such melting.

Fig. 9-32. Tractor trail bulldozed on the north slope of Alaska near the Canning River during the winter of 1967–68. The string of ponds formed during the summer of 1968, when the photograph was taken, and as thawing continues they will grow larger. Notice the patterned ground and vehicle tracks. *Averill Thayer, Bureau of Sport Fisheries and Wildlife*

Sewage disposal is perhaps the ultimate problem. Septic tanks and leach fields freeze, and in the absence of bacteria, decay does not dispose of waste as it does in warmer climates. At Point Barrow in Alaska the unsightly (but practical) solution is to heap everything atop an ice floe during the winter. In the summer the ice cake floats out into the Arctic Ocean, melts, and the waste sinks. Waste disposal is even more of a problem now that oil has been discovered in northern Alaska. The rapid development of the area will make that method of waste disposal and its attendant pollution intolerable.

Permafrost and the discovery of oil in the north slope of Alaska also precipitated a heated controversy over the Alaska pipeline, which extends from the petroleum fields south to the ice-free port of Valdez, a distance of about 1300 km (808 mi). The oil flowing through the pipeline, which is more than 1 m (3 ft) in diameter, has a temperature of 70–80°C (158–176°F). It was argued that if the pipe were buried, difficult problems would arise from melting of the permafrost. It has been calculated, for example, that within the first decade after burial a cylindrical area 6–9 m (20–30 ft) in diameter would be thawed around the pipe. In successive decades the thawing would continue, but at a diminishing rate.

The major construction problem was the condition of the permafrost before thawing. If it is dry, thawing should have only a slight effect. However, permafrost is in large part composed of fine-grained sediments with a high ice content, and thawing of the material forms a water-saturated slurry into which the pipe could sink. If the pipeline were on a slope, the slurry could flow out onto the landscape, allowing the pipe to settle even deeper—onto still-frozen ground—and the melting process would continue. Stress caused by such processes could cause the pipe to rupture, leading to oil spills that would rival in seriousness those that occur at sea. It was vital, therefore, to identify all potential problems before the pipeline was constructed.

After detailed investigations were made, it

Fig. 9-33. Insulated above-ground section of the Alaska pipeline north of Valdez. After the forest was cleared, the permafrost was protected from melting and damage from heavy equipment by a layer of gravel. *Alyeska*

was found that the pipeline could be placed below ground for about half of its length without serious consequences. Conventional burial procedures were used. Because of the problems associated with the melting of the permafrost and to insulate it against the arctic cold, the other half of the pipeline has been built above ground, on platforms about 15–21 m (50–70 ft) apart (Fig. 9–33). Areas disturbed during the construction will be revegetated, returning the environment to a stable condition in which the permafrost, protected by the vegetative cover, is prevented from melting.

SUMMARY

1. Gravity operates on all sloping ground, so that material moves to lower positions on the slope.
2. The transfer rate of material on a slope varies from millimeters per year to meters per second. Similarly, the amount of material involved in the transfer varies from small amounts to entire mountainsides.

3. *Landslides* and *earthflows* are the gravity movements that most often destroy property and dwellings. They can be initiated by natural or human-related causes, and once set in motion, they are difficult to stop. Slopes prone to landslides or earthflows should not be used as building sites.

4. A rock avalanche is the most spectacular kind of mass movement, because so much material is transferred downslope so quickly, perhaps on a cushion of air.

5. Mass movement takes place at a fairly rapid rate in areas of permafrost. Such terrain is extremely sensitive to human manipulation; any engineering projects on it must be undertaken with great caution.

SELECTED REFERENCES

Crandell, D. R., and Mullineaux, D. R., 1967, Volcanic hazards at Mount Rainier, Washington: U.S. Geological Survey Bulletin 1238.

Curry, R. R., 1966, Observation of alpine mudflows in the Tenmile Range, Central Colorado: Geological Society of America Bulletin, vol. 77, p. 771–76.

Ericksen, G. E., and Plafker, G., 1970, Preliminary report on the geologic events associated with the May 31, 1970, Peru earthquake, U.S. Geological Survey Circular 639.

Ferrians, O. J., Jr., Kachadoorian, R., and Greene, G. W., 1969, Permafrost and related engineering problems in Alaska, U.S. Geological Survey Professional Paper 678.

Kerr, P. F., 1963, Quick clay, Scientific American, vol. 209, no. 5, pp. 132–42.

Kiersch, G. A., 1965, The Vaiont Reservoir disaster, Mineral Information Service, vol. 18, no. 7, California Division of Mines and Geology, Sacramento.

Lachenbruch, A. H., 1970, Some estimates of the thermal effects of a heated pipeline in permafrost, U.S. Geological Survey Circular 632.

Péwé, T. L., 1966, Permafrost and its effect on life in the North, Oregon State University Press, Corvallis.

Price, L. W., 1972, The periglacial environment, permafrost, and man, Amer. Assoc. Geogr., Commission on College Geography, Resource Paper 14.

Sharpe, C. F. S., 1938, Landslides and related phenomena, Columbia University Press, New York.

Shreve, R. L., 1968, The Blackhawk landslide, Geological Society of America Special Paper 108.

Varnes, D. J., 1958, "Landslide types and processes," Chap. 3 in Landslides and engineering practice, Highway Research Board Special Report 29.

Washburn, A. L., 1973, Periglacial processes and environments, St. Martin's Press, New York.

Fig. 10-1. A river meanders across the Iranian desert in the land of Elam and flows past the ancient city of Susa, where Esther was chosen queen. *Aerofilms, Ltd.*

10

STREAM EROSION, TRANSPORTATION, AND DEPOSITION

Few natural phenomena are more intimately involved with human affairs than rivers. In centuries past such streams as the Nile, the Tigris, and the Euphrates literally were the givers of life as they threaded their way across a weary desert land (Fig. 10-1). Ancient civilization depended on such waters for irrigation; through that communal enterprise many of the attributes of modern urbanized society arose. The beginnings of mathematics, surveying, and hydraulics developed in the designing of dams and canals. One of the earliest projects was a long dike built about 3200 B.C. on the west bank of the Nile with cross dikes and canals to carry flood waters into basins adjacent to the river.

Boundary disputes and ownership problems logically led to a system of codes and usages that evolved into a pattern of laws and courts much like ours today. The Code of Hammurabi (c. 1900 B.C.) included a provision that if a landowner damaged his neighbor's land through neglect of a portion of a canal that was his responsibility, he was liable for all the damage.

Rivers have long played a role as natural barriers—and two that were of decisive importance in Roman times were the Rhine and the Danube. The crossing of the Danube by the barbarians commonly is cited as one of the events heralding the fall of the Roman Empire.

Contrasting with their role as barriers is the function rivers serve as communication routes. The Mississippi packet boat, with its flashing wheels and double columns of smoke, is gone forever. Its place is usurped by the vastly more powerful diesel-propelled towboat (which actually pushes its load) and its broad acreage of heavily laden barges driving against the current. The endless parade of diesel-powered barges that surges up and down the Rhine is an impressive sight to European travelers.

Rivers from time immemorial have been routes from the sea to the interior. Explorers have followed them; most of the world's leading cities are built on their banks. They are identified indissolubly with the history and national aspirations of almost all the lands that border them. It would be difficult to conceive of Germany without the Rhine, Vienna without the Danube, or Russia without the Volga or the Don.

In this chapter we will discuss first stream

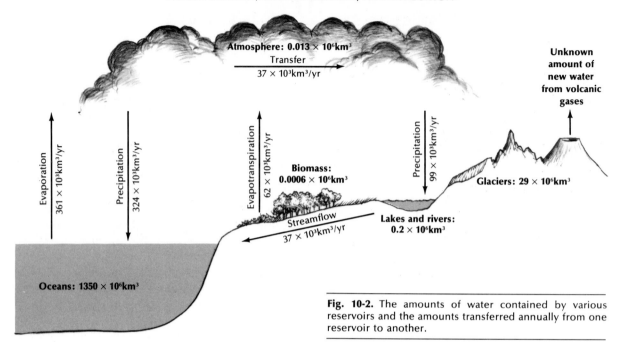

Fig. 10-2. The amounts of water contained by various reservoirs and the amounts transferred annually from one reservoir to another.

flow—how water flows down a valley—and then relate it to the work a river is capable of performing—most of which takes place during times of major flooding. We will then discuss the ways of deciphering river history and the role of rivers in landscape evolution. Throughout, the environmental problems involved in manipulating rivers will be pointed out. We have become aware of such problems through mistakes, and only by knowing how and why we erred can we avoid future problems.

THE HYDROLOGIC CYCLE

Before discussing stream flow, we should consider first the amount of water on the earth and how it moves from place to place (Fig. 10-2). There are several water reservoirs—the oceans, rivers and lakes, glaciers, the atmosphere, and the biomass—and of them the oceans contain by far the most water. Although the volume content of the various reservoirs has changed from time to time, as during major glaciations when considerable amounts of ocean water were transferred to glaciers,

most geologists believe that the total amount of water on earth has been more or less constant for the past billion years. New water may have been added to the earth's surface by steam condensation during volcanic eruptions, but the amount is insignificant in the overall picture.

Water is transferred from reservoir to reservoir each year, but, overall, the entire budget is in balance. That cycling process is known as the *hydrologic cycle*. The water that is annually evaporated from the ocean (approximately equivalent to 1 m, or 3 ft, in thickness) exceeds precipitation on the oceans, and the excess is transferred to the lands via the atmosphere. On land we see the opposite course—that is, more water reaches the ground as precipitation than leaves through *evapotranspiration*. The latter term encompasses water losses through evaporation as well as through transpiration from plants and animals. In fact, in well-vegetated areas the main loss of water to the atmosphere is through transpiration from plants. The cycle is completed and balanced as stream flow removes the excess from the land to make up for the deficit in the oceans.

It is stream flow to which we now turn our attention, because it performs work that continually changes the shape of the land. The geologist A. L. Bloom expresses the amount of available stream energy in a down-to-earth way:

The average continental height is 823 m above sea level. If we assume that the 37,000 cubic kilometers of annual runoff flow downhill an average of 823 m, the potential mechanical power of the system can be calculated. Potentially, the runoff from all lands would continuously generate over 12 billion horse-power. If all this power were used to erode the land, it would be comparable to having one horse-drawn scraper or scoop at work on each three-acre piece of land, day and night, year around. Imagine the work that would be accomplished! Of course, a large part of the potential energy of the runoff is wasted as frictional heat by the turbulent flow and splashing of water, but we will see that the "geomorphology machine" is really quite efficient, and in fact does erode and transport rock debris down to the sea almost as fast as if horse-drawn scrapers were hard at work on every small plot of land, over all the Earth.

STREAM FLOW

Although streams play so vital a role in our lives, and their control has engaged the efforts of people for centuries, many aspects of their behavior remain as mysterious today as they have always been. Great impetus has been given in recent years to the study of stream flow because of its importance in the design of high-head hydroelectric plants, dams and spillways, and increasingly complex irrigation systems. Every leading nation is actively engaged in research into the nature of stream flow, and the majority of them maintain large and well-equipped hydraulics laboratories. The largest in the United States is the U.S. Waterways Experiment Station, operated by the U.S. Army Corps of Engineers at Vicksburg, Mississippi. There, elaborate models of the Mississippi have been constructed and an immense amount of data collected and analyzed

in order to find ways to bring that unruly river and its tributaries under control.

From laboratory studies and field investigations around the world, and, as anybody who has rafted a wild river will testify, water flows mainly in a turbulent way (Fig. 10-3). In turbulent flow individual water particles thrash about in the most irregular fashion imaginable. Familiar examples of turbulent flow are the tumultuous rush of water through Niagara Gorge downstream from the plunge pool at the base of the Falls, or the maelstrom of white water at the bottom of the spillway at Grand Coulee Dam. Despite the random paths of the individual water particles, the main thrust of the water is forward, downslope in the direction that the stream is flowing. Sometimes the particles swirl upward like autumn leaves, or like dust devils in the desert—at other times they descend just as violently in the vortices of whirlpools and eddies. In part, it is the erratic flow pattern that makes the actual velocity of a stream so difficult to measure.

In very general terms, the velocity of a stream can be defined as the direction and magnitude of displacement of a portion of the stream per unit of time. Customarily we measure it in miles per hour (mph) or kilometers per hour (km/hr). Few streams, however, attain velocities in excess of 30 km/hr (19 mph), and velocities of less than 6 km/hr (4 mph) are more likely to be the rule. The velocity as much as any single factor is responsible for determining the size of particles that a stream can transport, as well as the way in which it carries its load.

Where is such a thing as an "average velocity" likely to be located within a stream? Different parts of the water in a stream advance at different rates and, like glacier flow, the center moves faster than the sides and the top faster than the bottom because of frictional retardation of flow along the channel perimeter (Fig. 10-4). The gathering of many data has shown that the average velocity of a river is approximated by that velocity at 0.6 of the distance from the surface of the river to its bed, about in midstream. How, then, do we obtain

Fig. 10-3. Turbulent flow patterns in the Colorado River. *Martin Litton*

that value, short of swimming? Actually, a reasonable figure can be obtained by throwing a stick into the middle of the stream, timing its travel over a known distance, and multiplying that velocity value by 0.8.

The average velocity of a stream depends on such factors as the gradient (or downvalley slope), the cross-sectional shape of the channel, the roughness of the channel surface, the discharge of the stream, and the quantity of sediment the stream is carrying. We will look at the influence of each factor separately, and then at how each can vary along the length of a river.

Increased gradient obviously speeds up a stream's flow. Where its gradient reaches zero (when it empties into another river or a lake, for example), its velocity becomes very low. Where slopes are vertical, as in a waterfall, the velocity approaches that of free fall. So it is in principle for almost all streams. Where the gradient is low, a stream lazes along; where the gradient is high, the water leaps and quickens in a headlong dash to the sea.

The cross-sectional shape of the channel has an effect on the velocity, too. In wide, shallow rivers much of the water is in contact with the channel perimeter; hence frictional retardation is great and velocity low. The channel shape that allows for the greatest velocity is something close to a semicircle—a shape that gives the maximum cross-sectional area and the shortest channel perimeter. The shorter the perimeter, the less the frictional retardation and the greater the velocity.

The effect of roughness on velocity is obvious. A smooth, clay-lined channel will promote an even, uniform flow, whereas an irregular, bumpy, boulder-filled one induces enough turbulence to retard significantly a stream's velocity.

Another important factor affecting stream velocity is *discharge*—the quantity of water that passes a point in a given interval of time, for example, in cubic meters of water per second. The discharge of most rivers is far from constant. The flow of northern rivers fluctuates with the melting of snow and ice. Tropical

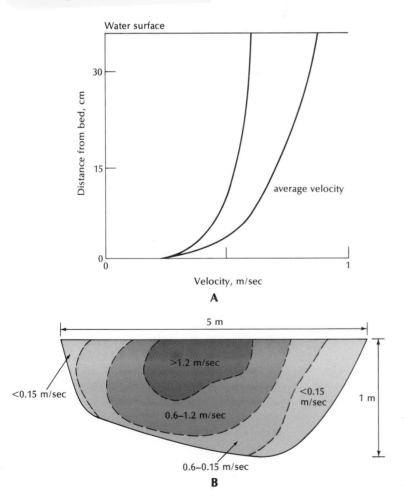

Fig. 10-4 A. Vertical velocity profile for two rivers in Wyoming. Notice that the velocity is zero at the bed of each stream. **B.** Velocity distribution in a cross section of a river channel in Wyoming.

rivers, too, especially those in monsoonal regions, show large seasonal variations, determined by the rhythm of wet and dry seasons throughout the year. Streams of the arid southwestern United States show as great a range as any. Throughout must of the year they may have no surface flow at all, but during a sudden cloudburst they can become raging torrents filling the channels from bank to bank.

The Amazon River is by far the world's largest, discharging an average of 156,000 m³/sec (5.5 million ft³/sec) to the sea. Just one day's volume would satisfy the needs of New York City for nine years! For comparison, the discharge of the Amazon is four times that of the Congo, the world's second largest river, and ten times that of the Mississippi.

Now, back to the influence of discharge on velocity. During a low discharge stage a river is shallow, much of the water is close to the channel perimeter, and the velocity is low. Increase the discharge, though, and less water per unit volume comes in contact with the perimeter. In such a situation, frictional retardation is less and the river can proceed downvalley at a higher velocity.

An increase in the sediment load and a corresponding decrease in the percentage of water has a strong braking effect on velocity. That is readily understandable because the more sediment a stream receives, the muddier it becomes, until finally it may become a mudflow. The ratio of solids in a muddy, viscous stream may be so high that the viscosity will increase to the point that the stream can no longer flow. Such an effect is demonstrated sometimes by the ephemeral streams produced by short-lived thunder showers in the desert.

All of the factors affecting velocity vary in a downvalley direction in a fairly systematic manner. It has to be so. For example, if the velocity of the upper part of a river were much faster than the lower part, it would overtake the lower part. Or, if the lower part ran much faster than the upper part, it might run away from the latter! Of course, such things do not happen, for rivers continually adjust themselves in order to remain intact ribbons of water flowing toward the sea.

Consider, for example, a river heading in high mountains and flowing across vast lowlands. In the mountains the gradient is steep but the velocity is checked by the boulder-strewn rough channel as well as by a relatively low discharge. Progressing downstream, the gradient commonly lessens, but not necessarily the velocity. Indeed it may remain constant, or even increase. Commonly discharge increases as more tributaries join the main stream, and the channel is smoother—both are factors that contribute to greater velocity. The tendency for greater velocity, however, could be offset by a lesser gradient, and the result could be constant velocity. There is no need to discuss further the interplay of factors. The point to be made here is that determining the velocity of a stream is a complex matter depending on many variables, none of which is easily measured.

STREAM TRANSPORTATION

The roiled cloud of sediment brought down by the Mississippi River stains the waters of the Gulf of Mexico far seaward of the river's mouth. In the Southwest, within a generation some fairly large reservoirs have silted up completely, and the lakes once backed up behind the dams have been converted into dreary expanses of muddy or dusty silt, depending on the season. After a heavy rain, many streets and sidewalks are slippery with a coating of mud, or are scattered with stones. Scores of other examples from everyday life are enough to convince an observant person that the land is inevitably wasting away, and that much of its substance is being swept to the sea.

That annual wastage can be an imposing amount is demonstrated by a single river, the Mississippi, which every day carries about one million metric tons of sediment to the Gulf of Mexico, and even more when in flood. That colossal drain of the central lowlands of the United States removes 5 cm (2 in.) of soil every 1000 years and has resulted in the construction within the past million years or so of a broad platform of sand, silt, and clay that covers an area of around 31,000 km² (11,966 mi²) at the river's mouth, with a central thickness of at least 1.5 km (1 mi).

Most people know that streams carry a heavy burden, but they may not know exactly how it is moved. Even experts in hydraulics are unsure of some of the dynamics involved. One problem is that most rivers transport most of their load during flood stages, and sediment sampling at such times can be rather perilous. However, geologists generally agree that a river moves its load in three major ways—in part by solution, in part by suspension, and in part by moving it bodily (rolling and sliding) along the bottom of the channel.

Dissolved load Dissolved material that rivers carry is supplied largely through the leaching out of soluble minerals in rocks and soils. It is the invisible dissolved load that gives some river water—especially the water in western U.S. rivers that cross arid or semi-arid regions—its distinctive taste. When such water evaporates it leaves behind alkali salts, a white residue that blankets the entire surface of the ground; from a distance extensive patches of it look like fields of snow. Such accumulations are fatal to nearly all crop plants and are the bane of many irrigation districts. Keeping soluble residues from accumulating in the soil is an unceasing struggle and one that has been lost in a number of reclaimed areas.

Dissolved loads go wherever river water goes, and only precipitate out if chemical conditions permit. They may be measured through analysis of the water. Error can creep into the interpretation of data, however, because pollution can account for as much as 50 per cent of the dissolved load of some rivers.

Suspended load The contrast between a clear trout stream in the high mountains and the muddy, roiled water sluicing through an arroyo as the aftermath of a desert cloudburst is largely a function of the suspended load: the visible cloud of sediment suspended in the water. If sediment is very finely subdivided it may remain buoyed up in the moving water almost indefinitely. That is likely to be true for such particles as clay-size grains; but not so

true for silt or sand, or for larger particles. How long such material keeps afloat depends on several factors: (1) the size, shape, and specific gravity of the sediment grains, (2) the velocity of the current, and (3) the degree of turbulence.

The effect of the factors is obvious. Flat mineral grains, such as mica flakes, will sift down through the water much like confetti, when compared with the more direct way in which nearly spherical grains settle out. Specific gravity is important, too, because denser substances such as gold nuggets, with a specific gravity of 16 to 19, are deposited far more rapidly than feldspar grains of the same dimensions, but with a specific gravity of about 2.7.

One of the most important factors in keeping particles in suspension is turbulence. If a particle is about to settle out and then is caught in an upward swirl of water, it may be whisked up suddenly in much the same way that tumbleweeds spiral upward in a swirling wind eddy in the desert. So it is most unlikely that individual grains of sediment will settle out at a uniform velocity along the entire course of a river. Rather, each particle follows a complex path, drifting down the river with the moving current, here and there swirling erratically—much like a sheet of paper caught in a vagrant wind.

Typically, not all of a stream's suspended load is distributed uniformly. Uniform distribution is likely to be true of the finer grain sizes, such as silt or clay, but the greatest concentration of larger grains, such as sand, remains closer to the bottom. Sand-size grains settle out more rapidly than do clay-size particles, and more turbulence and a stronger current are needed to lift sand from the bottom and keep it in suspension. In fact, the basal concentration of sand grains may be in suspension only briefly after which they sink back to the bottom of the channel to become part of the bed load.

Measurements of suspended sediments in a river are fairly easily taken. All that is necessary is to collect a bottle of river water and then to separate and weigh the solid sediment. If the average discharge is known, the annual suspended load can be calculated readily.

Fig. 10-5. A field of boulders near Manzanar, California, swept along as bed load by flash torrents from the distant canyon. These boulders may reside here for a long time and weather to smaller sizes before continuing their journey to the lower parts of the land. *Ansel Adams*

Bed load Part of a river's burden—the *bed load*—is transported by rolling or sliding along the bottom either as individual particles or collectively (Figs. 10-5 and 10-6). In terms of work accomplished by a stream in cutting down or widening its channel, the bed load does the lion's share. The bombardment of sedimen- tary particles against the sides and bottom of the channel wears it away as effectively as though it had been worked over by a harsh abrasive such as carborundum. In rocky streambeds smooth hollows called *potholes* may be formed (Fig. 10-7).

Individual particles may move by sliding, rolling, or *saltation* (from the Latin word, *saltare*, to jump). Saltation could be compared to the game of leapfrog. A sand grain may be rolling along the bottom, or may even be stationary when it is caught up by a swirling eddy of turbulence. Then it bounds or leaps through the water in an arching path. Should the velocity be great enough, it may be swept upward to become part of the stream's suspended load temporarily; if not, the particle sinks again, either to remain stationary or perhaps to continue downstream by leaps and bounds.

The bed load is a virtually impossible quan-

tity to measure in a natural stream as compared to the dissolved or suspended load. The latter are diffused throughout the main body of the river, but the bed load moves along the most inaccessible part of the stream—the bottom—and to add to the difficulty, it moves mainly during times of major floods, a time when the river's bed is least accessible for study. Bed loads can be trapped in the deltas of downstream reservoirs, and the simplest way to determine annual bed-load transport is to find such a delta and to measure its volume.

Fig. 10-6. This huge granite boulder, weighing 8 metric tons, stands in the atrium of the national headquarters of the Geological Society of America in Boulder, Colorado. The rock was once carried as bed load by the Big Thompson River, where it was rounded and smoothed during transport. *Jan Robertson*

Fig. 10-7. Bed load moving across a river bed not only polishes the bedrock but in places scours out potholes.
G.K Gilbert, USGS

Relative importance of dissolved vs. solid load
Rivers vary in the ratio of dissolved load to *solid load* (suspended load plus bed load) for a variety of reasons. It might come as a surprise, however, to hear that about one-half of the total world-wide load transferred to the oceans is in dissolved form. In short, much of the continents is dissolving away under the constant onslaught of rainfall.

Rivers in the United States show a complete spectrum in the ratio of the dissolved to solid load (Fig. 10-8). The Colorado River is basically a solid-load river, whereas those of the southeastern states carry approximately equal proportions of solid and dissolved loads. Climate and relief can explain many of the differences. The Colorado cuts across an arid to semiarid landscape, a terrain prone to rapid surface runoff that quickly picks up solid-load material. Also, because of the high relief of the region, much solid material is being constantly transported to the valley bottoms. In contrast, al-

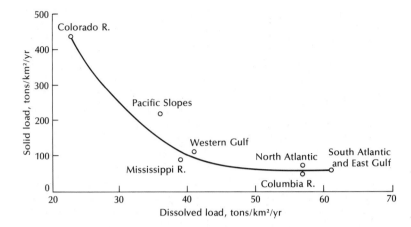

Fig. 10-8. Amounts of dissolved and solid load of various rivers and regions in the United States.

though solid material moves with more difficulty to the rivers in the southeastern states, which are well vegetated and lower in relief, high rainfall insures high rates of chemical weathering and thus high dissolved loads.

Putting it all together, we can estimate how long the United States and the world will last. The average rate of erosion of the United States before people began to disrupt the landscape has been estimated at 3 cm/1000 yr (1.2 in./1000 yr). Humans, through construction and agricultural practices, have caused that to increase dramatically in places—the increases can be one order of magnitude or greater (Fig. 10-9). Taking the United States as a whole, however, the present rate is about double the prehuman rate given above.

The prehuman world-wide rate of erosion has been calculated at slightly over 9 billion metric tons/yr, equivalent to a lowering of the total land surface at a rate of about 2.4 cm/1000 yr (1 in./1000 yr). Human intervention has increased the rate to some 24 billion metric tons/yr. At the natural rate, the continents would be swept to the sea in a bit over 300 million years, and humans are merely hastening the process along. It is doubtful that all the lands would be reduced to sea level, however, because the geologic record clearly indicates that it has never happened in the past. Highland areas are surely eroded down, but other lands, formerly low, are pushed up to form highlands and become good sources of sediment. How else, then, can the fairly young marine sediments high on Mount Everest be explained?

Competence and capacity The size of sedimentary particles a stream can transport, or its

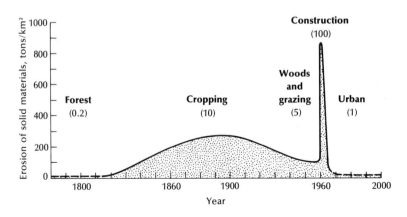

Fig. 10-9. Erosion rate related to land use in the Middle Atlantic region of the United States. The numbers in parentheses are the approximate erosion rate in cm/100 yr for the various times. The first increase in the erosion rate coincided with the clearing of forests for agriculture in the early 1800s. The rate decreased in the early to mid-1900s, when the land was used for grazing and partly returned to forests. Construction in the 1960s bared much of the land to runoff and erosion, but that was soon halted as people added vegetation, pavements, and other erosion-inhibiting covers to the landscape.

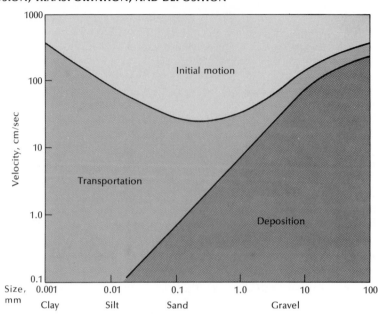

Fig. 10-10. Relationship of stream velocity to initial motion, continuing transportation, and deposition of particles of various sizes.

competence, depends primarily upon velocity. At low velocities many streams may run clear, and sediment grains on their beds may rest relatively undisturbed. With rising velocity the water becomes increasingly roiled, and larger and larger particles are set to moving. Finally, even such impressive loads as railroad locomotives may be swept along—several were rafted away, out of the roundhouse and into oblivion, during the Johnstown Flood (Pennsylvania) of 1889. In a cloudburst in the Tehachapi Valley of California in 1933, a behemoth of the rails, a Santa Fe steam freight-locomotive and its fully loaded tender, was swept several hundred meters downstream from the tracks and completely buried in the stream gravels.

The most stupendous historical example in the United States of the transporting power of running water probably was provided by the failure of the San Francisquito dam in California in 1928. When that 62.5-m (205-ft)-high concrete structure collapsed in the darkness, a wall of water 38 m (125 ft) high swept down the canyon with a velocity of perhaps 80 km/hr (50 mph), enough to raft individual 18-m (59-ft) blocks of concrete weighing as much as 10,000

metric tons almost 1 km (0.6 mi) downstream. Partially buried in gravel and boulders swept downriver by the same torrent, they much resembled gray icebergs stranded on an arctic beach.

The relationship of load size to stream velocity is not a simple one (Fig. 10-10). Research has shown that in order to initiate particle motion in sand or larger bed-load sizes a greater velocity is required than that needed to keep the particles in motion. It is the added force of momentum that helps to keep the particles moving. A somewhat surprising result of the research is that the velocity required to initiate motion of clay is similar to that required to move sand. The reason for the apparent anomaly is twofold: (1) clays rest at the bottom of the stream where velocities are nil and (2), clay particles are held tightly together by cohesive forces. High velocities are required to produce the conditions needed to break those forces, setting the fine-grained material in motion. Once in motion the fine particles stay in suspension over a wide velocity range, quite unlike the narrow velocity range under which bed-load material remains in motion.

Capacity is a term for the potential load that a

stream can carry, and, like competence, it is dependent in part upon the velocity of the current. A sluggish stream meandering across a swamp is capable of moving very little material in comparison with a boulder-rolling mountain torrent. Capacity is also dependent upon discharge. Obviously, a rivulet of water moving with the same velocity as the mile-wide Mississippi can move but a fraction of that river's immense burden.

If we remember that capacity is a measure of what a stream theoretically can do and that load is a measure of what a stream actually is doing, then we can keep the two terms straight. For example, a Lackawanna boxcar may have such a figure as "Capacity 100,000" stenciled on its side yet actually be carrying a load of only one-half that in automobile frames.

Graded streams and their disruption

A stream whose load is greater than its capacity is not likely to be spurred on by excess of zeal into carrying more than is expected of it. Instead, the overload is dropped as abruptly as a Peruvian llama deposits its burden if it is convinced that its carrying capacity of around 45 kg (99 lb) has been exceeded. When a stream deposits its excess load on the channel bottom, we say that it is *aggrading* its channel, and that may happen when too much sediment is supplied or when the particle size exceeds the competence. Conversely, an underloaded stream—one whose capacity is greater than its load—is likely to pick up an additional quantity of material by *degrading,* or eroding, its channel.

When a stream is balanced between the two extremes and has achieved equilibrium, so that its slope and discharge give it sufficient current to handle the load, it is *at grade.* A *graded stream* was aptly defined by the geologist J. H. Mackin in a beautifully worded statement:

A graded stream is one in which, over a period of years, slope is delicately adjusted to provide, with available discharge and with prevailing channel characteristics, just the velocity required for transportation of the load supplied from the drainage basin. The graded stream is a system in equilibrium; its diagnostic characteristic is that any change in any of the controlling factors will cause a displacement of the equilibrium in a direction that will tend to absorb the effect of the change.

A short discussion of Mackin's statement is in order. First, it is important to point out that equilibrium is reached only over a period of years. A river might put in only a few good days or weeks of work per year, during times of flood. The rest of the time it might just be loafing along with little tendency to work. So, it is necessary to look at the work record over the long haul.

Another important aspect is that if the equilibrium of a graded stream is disrupted, the stream will react quickly to counteract the disturbance. For example, if more load or larger particles are imposed on a stream, the stream might not be able to handle it. Like the llama, it will deposit the load in the stream bed where it will remain until a steeper slope that can handle the load is built and a new grade attained. A classic example occurred in the gold rush days in California, where hydraulic mining in the mountains greatly increased the loads streams had to carry (Fig. 10-11). The effects of the mining were felt far downstream in the Great Valley, where the rivers filled in or aggraded part of their channels. Because the channels then were narrower than before, floods became more frequent.

In the past, when major glaciers formed in the headwaters of mountain streams, they had a similar effect on stream equilibrium. Because glaciers are such effective erosive agents, they greatly increased the amount and size of the load in the rivers draining from their lower ends. The streams reacted by dumping part of their load and steepening their slopes. In that way they attained a new equilibrium.

The removal of part of a load from a river can bring about downcutting, or degradation. For example, when former glaciers melted, the loads of many rivers decreased, and the rivers were able to downcut enough to handle the post-glacial load.

Altering discharge also can greatly disrupt the equilibrium of a stream because of the

Fig. 10-11. Hydraulic mining for gold at Cherokee Flat, California, in the late nineteenth century. The excess sand and gravel were washed to the rivers, thus greatly increasing their loads. *J.S. Diller, USGS*

close correspondence between discharge and velocity. A decrease in discharge lessens both stream competence and capacity, and aggradation results, whereas the opposite conditions and degradation accompany an increase in discharge. How can discharge be altered? Nature can do it in the form of climatic change, or humans can do it in the form of water-diversion projects.

The side effects of damming a river vividly point out what happens when stream equilibrium is disrupted. Above the reservoir the stream carries its usual load, but when it meets the quiet reservoir waters the load is dumped to form a delta (Fig. 10-12). Because the delta has a low slope, some material is deposited on

its surface, and some is swept out and deposited on the delta front. As the delta continues to build out into the reservoir, the valley upstream must aggrade to maintain the equilibrium slope.

Things are not much better downstream from the dam. There the stream has been deprived of its load—part of which now forms the delta—and the river degrades. In any economic analysis of a proposed dam, such factors have to be considered. So also does the projected life-span of the reservoir, for surely the sediment will complete the task of filling it in one day.

Another common cause of stream disequilibrium is a change in *base level*. Base level is

Zion Narrows of the Virgin River, Zion National Park, Utah.
Deep canyon-cutting in Mesozoic sandstone in an arid cli-
mate.
Bill Ratcliffe

(Above) The Nile Delta, formed where the Nile meets the Mediterranean. In this mosaic from Landsat imagery, farmlands show as red, dramatically illustrating Egypt's dependence on the river. *General Electric Co.* (Below) The birds-foot shape of the Mississippi Delta is formed by the distributaries of the river as it meets the Gulf of Mexico. Present-day sediments show as light-blue in this Landsat image. *USGS*

the low point to which most streams flow—in short, the ocean. If ocean level changes up or down, the streams will meet the challenge by aggrading or degrading, respectively. Many coastal streams responded in such a way during the rapid fluctuations in sea level that accompanied the major Pleistocene glaciations.

FLOODS

When rains are heavy and fall day after day, people begin to consider the possibility of

floods, especially if they live along a river (Fig. 10-13). The ground will soak up a certain amount of water, but if rainfall continues, the ground's capacity to absorb any more is passed, just as is that of a sponge held under a flowing faucet. At such a point runoff is increased, and the water in the channel might swell to flood proportions (Fig. 10-14). Increased discharge means an increase in velocity, so that the river bed is scoured and the channel is enlarged. Part of the floodwaters is thus accommodated. But if discharge continues to increase, the ability of the channel

Fig. 10-12. Delta of the Virgin River in Lake Mead, Nevada. *John S. Shelton*

FLOODS

Fig. 10-13. On the evening and night of July 31 and August 1, 1976, 20.9 cm (8 in.) of rain fell in the canyon of the Big Thompson River, Colorado. The soil could not absorb that amount of water and the river flooded, reaching a maximum depth of almost 6 m (20 ft) and killing at least 139 persons. Such floods are estimated to occur every 100–500 years in that canyon. **A.** and **B.** (opposite) Before and after photographs of the town of Drake, where at least 19 lost their lives. *Hogan and Olhausen* **C.** (right) A 2-m (6.5-ft) boulder moved by the flood punched a hole into the side of this house. The largest boulder moved by the floodwaters is estimated to weigh 230 metric tons. *Dave Tewksbury*

to contain the rampaging river is exceeded, and the excess water spills out over the floodplain—that low-lying ground that periodically is inundated with floodwaters.

Earth scientists have ways of estimating how often floods of certain discharges will recur, and terms such as "5-year flood" or "25-year flood" are familiar to them. That means that it is estimated that floods of those discharges are thought to occur once in 5 years, or once in 25 years. They record the results of their work on flood maps, which show to what extent floods of various sizes might inundate a floodplain. Such maps are an integral part of wise land-use planning, and, if they are available, should be consulted if construction close to a river is being considered.

Flood maps must be updated fairly often. One reason is that each new flood provides new data with which to refine statistics on flood-recurrence intervals. Another is that urbanization tends to increase the incidence of floods of a particular size. The reason is simple:

the construction of roads and houses covers previous ground that once soaked up rainwater. The consequence is, of course, that runoff for a given-size storm is increased, which in turn increases the incidence of floods of a given size (Fig. 10-14B).

Few people driving through the eastern Washington desert on a hot summer's day realize that part of their route is along the course of an ancient flood, quite likely the largest on this well-watered planet! Called the Spokane Flood, it left such a clear mark on the Columbia Plateau that it appears in photographs taken from orbiting satellites, even though it took place thousands of years ago (Fig. 10-15). Geologist J. Harlan Bretz more than anyone else put together the geologic pieces on which the fascinating story is based.

The story begins less than 20,000 years ago when an enormous glacier pushed south along a wide front from Canada into eastern Washington, Idaho, and Montana. The ice blocked major drainages and formed a large lake known as Lake Missoula. Ice can hardly be said to make a stable dam, for it floats on water. Given the appropriate conditions, the dam burst and an amount of water equivalent to at least ten Amazons or one hundred Mississippis was unleashed on the basaltic plateau. Peak discharge lasted a day, and in a week or two the lake had drained and it was all over. Maximum water depths exceeded 200 m (656 ft), and velocities up to 75 km/hr (47 mph) seem possible.

What was left was devastation. Many square kilometers of basalt were reamed out into a landscape appropriately labeled the "channeled scablands." One of the most spectacular erosional features is Dry Falls, a cataract large enough to encompass several Niagara Falls (Figs. 10-16 and 10-17). Boulders up to 30 m (98 ft) were moved as much as 3 km (2 mi) by the floodwaters. Gravel bars formed, and some of the larger ones were more than 3 km (2 mi) long and 30 m (98 ft) high. Streams commonly form small sand ripples on their beds, but the Spokane Flood formed giant gravel ripples up to 7 m (23 ft) high and spaced more than 100 m (328 ft) apart.

The scabland features are so much out of line from the expected norm that it is no wonder that Bretz had trouble convincing his colleagues in the 1920s of a great prehistoric flood out West. He persevered, gathered more data, and eventually convinced the skeptics. Today few geologists doubt his interpretation. In fact, when in 1965 a group of geologists from many nations were taken through the flood area, they wired Bretz a message that concluded with the following words: "We are now all catastrophists."

Similar floods, but on a much smaller scale, occur when a dam fails. On 5 June 1976, the Teton Dam in southern Idaho failed and released 3.4 billion m³ (4.4 billion yd³) of water in eight hours (Fig. 10-18). A total of 480 km² (185 mi²) were inundated by the flood, in which eleven persons perished. Damage approached $1 billion. No one knows exactly what happened, but the early reports indicate that failure is related to highly fractured and porous

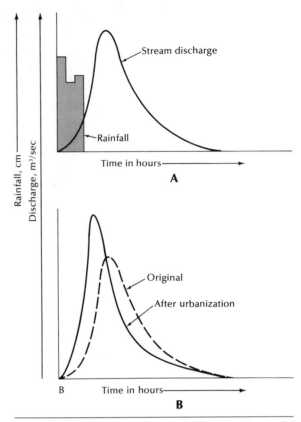

Fig. 10-14 A. Hypothetical relationship between the rainfall associated with an actual storm and the discharge of a stream. The lag between peak rainfall and peak discharge occurs because of the time required for the ground to become saturated. **B.** Alteration of the storm-discharge curve due to urbanization. Pavements and other impervious ground result in greater runoff and an earlier arrival of peak discharge, and floods may occur more frequently.

bedrock through which water could flow from the reservoir to the core of the earthfill dam, which it rapidly eroded away.

STREAM DEPOSITION

The broad plains bordering many of the large rivers of the world have been tempting sites for settlement since the beginning of history. Both the Egyptian and Babylonian civilizations were essentially riparian, and the life of their people was bound to the river, be it the Nile or the Euphrates, the giver of life in an arid land.

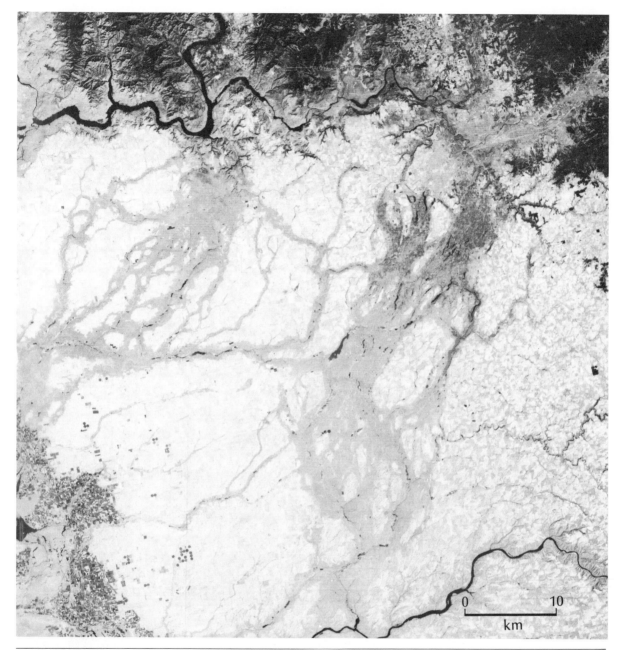

Fig. 10-15. Satellite photograph of the channeled scablands of eastern Washington, carved out by the ancient Spokane Flood. The path taken by the flood is shown by the gray braided pattern, whereas the present course of the Columbia River can be seen in the upper part of the photo. *Eros Data Center, USGS*

Fig. 10-16. Artist's rendering of the Spokane Flood with the glacier in the background. The cataract in the foreground is Dry Falls and measures 5.6 km (3.5 mi) wide and over 120 m (394 ft) high. *Washington State Park and Recreation Commission*

Fig. 10-17. Niagara Falls, as powerful as it is, possesses only a small fraction of the power put forth by Dry Falls at the time of the Spokane Flood. Photographed by George Barker in 1866. *Library of Congress*

Fig. 10-18 A. (above) Looking upstream to the Teton Dam, southern Idaho, a day after it failed. The debris adjacent to the stream below the dam was deposited by the flood. *U.S. Army Corps of Engineers* B. (below) The Teton Dam during the time of failure. The remaining section of the dam is to the right of the falls. The structure to the left is the dam's spillway. *Robert L. Jensen*

Fig. 10-19. River meandering across the plains of southern Wyoming. Meanders usually grow from the outside of the bends, and a cutoff course occurs where two out-ward-cutting bends meet. Oxbow lakes mark the course of the river cutoff.
Jan Robertson

Across the wide expanse of lowland border-ing a river, a stream such as the Nile is free to spread its waters in time of flood. In fact, the annual flood was an event of such importance to the survival of Egypt that a whole pantheon of deities centered around whatever divine force was responsible for the phenomenon. For centuries the Nile floods laid down sedi-ment—natural fertilizer—on farmlands ad-jacent to it. But the building of the Aswan Dam in the mid-twentieth century cut off that natural process. Even along the coast where the Nile meets the Mediterranean a decline in fish harvests and increased erosion may be tied to the fact that the dam checks not only the floods but the nutrients and sediments that flood-waters once carried to the sea.

In the pages that follow, we will discuss the characteristic features of floodplains, mention some problems in floodplain manipulation, and conclude with a description of sedimenta-tion and landform patterns where the rivers meet the sea.

Floodplain features From the nature of the term itself, we expect the surface of a flood-plain to be covered with deposits made by the river in time of flood. For example, the channel of the lower Mississippi River from a short dis-tance below its junction with the Ohio down to the Gulf of Mexico lies wholly upon its own alluvium.

Most of the more common floodplain features are along rivers that swing in large curving bends, or *meanders,* as they progress toward the sea (Fig. 10-19). The word "meander" comes from a river in western Turkey, the Menderes, and the derivation of the word is from the Latin *maendere,* to wander. Meanders are one of two most common river-channel patterns, but we will leave a discussion of their origin until later.

Floodplains sometimes are bordered by low bluffs marking the outer limits of the band of swampy ground across which the river has been free to wander, or meander (Fig. 10-20). Most such rivers—and the lower Mississippi is typical—are confined by low embankments, or natural *levees,* which slope gently away from the river. A section of low-lying ground, the *backswamp,* commonly is found between the bluff and the natural levee. It is a boggy section whose surface waters cannot flow back into the river because the slope of the natural levee is against them.

Natural levees are built by the river when it overtops its banks during a flood. Rather than surging violently over its banks at a single outlet, a typical flood moves away from the river toward the backswamp as a sullen, tawny, inexorably spreading tide of muddy water. The greatest check to the velocity of floodwaters comes when they leave the stream channel and first encounter the lake-like expanse of floodwaters that have inundated the backswamp. The bulk of the suspended sediment is deposited forthwith at the channel's edge where

the sudden drop in velocity occurs. The result, then, is that a narrow embankment, chiefly of fine sand and silt, builds up on either side of the lower Mississippi. Because their water content is less than the immediately adjacent, ill-drained backswamp, and the grain size of their sediment is larger than that of the swamp muck, the natural levees provide the only solid ground in such a saturated region as the lower Mississippi floodplain. For that reason roads, settlements, and farms are clustered along the higher and firmer ground of the levee immediately adjacent to the river. The backswamp, with its intricate pattern of bayous, branching channels, and lakes is virtually uninhabited, with the exception perhaps of birds and muskrats and their hunters.

In some places, tributary rivers can be caught in the low areas between the natural levee and the bluff (Fig. 10-20), running parallel to the mainstream for many kilometers before they are able to join it. Geologists have termed such rivers *yazoos* after a stream of that type—the Yazoo River in Mississippi.

In meandering rivers the main current clings to the outer side of a bend, where it is shunted by centrifugal force. There the current is strong, and the bottom of the channel is scoured more deeply than elsewhere. When the current leaves a bend, it normally does so on a tangent and may occupy a variety of positions before reaching the next bend. Upon entering the next bend it crosses to the opposite bank. The *crossings,* as they are called, are beset by shoals and shallows, or sand bars.

If we draw two profiles of a river channel, the first between two bends, where the current

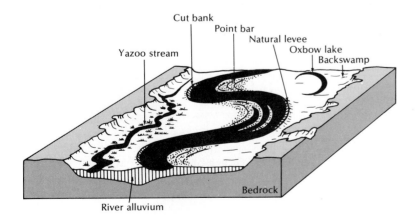

Cut bank
Point bar
Natural levee
Oxbow lake
Backswamp
Yazoo stream
Bedrock
River alluvium

Fig. 10-20. Characteristic features of the floodplain of a meandering river.

is approximately in midstream, and the second in a bend, the two cross sections will be quite unlike. In the first profile the river flows in a broad, shallow, nearly flat-floored trough. But in the bend, the outside of the curve is deep and the inside is shallow. In other words, the cross-sectional pattern is wedge shaped, with the thicker part of the wedge on the outside (Fig. 10-21).

How meanders enlarge is a question that has been long debated. The most generally accepted theory is that the outside bank is undermined by deep scouring, especially when the river is in flood, thus over-steepening the bank and slicing away its foundations. Commonly, the bank fails by slumping into the stream through the removal of support at the base, rather than being sawed horizontally by the river. No wonder, then, that the outside bank is called the *cut bank*. Much of the material derived in that way is deposited on the inside of the next bend downstream as a *point bar*. Thus, meanders can migrate laterally yet keep the same cross-sectional area, for generally the same amount of material that is taken from the cut bank of a bend has been added to the point bar of the next bend downstream.

Meanders migrate in the direction of the cut bank. In places, the neck of land between the bends has been eroded through and a new, more direct channel—called a *cutoff*—is followed by the river. The old channel then contains a crescent-shaped lake, termed an *oxbow lake*, that eventually fills in with sediment. All told there have been about 20 naturally occurring cutoffs on the lower Mississippi since 1765. About 15 have been made artificially by the Mississippi River Commission since 1932 to straighten out the river's course, thereby increasing the gradient and thus the velocity, and as a consequence theoretically diminishing the flood hazard by improving the hydraulic efficiency of the channel. As we shall see later, however, straightening channels can have some disastrous side effects.

Historically, one of the more interesting cutoffs of the Mississippi occurred at Vicksburg in 1876. Before that date the river made a

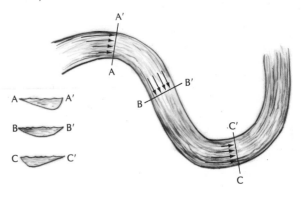

Fig. 10-21. Plan view of a meandering river, showing the change in channel shape with position along the river. The length of the arrows depicts relative river velocity.

broadly sweeping curve past the city (Fig. 10-22). In 1876, however, the river formed a cutoff south of Vicksburg that isolated that town from the river. Ironically, in 1862, General Ulysses S. Grant tried unsuccessfully to construct an artificial cutoff at about the same place so that Union river traffic could by-pass Confederate guns at Vicksburg.

Let Mark Twain, in his role of steamship pilot, have the last word on meandering streams. In *Life on the Mississippi* he writes about the river meanders and how the river is shortened when it cuts through the narrow neck of a meander. He grossly misuses the principle of uniformitarianism and is not very kind to science in general, but who will argue?

Therefore, the Mississippi between Cairo and New Orleans was 1215 miles long 176 years ago. It was 1180 after the cutoff of 1722. It was 1040 after the American Bend cutoff. It has lost 67 miles since. Consequently its length is only 973 miles at present.

Now if I wanted to be one of those ponderous scientific people, and "let on" to prove what had occurred in the remote past by what had occurred in a given time in the recent past, or what will occur in the far future by what has occurred in late years, what an opportunity is here! Geology never had such a chance, nor such exact data to argue from! . . . Please observe:

In the space of 176 years the Lower Mississippi has shortened itself 242 miles. That is an average of a trifle over one mile and a third per year. Therefore, any calm person, who is not blind or idiotic, can see that in the Old Oolitic Silurian Period, just a million years ago next November, the Lower Mississippi River was upwards of 1,300,000 miles long, and stuck out over the Gulf of Mexico like a fishing rod. And by the same token any person can see that 742 years from now the Lower Mississippi will be only a mile and three-quarters long, and Cairo and New Orleans will have joined their streets together, and be plodding comfortably along under a single mayor and a mutual board of aldermen. There is something fascinating about science. One gets such wholesale returns of conjecture out of such a trifling investment of fact.

In places, meandering rivers are considered a real nuisance, especially if the lateral migration is at the expense of valuable land. Remedial measures to thwart such wayward rivers vary, but one common one is to line the outside bend with that symbol of American affluence—the junked car.

If rivers do not have a meandering pattern, chances are that the pattern will be *braided*, the only other major river pattern. Rather than flowing in a single, rather narrow but deep channel, the river follows a braided pattern—a series of wide, shallow anastomosing channels that continually join and part from one another in a downstream direction (Fig. 10-23). The individual channels may change position hourly, daily, or seasonally—in short, they are quite active.

The reason that some rivers are braided and others meander lies mainly in the type of sediment the stream carries. Meandering streams require a rather tough bank material to restrict the river to a single channel. The toughest material is cohesive silt and clay, and so it is that meandering rivers are those with a relatively high suspended load. In contrast, a stream can easily spread into many shallow channels if the bank material is loose and noncohesive. The weakest bank materials are sand

Fig. 10-22. Major changes in the courses of the Mississippi and Yazoo rivers over a 16-year period.

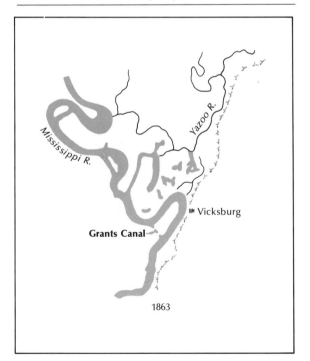

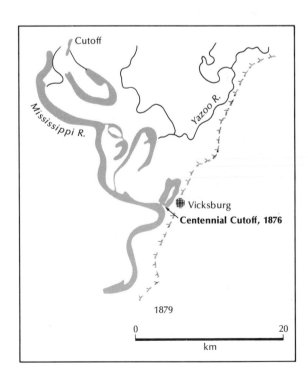

Fig. 10-23. Braided channels of the Nelchina River in the Copper River region of Alaska. *J.R. Williams, USGS*

and gravel; hence, braided streams are characteristically bed-load streams.

Stream pattern is all a part of the equilibrium of a stream, which is why Mackin talked of "prevailing channel characteristics" in his definition of a graded stream. A high-gradient, boulder-carrying braided stream can be as much at grade as a low-gradient, low-velocity meandering stream burdened with silt. In each case the stream may be doing only the work asked of it, no more or no less.

Deltas Herodotus, in the fifth century B.C., impressed by the branching pattern of the distributaries of the Nile, compared the form of the watery, muddy region between Cairo and Alexandria to the Greek letter *delta*, Δ. The comparison is so apt that it has won acceptance in most of the languages of western Europe.

The Nile delta is a nearly ideal example of that particular landform, so much so that few others measure up to its perfection. A high-altitude photograph shows how the main channel of the river separates into a host of branching lesser arms called, quite appropriately, *distributaries* (Fig. 10-24). Another feature of the Nile delta is the bordering bays and lakes, of which one, Abu Qir Bay, is a good example. Abu Qir was the site of the Battle of the Nile, in which the French fleet was destroyed by Nelson, thus ending Napoleon's hopes for an Eastern empire. Similar *delta-flank depressions*, as such water bodies are called, border many of the other deltas of the world. Well-known ones are Lake Ponchartrain by the Mississippi delta, the Zuider Zee (IJsselmeer) and marshes of Zeeland adjacent to the Rhine, and the lagoon surrounding Venice at the mouth of the Po.

Not all of the world's rivers form deltas where they enter the sea. Two large ones that do not are the St. Lawrence and the Columbia —the St. Lawrence because it has little chance to pick up much sediment in the short run between Lake Ontario and the Gulf; the Columbia because it discharges directly into the open sea whose powerful waves and currents quickly redistribute the rivers's burden of sand and silt.

The best-developed deltas are most likely be constructed where a river moves a large load of sediment into a relatively undisturbed body of water. Examples of such impressive accumulations of riverine deposits are the great deltas at the mouths of the Ganges-Bramaputra in India and East Bangladesh, the Indus in Pakistan, the Tigris-Euphrates in Iraq, the Niger in Nigeria, the Yangtze Kiang and Hwang Ho in China, the Mississippi in the United States, the Danube in Romania, and the Volga in the USSR. One of the most picturesque of the deltaic worlds is the Camargue, the land of cowboys and semi-wild cattle at the mouth of the Rhône in southern France.

In North America, the Mississippi is by far the best known geologically, not only because of its long record of channel changes but because about 90,000 oil wells have been drilled in its sediments and because repeated geophysical surveys have been made up and down its length. All of that information combined gives us a uniquely detailed, three-dimensional picture, not only of the 29,000 km² (11,194 mi) of delta surface but of the over 1.5-km (0.9-mi)-thick sediments below the waters of the Gulf of Mexico as well.

The great delta of the Mississippi is not a

Fig. 10-24. Satellite photograph of the Nile Delta. *NASA*

Fig. 10-25. At Head of Passes, the Mississippi River divides into several distributaries which lead out to the Gulf of Mexico. *Humble Oil and Refining Co.*

simple structure, but is a compound feature built up through a complexly overlapping pattern of *subdeltas* that have been constructed over the last 5000 years. From the older deltas it can be seen that the river established its present course only very briefly before the arrival of the first European explorers, Cabeza de Vaca in 1528(?), or Hernando de Soto's followers in 1544.

Earlier than 5000 years ago, sea level was lower than it is at present as a result of the stockpiling of oceanic waters on land in the form of glaciers. At the height of the glaciation the Mississippi flowed in a channel en-

trenched in older deltaic sediments. It reached the Gulf of Mexico at the head of a submarine canyon located south of New Orleans and southwest of the present birdsfoot delta, so named for its obvious shape. Melting glaciers returned water to the sea, the sea level rose, and the present delta began to take shape.

Head of the Passes is where the present-day river breaks up into three major and a number of minor distributaries (Fig. 10-25), and there the river depth is around 12 m (39 ft). The slope of the channel bottom in the distributaries is upstream against the gradient of the river surface because most deposition occurs at the mouth of the distributaries; there the river builds up shallow sand bars as it enters the Gulf of Mexico. Thus, the normal low-water depth at the mouth of most distributaries is as little as 3 to 4.5 m (10 to 15 ft)—without the interference of dredging and jetty construction—and in some of the minor, less frequently used channels, as little as 1 to 1.5 m (3 to 5 ft).

Natural levees border the channels of distributaries that are growing outward into the Gulf. The height of the natural levees increases gradually inland from the Gulf until they are around 4 m (13 ft) high at New Orleans, a distance of 165 km (103 mi) from Head of the Passes.

This strange deltaic world—half water and half land, the abode of water birds of great variety, with a continuously changing pattern of lakes, swamps, marshes, and constantly shifting streams—is one of the most important environments for human beings. From the beginnings of historic time their fertile soil, their network of waterways, and their position as a meeting ground between seafarers and those who make their living along the rivers of the world have made deltas tempting sites for coastal cities. Such a bewildering maze of channels, large and small, was made to order for piracy, as witness the successful operations for many years of the brothers Lafitte at Barataria in the Mississippi delta.

Quite a price is exacted, however, from a delta city in return for its communication advantages. Such a city is under constant threat of inundation by flood, and building foundations are insecure—a visit to Venice is an impressive illustration of what differential settling can do to structures built on delta mud. Even such a shallow excavation as a grave may fill with water (the vaulted sepulchers of New Orleans are an answer to that problem), and the development of an unpolluted local water supply is a difficult task. Over-riding the many adversities is the greatest threat of all, that the river may change its course completely or the channel may silt up. A good example of the latter fate is the ancient city of Ravenna; in the days of the Emperor Justinian (483–565) it

was a leading seaport on the Adriatic; now it is about 10 km (6 mi) from the coast.

The low country bordering the mouths of the Rhine (the Randstad) provides an impressive example of the problems besetting the inhabitants of a delta. The Randstad is one of the most densely inhabited regions of the world, and one where an immense volume of seaborne and riverborne trade moves through such ports as Rotterdam and Amsterdam. Those cities, already below sea level, throughout their entire existence have not only had to endure attacks overland—a hazard to which delta cities are peculiarly vulnerable—but have had to labor valiantly to hold back the sea. The explanations of all the difficulties faced by the people of the Rhine delta are complex, but they certainly include the currently rising level of the sea, the compaction of the water-logged clays upon which the cities stand, and the apparently geologically active subsidence along that part of the North Sea coast.

Many of the same difficulties plague New Orleans, not the least being the problem of subsidence. Indications of a relative lowering of the land with respect to the sea are everywhere. Among them are the remains of Indian settlements far out on the bottom of Lake Ponchartrain, drowned cypress trees and inundated farmland around the lake's margin, and the sunken streets and graves of the deserted settlement of Balize near Head of the Passes. The now-silent, shell-paved streets are buried under the marsh about 1 m (3 ft) below sea level.

There is little disagreement over the evidence of subsidence in many of the world's deltas, but there are strong differences of opinion as to its cause. There are those who believe that the addition of as much as 1.8 million metric tons of sediment a day, as in the Mississippi delta, makes for an excess load on the earth's crust, which bows down as a consequence. Opponents of that belief point out that the excess load is not so great as might appear at first, since the weight of the displaced sea water has to be considered, too. Thus, the material added to the crust may have

Fig. 10-26 A. (opposite, top) Flight of river terraces along the Madison River, Montana. *W. C. Bradley* **B.** (opposite bottom) River terraces along the Rangitata River, Canterbury, New Zealand. Most of the terraces were formed during the time of the last glaciation. *New Zealand Geological Survey* **C.** (above) Rounded river gravel overlying a dark-colored bedrock surface cut by the river. Before a flat adjacent to a river can be called a river terrace, deposits such as these must be present. *W. C. Bradley*

a specific gravity of about 1.8. How the lighter material displaces heavier sub-crustal material with a specific gravity of perhaps 3.3 is a tricky question to answer.

The answer probably is not a simple one, but certainly involves many of the factors operating in the Netherlands. The addition of an extra burden of sediment, as in the Mississippi delta, may not be adequate in itself to bring about a broad crustal downwarping, but if a load such as the Mississippi's accumulates in a region where subsidence is the dominant geologic process, the additional weight certainly is not going to work in an opposite direction.

Stream terraces—a case for disequilibrium

Along many streams are found extensive flat areas that slope downstream more or less parallel to the slope of the stream (Fig. 10-26). If the "flats" are underlain by river deposits, the landform is called a *river terrace*. In essence, such terraces are ancient floodplains, long abandoned and stranded high enough so that present-day floods no longer overtop them. River terraces and their associated deposits are important in deciphering the history of a river—a task, as we shall see, that is far from easy.

Two major kinds of terraces are recognized (Fig. 10-27). One is a *cut terrace,* so named because it consists of a thin veneer of gravel resting upon a fairly smooth surface cut from bedrock. How thin is thin? Geologists have their opinions, but most would say that the layer of gravel can amount to no more than the thickness moved during the deep scour that accompanies major floods. For small rivers, that means a depth of about 5 m (16 ft) or less, and for large rivers about 10 m (33 ft). The other kind of river terrace is a *fill terrace.* It too rests on bedrock, but the river deposits are relatively thick. It is difficult to generalize, but the deposits are thicker than the depth of scour and hence thicker than the gravels of cut terraces. Geologists find such qualitative rules of thumb indispensable, and that is just one of many.

We can learn something about former river behavior through the study of terraces and deposits. First, consider cut terraces. In some places they might be 1 km (0.6 mi) or more wide. In order to develop such an extensive former floodplain, a river must be in near-perfect equilibrium for a long time, swinging back and forth across the floodplain and nipping away at valley walls. Imagine a horizontal saw cutting through the landscape. Although the rates at which rivers cut through bedrock are hard to come by, we can estimate that surfaces of 1 km (0.6 mi) or more found in many parts of the American West may have required near-equilibrium conditions for hundreds of thousands of years. Something then changed the equilibrium of the river so that it had no choice but to downcut, leaving the abandoned floodplain as a cut terrace.

Fill terraces represent a more complicated

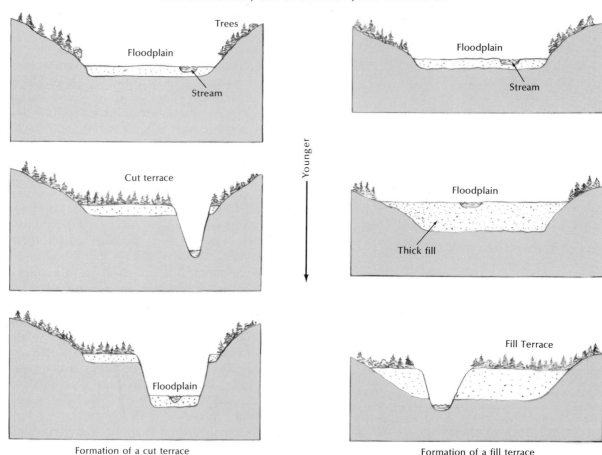

Fig. 10-27. The formation of a cut terrace (left) and a fill terrace (right).

history. Initially, a river flows at a low level in its valley, perhaps swinging back and forth on a floodplain. Aggradation then follows as river sediment is stored on the floodplain rather than being carried farther downstream. Next, the floodplain becomes a terrace as the river downcuts. The sequence of events, therefore, is aggradation followed by degradation. Some valleys have many fill terraces inset into one another (Fig. 10-28)—indicating that the aggradation-degradation cycle has taken place many times.

It takes nothing short of a good detective to come up with a reason for terrace formation, and, like so many other geologic solutions,

the best one usually seems to be a list of possibilities. In general, geologists look for "downstream" or "upstream" causes. Downstream variation in base level always is a possibility because a river responds quickly to variations in sea level at its lower end, aggrading as sea level rises and degrading as it falls. An upstream cause could be the reaction of a river to changes in the size and amount of load or in the volume of discharge—changes that might have resulted from a former climatic change. Whatever the correct interpretation, it does not come easily. The slope of a terrace, the characteristics of its deposits, and any remaining ancient channel patterns etched on its surface all have to be compared with similar features of the present-day river before any one hypothesis is preferred over others.

Tectonism also can affect river-terrace formation. For example, rivers usually entrench in areas undergoing uplift but aggrade in areas undergoing tectonic flattening. Some thousands of years after the act, an uplifted area might be characterized by cut terraces, and a flattened area by fill terraces. In the latter case, faults might not be recognizable in the loose gravels, and river terraces may be the only clue to former tectonic activity.

Artificial stream disequilibrium   Rivers can be in or out of equilibrium as part of their own doing, or they can be helped along by the works of humans. The problem here is that people chose to live along rivers but did not always build in the right places—suitable sites could only have been chosen after long-term study of the behavior of the river, and nobody had time for that. So when catastrophic events took place—catastrophic to landowners but run-of-the-mill events in the life of the river—the people affected wanted somebody to do something. In many places somebody did do something, and in many places the results had unforeseen side-effects.

Examples of river disturbances caused by humans abound. Because there are so many, we will cover only one fairly "innocent" one here, innocent at least at first glance. It should be pointed out that, as with any patient, before remedial measures are prescribed it is not

Fig. 10-29. (Overleaf) The Grand Canyon at the foot of the Toroweap. The Colorado River was mainly responsible for cutting the canyon vertically, but its breadth is largely the result of slope processes that etched out the rocks in delicate relationship to their resistance to erosion. Rocks that are readily eroded form the gentle slopes whereas more resistant rocks form the cliffs. *Drawing by William H. Holmes, from J.W. Powell, The exploration of the Colorado River, 1875.*

a bad idea to try to find out how a river reacted to natural changes in the past before imposing new changes on it. As was emphasized earlier, however, many rivers, like people, simply do not divulge their past histories readily.

An example is that of the meandering Blackwater River in Missouri. Local flooding had been a problem before 1910, and controlling measures seemed in order. The parameters of the original river in the headwaters reach were: 54 km (34 mi) long, 1.7 m/km (9.3 ft/mi) gradient, with 1.8 meanders/km (3 meanders/mi). Bridges 15 to 30 m (49 to 98 ft) wide spanned the river. Because meandering rivers are slow moving, they cannot get rid of the rapidly increasing discharge that accompanies storms, and the result is flooding. The common remedial practice is to cut a straighter channel, which increases both stream gradient and velocity and disperses the floodwaters more quickly.

In 1910, the Blackwater was straightened and channelized. As Mark Twain points out so vividly, straightening also results in shortening. The river was shortened by about one-half at the expense of increasing the gradient to 3 m/km (17 ft/mi). The artificial channel was cut 9 m (30 ft) wide at the top, 1 m (3 ft) wide at the base, and 3.8 m (12 ft) deep, for a cross-sectional area of 19 m² (23 yd²).

The floods were contained in the channel, but the latter had a real zest for growth. It enlarged greatly over the following 60 years, and the river sector that enlarged the most measured 71 m (233 ft) across at the top, 13 m (43 ft) at the bottom and 12 m (39 ft) deep, for a cross-sectional area of 484 m² (579 yd²)! In any river system the tributary streams follow the lead of the main stream, and drainage was no excep-

Fig. 10-28. Three fill terraces, each underlain with stream deposits of a different age. The highest terrace is the oldest and the lowest is the youngest.

tion. The tributaries of the Blackwater downcut just as fast as the main stream, and gullies began to spread over the landscape.

Bridges were built across the straightened channel, but as the channel grew wider and deeper, the bridges had to be lengthened and vertical pilings extended to avoid collapse. In spite of those efforts, many could not withstand the stress and did collapse.

The effects of channel straightening were also felt downstream beyond the limits of the artificial channel, for sediment derived from the enlarging channel was deposited there. Deposition was so rapid that two successive generations of fence posts were buried in the flood debris.

The story of the Blackwater is an excellent small-scale example of what can happen when an attempt is made to regulate a river. Once the forces are set in motion, they are difficult to check. We now have enough case histories to intelligently forecast the various side effects of most regulatory measures. So it is hoped that the arguments of cost versus benefit can be properly laid before those individuals charged with making decisions.

Role of streams in landscape evolution

Today virtually everyone recognizes that valleys, even such imposing ones as Hells Canyon in Idaho, the Grand Canyon in Arizona (Fig. 10-29), or Kings Canyon in California—all of them more than 1 km (0.6 mi) deep—were cut by the narrow ribbon of turbid water in the stream at the bottom, barely visible hundreds of meters below the canyon rim.

This belief, so obvious now, was disputed by our predecessors, and was not fully acceptable even to such illustrious figures as Sir Charles Lyell and Charles Darwin. As late as 1880 many geologists were content to believe that while streams were capable of some downcutting, nonetheless deeper and more impressive gorges, such as Grand Canyon, resulted from a violent sundering of the earth's crust. The presence of such a river as the Colorado was purely fortuitous—instead of cutting the can-

yon it simply followed the course it does because that was predetermined as the lowest and easiest route to follow.

Although a belief in a cataclysmic origin unquestionably has greater appeal to the imagination, it simply means that many geologists of a century ago had overlooked the succinct and eloquent statement made in 1802 by a Scottish mathematician and amateur geologist, John Playfair (1748–1819). Although his essay is now widely quoted, it is worth repeating because of the clarity of style—an attribute not always characteristic of scientific writing today—and because his choice of essentially simple, straightforward words made the point so long ago that streams do in fact excavate the valleys they occupy.

If indeed a river consisted of a single stream, without branches, running in a straight valley, it might be supposed that some great concussion, or some powerful torrent, had opened at once the channel by which its waters are conducted to the ocean; but when the usual form of a river is considered, the trunk divided into many branches, which rise at a great distance from one another, and these again subdivided into an infinity of smaller ramifications, it becomes strongly impressed upon the mind, that all these channels have been cut by the waters themselves; that they have been slowly dug out by the washing and erosion of the land; and that it is by the repeated touches of the same instrument that this curious assemblage of lines has been engraved so deeply on the surface of the globe.*

The Colorado River is responsible for cutting its channel to the bottom of the Grand Canyon, but it is not responsible for the entire excavation of the gorge, 21 km (13 mi) wide at the top. The widely flaring upper portion of the canyon, outside of the narrow slot actually cut by the Colorado, is the result of mass movement downslope, of weathering, and of slope wash. The river has served as a gigantic conveyor belt, running without cessation and carrying away most of the debris supplied to it while continuing to cut downward.

* From Illustrations of the Huttonian theory of the earth, by John Playfair, Edinburgh, 1802.

The long-term goal of all rivers is to help wear the land down to a nearly featureless plain near sea level where stream erosion virtually ceases. Sea level was called the *base level of erosion* years ago by Major John Wesley Powell (1834–1902), pioneering geologist of the far western United States and leader of the first party to explore the Grand Canyon of the Colorado (Fig. 10-30). In his thinking the formation of such a plain included much more than the carving of a narrow stream channel down to sea level; it also involved the wearing down of all interstream areas until an entire region was nearly at sea level.

The effect of the slow wasting away of the area between streams is shown in the accompanying diagram (Fig. 10-31). A once broad upland, trenched by narrow canyons, may be reduced over many years to a nearly level plain not very far above sea level. To such a broad, nearly featureless plain the term *peneplain* is applied, from the Latin *paene*, meaning almost, and the English "plain."

A peneplain, because it is a product of widespread degradation by streams and mass wasting, can never have a perfectly level surface, although it may come very close to it. Being a product of erosion, the surface of the plain truncates the underlying bedrock. The bedrock surface may not be worn down everywhere to the same monotonous level; portions underlain by more resistant rock may stand higher than their surroundings. Such isolated, residual hills, or even mountains, are called *monadnocks*, after Mount Monadnock in New Hampshire (Fig. 10-31).

Few incontestable examples of peneplains have been described from round the world, although many partial peneplains, or partially convincing examples, have been described. Many expanses of nearly level plains of barren rock, such as the Hudson Bay region of Canada or much of the Scandinavian Peninsula, have complex histories, and each of those cases involves widespread glacial stripping.

Some geologists propose that a few of the flattish uplands in the western mountains of the United States and other areas might be up-lifted remnants of former low-relief erosion surfaces. In such cases erosional flattening would have preceded any mountain-building activity. Because such surfaces occur in high mountains that themselves are eroding rapidly, their days of preservation are numbered. In geology, however, numbers run large, and the high flatlands may be a part of the landscape for several million years to come.

Actually, there is less than general agreement among geologists about the way in which a combination of stream erosion and mass wasting operates to produce peneplains (Fig. 10-32). One way is through *downwasting*, a process in which the steepness of slopes adjacent to a stream valley gradually diminishes through soil creep, gravitative transfer, and the decomposition of rocks. The process might be regarded as a point of view based on the work of William Morris Davis (1850–1934), a pioneer American scientist. A different solution was advocated by an Austrian geologist, Walther Penck (1858–1945). According to Penck, the landscape evolves through *backwasting*, a process in which valley slopes retreat essentially parallel to themselves. An equilibrium slope is first established, as in Figure 10-31. Once equilibrium has been reached for that particular environment, the slope will continue to retreat. Eventually a graded, stripped rock surface develops at the base, and, as the slope retreats, widens. Ultimately, all land above the level of the widening platform will be stripped off as the separate, retreating slopes meet, and an entire region will have been worn down to a base level of erosion with a very gentle gradient toward the sea.

That geologists still argue the merits of backwasting versus downwasting is testimony to the impact Davis and Penck had upon geological thought. However, a final choice between the two explanations would be premature in view of the state of knowledge today. The answer to the problem of the wearing away of the land between the rivers is complex, and probably involves elements of each explanation. If any differentiation can be made, the possibility seems strong that backwasting is the leading

Fig. 10-30. Major John Wesley Powell led the first expedition into the Grand Canyon in 1869. His own words describe well the mood of the small band as it moved into the chasm:

August 13. We are now ready to start on our way down the Great Unknown. Our boats, tied to a common stake, are chafing each other, as they are tossed by the fretful river. They ride high and buoyant, for their loads are lighter than we could desire. We have but a month's rations remaining. The flour has been resifted through the mosquito-net sieve; the spoiled bacon has been dried, and the worst of it boiled; the few pounds of dried apples have been spread in the sun, and reshrunken to their normal bulk; the sugar has all melted, and gone on its way down the river; but we have a large sack of coffee. The lighting of the boats has this advantage; they will ride the waves better, and we shall have but little to carry when we make a portage.

We are three quarters of a mile in the depths of the earth, and the great river shrinks into insignificance, as it dashes its angry waves against the walls and cliffs, that rise to the world above; they are but puny ripples, and we but pigmies, running up and down the sands, or lost among the boulders. . . .

With some eagerness, and some anxiety, and some misgiving, we enter the canyon below, and are carried along by the swift water through walls which rise from its very edge. They have the same structure as we noticed yesterday—tiers of irregular shelves below, and, above these, steep slopes to the foot of marble cliffs. We run six miles in a little more than half an hour, and emerge into a more open portion of the canyon, where high hills and ledges of rock intervene between the river and the distant walls. Just at the head of this open place the river runs across a dike; that is, a fissure in the rocks, open to depths below, has been filled with eruptive matter, and this, on cooling, was harder than the rocks through which the crevice was made, and, when these were washed away, the harder volcanic matter remained as a wall, and the river has cut a gateway through it several hundred feet high, and as many wide. As it crosses the wall, there is a fall below, and a bad rapid, filled with boulders of trap; so we stop to make a portage. Then we go, gliding by hills and ledges, with distant walls in view; sweeping past sharp angles of rock; stopping at a few points to examine rapids, which we find can be run, until we have made another five miles, when we land for dinner.

Then we let down the lines, over a long rapid, and start again. Once more the walls close in, and we find ourselves in a narrow gorge, the water again filling the channel, and very swift. With great care, and constant watchfulness, we proceed, making about four miles this afternoon, and camp in a cave.

A. Powell and his companions before the journey. *Library of Congress* **B.** Powell with Tau-gu, a Paiute Indian chief. *Smithsonian Institution* **C.** Powell sat in a captain's chair lashed to his boat. Making up the party were four other boats and nine men. Not all of them completed the journey; three men who climbed to the canyon's rim met their deaths at the hands of Indians who did not believe that the explorers had actually navigated the river. *J.K. Hillers, 1869, USGS*

Fig. 10-31. The successive cross profiles of a valley, beginning with a narrow canyon and progressing toward a peneplain. On the right, the slopes are diminished by downwasting; on the left, the slope angles are parallel, and the valley walls retreat through backwasting.

process in the arid lands of the world, while downwasting is paramount in humid lands, where vegetation blankets the slopes, soil is deep, and mass movement plays a strong supporting role.

Regardless of the details involved, geologists today generally agree that the running water of streams combined with mass wasting—given time—can wear away the highest mountains made of the most resistant rocks, until they are reduced to a nearly featureless plain (Fig. 10-33). That so much of the land surface of the world fails to conform to that drab description indicates in itself the recency and continuing nature of the forces of deformation affecting the rocks of the earth's crust. Add to that crustal unrest the fluctuating level of the oceans during the ice ages, and no wonder we cannot appease the skeptics. To them we can only say "It could happen, given more time."

Low-relief erosion surfaces probably have been formed again and again through the long antiquity of the earth. Some ancient and buried unconformities could be former peneplains graded to an unknown base level.

RATIONAL USE OF RIVERS

In the early days of our country's history, before the transcontinental railroad was completed and before the present web of highways was constructed, rivers provided the principal path of transportation for both goods and people. Now, of course, they are no longer used for transport to any great extent, but certainly they are used in other ways. One of the most important is as a supply of water for

Fig. 10-32. Mount Monadnock, New Hampshire, rises above a low-relief erosion surface that was subsequently modified by glaciation. *John S. Shelton*

both individuals and industry, the natural outcome of which is the use of rivers as a means of disposal of our monumental amount of waste products and sewage (Fig. 10-34).

In an attempt to harness the forces of nature for our own benefit, we have built and propose to build a great many dams across our rivers.

The water stored behind the dams is being used for a number of purposes: production of electric power, maintenance of a steady supply of water for irrigation, and flood control, among others. At the same time, the number of rivers now preserved in national parks and monuments, or designated as Wild Rivers in-

Fig. 10-33 A. (above) Flat erosional remnants form the summit plateau of the northern Wind River Mountains, Wyoming. *Jan Robertson* **B.** (below) Peneplain cut in schists and subsequently dissected by streams, west of Dunedin, New Zealand. *W. C. Bradley*

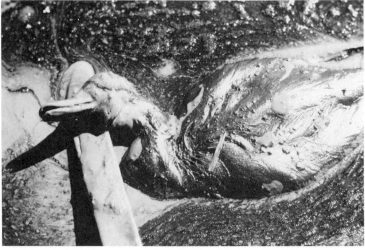

Fig. 10-34. (A) (above) Assorted junk and **(B)** (right) a dead duck in a pond filled with acid water and oil near Ogden, Utah. It was later cleaned up under supervision of the Environmental Protection Agency to prevent possible contamination of Great Salt Lake and a nearby wildlife preserve. *Docuamerica*

Fig. 10-35. Hell's Canyon, cut by the Snake River in Idaho averages 1680 m (5510 ft) in depth and is deeper than the Grand Canyon. Layers of dark basalt make up the walls—a marked contrast to the play of colors in the Grand Canyon. *Oregon State Highway Department*

dicates that their aesthetic and recreational value is receiving more and more recognition.

That there is a mighty conflict among the users is clear from even minimum attention given by the mass media. Choices must be made, but too often they are made by a small group of persons who do not necessarily represent the views of the majority of those affected. The intent of the decision-makers may not be to deceive deliberately; they may simply have neglected the fact that the views of the community and the country at large may differ from their own.

Even when plans for development of a river are disclosed, the proposers are often remarkably stubborn, refusing to listen to the opposition's arguments or to consider alternatives. One prominent hydrologist, Luna Leopold, believes that a reason for such behavior is that while proposed benefits can be *quantitatively* stated, the "nonmonetary values are described either in emotion-laden words or else are mentioned and thence forgotten." In an effort to remedy such situations, Leopold devised a chart to evaluate quantitatively some of the aesthetic factors of river sites. Disclaiming any personal bias, he applied his method to twelve river sites in Idaho in the vicinity of Hells Canyon of the Snake River, an area in which the Federal Power Commission wants to construct one or more additional hydroelectric dams.

After comparing the Hells Canyon site with the other sites in Idaho which were capable of hydropower development, he found Hells Canyon the most worthy of preservation (Fig. 10-35). Leopold then compared the site to other river valleys that lie within national parks: (1) the Merced River in Yosemite, (2) the Grand Canyon of the Colorado River, (3) the Yellowstone River near Yellowstone Falls, and (4) the Snake River in Grand Teton National Park. His conclusion was that "Hells Canyon is clearly unique and comparable only to Grand Canyon of the Colorado River in these features." It remains to be seen whether his method will be a valuable tool in determining the rational use of our river resources.

SUMMARY

1. Streams are the main agents that transport material from land to sea.
2. Water in a stream flows in a turbulent fashion, and turbulent flow aids greatly in the transportation of the stream's *solid load* (*bed load* and *suspended load*). The *dissolved load* quantitatively makes up an important percentage of the total load carried by a stream. It is derived from chemical weathering.
3. Stream velocity is determined by channel shape and roughness, discharge, and stream gradient. The greater the velocity, the greater the size of material a stream is capable of moving.
4. Streams strive for an equilibrium in which the load added to the stream is balanced by the material carried by the stream. A stream which has reached equilibrium is called a *graded stream*.
5. Stream terraces reflect former stream disequilibrium so that the stream either *aggraded* or *degraded*. The two main kinds of stream terraces—*cut terraces* and *fill terraces*—each reflect a different stream history. Human intervention can bring about stream disequilibrium, and often the effect is detrimental to people and structures along the river.
6. Floods are natural to all streams and the identification of the extent of inundation for floods of various magnitudes is an important part of land-use planning. People can cause more frequent flooding through building practices.
7. The two main kinds of stream patterns are largely related to the kind of load. Braided patterns reflect a high bed load, and meandering patterns a suspended load.
8. High-relief landscapes, given sufficient time and tectonic stability, can evolve to a low-relief rolling surface—a *peneplain*. Slope processes mainly are responsible for the lowering of the original landscape, especially in a humid climate, whereas the main role of the rivers is to transport the material out of the drainage basin.

SELECTED REFERENCES

Baker, V. R., 1973, Paleohydrology and sedimentology of Lake Missoula flooding in eastern Washington, Geological Society of America Special Paper 144.

Bloom, A. L., 1969, The surface of the earth, Prentice-Hall, Englewood Cliffs, New Jersey.

Bretz, J. H., 1969, The Lake Missoula floods and the channeled scabland, Journal of Geology, vol. 77, pp. 505–43.

Easterbrook, D. J., 1969, Principles of geomorphology, McGraw-Hill Book Co., New York.

Emerson, J. W., 1971, Channelization: a case study, Science, vol. 173, pp. 325–26.

Garner, H. F., 1974, The origin of landscapes, Oxford University Press, New York.

Leopold, L. B., 1969, Quantitative comparison of some aesthetic factors among rivers, U.S. Geological Survey Circular No. 620.

Leopold, L. B., Wolman, M. G., and Miller, J. P., 1964, Fluvial processes in geomorphology, W. H. Freeman and Co., San Francisco.

Mackin, J. H., 1948, Concept of the graded river, Geological Society of America Bulletin, vol. 59, pp. 561–88.

Morisawa, M., 1968, Streams, their dynamics and morphology, McGraw-Hill Book Co., New York.

Ruhe, R. V., 1975, Geomorphology, Houghton Mifflin Company, Boston.

Thornbury, W. D., 1969, Principles of geomorphology, 2nd ed., John Wiley and Sons, New York.

Fig. 11-1. Apollo 7 photograph of the arid Zagros Mountains of Iran. The circular structures are salt domes, which can be preserved at the surface only because of the extreme dryness. *NASA*

11

DESERT LANDFORMS AND DEPOSITS

Least familiar of land areas, aside from the extreme Arctic, are the world's deserts (Fig. 11-1). Perhaps their seeming mystery lies in their relative remoteness from lands such as western Europe and the Atlantic coast of North America, where the modern patterns of Western civilization developed. Had Western life remained centered on the Mediterranean, deserts would have been much closer to our daily lives because the limitations imposed by aridity bear heavily on such bordering countries as Spain, Morocco, Algeria, Libya, Egypt, and Israel.

In earlier days much of the southern shore of the Mediterranean was the granary of Rome, and once-flourishing cities, such as Leptus Magnus in Libya, are now stark ruins half buried in the sand. One of the problems in studying deserts is that the boundaries are not fixed inexorably, but may expand or contract through the centuries. In fact, there is much evidence from *paleobotany*—the study of fossil plants—that deserts are relatively late arrivals among the earth's landscapes. Deserts require a rather specialized set of circumstances for

their existence, and in a moment we shall inquire into what some of them are.

First of all we need a working agreement as to what constitutes a desert. Temperature is not the only factor; some are hot almost all of the time, others may have hot summers and cold winters, and some are cold throughout much of the year. Drought is their common factor; in a general way we might call those regions deserts where more water evaporates than actually falls as rain. Because drought is their prevailing characteristic, deserts notably are regions of sparse vegetation. Few are completely devoid of plants, but some come very close to that ultimate limit. Typically, desert plants are widely spaced (Fig. 11-2). Their colors tend to be subdued and drab, blending with their surroundings. Their leaves may be small and leathery in order to reduce evaporation. In fact, some, such as the saguaro of Arizona or the barrel cactus of the Sonoran Desert, may have no leaves at all. Other desert plants, such as the ubiquitous sage, may develop an extraordinarily deep root system in

Fig. 11-2. Desert vegetation in Anza Borrego State Park, California. *Jeffrey House*

proportion to the part of the plant that shows above ground. Plants with such adaptations of extensive roots, leathery leaves, and large water-holding capacity are called *xerophytes*, from a combination of Greek words meaning dry and plant.

Every gradation exists in deserts, from those that are completely arid and essentially barren expanses of rock and sand, devoid of almost all visible plants, to deserts that support a nearly continuous cover of such plants as sagebrush and short grass. Dry regions with such a characteristic seasonal cover are best referred to as *steppe,* and commonly are marginal to more desolate areas.

Dry regions occupy one-third of the earth's

Fig. 11-3. Distribution of the earth's hot desert regions.

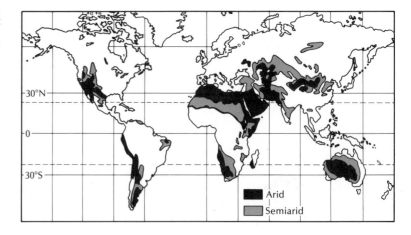

land area and are concentrated in subtropical and in middle-latitude regions (Fig. 11-3). For example, there is nearly continuous desert from Cape Verde on the west coast of Africa, across the Sahara, the barren interior of Arabia, the desolate mountains of southern Iran, and on to the banks of the Indus in Pakistan. The preponderance of the dry areas of the earth—exclusive of the Arctic—are on either side of the equator, chiefly around latitude 30°, and they tend to favor the western side of continents. Polar regions also harbor deserts, but such places are cold and will be discussed at the end of this chapter.

Contrary to the popular image, most deserts are not vast shimmering seas of sand across which sheiks on spirited stallions ride. Although many deserts are sand covered, the majority are not. They are more likely to be broad expanses of barren rock or stony ground with only a rudimentary soil profile developed. Ground colors are largely those of the original bedrock. For example, the bright red color that we associate with such places as Grand Canyon and Monument Valley comes in large part from coloring matter within the rocks themselves, rather than from red-soil forming processes active there today.

It is typical of many desert regions, especially those in continental interiors, that streams originating within the desert often falter and die within the desert's boundaries. A pattern of streams that do not reach the sea is called *interior drainage*, an unusual feature to visitors from well-watered regions with through-flowing streams. Some desert streams simply wither away and sink into the sand. Others may carry enough water to maintain a lake in a structural basin at the end of their course. Such lakes have no outlets and are almost universally salty or brackish—the Dead Sea, about 396 m (1300 ft) below sea level at the end of the River Jordan, is a renowned example. A larger water body without an outlet is the Caspian Sea, and even though it is supplied by the mighty Volga, not enough water reaches it to overcome the inexorable losses of evaporation and to allow the lake to spill over the low divide separating it from the Don River and the Black Sea. A third notable example is Lake Chad, covering some 22,000 km² (8494 mi²) in the southern Sahara.

In many arid parts of the world, where the water brought in by streams cannot hold its own against evaporation, desert lakes may be only short-lived seasonal affairs, or may be completely dry for decades. Such ephemeral lakes, so characteristic of drought-burdened lands, are called *playa lakes* in the southwestern United States, and they evaporate quickly, leaving a *playa* (Fig. 11-4). Some playas may be glaring expanses of shimmering salt, such as the Bonneville Salt Flats near Great Salt Lake in Utah, or they may be broad, dead-flat, clay-floored dry lakes—seemingly created for landing fields, they have such an ideally level surface.

Before going into the causes of deserts, it might be appropriate to give data on some of the extreme conditions that characterize them. A temperature of 57°C (135°F) has been recorded in Algeria. No wonder people talk of frying eggs on rocks! Rock surfaces reach even higher temperatures, so that it is not surprising that many geologists attribute some rock-weathering features not only to high temperatures but also to day-night, hot-cool cycles in temperature. Although the lowest recorded mean annual precipitation of 0.4 mm (.016 in.) comes from the Sahara, some of the larger Saharan storms can drop rainfall at a rate of 1 mm/minute (0.04 in./minute). Other areas can go years without rainfall, the record being 15 years in Southwest Africa.

CONDITIONS CAUSING DESERTS

What special circumstances are responsible for causing some parts of the earth's surface to be deprived of normal rainfall? Omitting the polar regions of deficient precipitation, there are three major types of arid regions: (1) Horse Latitude deserts, (2) rain shadow deserts, and (3) deserts produced by cold coastal currents in tropical and subtropical regions.

Most of the world's deserts are located about

Fig. 11-4. Playa in Fish Lake Valley, at the foot of the Pinto mountains. *John Haddaway*

30° north and south of the equator, in the so-called Horse Latitudes (Fig. 11-5). The origin of such low latitude deserts is intimately connected with the circulation of the earth's atmosphere. Near the equator, the sun's rays strike the earth's surface more directly than at other places, resulting in greater solar radiation which heats the air. The hot, moist air rises, cools, and loses its ability to hold water. The excess water falls as rain in the equatorial regions. The cooled air spreads northward and

southward from the equator into the areas of the Horse Latitudes where it descends and is warmed, enabling it to hold much more water. The Horse Latitude deserts, then, are deserts because they are continually parched by warm, dry winds that suck up any moisture they can find.

A second cause of aridity is the interposition of a mountain barrier in the path of a moisture-bearing air current. A striking example may be found in the western United States, where the desert stretches eastward in the so-called rain shadow of the Sierra Nevada of California. As moisture-laden air from the ocean rises along the west side of the Sierra Nevada, it is cooled and is forced to give up the water that it can no longer hold. The dry, cooler air flows down the eastern slope, is warmed again, and evaporates any water it comes across.

The third type of deserts, those formed because of cool coastal currents, are perhaps the least familiar to North Americans. Such deserts flourish along the mid-latitude coasts of continents whose shores are bathed by cold coastal currents, such as the Humboldt Current off the coast of Chile and Peru, or the Benguela Current off the Kalahari Desert of southwest Africa, both of which run northward along the coasts.

The latter deserts are exceptionally impressive for the dramatic climatic contrasts encountered within extremely short distances. The desert of southern Peru and northern Chile, for example, is one of the driest lands on earth, yet its seaward margin is concealed in a virtually unbroken gray wall of fog. Winds

blowing across the cold waters of the coastal current are chilled, their moisture condenses, and thus a seemingly eternal blanket of fog stands over the sea. Once the fog drifts landward to where the air temperatures are higher, the fog burns off almost immediately, and the water-holding capacity of the air current increases rather than decreases as it moves across the heated land.

Few deserts, however, are the product of a single cause operating to the exclusion of all others. For example, the Atacama Desert on the western slope of the Chilean Andes is not only affected by the cold Humboldt Current; it is also located in the Horse Latitudes. Furthermore, cold air sweeps down from the Andes, so that all three conditions causing deserts operate at one place.

In summary, we can say that there are no simple explanations for the existence of deserts. Their origins are complex, but they are well worth trying to understand, not only for the intellectual challenge but also because so much of the future of the human race is dependent upon the utilization of arid lands. An equally intriguing question is whether the boundaries of deserts are stationary, or whether they are contracting or expanding. As we will see, the climate of the world has slowly changed in historic time. That change has been strikingly true of deserts, and much of the most compelling evidence comes from the Sahara and the Middle East. Artifacts, stone implements, and rock paintings of extraordinary subtlety and sophistication testify to the presence of early man in what today are desolate expanses of the Sahara (Fig. 11-6). At a later date much of that barren region was the granary of Rome, and colonial cities of that day as well as roads along which the legions marched are now overrun by drifting sand. The expansion of the desert broke the slender thread of communication linking the cultures of the Mediterranean and African worlds. Apparently thereafter the two had a vague and uncertain awareness of each other over the centuries, but it was not until the introduction of the camel caravan that the land connection

Fig. 11-5. Generalized diagram of atmospheric circulation.

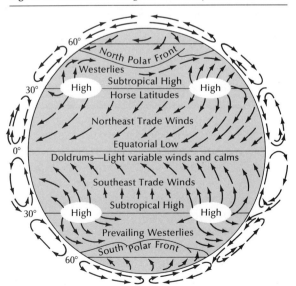

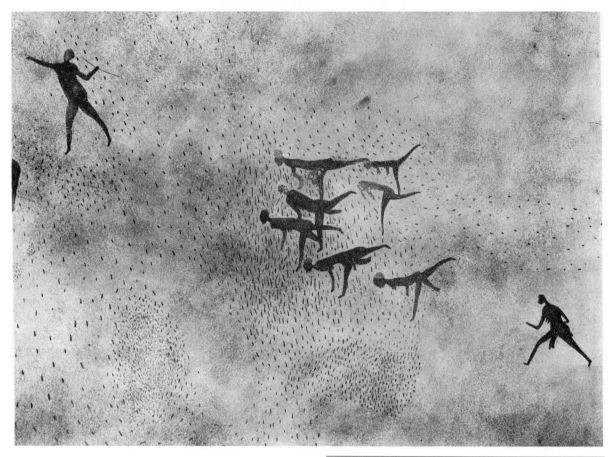

Fig. 11-6. A Stone Age rock painting in the African desert depicts people gathering grain in a graceful ballet. This and other similar paintings suggest a wetter climate in the past. *Eric Lessing, Magnum*

was re-established. By that time each culture had evolved along a different path.

SURFACE FEATURES

In many parts of the desert the land is simply barren rock outcrop or is strewn with rocks and boulders. Commonly the rock is coated with *desert varnish,* a shiny, bluish-black substance peculiar to desert environments. It can coat entire stones or form patterns on high vertical cliffs (Fig. 11-7A). In the Southwest, Indians have scratched off the very thin coat-

ings in making their very picturesque petroglyphs (Fig. 11-7B). Chemical analyses indicate that the varnish is a mixture of iron and manganese oxides derived from weathering—either from the rock on which they form or from the adjacent soil. Archeologic and geologic evidence suggests that several thousand years are required before desert varnish becomes a prominent desert feature.

Another unique feature of deserts is *stone pavement*—a thin veneer of stones one to two stones thick. Looking closely at the surface, one can see that the stones form a tightly packed mosaic, as if put together by a mason— hence the term stone pavement (Fig. 11-8). Typically the layer of soil directly beneath the surface has a much lower concentration of stones. How, then, are stone pavements to be explained?

Two popular hypotheses help explain the origin of most stone pavements. One is that wind erosion is responsible. Consider a sand and gravel deposit exposed at the surface with little vegetation for cover. Winds sweeping across the surface can pick up sands and finer-grained materials and carry them away, but not the gravel. If the process goes on long enough the remaining stones will form a tightly packed lag concentrate. The other theory is that stones move from the shallow depth to the surface. If the gravels were originally scattered throughout a somewhat clayey matrix, the shrinking, cracking, and swelling that accompanies drying and wetting cycles can actually move gravel toward the surface, and there concentrate it into a pavement. Whatever their origin, stone pavements form an armor that protects the surface from further wind erosion. And, like desert varnish, they take a long time to form; commonly those gravel surfaces with the best development of stone pavement also will have a well-developed coating of varnish.

STREAM EROSION AND DEPOSITION

Paradoxical as it may seem, the leading agent responsible for sculpturing desert landforms is running water, perhaps more so than in humid regions. Puzzling indeed is how the enormous volumes of rock that once filled desert canyons or shaped the mountains themselves were removed or worn down to bare rock plains when there appear to be no streams at all in that seemingly timeless land.

Part of the answer may be that not all landforms in all deserts were formed under the climatic regime we see today; they may be fossil landscapes in a sense, survivals of erosional patterns carved during a time more humid than ours. We shall see one result of a recent climatic shift a little farther on when we

talk about the lakes that formed in desert regions during times of recent glaciation. Not all the erosion of desert landscapes, however, took place under climatic controls alien to today's world. Almost all deserts have some rainfall, even though 10 to 15 years may elapse between showers. So it could be that the infrequent storms, spread over a long enough period of time, provide enough water to get the job done.

The nature of runoff

Contrary to popular belief, cloudbursts are relatively rare in arid lands—they are much more common where rainfall is greatest; for example, in the rainy tropics or the southern Atlantic coastal states. When a moderate rain does fall on the desert, it is likely to assume the proportions of a cloudburst in a more humid region and also to do a very effective job of erosion. The reason is that little or no vegetation protects the slopes from the spattering effect of raindrops or from water running across the surface, rapidly cutting ravines and arroyos. *Arroyo* is a Spanish word for stream or brook, and in the southwestern United States it is used for the steep-sided, flat-floored river washes that are so common there (Fig. 11-9). *Badlands,* or slopes scored with great numbers of gullies, large and small, are also characteristic desert landscape elements caused by runoff.

Anyone caught in a sudden desert downpour is likely to have an unforgettable experience. In a matter of moments a dry, sandy arroyo, bordered by low but steep cliffs, is filled with a surging, mud-laden flood. The stream swirls and churns violently forward, sweeping along a great mass of debris. Boulders of all sizes can be moved in the torrents. Such flash floods make deserts impassable until the arroyos drain. And they do that almost as rapidly as they fill since there is no continuing source of water as there is in regions of plentiful rain and perennial springs. After only a few hours beneath the desert sun, an arroyo floor covered by 3 m (10 ft) of water may become dry

Fig. 11-7 A. (above) Dark-colored desert varnish coats these sandstone cliffs along the Grand Canyon of the Colorado River, and is responsible for the reflective sheen between the major vertical joints. *W.C. Bradley* **B.** (opposite) Desert varnish on sandstone in the Glen Canyon of the Colorado River. Shamans of the prehistoric Indian tribes inscribed petroglyphs by scraping away the dark coatings. Notice the bighorn sheep, important prey of the Indians, and the self-portraits of the shamans with headdresses of bison horns. *Utah Museum of Natural History*

sand again, interrupted by only occasional pools of muddy water. The evidence of the event, however, remains for a long time.

Occasionally the short-lived torrential floods overflow the low banks of dry washes, spreading a sheet of muddy, turbulent water over the desert floor. Such a *sheet flood* is vastly effective in picking up loose sediment and shifting it around the landscape.

Deserts vary in their rates of erosion. In small drainage basins in the United States, the maximum amount of erosion seems to occur in

Fig. 11-8. Well-developed stone pavement in the Mojave Desert, California. *W.C. Bradley*

Fig. 11-9. Arroyo incised into the floor of San Timoteo Canyon, California. The flat river bed and steep walls are typical of many arroyos. *Richard Reeves*

semiarid regions, with lesser rates in both arid and humid regions. Erosion rates in more arid regions are less because rainfall is less, whereas in humid regions the stabilizing effect of vegetation lowers the rate.

Depositional landforms

Wherever erosion occurs, there is deposition close by. Arid regions are especially prone to deposition because ordinary streams cannot escape beyond the confines of the desert. Their water sinks underground into the sandy stretches of their own normally dry stream beds, or it evaporates, or it may be withdrawn by fiercely competitive water-seeking plants, such as the mesquite and tamarisk that line the banks of many desert watercourses. Much of the desert landscape is dominated by stream deposits, in large part because of the inadequacy of running water to move great quan-

tities of debris out of a desert basin and on to the sea.

One of the most characteristic desert landforms is the *alluvial fan* (Fig. 11-10). It takes its name from its shape, which most approximates a portion of a cone that enlarges downslope from the point where a stream leaves the mountain front. Fans commonly are composed of accumulations of gravel and sand, and deposition on their surfaces is attributed to two main causes, both of which reduce the velocity of the stream. One is that much of the water sinks into the porous, sandy subsurface layers of the fan, thereby decreasing the discharge, which is one of the factors controlling the velocity. The other is that the main channel, upon leaving the mountain front and entering the realm of the fan, soon separates into a score of distributaries and acquires a braided pattern. The increase in length of channel perimeter also helps to cut down the velocity and to bring about deposition.

Fig. 11-10. Alluvial fans in Death Valley, California.
John S. Shelton

Fig. 11-11. Alluvial fans form bajadas on both sides of Death Valley, California. The view is toward the west side of the valley, where the largest fans have formed. The present stream channels are light in color, in contrast to the inactive surfaces, which are darkened by desert varnish. The white material on the valley floor is precipitated salt. *USAF*

Alluvial fans come in various sizes and degrees of slope, but there is consistency in the variation. In a general way, the area of a fan is proportional to the area of the drainage basin from which the fan debris is derived; in short, small drainage basins produce small fans and large drainage basins, large fans. The slope of a fan is related to the parameters that control the slopes of streams. Fans derived from large drainage basins tend to have relatively low slopes, as do those constructed by streams with large discharge. Finally, it should be obvious that sediment size helps to control the slope, for gravels are usually associated with fans having steeper gradient.

Streams may be able to cross the fan in time of flood, but under normal conditions they sink into the sandy ground almost as soon as they leave the bedrock of the mountains. Visitors to a dry country almost always are surprised to see a stream waste away, growing thinner and thinner downstream and then vanishing completely—quite the reverse of humid-climate streams, whose volume commonly increases downstream. Actually the water does not vanish but percolates slowly through pore spaces in the fan. Far down the fan, the same water may, for a variety of reasons, be forced to the surface and seep out in springs—commonly marked in the desert by clumps of mesquite trees, one of the best guides to water in the arid Southwest.

If there is a one alluvial fan at the base of a desert range, there will probably be others—in fact, there probably will be one at the mouth of each principal canyon. Overlapping like palm fronds, the fans form a nearly continuous apron sloping away from the mountain front to the basin floor. The apron is called a *bajada*, from a Spanish word meaning a gradual descent (Fig. 11-11).

Many desert basins, especially those in the Great Basin, are closed areas with interior drainage. Debris shed from the mountains is slowly filling in the basins, but the task has not been finished. In a humid region, where precipitation is greater than evaporation, such places would be occupied by a lake whose waters would rise until they spilled over the lowest point of the basin rim.

Desert lakes do exist however, as demonstrated by the Great Salt Lake in the eastern part of the Great Basin and Pyramid Lake in the western part. Many desert lakes fluctuate in depth with climatic variations, and their waters are saline. Accounting for the salinity are salts brought to the lake as dissolved stream load and concentrated as pure water evaporates from the lake surface. Because everything soluble in desert rocks may be carried into such a lake, it is no wonder that some desert lakes are natural chemical factories. Saline lakes in volcanic areas are likely to be especially prolific sources of unusual substances, including potash, potassium salts, and in the eastern desert of California, boron compounds.

During a dry climatic cycle some saline lakes may evaporate completely, leaving an achingly white residue of salts, as complex chlorides, sulfates, and carbonates. Called *salt playas*, such surfaces are most likely to be found at the end of a long and integrated drainage system. The incredibly jagged terrain of salt pinnacles on the floor of Death Valley (Fig. 11-12), nearly 90 m (295 ft) below sea level, was a fearsome stretch for the first party of emigrants to cross, in their battered wagons drawn by plodding oxen. That playa receives some of its water and salt from the Amargosa River, which name appropriately enough means bitter, after its long and lonely journey across the hottest desert in the United States.

Clay playas are more likely to be found in smaller basins. Most of the year they are dry and their surface is baked as hard as a brick; in fact, they make ideal emergency landing fields. One such lake floor was converted into a gigantic multidirectional landing ground— 24-km-long (15-mi) Rogers Dry Lake at Edwards Air Force Base in eastern California.

When a typical short-lived downpour is over, desert streams waste away as suddenly as they sprang into being. Then the arroyo bottoms quickly return to their seemingly unchanging state of dry, shifting sand enclosed between steep arroyo walls. The playa lake may endure a bit longer, but ultimately its murky, dark-hued water evaporates or sinks underground. Before that happens, the suspended load of finely divided silt and clay particles is disseminated throughout the ephemeral lake. Thus, when the lake has vanished, its newly revealed floor is a dead-flat expanse of uniformly distributed fine-grained tan sediment. Very often, too, surface layers of playa clay shrink as they dry, cracking into thousands of small polygonally shaped blocks.

Erosional landforms

In addition to gullies, valleys, and canyons, there is in the desert another erosional landform, the *pediment*. Pediments are bedrock surfaces stripped bare that slope gently away from low desert mountains toward the lower part of an intermontane basin (Fig. 11-13). The surface usually meets the mountain front at a sharp angle. From a distance the broad, encircling surfaces look like a uniformly sloping bajada rather than the product of long-continued degradation and the removal of many thousands of cubic meters of bedrock (Fig. 11-14). In fact, no one yet has come up with an easy way to differentiate fans, which are areas of deposition, from pediments, which are areas of erosion.

Pediments were first described in this coun-

Fig. 11-12. The Devil's Golf Course, Death Valley, California, is composed of salts—complex chlorides, sulfates, and carbonates—which were concentrated in Lake Manley at the end of a long drainage system. The salts were then precipitated as the lake disappeared. *Ansel Adams*

Fig. 11-13. Pediment cut in granite sloping away from the mountain front in the Mojave Desert, California. In places, low, resistant bedrock hills rise above the surface. *W.C. Bradley*

Fig. 11-14. Pediments near Phoenix, Arizona. It is believed that the mountain fronts all retreated parallel to their trends, leaving the broad pediments behind. *W.C. Bradley*

try in 1897 by an early American exploring geologist, W. J. McGee. He was as surprised as anyone by their true nature:

During the first expedition of the Bureau of American Ethnology [1894] it was noted with surprise that the horse shoes beat on planed granite or schist or other rocks in traversing plains 3 or 5 miles from mountains rising sharply from the same plains without intervening foothills; it was only after observing this phenomenon on both sides of different ranges and all around several buttes that the relation . . . was generalized.

The process by which pediments are formed has long been a puzzle to geologists who have vigorously debated the topic since McGee first described them. Very briefly, there are two principal schools of thought on the subject. Some geologists think that a pediment is formed by lateral planation of streams; that is, erosion caused by a stream as it swings back and forth over a rock surface. The materials carried by the stream slowly cut away at the bedrock, and, where the stream touches the mountain front, it causes it to retreat ever so slightly. Other geologists believe that pediment surfaces weather, and that the products of weathering are removed by an unconcentrated flow of water sweeping across the pediment during rainstorms. The mountain front also weathers, and various processes, including gravity and running water, deliver debris to the upper end of the pediment. If the material is the appropriate size, it will continue on its journey across the pediment; if not, it will reside at the mountain front until weathering reduces it to a size that can be transported. Between those two extreme views is one that says that both processes are important in pediment formation and that, depending on special local circumstances, one process dominates the other.

The most recent theory on the origin of pediments was proposed to explain those in the Mojave Desert. Evidence suggests that the Tertiary landscape, say eight million years ago, was one of soil-covered hills with fronts retreating in a parallel fashion (Fig. 11-15). A cli- ·

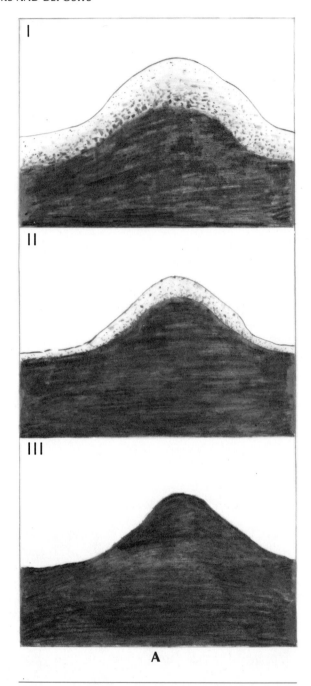

Fig. 11-15 A. Landform evolution in the Mojave Desert. Stage I is the soil-covered granitic landscape in Tertiary time. The bedrock becomes more weathered toward the surface, and jointing controls the blocky weathering forms. Stages II and III represent progressively greater stripping of the landscape to the bedrock that predominates today. The low-lying flat areas would be the present-day pediments.

matic change toward greater aridity increased the erosion rate and stripped the land of its soil cover a few million years ago. Bedrock pediments are only part of the stripped surface. In the case of the Mojave theory, the pediment form is inherited essentially from the past.

Certainly the last hypothesis has much merit, and can be called upon to explain some pediments. Whether it explains all, however, is doubtful.

Erosional cycle in an arid region

Geologists have made a considerable effort to determine if landforms evolve through time, from one characteristic form to another. Some, for example, envisage the landscape in a humid climate as initially hilly, only to be converted into a peneplain after millions of years of erosion, dominantly backwasting. And, so the argument goes, perhaps an erosion cycle can also be deciphered from desert landforms.

As an example, we shall use the Southwest (Fig. 11-16). We start with a recently uplifted mountain range, bounded by normal faults. Its upland surface is cut by narrow gullies separated by broad, flat divides. Down-dropped blocks form basins that separate adjacent ranges. As weathering and erosion proceed, the mountains slopes recede nearly parallel to their original slopes; the flattening and rounding caused by soil creep in humid regions is not to be found. Also, waste material is not carried out of the area but accumulates in the basins. Because streams are few and flow only for short distances, there is no chance for an integrated drainage system to develop. As a consequence, the valley floor is occupied by a playa. Alluvial fans lead out to the playa. The basin gradually fills with sediment and drainage eventually can spill into the next basin, becoming "through-going."

Provided the area remains tectonically stable for a long time and through-flowing drainage is established, slopes eventually can retreat from the original mountain front to form pediments. The pediment continues to expand at the expense of the mountains until only mere fragments of the mountains—called *inselbergs* (German for island mountains)—remain. Finally, the inselbergs are consumed and low bedrock domes are all that remain of the mountains. These extensive coalescing pediments are called *pediplanes* and correspond to the peneplains of a humid region. Thus, theory has it that gently rolling terrain of very low relief is

Fig. 11-15 B. Mountain front in the Mojave Desert made up of huge granitic boulders similar to those in Stage III of (**A**). *W.C. Bradley*

the end result of erosion in any region. Few areas remain tectonically stable long enough for such landscapes to develop, however.

Active faulting and climatic variation can alter the processes that form alluvial fans and pediments. In a very general way, at least in the Great Basin, alluvial fans are found in the north, where active faulting is taking place, and pediments occur in the south, where faulting processes have long been dormant. So, without belaboring the point, tectonic stability is essential to pediment formation.

Within most major alluvial fans, several different surfaces can be recognized (Fig. 11-17). The stream has incised the old fans, forming a new one at a lower level. The process can happen again and again. In places such as the west side of Death Valley, the incision may have been caused by the upward tilting of the mountain block. That tilting could also explain the drastically different sizes of fans on either side of the valley. The west-side fans on the uptilted block are large, whereas those on the east side (the down-tilted part of the valley) are continually being buried by playa sediments, which keep the fans small. A close study of fans, therefore, is an important clue to tectonic history. Such studies become more relevant as desert terrain is increasingly encroached upon by humans, with some areas being considered as sites for nuclear power plants.

Before one opts for the interpretation that multi-fan surfaces indicate tectonic activity, climatic change also should be considered as a cause of such surfaces. Streams that build fans should respond to climatic changes by degrading at some times and aggrading at other times. For example, with a change to a wetter climate and greater discharge would a stream degrade, only to be followed by aggradation during a drier climate? Such questions are tough and there are no pat answers. We know climatic changes have taken place in the deserts, and we keep trying to read the changes in the fan deposits. But the answers are not readily divulged.

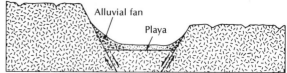

A   Initial block faulting forms a basin-and-range topography. Undissected uplands may be present in the mountains.

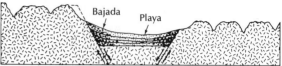

B   The basin fills with sediments—fine grained playa sediments in the center and coarse-grained fan deposits around the margins. Drainage becomes through-going.

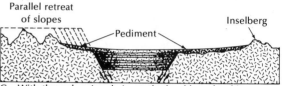

C   With through-going drainage the local base level is relatively stable. Slopes retreat parallel to themselves, forming pediments with a thin gravel veneer.

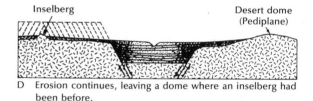

D   Erosion continues, leaving a dome where an inselberg had been before.

Fig. 11-16. The erosional cycle in the semiarid to arid Southwest.

WIND

Wind erosion

Over many of the dry lands of the world the wind blows seemingly without restraint, adding a note of harassing melancholy to an already unbearably lugubrious scene. Such a desolate region, with the wind keening ceaselessly across it, is the drier part of Patagonia of the far southeastern reaches of Argentina. Other deserts are perhaps as windy on occasion, but in most of them times of extreme windiness alternate with times of calm.

Oldest fan   Youngest fan remnants   Intermediate-age fan remnants

Fig. 11-17. View westward across Death Valley to the Panamint Range. Several ages of fans can be seen, based on position above the stream (the higher fans are older) and on tonal differences caused by desert varnish. *H.E. Malde, USGS*

During intervals in which wind blows, its erosional effectiveness is likely to be greater than in humid regions. In the latter, the ground surface is protected by vegetation and by a more tenacious mantle of weathered soil, which also may be damp throughout most of the year. Vegetation plays an important role indeed because the velocity of the wind drops off rapidly at the ground because the plants act as an extremely effective baffle.

The wind is a very effective agent of transportation for certain sizes of particles, and scarcely at all for others. It is that high degree of selectivity that makes the wind such an unusual agent of erosion. Obviously there is an upper limit to the size of particles that it reasonably can be expected to move. Very few boulders of the size that are handily transported by streams, waves, or glaciers will be moved by the wind—even by a tornado. Such

Fig. 11-18. Gravel moved during a 1969 windstorm in Boulder, Colorado. These grains bounded down an asphalt road driven by winds exceeding 200 km/hr (120 mi/hr). *University of Colorado Daily*

Fig. 11-19. Paths that saltating sand grains take across two different surfaces.

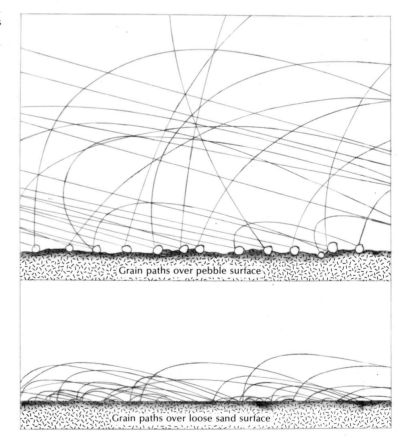

Grain paths over pebble surface

Grain paths over loose sand surface

ruptly, the force of their impact setting other grains in motion. The height to which the grains bound depends on the characteristics of the surface. Grains reach greater heights when rebounding off a hard surface, such as a pebbly desert pavement or a road, than from a loose sand surface, which absorbs a lot of the impact energy. Saltating sand particles seldom rise to heights greater than a meter (3 ft) or so above the surface (Fig. 11-20).

Saltating grains have a marked effect on the shape of rock outcrops (Fig. 11-21) or gravel

Fig. 11-20 A. (above) Erosion of this road marker has been the most effective close to the ground, where most of the sand grains are moving. **B.** (right) Telephone poles sheathed in metal to protect them from saltating sand particles. *D.C. Strong*

immobility is true for loose material down to the size of large pebbles—although the smallest pebbles, about 4 mm (0.16 in.) in size, commonly are moved (Fig. 11-18).

Sand grains move along the ground in a series of hops—a motion known as *saltation* (Fig. 11-19). Grains may bounce off other grains making up the surface, or they may stop abthat they beat against. In general, the upwind side of a gravel particle will be cut gradually to form a low-angle planar surface called a *facet* (Fig. 11-22). Facets form at approximately right angles to the wind, and attempts have been made to deduce ancient wind directions from faceted rocks, or *ventifacts*. But a stone can shift position, presenting a new face to the driving winds, and a new facet is cut. In fact, multi-faceted stones are more common than not. Stones can be shifted about by wind scouring near their bases, or by other means, not the least of which are frost heave or animal disturbance.

Odd as it may seem, very fine-grained particles, such as silt or clay flakes, are not easily started in motion by the wind. In general, the same principles apply here that apply to streams. Silt and clay particles, with their small dimensions and strong cohesiveness, cannot be picked up readily by a moving current of air because wind velocity diminishes close to the ground (Fig. 11-23). Their tenacity when in place is demonstrated when the wind blows full strength across the surface of a clay playa. Very little dust is stirred up as a rule, and the hard-packed clay particles hold firm. Some of the looser sand and silt around the margin, however, may be picked up, especially if the playa surface is sun cracked.

Once silt- or clay-size particles are picked up by the wind, however, they remain in suspension much longer than sand. There are no restraints imposed on their travel comparable to those of their waterborne equivalents, which are limited to the drainage pattern of a stream or to a glacier's course. Windborne dust swept from the fields of Colorado in the Dustbowl days of the 1930s was carried as far east as the New England states. In fact, some dust was blown far beyond, out over the North Atlantic.

There are scores of other examples of the efficacy of air currents in moving fine particles over vast distances. Volcanic eruptions of the explosive sort are the most telling kind because they constitute a point source for the volcanic dust. Also, because of the unique nature of the dust, it can be traced for great distances. In

Fig. 11-21 A. (above) This granite outcrop in the Atacama Province of Chile has been undercut by the abrasive action of windblown sand. *K. Segerstrom, USGS* **B.** (below) Wind-driven sand etched this rock photographed near Fortification Rock, Arizona, in 1871, by T.H. O'Sullivan. *Library of Congress*

Fig. 11-22. Ventifact of granite in Wyoming showing several faceted faces. *W.H. Bradley, USGS*

Fig. 11-23. Variation in wind velocity for two different events with proximity to the ground surface.

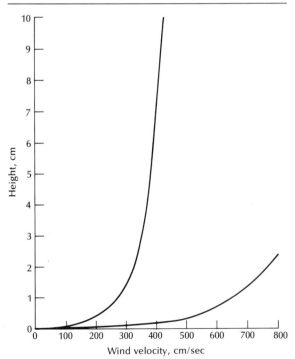

1883 Krakatoa, near Java, hurled dust into the upper atmosphere to circle the earth several times and to produce an appreciable fall in regions as remote from the source as western Europe. Icelandic eruptions have made appreciable deposits in Europe, too, as well as in eastern North America.

More typical, however, of the aspect of dust transport with which we are concerned here is the dust whirled high into the air by turbulent winds of the Sahara and broadcast far and wide over the Mediterranean, southern Europe, and on occasion as far as England, 3200 km (1988 mi) away.

The sheer quantity of material that can be in the air at any given time is surprising. In a dust storm of average violence a cube of air 3 m (10 ft) on a side might well have 28 gm (1 oz) of dust suspended in it. Such an amount seems trifling, but if we increase the size of our cube of air until it is 1.6 km (1 mi) on each side and maintain the same saturation of dust, then the air current is supporting 4000 metric tons of solid material. Thus, a storm 500 to 600 km (311 to 373 mi) across might well be sweeping 100,000,000 metric tons of solids along with it!

Such facts are strong evidence that the wind can be an effective erosional agent and one of especially great significance in arid regions. In fact, it is the only agent that can transport material beyond the confines of a typical desert. Some dust from Africa, for example, has been recognized in the Caribbean region.

Some of the closed basins in the desert excavated, or as earth scientists say, *deflated*, by the wind reach very large dimensions (Fig. 11-24). A famous example is the great Saharan oasis of Kharga west of the Nile, which is about 190 km (118 mi) long by 19 to 80 km (12 to 50 mi) wide and 180 to 300 m (590 to 984 ft) deep. A series of longitudinal sand dunes trails off downwind from the oasis, whose sandy floor has almost certainly been the source of supply. That is not to say that the wind hollowed out the whole basin; its origin unquestionably is more complex. Rain wash and streams may erode the friable sandstone and sweep the loose sand down to the floor of the trough.

Fig. 11-24. Deflation basins in the Sahara, cut in folded sedimentary rocks. Although basins can have many origins, those cut in desert bedrock often are best explained by the removal of material in an uphill direction. *Eros Data Center, USGS*

There the sand rests until it is picked up by the wind to accumulate in dunes to the leeward of the basin.

Ground water sets a lower limit to wind erosion. Once the desert surface is deflated to the level of the water table, so that the ground is kept damp, the wind can no longer pick up loose material with the same ease, and deflation slows to a virtual halt. In Kharga the beneficial effect of deflation has been the appearance of springs around the margins of the great depression. Such springs are the source of water for the true oases, for with only a little ground water to draw on, date palms flourish and make a startlingly green contrast to their stark surroundings.

Another impressive closed basin in the Sahara is the Qattara, about 300 km (186 mi) long with a floor 134 m (440 ft) below sea level that is a searingly forbidding quagmire of salt

and shifting sand covering about 18,000 km² (6950 mi²). Whatever the origin of the Qattara Depression, the wind almost certainly played a prominent role in enlarging it and in deepening it to the water table. Evaporation of the water through centuries has been responsible for the accumulation of the salt deposits. That immense, impassable saline trough acted as a barrier in World War II, making it impossible for Rommel's Afrika Korps to turn Montgomery's flank compelling them to attack the concentrated British forces at El Alamein on the

Fig. 11-25 A. (opposite, above) Terraced landscape in loess, Shensi Province, northern China. **B.** (opposite, below) Cave houses dug in deposits of loess, Shansi Province, northern China. Loess is not only easily excavated, but it stands in a vertical face without failing. *H.E. Malde, USGS*

narrow neck of land separating the Qattara from the Mediterranean.

At some future date the Qattara Depression may have surprising utility if an unorthodox proposal should be developed leading Mediterranean water to it through canals and a tunnel, and then using the 460-m (1509-ft) drop available from the rim of the depression to generate power from the unlimited supply of water available in the sea. Gradually the depression would fill with water to form a concentrated saline lake whose surface, it is estimated, would stabilize around 46 m (151 ft) below the level of the Mediterranean when the input of sea water balanced the loss of lake water through evaporation beneath the Saharan sun.

Wind deposition

Loess Dust lofted out of the desert by winds may be swept for scores of miles before it sifts down to accumulate in a tawny blanket of sediment that may be enormously distant from the source. Such material, in which silt-size particles predominate, is called *loess*.

The most renowned of such deposits are those of northeast China, and it was to them that Baron von Richthofen (1833–1905), a leading German geologist, gave the name of *löss* while on an exploring expedition to the outermost parts of the Russian and the Chinese empires (Fig. 11-25). The loess of China has been exported by the wind out of the Gobi and across the Kalgan Range, upon whose barren ridges the Great Wall was built. Deposited on the North China plain by the dust storms of centuries, loess lies deep in the valleys of that ancient land, often to thicknesses of hundreds of meters, although it lies much thinner on the crests of divides. The tan-colored silt gives the Yellow River its name, as well as the Yellow Sea, whose waters are stained for hundreds of kilometers from shore by suspended dust particles. Loess is also found in other regions where it is closely associated with glaciation (see Chap. 12).

Sand dunes Where sand is plentiful in arid regions, winds move it and pile it into characteristic heaps called sand dunes. Such features are not, however, limited to deserts. Many of the larger and more renowned of the world's dunes are along shorelines, such as those on the eastern shore of Lake Michigan, the length of Cape Cod, and the coast of Somalia. Dunes also border the sandy plains of some large rivers—the Volga is an outstanding example.

Few deserts are completely sand covered. Nevertheless, there are sandy areas in almost all deserts; the south-central part of Arabia and the western part of the Sahara are among the best known. Broad dune-covered areas in the Sahara and elsewhere are called *ergs* (sand seas) because they emulate a wave-tossed sea so well (Fig. 11-26).

Sand dunes consist dominantly of sand-size grains, which bear testimony to the extraordinary sorting ability of the wind. Finer material such as silt may be blown far away, whereas coarse rock fragments, such as pebbles and gravel, may lag behind the sand.

Dunes are neither stable nor permanent features of the landscape, and may be continuously on the march. Usually, they have a gentle side facing toward the wind and a steep side facing away from the wind. Wind-drifted sand blows up the gentler slope of a dune, and when it reaches the crest it may be carried a short distance over it—the tops of dunes sometimes seem to be smoking when the sand is driving across them. Behind the crest the sand drops out of the wind stream to accumulate on a steeper slope, the *slip face*. When the sand is dry and well-sorted, the inclination of the slip face may be as much as 34°. Should the slope become steeper, the sand becomes unstable and shears along a slightly gentler plane, with the result that a small avalanche of dry sand glides to the base of the dune (Fig. 11-27). When new sand falls on the slip face, the slope steepens once more, and so the process repeats itself again and again. The net result is a transfer of sand from the upwind to the downwind side of the dune. Thus the dune slowly

Fig. 11-26. Great sand seas, or *ergs*, shown here in the eastern Rub' al-Khali, are typical of the desert of Saudi Arabia, as well as parts of the Sahara and other deserts of the world. In the foreground is the camp of a geophysical party exploring for petroleum. *Aramco*

migrates—in a sense rolling along over itself.

Dunes show a fascinating variety of shapes and patterns. Where the wind holds more constantly from a single direction, as it may along a sea coast, they are likely to have a more persistent geometry. For example, they may be aligned at right angles to the wind, in which case they are called *transverse dunes*. Such dunes are likely to be quite short and to flourish where an abundant supply of sand is available and where the winds are strong. Typically, many coastal dunes are in that category.

If dunes are lined up parallel to the prevailing wind they are known as *longitudinal dunes*. The latter are most likely to form where the

Fig. 11-27. Avalanche tracks on the slip face of a miniature active sand dune. *W.C. Bradley*

wind blows strongly from a single quarter, the supply of sand is sparse, and vegetation is virtually absent. Such conditions are fulfilled admirably in the remote Great Sandy Desert of Australia. Some of the individual longitudinal dunes there are said to be more than 95 km (59 mi) long.

A curious and aesthetically appealing sand dune is the *barchan* (Fig. 11-28). Barchans are beautifully symmetrical, crescent dunes—sometimes as perfectly proportioned as the crescent moon, symbol of Islam. Pointing downwind are the horns of the crescent, and the steeper slip face lies between them. Barchans, too, require winds that are constant in direction and not too high in velocity. Such

dunes flourish in the tradewind deserts or in coastal deserts such as the Atacama. As a rule, barchans are migratory and may march across the desert landscape at a rate of as much as 25 m (82 ft) per year. Sand blows up the gentle slope of the crescent and slides down the steeper slip face—a procedure typical of most dunes. The chief variation here is that sand swept around the ends of the dune tails off downwind to form the horns of the crescent. As the dune migrates those points continue to pace its progress.

The desert surface surrounding a barchan is more likely than not to be barren bedrock, almost completely devoid of sand. The wind is a remarkably tidy housekeeper, whisking up loose sand from the bedrock surface of the desert between the dunes. Part of the reason seems to be that grains bounce along much more readily over a bare rock surface than they do over sand. Sand has a retarding effect on bouncing grains, and once they start to accumulate, as in a dune, their independent, free-roving days temporarily, at least, are ended.

DESERT LAKES

Among the many distinctive features of such a dry and furrowed landscape as our own Southwest are the desert lakes. They owe their existence in large part to the inability of desert streams to develop through-flowing courses. Where such streams are blocked, even though by no more than the advancing toe of an allu-

Fig. 11-28. Barchan dunes along the west shore of the Salton Sea, California. The form of these dunes indicates that they are moving from the upper left to the lower right. *John S. Shelton*

Fig. 11-29. Late Pleistocene lakes in the Great Basin, relative to present lakes. Arrows indicate direction of stream flow from one lake basin to another. In a few places, the lakes rose so high that drainage spilled out of the Basin. The most famous outlet is Red Rock Pass; through it Lake Bonneville spilled north, sending a catastrophic flood down the Snake River that approached depths of 100 m (328 ft) in some places.

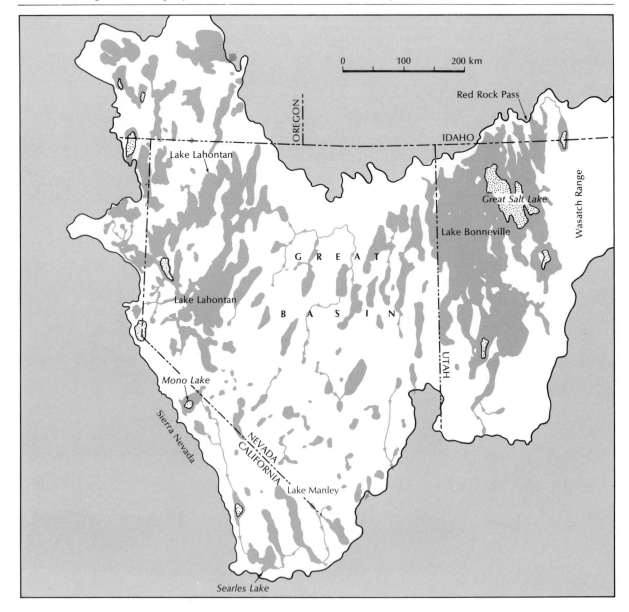

Fig. 11-30. Nineteenth-century etching of the Lake Bonneville shorelines on the northern end of the Oquirrh Range, Utah from G.K. Gilbert, 1890, *Lake Bonneville: U.S. Geological Survey Monograph 1.*

vial fan, their water is ponded and a lake results. Some desert lakes, like Great Salt Lake with a surface area of about 3800 km² (1467 mi²), are quite large. Others are little more than saline ponds.

Almost all desert lakes have a common attribute: their water is brackish or saline to a greater or less degree. The concentration of salt in Great Salt Lake ranges from as little as 14 per cent to as much as 27 per cent of the weight of the water, as compared to 3 per cent for the oceans. So Great Salt Lake is one of the world's most saline lakes—one in which swimmers have little problem floating.

Desert lakes are extremely sensitive climatic indicators. In a dry cycle their water wastes

away through evaporation, and the drought-diminished streams are not able to hold their own against the loss. The lake level drops and the shores are bordered by an ever-widening band of salt. Most famous of such expanses is the Bonneville Salt Flat adjacent to Great Salt Lake, the scene of many a determined assault on land speed-records.

Another interesting feature of deserts is the evidence that they were once the site of far larger lakes than the shrunken remnants of today (Fig. 11-29). The most redoubtable of the now vanished inland seas in the United States was Lake Bonneville. It was the precursor of present-day Great Salt Lake, and shorelines of the one-time inland sea now scar the higher slopes of the Wasatch Mountains more than 300 m (984 ft) above the modern lake. The area flooded by Lake Bonneville was more than 51,000 km² (19,691 mi²). During part of its history, Lake Bonneville had an outlet at Red Rock Pass north to the Snake River and thence to the Pacific by way of the Columbia River.

A contemporary of Lake Bonneville was Lake Lahontan, located mostly in western Nevada not far from Reno. In that ruggedly mountainous area, all the intervening valleys were filled with long, narrow arms of the lake. Pyramid, Walker, and Winnemucca lakes are the chief remnants of Lake Lahontan, but both Lahontan and Bonneville left their imprint on the landscape in an impressive array of wave-cut and wave-built landforms (Fig. 11-30). Among them are wonderfully well-preserved beaches, gravel bars, sea cliffs, deltas, and limy tower-like deposits known as *tufa* (Fig. 11-31). The latter deposits are built up underwater by calcareous algae.

A remarkable set of lakes briefly was a part of the California landscape in the desert east of the Sierra Nevada. Individually they were far smaller than such giants as Lahontan and Bonneville, and formed part of an extensive network of connected lakes and streams. One series extended north from the site of modern Lake Arrowhead in the San Bernardino Mountains across one of the driest parts of North America, the Mojave Desert, to Death Valley. The now-desiccated depression then held a lake perhaps 145 km (90 mi) long and 183 m (600 ft) deep, to which the name Lake Manley is given—in honor of Lewis Manley, a mountain man of great strength and resolution who saved the first party of pioneers to reach Death Valley. In a period of six weeks he hiked all the way to the coast and back again, and then back to the coast, in order to bring supplies and to lead the survivors out of their trap.

To the west of Death Valley a similar set of lakes and streams led from Mono Lake at the base of the Sierra Nevada, down the length of Owens Valley to the basin of Searles Lake and on to Death Valley. Searles Lake acted as a gigantic chemical processing plant, concentrating an enormous tonnage of dissolved ma-

Fig. 11-31. Tufa mounds in Mono Lake, California, photographed in 1886. Many such features are still being formed today. Ancient weathered mounds are found related to higher and older shorelines here as well as along the shorelines of other desert lakes. From the *Eighth Annual Report* of the U.S. Geological Survey, 1887.

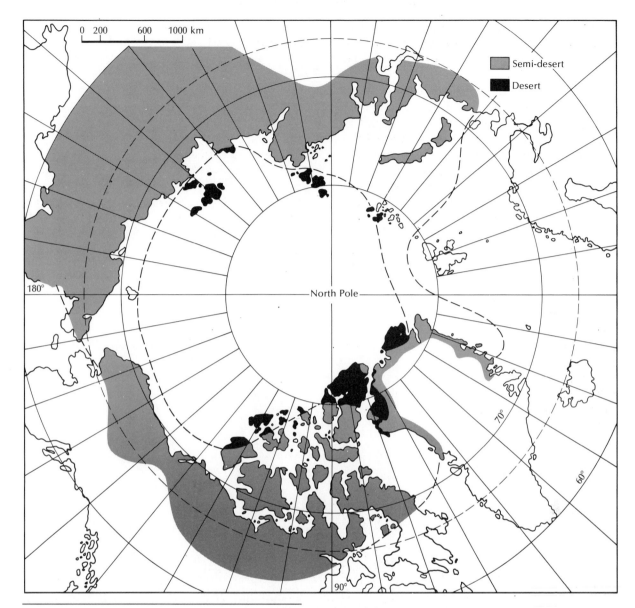

Fig. 11-32. Probable distribution of desert and semi-desert regions in the Arctic.

Fig. 11-33. Cavernously weathered granitic boulders. **A.** (above) in Death Valley National Monument, California. *R.M. Burke* **B.** (below) in Taylor Dry Valley, Antarctica. *T.L. Péwé*

terial which is now being recovered from the dazzlingly white expanse of the saline playa.

Desert lakes are equally characteristic of other arid regions throughout the world. Among well-known examples are Lake Chad in Africa, Lake Eyre in Australia; and Lop Nor, Lake Balkhash, and the Aral Sea in Central Asia. Not only were the Aral and Caspian seas larger in the recent geologic past than they are today, but they were connected with one another as well as with the Black Sea. Many others, such as the Dead Sea, are rimmed by

abandoned shorelines that scar the barren slopes of the bordering desert hills much like gigantic flights of steps.

The obvious recency of the expanded lakes, coupled with the fact that in a few locations, such as along the flanks of the Sierra Nevada and the Wasatch Range in Utah, the shorelines actually cut deposits laid down by ancient glaciers, shows that the last "high stand" of the lake generally coincided with the time of ice advance. Furthermore, the record is very clear that such events happened more than once,

Desert in bloom near Bahia San Louis Gonzaga, Baja California, Mexico. Because so little of the ground is protected by vegetation, infrequent torrential rains can cause a considerable amount of erosion in a short time. *Eliot Porter*

(Above) Ayers Rock, an erosional remnant of arkosic sandstone rises dramatically from the desert plain of central Australia. *Pan American World Airways* (Below) Lake Powell, formed by the Glen Canyon Dam on the Colorado River, contrasts markedly with the surrounding arid landscape. *American Airlines*

Fig. 11-34. Harmful effects on landscape of cross-country motorcycle travel near Los Banos, California. **A.** (above) General view of erosion and vegetation removal. **B.** (left) Small gullies in a heavily used area. Motorcycles compacted the soil to a depth of 1 m (3 ft), and increased runoff and erosion resulted. The area was closed to motorcycles in 1969; the photograph was taken in 1975. Notice the revegetation of part of the disturbed slope. *R.F. Hadley, USGS*

and that all the high stands occurred during glacial times and that the "low stands" (times of near-complete desiccation) coincided with times in which the glaciers disappeared. Climatic change and the increase of meltwater appear to have been the causes of lake expansion, and, as the glaciers of the world receded to their present diminished extent, the level of North America's desert lakes fell. Many of them vanished almost entirely, leaving a barren expanse of salt or shrunken alkaline ponds as relics of what once was an inland sea.

POLAR DESERTS

Another desert environment that we are getting to know better is the polar desert—the cold dry areas that circumscribe the North Pole (Fig. 11-32) and that occur in the ice-free areas of Antarctica. The main differences between polar deserts and hot deserts is that the former receive most of their precipitation in solid form and are characterized by permafrost (see Chap. 9).

Although some features of polar deserts are closely linked to the presence of permafrost, other features are quite similar to those of the hot deserts. So striking are some of the similarities that, were it not for the temperature, visitors might think they were in a hot desert! Rock weathers into rather weird forms in both cold and hot deserts even though the processes of weathering might differ (Fig. 11-33). For example, salt weathering and wind erosion occur in both environments, and the little precipitation that does fall has some effect in both. Temperature alone, however, is not an effective weathering agent in polar deserts. Soils in both environments can be quite similar, with thin horizons, and salt accumulations in the subsurface. Information on the rates of weathering and soil formation is somewhat scanty, but so far it seems that those processes take place more slowly in the cold deserts. Ventifacts occur in both environments, and even salt lakes, the trademark of the hot deserts, are

found in some polar deserts. Lakes in the Dry Valleys of Antarctica are an example.

Some hot-desert features, however, seldom occur in polar deserts. Although in places there seems to be an adequate source of sand in the ubiquitous glacial outwash plains, for some reason the winds have not picked up the sands and shaped them into the sand seas so often characteristic of hot deserts. Although alluvial fans are seen in the polar deserts, pediments are not. Perhaps, as with so many other geologic anomalies, we can call upon time to explain their absence. Many polar deserts only recently were covered by glaciers. Given more time, might pediments form? Nobody knows, and there is always the possibility that processes associated with permafrost will tend to negate pedimentation processes.

A FRAGILE LANDSCAPE

In many desert areas in the United States, population is burgeoning and with it problems in land use and abuse. In well-watered areas, for example, vegetation grows rapidly after any disturbance and quickly hides scars. In deserts, however, the works of humans remain in plain view for many years.

A recent problem in desert landscapes has been the damage caused by recreational vehicles. The ground is often disturbed to the extent that tire tracks can be seen crisscrossing the desert from miles away (Fig. 11-34). The tracks not only mar the landscape, but they can compact the soil or disturb the vegetation to the point that increased runoff and gullying become a real problem. Motorcycles have now made their way into the Arctic, where they are run across the tundra—only time will tell how they will affect that fragile terrain. We do know that the scars in both deserts will last a long time, however, because deserts heal very slowly. Marks made today will undoubtedly last for the next generation to view, if not longer. Deserts should be treated with care so that future generations may admire and enjoy their special beauty.

SUMMARY

1. Hot deserts occupy about one-third of the earth's land area, and are characterized by rates of evaporation and plant transpiration that greatly exceed precipitation. Vegetation cover, therefore, is sparse, and that plus the torrential but localized nature of the rainfall help account for many desert features.

2. Many surfaces in the desert are characterized by *desert varnish* and a *stone pavement*. The former results from weathering and the latter from either wind action or the upward movement of stones from shallow depths.

3. The main depositional landform characteristic of hot deserts is an *alluvial fan,* formed from material deposited by a stream as it leaves the mountains and enters an intermontane basin.

4. The main erosional landform in hot deserts is a *pediment*. Its surface form is similar to that of a fan, but it is bedrock rather than deposited material. The origin of pediments is still debated.

5. Wind is a very effective agent in shaping some deserts. The sand it carries cuts away at stones and bedrock alike, and in some places it cuts and excavates bedrock basins.

6. Wind-blown sand is deposited into a variety of dune forms.

7. Many deserts were wetter in the past, as shown by evidence that lakes existed in many present-day desert basins during times of past glaciation.

8. Polar deserts are cold deserts, and, with the exception of dune fields and pediments, have many features in common with hot deserts.

SELECTED REFERENCES

Bagnold, R. A., 1941, The physics of blown sand and desert dunes, Methuen and Co., London. (Repr. 1965, Halsted Press, New York.)

Bull, W. B., 1968, Alluvial fans, Journal of Geological Education, vol. 26, no. 3, pp. 101–6.

Cooke, R. U., and Warren, A., 1973, Geomorphology in deserts, University of California Press, Berkeley.

Glennie, K. W., 1970, Desert sedimentary environments, Elsevier Publishing Co., New York.

Hadley, Richard F., 1967, Pediments and pediment-forming processes, Journal of Geological Education, vol. 15, no. 2, pp. 83–89.

Leopold, A. S., and the Editors of Life, 1962, The desert, Time Inc., New York.

Morrison, R. B., 1968, Pluvial lakes, pp. 873–83 in The encyclopedia of geomorphology, R. W. Fairbridge, ed., Reinhold Book Corp., New York.

Oberlander, T. M., 1974, Landscape inheritance and the pediment problem in the Mojave Desert of Southern California, American Journal of Science, vol. 274, pp. 849–75.

Smiley, T. L., and Zumberge, J. H., eds., 1974, Polar deserts and modern man, The University of Arizona Press, Tucson.

Fig. 12-1. The glaciated peak of Mount Hayes in the Alaska Range, Alaska. *Austin S. Post, USGS*

12

GLACIERS AND EFFECTS OF GLACIATION

Scenically, the world is more indebted to glaciation than to any other process of erosion (Fig. 12-1). Without glaciation there would be few of the jagged peaks that stand in isolated splendor along the crest of many of the world's lofty mountain ranges. Such a resplendent peak as the Matterhorn is an example, so familiar through endless repetition in calendars and travel posters as to verge on the trite—until it is actually seen.

The formation of steep cliffs in the valley heads and along the valley walls, so challenging to rock climbers in the Alps, the Rocky Mountains, Alaska, and the Sierra Nevada is but a single aspect of glaciation. That a glacier can erode more deeply in some parts of its channel and less deeply in others, and that the material it deposits has an irregular surface, is responsible for the multitude of lakes that add such interest to the landscape of alpine regions

throughout the world as well as to such lower-lying areas as the Great Lakes region, northeastern Canada, and Scandinavia. Unfortunately, however, the same lakes are mainly responsible for a great summer flourish of insects. Deep valleys whose outlines have been sharpened by the glacial file—such as Yosemite Valley, the Lauterbrunnenthal of Switzerland, and the Norwegian fiords—are sufficiently spectacular to support large and flourishing tourist industries. These and many other testimonials to the effectiveness as well as the uniqueness of glacial processes are widespread throughout the high latitude and high altitude countries of the world.

The ice age just ended provides one of the most stirring chapters in earth history, one that to a greater or lesser degree has affected the lives of us all. Soil and loose rocks were stripped from vast land areas, leaving barren

Fig. 12-2. Enormous boulder resting on glacial ice, Baffin Island, Northwest Territories, Canada. Once the ice melts the boulder will be known as an *erratic*. *J.D. Ives*

rock behind. The load of stripped material was deposited toward the glacier margins. In addition, glacially produced silt and clay were blown off the floodplains of the world's major glaciated rivers and deposited as a blanket across the landscape for many kilometers downwind. Such deposits constitute some of the world's finest agricultural lands.

Large lakes were created where none so extensive had existed before. Lake Agassiz located just west of the present Great Lakes was one of them, and its remnant, Lake Winnipeg, is a giant in its own right. The Great Lakes, also in large measure a product of glaciation, are

still with us. Great Salt Lake in Utah is a remnant of a much larger lake. And the rapid emptying of a glacier-dammed lake in the northwestern United States unleashed a prehistoric flood that could be the largest ever recorded.

Many of the areas blanketed by glacial ice were bowed down under its weight, and when the burden was lifted through the disappearance of the ice, the land rebounded. In Canada, uplifted wave-cut features show that the Hudson Bay region has risen 300 m (984 ft) or more. Historic records and various shoreline structures such as ancient landing places show that the rise still continues. It reaches the unusually high rate of about 2 m (6.5 ft) per century at the southern end of Hudson Bay and decreases to about zero in the vicinity of the southern Great Lakes.

World-wide, the sea level swung in rhythm with the waxing and waning of the ice sheets. When the ice sheets expanded, tens of millions of cubic kilometers of water were withdrawn from the sea via the atmosphere and locked up on land as ice. Sea level was lowered as a consequence, perhaps by as much as 140 m (459 ft). That may not seem like much, but it was enough to alter profoundly world geography. Land areas now separated were then connected, and migrations of whole populations of animals were made possible by the existence of *land bridges*, as they are called. Among the natural causeways were such links as those that once connected Tasmania and Australia, Ceylon and India, New Guinea and Australia, and some of the islands of Indonesia. Most renowned of all was the land bridge that joined Alaska and Siberia, areas now separated by the 55-m (180-ft) waters of Bering Strait. In a way, that link must have been a veritable freeway, with all sorts of creatures, including human beings, pattering to and fro. Westbound from the New World, migrating into the Old, went zebras, camels, tapirs, and horses. Eastbound immigrants were elk, musk oxen, bison, elephants, mountain sheep, and mountain goats. Not the least were human beings.

There are many reasons why geologists study glaciers and the effects and deposits of former glaciations. Perhaps foremost is that they indicate rather dramatic past climatic changes. If we can determine when such changes occurred, we might be able to predict the frequency of future climatic change. Another reason for studying glaciers is that glacial deposits often create special engineering problems. Further, monitoring the amount of ice on land helps us to determine if the rise and fall of sea level along some coasts is due to world-wide changes in the volume of ice or to local tectonic causes.

THE GLACIAL THEORY

No wonder much of the evidence of past glaciation attracted the attention of observant men in Europe and in New England in centuries past. The lavish supply of boulders on New England farms was a source not only of wonderment to the early settlers but also of wearisome, backbreaking toil. So much labor was involved in clearing fields strewn with glacially transported stones that more than one young man was readily convinced that a life at sea could be no harsher—even on a New Bedford whaler.

For many, the presence of those stranger-stones, found far from their place of origin and very often completely out of harmony with their new environment—for example, granite blocks resting on a limestone terrain—was adequately explained as the work of that "vindictive affliction," the Great Flood of Noah.

In Great Britain, much of which was covered by only recently vanished glaciers, the widely scattered glacial deposits were called *drift*, a name betraying the belief that it originated as a deposit spread far and wide by an all-encompassing sea. Some even accepted the idea that icebergs and ice floes may have transported the *erratic boulders* (Fig. 12-2), as such out-of-place rocks are called, for British whalers working off the coast of Greenland had seen such debris embedded in sea ice there, as had explorers elsewhere in the Arctic.

Persuading English geologists that glaciers had scoured the inland surface and transported rocks the size of small houses for scores of kilometers was a difficult task, for there were no existing glaciers to serve as models. It is not surprising, therefore, that the most eloquent advocates of the notion that ice could perform prodigies of work in shaping mountains and excavating valleys were the Swiss. There are about 2000 glaciers in the Alps, and through the centuries their snouts have advanced or retreated, and alpine passes have been alternately ice free or ice blocked. Some villages occupied in medieval times are now buried by ice. The silver mine of Argentiere, active in the Middle Ages, is now covered by the glacier of

Fig. 12-3. Louis Agassiz (1807–73). *Harvard University Archives*

Mont Blanc near Chamonix in the French Alps. Many alpine villagers must have been aware that when a glacier receded it left behind it a trail of barren, stony ground interrupted by low, rocky ridges, diversified by lakes and ponds, and strewn freely with rock fragments, large and small.

In the early nineteenth century, several European geologists became convinced that an ice sheet had covered much of northern Europe, an idea that aroused the curiosity of one of the leading Swiss naturalists of the day, Louis Agassiz (Fig. 12-3). He persuaded one of the geologists to take him on an expedition in the Alps. Agassiz became a believer, and when in 1837 he came to the United States and to a professorship at Harvard University, he spread the word far and wide of a "Great Ice Age" that had once refrigerated most of the Northern Hemisphere. A concept so novel aroused opposition, for some accepted the far more labored explanation that (1) the land sank; (2) the sea spread inland, and boulders and other detritus were rafted by icebergs far and wide across its waters; and (3) the land rose, thus shedding its oceanic waters which left

behind a residue of rocks and boulders scattered over the landscape.

Although Agassiz deserves full honor for carrying the word from Europe to North America, he was not the first to believe that glaciers had once advanced across the European landscape. The problem of discovering who had the original idea is one of the manifold difficulties confronting the historian of science. So it was with the glacial theory. A number of remarkably perceptive persons had glimpsed the truth and then almost immediately were forgotten. As early as 1802, erratic boulders in the Jura Mountains were recognized for what they were—glacially transported rocks. In Germany a professor of forestry, Bernhardi, employing the same reasoning, wrote a paper in 1832 stating his belief that the ridges of drift and erratic boulders that are significant terrain elements of the northern plain were evidence that glaciers had advanced southward from lands far to the north. He suffered the familiar fate of a prophet in his own country in that few people paid the slightest attention to him. Not until 1875 did German scientists accept the singular idea that their homeland had once been overridden by a sheet of ice.

Forward steps in science very often are the work of many, rather than the brilliant inspiration of a single genius. Given a similar environment, or similar evidence, different people working quite independently of one another may come to the same general conclusion. The virtually simultaneous announcement of a mechanism of evolution by Charles Darwin and Alfred Russel Wallace is a classic example.

DISTRIBUTION AND FORMATION OF GLACIERS

Glaciers at present occupy about 15,000,000 km² (5,791,500 mi²) of the earth's surface—about 10 per cent of the total land area. Most of the ice is locked up in two ice caps: Antarctica, which accounts for about 84 per cent of the present ice in the world, and Greenland. The rest of the ice is scattered around the world, generally in mountainous areas. Glaciers are especially prominent features in the ranges that parallel the coasts of Alaska, British Columbia, and Washington; in the Rocky Mountains of Canada and the northern United States; on many islands of the Arctic region, including Greenland, and in Scandinavia, the

European Alps, the Southern Alps of New Zealand, and the Andes.

The amount of water locked up in glacial ice can be converted to depth of ocean water; thus we can appreciate its quantity and gain some idea of what might happen if all the ice on earth were to melt. It is no simple task to measure the amount of existing ice because the configuration of the bases of glaciers are not well known. The best estimates, however, are that if all the present ice were to melt, sea level would lie some 100 m (328 ft) above its present level. A glance at a map of the United States shows that many coastal cities would lie under 100 m of water, whereas many inland cities would become seaports (Fig. 12-4). There is no need for alarm, however, because the major ice caps are fairly stable, and any changes that might take place would do so gradually, over many thousands of years. The glacial recession since the 1890s, for example, has resulted in a general sea level rise of 12 to 30 cm (5 to 12 in.)—a change perhaps not important to many of us, but certainly important to people living on low-lying coasts. In low-lying areas of Holland, for example, dikes have been built to keep the rising sea from submerging the land. It should be pointed out, however, that a rising sea level is not the only reason that some areas are going

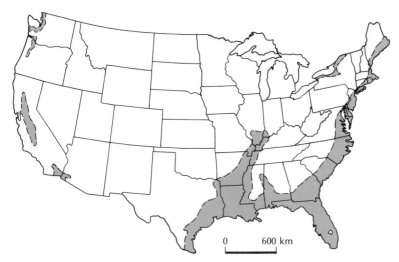

Fig. 12-4. Map showing the approximate location of the shoreline in the United States, if all glaciers in the world were to melt.

Fig. 12-5 A. (above) Newly fallen snowflake. The specific gravity of new snow is low because much of its volume is occupied by air. **B.** (below) With time, snowflakes melt and refreeze into rounded snow particles. *E.R. LaChapelle*

under water. Other reasons are tectonic subsidence and compaction of the sediments that make up the land. More will be said of those processes later.

Change of snow to ice

Without snow there would be no accumulation of extensive bodies of ice, and without ice there would be no glaciers. True glaciers, however, must display evidence of flowage. Glaciers may develop in those parts of the world where the combination of sufficient winter snowfall and low summer temperatures results in some snow remaining year around. Thus, glaciers are active today in high mountains over much of the globe, and in the farther reaches of the Northern and Southern hemispheres. On the high upper slopes of such equatorial mountains as Kilimanjaro in Africa, the summits of the Andes in South America, and the Carstenz Toppenz Range in New Guinea ice fields lie at altitudes of 4800 to 5400 m (15,744 to 17,712 ft). In mid-latitudes, as in the Sierra Nevada of California and the Swiss Alps, they lie close to 3000 m (9840 ft). Finally, they form near sea level in Antarctica (55°S), and at about 600 m (1968 ft) in lands bordering the Arctic Ocean.

More than cold temperatures are needed for snow to accumulate and remain from year to year. We can demonstrate that readily by pointing out the vast areas of the cold Arctic that are not covered with glaciers. Sufficient snowfall also is required to maintain glaciers, but the exact amount will vary from place to place. In maritime regions with high summer snowmelt, about 4 m (13 ft) of snowfall (on a water-equivalent basis) are required to maintain glaciers, whereas only a fraction of a meter is required in the polar deserts of the northern Arctic and Antarctica, where summer snowmelt is at a minimum.

Local environment also plays a significant role in determining positions of glaciers. North-facing slopes are shadier and receive less solar radiation than south-facing slopes, and, under the right conditions, ice can persist on such northerly exposures but not on those facing south. In many North American mountain ranges the western side may be the windward slope, where precipitation is greater and cloud cover is more persistent. There glaciers commonly lie at lower altitudes than they do on the eastern side. Wind is also a determining factor in some places, as it redistributes winter snowfall. In parts of the Rocky Mountains of Colorado and Wyoming especially, dry snow falling on the alpine tundra slopes is picked up by winds from the west and redeposited on the heads of east-facing valleys, eventually to be converted to glacier ice.

Before it becomes a glacier, snow must convert to ice, which is no simple process. Snow, strictly speaking, is not frozen water, as is ice, but is frozen water vapor. In other words, it is water that has crystallized directly from water vapor in the atmosphere. Since snow is a crystalline substance, snowflakes grow in regular geometric patterns (Fig. 12-5A). Although they seem to show an infinite variety, it probably is not strictly true that no two snowflakes are ever alike. Actually, crystalline water behaves like other mineral crystals in that there is an established crystal form for the compound —a variant of the hexagonal system—the same system of which quartz is a member. The specific gravity of snow is much less than that of water, so that 1 cm (0.4 in.) of snow is frequently equal to 1 mm (0.04 in.) of rain water.

After snowflakes lie on the ground for a short while, they ordinarily undergo a change (Fig. 12-5B). Individual flakes may *sublimate* (pass directly from a solid to a gaseous state), they may melt and refreeze into granules of snow. That gritty, granular snow, with a texture much like coarse sand, is a familiar phenomenon in snowbanks that survive for a fair share of the winter behind a building, in the shade of a forest, or in the lee of a cliff. Such granular, recrystallized snow is called *firn* (from a German adjective meaning of last year) in the German-speaking parts of Switzerland; *névé* (a word going back to the Latin stem of *nix* for snow) is used in French-speaking areas. Both terms are used in English as well.

Firn, which typically accumulates on the upper slopes of alpine mountains, goes through a gradual transition into glacier ice. Firn is usually white or grayish white, and the spaces between the granules are filled with trapped air. At a depth equivalent perhaps to an accumulation of three to five years of firn, the pore spaces become smaller, or even may be lacking, and the transition into blue glacier ice made up of interlocking ice crystals is completed (Fig. 12-6). The process is accompanied by an increase in the specific gravity from perhaps 0.1 in newly fallen snow up to 0.9 in solid ice. The change from firn to ice is aided by an increase in pressure resulting from the weight of the overlying snow and ice. As a glacier moves downvalley the ice crystals undergo recrystallization and may reach diameters as large as 7 to 10 cm (3 to 4 in.).

The conversion of snow to ice is an excellent example of present-day, rapid metamorphism. Snow falls on the ground and builds up sedimentary layers (Fig. 12-7). As temperature and pressure conditions change, the snow is metamorphosed into ice, which moves downvalley, developing many flow patterns that are akin to patterns made by folding in metamorphic rocks. As the snow becomes ice, much of the glacier remains solid, just as rocks do during metamorphic change. If the formation of glaciers were not a solid-state change, the ice would melt and the glacier would vanish before our very eyes.

Mechanism of glacier movement

Nobody doubts that glaciers flow. A common proof is that the rocks being deposited at the front of alpine glaciers are derived from the cliffs that flank the upper end of the glacier. Gruesome evidence of such movement comes from the disappearance of an early mountaineer in the European Alps. Efforts to find him failed, but his body did show up decades later at the lower end of the glacier. In the section that follows we will look first at glacier flow

and then try to explain how such motion comes about.

The cross-valley velocity of a glacier in an alpine valley is very similar to that of water in a stream. The flow is greatest in the center, and the least on the sides and bottom (Fig. 12-8). The diminishing velocity toward the valley sides and bottom is the result of friction between the ice and the bedrock. Average velocities vary from glacier to glacier, but most fall between 3 and 300 m (10 and 984 ft) per year.

Occasionally a glacier becomes decoupled from the rock floor and sides and advances downvalley at truly fantastic velocities, some approaching as much as 6000 m (19,680 ft) per year. Such movements are known as *glacial*

Fig. 12-6. Interlocking crystals of glacier ice, as seen under a microscope. *Chester Langway, Jr.*

Fig. 12-7. Sedimentary layers in glacial ice are well shown at the edge of this glacier in Vatnajökull, southeast Iceland. *J.D. Ives*

surges and are usually characteristic of valley glaciers, although ice caps have also been known to surge. Surges generally take place after a glacier has been stagnant or even receding; suddenly it moves several kilometers in a few months. In 1953, for example, the Kutiah Glacier in the Himalayas surged 11 km (7 mi) in three months. Various theories have been suggested to account for glacial surges. Some geologists have thought that earthquakes might shake great quantities of snow and ice down onto a glacier, or that increased snowfall for a few years at the head of a glacier could cause surging. Others have suggested that an increase in the amount of meltwater percolating down through a glacier might help it to slide along its bed. A credible explanation recently put forth is based upon the formation of a block of stagnant ice at the end of a valley glacier. Such a block would act as a dam until the pressure of the flowing ice behind it forced it to give way.

The way ice flows is complex. Generally, the lowest part of most glaciers moves by actually sliding over the rocks, as shown in Figure 12-8B by the displacement of the base of the pipe from X' to Y'. Evidence for such movement, which is called *basal slip*, are the polished and scratched rocks left behind when glaciers melt (Fig. 12-9). In contrast, the lower part of a glacier (Fig. 12-8B, between Y' and Z) creeps downvalley through *plastic flow* because the ice crystals deform under the pressure of the overlying ice. The upper part of a glacier (above Z in Fig. 12-8B) consists of brittle ice since there is not enough overlying ice and snow to cause it to deform plastically. Such ice rides piggyback on the lower ice, which is continually undergoing deformation. Deep flowage, combined with the valley-side friction and the low strength of the brittle ice, causes innumerable cracks or *crevasses* to form in the brittle ice (Fig. 12-10). Such crevasses can extend to 30 m (98 ft) or more in depth; at greater depths plastic flow of the ice seals them off.

The above discussion pertains to most glaciers, but it is not quite accurate for those located in polar regions. There, temperatures are so low that glaciers are most likely frozen to their bedrock bases. Basal slip, therefore, is not so important in the motion of polar glaciers as it is in the motion of temperate-climate glaciers, which are not frozen to their bases.

The glacier budget

Glaciers are continually gaining and losing mass, and by a series of detailed measurements we can determine those gains and losses; in short, we can determine a budget. But while we can manipulate a budget, a glacier cannot easily hide its surpluses or deficits. If, for example, it has a surplus year, the front may advance downvalley, whereas in a deficit year the front may retreat.

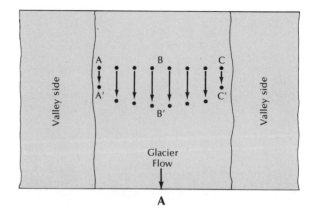

A

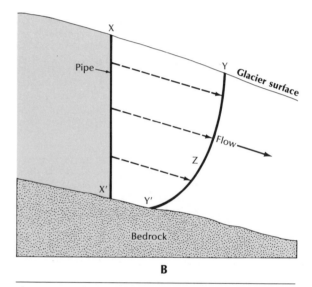

B

Fig. 12-8 A. Plan view of an alpine glacier, showing how velocity measurements are made. Stakes are set in the ice in a straight line (A-B-C). Eventually the movement of the ice will cause the line to curve (A'-B'-C'). The distances between the original and present positions of the stakes indicates the amount of displacement, from which the velocity is calculated. **B.** Vertical profile of an alpine glacier, showing the velocity distribution with depth. A vertical pipe extending down to bedrock is placed at X. After several years the pipe will bend into a curved form (Y-Y'), its top moving from X to Y and its base from X' to Y'.

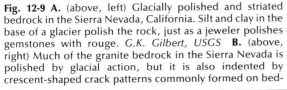

Fig. 12-9 A. (above, left) Glacially polished and striated bedrock in the Sierra Nevada, California. Silt and clay in the base of a glacier polish the rock, just as a jeweler polishes gemstones with rouge. *G.K. Gilbert, USGS* **B.** (above, right) Much of the granite bedrock in the Sierra Nevada is polished by glacial action, but it is also indented by crescent-shaped crack patterns commonly formed on bed-rock overridden by ice. In some places crack orientation can be used to determine the direction of ice movement. *G.K. Gilbert, USGS* **C.** (below) Glacial action grooved this limestone bedrock now exposed in a quarry in Vermilion County, Illinois. The long scratches are called *striations*. *Illinois Geological Survey*

Fig. 12-10 A. (right) Crevasses in a glacier on Mount Rainier, Washington. Irregular bedrock topography sets up complex flow patterns in the plastically deforming ice near the base of the glacier, causing the brittle uppermost ice to crack and form crevasses. *Peter W. Birkeland* **B.** (below) Crevasses as a tourist attraction, 1895. Their patterns change with the constant glacier movement, and in places they are covered by weak snow bridges—a hazard to mountaineers. Note the depth of this crevasse in Muir Glacier, Alaska. *Library of Congress*

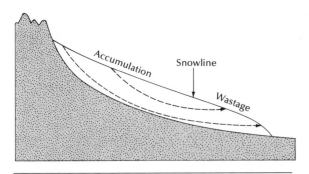

Fig. 12-11. Cross section of a valley glacier, showing the areas of accumulation and of wastage or ablation. The dashed lines are the approximate flow lines of the glacier particles. Notice how they descend and then rise toward the surface in the ablation area. A snow particle in the accumulation area will follow such paths, convert to ice on the way, and eventually melt when it reaches the surface.

In order to understand a glacier's budget, we need to determine where the gains and losses are taking place. Every glacier has a fairly narrow zone on its surface known as the *snowline* (Fig. 12-11). That zone is identified late in the melting season; above it, some snow lingers from year to year and accumulates, whereas below it the snow from the previous winter (and some of the underlying ice) melts and is lost. Thus there are distinct areas of accumulation and wastage, or *ablation*, on glaciers (Fig. 12-12). In a very general way, the area of accumulation makes up about two-thirds of the total surface area of a valley glacier.

With the glacier budget in mind, we now can examine how a glacier advances and retreats. We will focus our attention on the front of a stable glacier; that is, one that is neither advancing nor retreating because the accumulation of snow is balanced by ablation (Fig. 12-13, position A). The glacier maintains the same form over the years, so that all losses in the ablation area are precisely balanced by the flow of ice into that area from up-valley. To illustrate the point, we might compare a glacier to a side of bacon being fed into a slicer. Al-

though the bacon is continuously being shoved forward, it never advances beyond a fixed point because it is always being cut off by the oscillating blade. In contrast, if the snowline were to lower (Fig. 12-13, position B), the area of accumulation would be enlarged. The amount of ablation would then be less than the gains, and the front would advance. Advance takes place until, for that snowline position, a balance between gains and losses is reached. Similarly, if the snowline rises, the opposite effects are seen—the area of wastage is enlarged, more ice melts than can be compensated for by accumulation, and the front retreats until a balance with respect to the new snowline (Fig. 12-13, position C) is reached. If the snowline continues to rise until it no longer intersects the land surface, the glacier melts away.

ALPINE GLACIATION

Although individual glaciers and ice fields are tremendously diverse, they can be placed in two broad categories: *alpine glaciers* and *continental ice sheets.* Alpine glaciers have their origin on mountain slopes and summits that rise above the snowline. They advance downslope under the urging of gravity, and more often than not take the path of least resistance by following a pre-existing stream valley. In contrast, continental ice sheets are irregular in shape, cover a large land area, and are not necessarily guided in their flow pattern by the underlying topography.

Continental ice sheets were much more important in the recent geologic past than they are today. During the ice ages, North America as far south as the Ohio and Missouri rivers was buried under at least several kilometers of ice, as was most of northern Europe. Fortunately, such vast expanses of frozen water are gone, and surviving relics such as the Greenland ice cap, although of tremendous extent, are greatly subordinate to the vanished titans. Continental glaciers override the terrain and

Fig. 12-12. The snowline on these glaciers is marked by the transition from clean snow at their upper ends to streaked, "dirty" ice at their lower ends. The accumulation area lies above the snowline; the ablation area below it. Clyde Fiord on Baffin Island, Northwest Territories, Canada. *J.D. Ives*

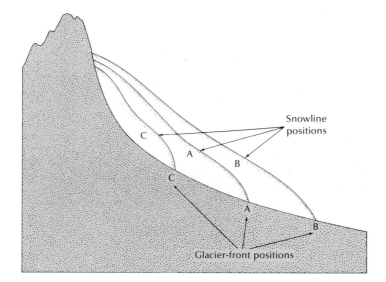

Snowline positions

C

A

B

C

A

B

Glacier-front positions

Fig. 12-13. Cross section of an alpine glacier, showing how its size is governed by the snowline position. An adjustment in size comes only through frontal advance or retreat because the upper end of the glacier is always anchored to the headwall.

subdue many of the irregularities they encounter. Alpine glaciers are more likely to accentuate irregularities in the landscape, making bold peaks even more jagged, and steepening the walls of already deep canyons.

Although alpine glaciers are conspicuous elements of the world's snow-capped mountain ranges, they, too, are much smaller than they were in the ice ages. At that time, glaciers in the Alps advanced northward far beyond the foothills of the Alps onto the lowlands near Munich and southward into the low hills marginal to the Po Valley of Italy. Reduced though they may be, some of the surviving alpine glaciers are impressively large. Several of the Himalayan ice streams are 40 km (25 mi) long, and the Great Aletsch in Switzerland has a length of 22 km (14 mi). The Seward glacier in Alaska with its tributaries has a total length approaching 80 km (50 mi).

In the next section we will discuss glacial erosion and deposition, emphasizing alpine glaciation. The discussion should help us to understand the effects of continental ice sheets on the landscape.

Glacial erosion

There appear to be two leading ways in which alpine glaciers shape the land surface upon which they rest: *glacial quarrying* and *glacial abrasion*. In the first process, rocks are sprung, or pried, out of place in much the same way that they are in commercial rock quarries—except, of course, at a far slower rate. Abrasion takes places when the rock surface on which the ice rests is scoured or worn down—in much the same way that a wood surface may be sandpapered.

The way in which quarrying operates is not completely understood—scarcely surprising, because it is a process that takes place beneath the glacial ice in an environment inaccessible to us by ordinary means.

The nearest access we have to that frigid, subglacial world is through the crevasses that extend down through the ice to the bottom of the glacier. Such deeply penetrating fractures are rare, but are reasonably common in the accumulation area at the headward end of a glacier. In 1899 in the Sierra Nevada of California one of the leading topographers of that day, W. D. Johnson, had himself lowered on a line to a depth of 45 m (148 ft) into a crevasse, certainly no venture for the fainthearted—and saw that glacial ice indeed was capable of prying strongly jointed rocks loose from their foundations.

Glacial crevasses provide an avenue by which *meltwater,* supplied by melting of snow, névé, and surface layers of ice, streams down into the inner recesses of a glacier. If meltwater penetrates to the rocky headwall or to the glacier floor, it percolates into the cracks and joints of the bedrock. As the water freezes it expands, exerting a tremendous leverage against the enclosing rocks, which may then be pried loose.

Should those rocks be frozen into the glacier, they are carried along with the glacial ice as it moves downslope, and a new surface is bared for the process to be repeated. The more often glacial meltwater freezes in the subglacial rock joints, the more effective the process of rock quarrying will be.

Rocks ranging from blocks the size of boxcars down to fragments the size of flour grains are embedded in the ice as the result of glacial quarrying and mass wasting from the valley walls. The rocky debris is dragged along with the glacier, and as it moves downslope, it acts as an abrasive. Where the embedded rock fragments are large, they may gouge out long grooves and scratches, called *glacial striations,* in the bedrock. Where the rock fragments are fine grained, they may polish the surface of the overridden rocks, much like a lapidarist's fine emery powder does. Visitors to the high parts of Yosemite National Park are impressed by the broad expanses of smoothly polished, shining granite, as fresh looking as though it had been given its bright sheen only yesterday (see Fig. 12-9). The process of abrasion works both ways; many of the rock fragments embedded in the ice develop scratches on their surfaces.

Landforms produced by glacial erosion

As the debris-laden ice quarries and grinds away the surface over which it moves, characteristic landforms are produced. Most commonly it smooths down irregularities, so that

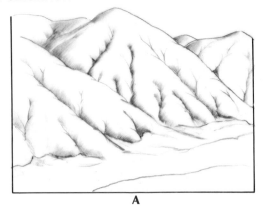

A

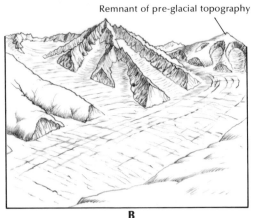

Remnant of pre-glacial topography

B

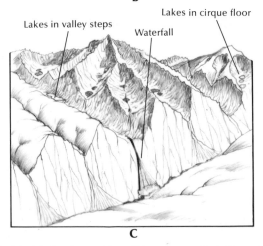

Lakes in cirque floor
Lakes in valley steps
Waterfall

C

Fig. 12-14. The alteration of a stream valley by alpine glaciation. **A.** Terrain before glaciation with smooth, rounded hillslopes and a stream meandering through a V-shaped valley. **B.** Maximum extent of glaciation. Ridges and peaks close to the active glaciers are steepened and take on a jagged appearance. Not all of the rounded mountain summits are consumed, however. **C.** After the ice melts, the characteristic glaciated landscape appears: U-shaped valley, hanging valleys, cirques, and lakes.

on a typically ice-abraded landscape rock knobs are rounded off and the lower ends of spurs and ridges are blunted or even worn away. Valleys are deepened, made more linear (Fig. 12-14). In the most striking examples their sides are steepened until they are almost vertical, as in Yosemite (Fig. 12-15) or the Lauterbrunnenthal in Switzerland.

Of the features characteristically resulting from glacial quarrying, one of the most impressive is a *cirque* (Fig. 12-16). A cirque is a horseshoe-shaped, steep-walled, glaciated valley head; it is such a distinctive landscape element that a name for it exists in the language of every western European country in which it is found. A French word meaning circus, the term "cirque" is not completely appropriate in the glacial connotation because a true circus is a fully round figure. Actually, a glaciated valley head is a half-round feature and is more nearly comparable to the traditional form of a Greek theater. To return to the other names for a cirque: in German-speaking lands it is a *kar;* in Wales, a *cwm;* in Scotland, a *corrie;* and in Scandinavia, a *botn* or a *kjedel*. This brief linguistic excursion illustrates that geology is a truly international science; no single country has a monopoly on all the names for landforms.

The origin of cirques is still somewhat of a mystery. They seem, however, to result partly from active plucking or quarrying at the head of a glacier, probably as the result of the downward percolation of water and frost-riving of the rocks in the headwall that towers above the glacier surface. Another process that seems important is the rotational movement of the ice, which gouges out closed basins in the floors of cirques. Should the glacier disappear, as so many have since the end of the ice age, then the newly bared cirque is an impressive alpine sight. The rock wall at the upper end of such a titanic box canyon may be a cliff 1000 m (3280 ft) high, and the base of the cliff is usually completely free of the long and sloping apron of talus blocks so common at the base of other cliffs. Commonly a lake occupies the closed basin on the cirque floor (Fig. 12-17).

Scores of clear, often dark blue, rock-basin lakes add immeasurably to the beauty of alpine scenery, and many of them result from differential glacial scouring. Unlike streams, glaciers dig deeply in some places and much less so in others. The floor of a glacial trough may consist of closed basins, in which water collects after the ice has melted, and of intervening barren ice-smoothed ridges. Such an irregular floor, sometimes deep beneath the ice, sometimes shallow, can be excavated so long as there is a forward slope to the surface of the glacier. In the same way, lava flows are able to cross a corrugation of ridges and valleys athwart their path.

Although glacial erosion is responsible for the extremely rugged alpine scenery, it is also responsible for the passes that allow for travel through such forbidding terrain. When two glaciers exist in parallel valleys close to one another, then the divide separating them may become progressively narrowed through oversteepening until it is reduced to little more than a rock screen (see Fig. 12-14) called an *arête* (from the French word for ridge or fishbone). Similarly if two glaciers are on opposite sides of a divide and flowing away from each other, the headwalls of the cirques may intersect until only the narrowest sort of rock partition separates the two cirques. The top of such a rock screen may be almost razor sharp, and commonly is surmounted by jagged pinnacles or spires. If glacial quarrying continues actively, in very little time the intervening ridge may be stripped away completely. When that happens, the two glacial troughs intersect, and a sometimes strategically important transmontane pass may result. Some of the famous alpine passes, such as the St. Bernard, St. Gothard, and Simplon, are of that type, as are Berthoud Pass in the Rockies and Tioga Pass in the Sierra Nevada. In some glaciated mountains glaciers radiate away from the summit area like spokes of a wheel. Should the glaciers continue to quarry actively at their upper ends, then the mountain may be whittled away by the concerted attack until only a jagged, saw-toothed pinnacle called a *horn* survives. The Matterhorn is the world's most familiar example of such a glacially accentuated peak, and Mt.

Assiniboine in Canada is well known to North Americans (Fig. 12-18).

Downvalley, a primary result of glacial erosion is the overdeepening of valley floors and the steepening of valley walls. The part of an alpine valley occupied by a glacier resembles the channel occupied by a river, but on a far grander scale, because the volume of water in the glacier is far greater.

The statement commonly is made that such a glacially occupied, ice-scoured valley has a *U-shaped* cross section (Fig. 12-19), in contrast to the *V-shaped* form of a stream valley. In reality, a cross section of a glaciated valley more nearly approximates a *catenary* (a curve produced by a wire or chain hanging from two points not in the same vertical line). Such curves are seen in the gracefully sweeping loops of the wire between the towers of a transmission line. If the points of suspension are close together, the curve is narrow and the slope is steep; if they are far apart, the curve is open and the side slopes are gradual. Glaciated valleys also commonly are characterized by series of gigantic

Fig. 12-15 A. (below) Distant view of Yosemite Valley, California, looking east. Glacial erosion extensively modified its original sloping walls, creating a U-shaped valley. Yosemite's numerous vertical walls challenge rock climbers from around the world. *USGS* **B.** (opposite) Within Yosemite Valley. The lower end of the stream valley on the right was oversteepened by glacial erosion, producing a hanging valley and the picturesque waterfall. *Ansel Adams.*

cliffs separated by flat stretches that resemble cyclopean steps. They owe their origin to a combination of the variabilities of glacial flow (and thus the ability to erode) and the relative resistance of the bedrock to glacial erosion.

Among the more photogenic results of alpine glaciation are waterfalls that plunge over the rims of glaciated valleys and cascade in long streamers down the shining walls (Fig. 12-20). There are a number of explanations for the origin of such dramaticaly discordant side streams. A widely accepted view is that they were unable to cut down rapidly enough to keep pace with the main canyon, which was being deepened by the glacier. Thus their valleys are called *hanging valleys*.

Other spectacular landforms of glaciated terrain are the *fiords* that flank many high-latitude coasts, such as in Alaska, the Canadian Arctic, Norway, Chile, and New Zealand (Fig. 12-21). They are narrow troughs that differ from land-based flat-floored glacial valleys (compare with Fig. 12-15) mainly in that they are submerged by the sea, but also in their truly

Fig. 12-16. Cirques at the heads of glaciated valleys in the Wind River Mountains, Wyoming. The cirques are cut into an old erosion surface, the dissected remnants of which now form rolling plateau-like uplands. Lakes commonly occur in the cirque basin, but they also occur downvalley in bedrock rimmed basins and are a characteristic feature of glaciated terrain. *Austin S. Post*

Fig. 12-17. Iceberg Lake, Glacier National Park, Montana, occupies the floor of a cirque. *Austin S. Post*

Fig. 12-19. U-shaped valley formed by a glacier flowing on granite, Wind River Mountains, Wyoming.
Peter W. Birkeland

fantastic depths. One in Norway, for example, is 1300 m (4264 ft) deep; another in Antarctica is more than 1900 m (6232 ft) deep. Not uncommonly, there are fairly shallow submerged rock barriers that lie 100 to 200 m (328 to 656 ft) below sea level at their mouths. Geologists have long speculated upon the origin of fiords, and the theories range from one of tectonic

Fig. 12-18. Mount Assiniboine in the Canadian Rockies is a classical example of a horn formed by glacial erosion.
Austin S. Post

origin to one of glacial scouring and over-deepening of former stream valleys. Surely, many fiords owe much of their origin to glacial erosion and stand as testimony to the tremendous work a glacier can perform.

Glacial deposits and depositional landforms

Much of the immense quantity of debris carried by glaciers comes to rest beneath or along the periphery of the ice downvalley from the snowline. The sheer volume of debris brought down by the ice, as well as the size of

Fig. 12-20. (opposite) When Yosemite Valley (foreground) was being deepened by glacial erosion, a small sidestream unable to keep up with erosion was left hanging when the glaciers melted; thus Yosemite Falls was created. Photographed in 1866 by Carleton E. Watkins. *Metropolitan Museum of Art*

Fig. 12-21. (below) View east along a fiord cut into a high mountain plateau. The waters are 1000 m (3280 ft) deep, and its valley walls are 1500 m (4920 ft) high. *Geodetic Institute, Copenhagen*

Fig. 12-22. Terminal moraine forming along the northeast margin of the Barnes Ice Cap, Baffin Island, Northwest Territories, Canada. *J.D. Ives*

many fragments, is more than the meltwater streaming away from a glacier can remove—at least not as quickly as it is supplied. Consequently, the sides and lower end of the glacier are nearly always smothered under a heavy rock burden (Fig. 12-22). The mass of ice-transported debris may accumulate as a hummocky, crescent-shaped, rocky ridge, looped around the snout of the glacier. A *terminal moraine* (from the Provençal French word *morena,* or heap of earth), or *end moraine* forms at the end of the glacial lobe. Ridges of debris that continue up-valley along the sides of the glacier form *lateral moraines* (Fig. 12-23). As a rule they are higher and bulkier than terminal moraines, and their crests slope forward with about the same inclination as the glacier surface. Should two glaciers join,

Lateral moraines

c

b

Terminal moraines

a

Fig. 12-23. Terminal and lateral moraines, Mount Jacobsen, British Columbia, Canada. The moraines mark the former extent of the glacier, perhaps during the early 1900s. A change in climate caused the ice to retreat, leaving behind the moraines, a moraine-dammed lake, and ice-scoured bedrock. *Austin S. Post*

Fig. 12-24. Medial moraines formed by the joining of lateral moraines, Kaskawulsh Glacier, St. Elias Range, Yukon Territory, Canada. The number of medial moraines is a good indication of the number of tributary glaciers that feed a main glacier. *Austin S. Post*

then the lateral moraines that meet at their intersection may unite and continue together down the middle of the ice stream as a dark band of rocky debris known as a *medial moraine*. In fact, there may be as many bands as there are unions of trunk and tributary glaciers, resulting in a wonderfully banded, candy-cane effect of strips of white ice interspersed with darker morainal layers (Fig. 12-24).

Lateral and terminal moraines form by several processes. Most of the material in lateral moraines is derived from the valley walls above the ice. Material that avalanches or mass wastes down the slopes is caught in the trough between the ice and the valley wall and forms a ridge as it is dragged downvalley by the moving ice. There are several opinions as to the origin of terminal moraines, but we will touch on just two here. One is that glaciers, as they advance over terrain covered by a lot of loose debris,

push or bulldoze the material into a moraine. The other theory has to do with the melting of the ice at the lower end of a glacier. Such ice carries a fair amount of debris in it, material left behind as the ice melts which is not carried away by the meltwater streams.

The combination of end moraines and lateral moraines that once looped in a semi-circular festoon around the terminus of a glacier often plays a scenically significant role after the ice has disappeared. For a time morainal embankments may serve as natural earth-fill dams, and quite successfully, even though in a sense they are pointed the wrong way. They are concave upstream, instead of being arched convexly against the reservoir as a well-designed dam is. Nonetheless, such morainal dams effectively impound the waters of some of the world's most scenic lakes. The best known probably are those bordering the Alps, such as Como,

(Above) In the lower parts of a glacier where ablation is dominant, the ice is commonly debris-laden and very rough as in Turamis Glacier in the Pamirs, Tadzhikskaya SSR. (Below) Detail of the topography in the ablation region of North Enilchek Glacier in the Tien Shan of central Asia. Note that the layering in the ice is near-vertical, no doubt due to deformation in the ice as it moves downvalley. *Both photos, Nicolai Gridin*

(Above) The braided pattern of the Valannuker River, Iceland, photographed under the midnight sun. This stream heads at the terminus of an active glacier. *Barbara Jarvis*

(Below) The Lauterbrunnen, a U-shaped glacial valley in the Bernese Oberland, Switzerland.
Swiss National Tourist Office

Fig. 12-25. Glacial till in the Sierra Nevada, California. Typical of many alpine areas, the till is very bouldery and boulders occur in a wide range of sizes.

Maggiore, and Garda on the Italian side and Neuchâtel, Geneva, Luzerne, and Constance on the north. North America has such moraine-blocked water bodies, too; Lakes Mary and MacDonald in Glacier National Park are two of the most striking examples.

Material laid down directly by ice is called *till* and consists of poorly sorted debris in which the size of particles ranges from clay to huge boulders (Fig. 12-25). Till underlies the moraines, a term saved for the resulting landforms. Few geologic processes leave such a jumble of debris, with, perhaps, the exception of desert and volcanic mudflows. In fact, in places it is difficult to determine if a deposit is a till or a mudflow. The material can be identified by examining the boulders it contains for striations and the bedrock on which it rests for polishing. Differentiation of the deposits is not merely an academic exercise, however, be-

cause assessment of geologic hazards on the high, glaciated volcanoes of the Northwest, for example, demands that one be able to tell one deposit from another. One might feel rather foolish in identifying as a mudflow—and thus assigning to an area a high-hazard potential— what was actually a till laid down by a sluggish glacier 20,000 years ago. Yet many such errors have been made by geologists.

Rock glaciers

In some areas of a continental climate, such as the Rocky Mountains of the United States, ice glaciers are not too common—probably the result of a combination of light winter snowfall and relatively high summer snowmelt. In their place we commonly see rock glaciers, tongue-shaped masses of rocky debris and ice slowly moving downvalley (Fig. 12-26). Because of the inherent problems of digging into a rock glacier, we have little data on their interiors. Some seem to consist of essentially clean glacier ice overlain by a thin mantle of rocky debris, whereas others probably contain rocks throughout cemented together by ice. Rock glaciers move downvalley just as ice glaciers do, through basal sliding and plastic flow. As they advance their fronts continually are be-ing oversteepened, and rock avalanches are common there.

Apparently special conditions are necessary for their formation. One is that altitudes must be high enough for some snow to linger into the late summer. The other is that the sur-rounding cliffs must supply rocks rapidly enough to bury the summer snow and protect it from further melting. In fact, the surface

Fig. 12-26. Rock glacier on Mount Sopris, Colorado. Rockfalls from the high cliffs continually feed debris to the glacier. The slow movement downvalley results in transverse furrows and ridges on its sur-face. Rock avalanches are common on the front of the glacier, steepened by a con-stant push from behind. *Edwin E. Larson*

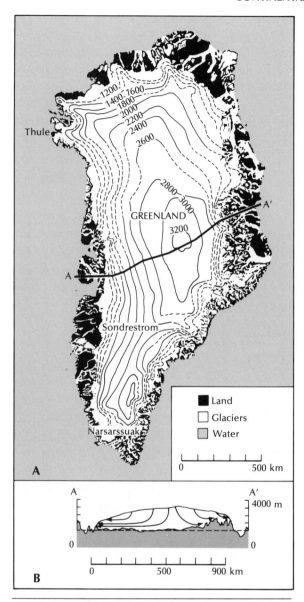

Fig. 12-27 A. Map of Greenland showing the areas of land and ice, with contours depicting areas of equal altitude on the latter. **B.** Cross section of Greenland along line A-A' of the map. Arrows depict approximate flow lines of the glacial ice.

mantle of rock protects or insulates the underlying ice so well that rock glaciers persist at lower altitudes than those at which some present-day ice glaciers could survive.

Ice and rock glaciers have different sensitivities to climatic change. A warming climate during the last century has caused many ice glaciers to retreat. In contrast, some rock glaciers behave as if they have never heard of the climatic change! Many of the larger ones are still advancing down valleys, knocking over trees in their path. We need not send out a rock-glacier alert, however, as their movement is very slow—about 1 m (3 ft) per year.

CONTINENTAL GLACIATION

Continental ice sheets, unlike alpine glaciers, did not move under the impetus of gravity in tongue-like masses through pre-existing valleys countersunk in the flanks of mountain ranges. Instead, during the ice ages they sprawled broadly as huge disk-shaped masses of ice over much of northern Europe and northern North America, covering more than 20 million km² (7.7 million mi²) of land surface. In North America the ice mass must have been a wall of nearly unimaginable size, extending almost 6400 km (3977 mi) across the entire width of Canada. That frozen tide spread southward into the United States roughly to the line of the pendant loop marking the courses of the Ohio and Missouri rivers today. In order to find a partial replica of the vanished ice sheets we need to turn to the ice sheets of Greenland and Antarctica, both of which resemble gigantic domes.

The Greenland ice cap covers some 2 million km² (0.8 million mi²) and is shaped like an elongate dome that runs parallel to the trend of the island (Fig. 12-27). The ice dome reaches an altitude of 3.3 km (2 mi), just slightly lower than the highest peaks that flank the east coast, and it extends to below sea level. The landscape along both coasts is quite irregular, and in places the ice overtops all the topographic barriers in its way. Such behavior is typical of a continental glacier—it flows in the direction of the surface slope, and if a mountain range is in the way, it simply rides over it. If a mountain is too high to be overridden, the ice flows around it. In general, the ice from the central part of the island moves down through deep, narrow fiords to the sea. There the ice may break off, a process known as *calving*, and drift away as *icebergs* into the sea lanes of the North Atlantic (Fig. 12-28). Such an iceberg sank the *Titanic* in 1912 with a loss of 1489 lives. The same icebergs carry entrained glacial debris, dumping it far out at sea as they melt. Such material rains down on the ocean floors, delivering sediment to those regions.

The Greenland ice cap is similar to alpine glaciers in that there is a central area of accumulation surrounded by an area of wastage. As shown in Figure 12-27, ice flow in the central

Fig. 12-28. A birthplace of icebergs. Storstrømmen, east Greenland. *Geodetic Institute, Copenhagen*

Fig. 12-29. Portion of the Antarctic Ice Sheet near McMurdo, looking northward. In the foreground are the dry valleys with the tongue-shaped Victoria Upper Glacier. The mountains in the middle ground are part of the Transantarctic Mountains; beyond is the Mackey Glacier, which drains part of the Antarctic Ice Sheet. *USGS*

area is inward, whereas in the surrounding area the flow lines intersect the ice surface. Unlike glaciers located farther from the sea, however, a larger proportion of the loss is through calving than through surface melting.

The Antarctic Ice Sheet is more than seven times the size of Greenland, but it has a lot in common with it (Fig. 12-29). For example, the Antarctic sheet also extends below sea level, and in places it flows against the "topographic grain" to reach the sea. It differs from the Greenland sheet in that its outer edge is mostly grounded on the continental shelf, and, because of the extremely cold climate and the

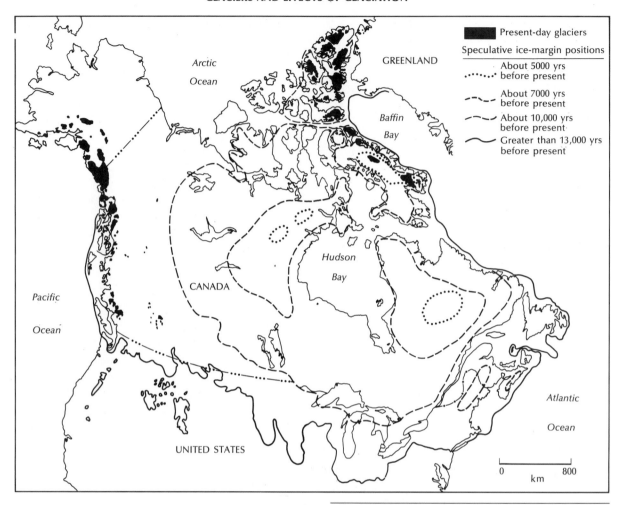

Fig. 12-30. Area covered by major continental glaciers in North America. The lines denote various positions of the ice front at different times (in years before present).

general absence of ice melting, in that most of the wastage is through calving. Thus an ice advance could only occur if the sea level were to drop, exposing a larger expanse of continental shelf. Parts of the ice sheet receive so little precipitation (less than 5 cm, or 2 in., per year) that the flow of ice is extremely slow and the area is a desert, but of the cold variety. Indeed, landforms and soil and weathering features in many areas in Antarctica closely resemble those found in hot deserts.

Today interest in the Antarctic Ice Sheet is high. Because the amount of water locked up in it is so great, small changes in its volume will strongly effect world-wide sea level. Most likely, the changes will take place very slowly, but some earth scientists feel that parts of the sheet could surge, just as alpine glaciers can surge. If that happened over a short enough time period and involved large masses of calving ice, the effect surely could be catastrophic to many coastal areas.

Enough data are now available to reconstruct the last major continental glaciers with some accuracy (Figs. 12-30 and 12-31). The generating centers for much of the North American ice appear to have been Baffin Island, the Labrador Peninsula, the country west of Hudson Bay, and the islands north of the Canadian mainland. Ice from those centers coalesced into one enormous ice sheet, which began to spread to the south and to the west. At times it joined the Cordilleran Ice Sheet, which was centered over the mountains of western Canada.

Both geologists and archeologists are interested in the placement and timing of the joinings, because it was a major migration route for human beings and animals from Alaska southward. The Cordilleran Ice Sheet also sent a thick tongue of ice southward into the Puget Lowland of Washington, covering areas where the cities of Bellingham and Seattle now stand. In Europe the ice sheets spread outward from the backbone of the Scandinavian Peninsula, and to a lesser degree from Britain and Ireland, traveling south beyond the present-day cities of Warsaw and Kiev and almost to the gates of Volgograd. Hills and mountains that were overridden by those great masses of ice are more likely to be rounded off than to be surmounted by the spires, castellated divides, and horns so typical of alpine glaciation. Mountains of surprising height have been buried beneath the ice of continental glaciers. Among those in the northeastern United States are the Catskills (1280 m; 4198 ft), the Adirondacks (1615 m; 5297 ft), and the Presidential Range of New England, whose highest point is Mount Washington with an altitude of 1917 m (6288 ft).

Landforms of continental glaciation

The landforms associated with continental glaciation are complex, and, although many are similar to those produced by alpine glaciation, they occur on a much larger scale. Cirques, of course, are not present because the ice does not flow from a jagged mountain top, but in-

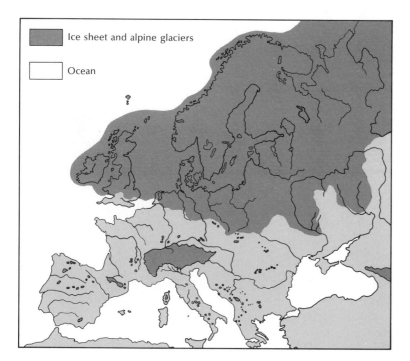

Fig. 12-31. Limits of glaciation in Europe. A continental ice sheet spread south from Scandinavia and the British Isles into northern Europe and Russia. South of the ice sheet, alpine glaciers covered many parts of the Alps, but the two main ice masses did not join. Other local glaciers also are shown. The coastline shown is the present one. In glacial times, however, sea level was lower and the coastline was seaward of its present position.

Fig. 12-32. Glacial scouring has been such in this granite terrain that joints and other structural features show up very clearly. Melville Peninsula, Northwest Territories, Canada. *Dept. of Energy, Mines, and Resources, Canada*

Fig. 12-33. Hummocky ground moraine near Cypress Hills, Saskatchewan, Canada. *Geological Survey of Canada*

stead from the central portion of an enormous ice dome.

In the central part of the vanished ice sheets, such as in northeastern Canada, the ice scraped and scoured the landscape. We find, therefore, vast stretches of smoothed bedrock that is stripped of its soil and in which all minute bedrock features, such as joint patterns, have been etched into relief (Fig. 12-32). Weaker portions of the bedrock were scooped out, forming closed depressions that now hold lakes, such as the uncountable ones that stretch across large areas of Labrador. Telltale striations in the bedrock give us the only evidence of the direction in which the ice flowed during the latter stages of the ice sheet, because the land is so flat and the ice was so thick.

In contrast to the scrubbed interior areas covered by the ice sheets, large areas of till

Fig. 12-34 A. (above) Depression formed by the melting of a block of glacial ice in river deposits fronting the Hidden Glacier, Yakutat District, Alaska. *G.K. Gilbert, USGS* **B.** (below) Closed depression in till in the Matapedia Valley, Québec, that probably formed when a block of glacial ice melted. *Geological Survey of Canada*

were deposited around its margin, leaving an undulating terrain known as *ground moraine* (Fig. 12-33). Indeed many of our major northern cities are built on till. Where it is sandy and bouldery, till makes a good base for buildings and freeways, but where it is rich in clay, it makes for very slippery foundations. Tills and associated deposits often change properties over short distances laterally; thus geologists find it difficult to predict the kind of surface materials that might be encountered during a construction project. Extensive field mapping and drill-hole data are essential to builders working in glaciated areas.

In places, blocks of ice may have been entrapped in the till, which upon melting produce a depression called a *kettle* (Fig. 12-34). It is no wonder then, that glaciated areas abound with lakes. Anyone caught in a traffic jam behind a Minnesota car can read on its license plates that back home there are 10,000 lakes. If the license plate were larger and the governor was an ex-geologist the true story of their origin could be told!

One major landscape feature of North America—the Great Lakes—resulted from the advances and retreats of the last major glaciation, called the Wisconsin Glaciation. In part the lake basins are ice-gouged, and their floors lie far below sea level; the bottom of Lake Superior is 213 m (699 ft) below sea level and that of Lake Michigan 104 m (341 ft). They are partially blocked by moraines, especially around the south end of Lake Michigan, whose pendulous, lobate form is a close counterpart to the moraine-outlined glacial lobe whose place the lake has usurped. The Great Lakes had a different pattern during the last part of the ice age than they have today; for one thing they were dammed to the north by the retreating wall of the receding ice sheet; for another, their levels were higher than those of the present-day lakes, and their outlets were quite different. One outlet was via the Mohawk Valley and the Hudson, while another was down the course of the Illinois River to the Mississippi and thence to the Gulf of Mexico, rather than to the Atlantic by way of the Gulf of St. Lawrence, the present route.

Some glacial forms on ground moraine are not random heaps of till but show a regular geometry. Among the shaped features are swarms of curious elliptical, rounded low hills resembling a whale, or the bowl of a teaspoon turned upside down. Such hills are called *drumlins*, from an Irish Gaelic word *druim* which means the ridge of a hill (Fig. 12-35). Of those curious features, certainly the most renowned is Bunker Hill, although there are many others in the Boston region, including some of the islands in the harbor. Drumlins vary widely in size and shape, but few are more than 1 km (0.6 mi) long or more than 30 m (98 ft) high. In general, they appear to have a high percentage of clay in their makeup. Although details of their origin are uncertain (no one ever saw a drumlin being made), there is little doubt that they originated beneath moving ice. We can say that with some certainty because they occur in groups, parallel to one another and to the known directions of ice transport. Proposals for their origin range from erosion of pre-existing till or bedrock, to a purely depositional landform feature related to the mechanics of till deposition.

Eskers, from a Gaelic word used in Ireland, are elongate, narrow, sinuous ridges of stratified sediment that commonly wander across the countryside, much like a canal levee or a railroad embankment laid out by a mildly inebriated surveyor. Generally their crests are rounded, their side slopes are moderately steep, and their longitudinal slope is gentle (Fig. 12-36). Like conventional streams, they may meander; occasionally they are joined by tributaries, but unlike ordinary streams they may climb up hill slopes, especially where they cross low ridges through passes. Seldom do their crests stand much more than 30 m (98 ft) above their surroundings. Some may be as much as 500 km (311 mi) long, although most are a great deal less. The consensus today is that eskers probably are deposits made by streams flowing in ice tunnels at the bottom of a glacier.

Along the terminal portion of the former continental ice sheets are crescentically looped morainal ridges that are characteristic of landscapes in the North American mid-continent (Fig. 12-37). The height of such ridges rarely exceeds 30 m (98 ft), and in some places wide gaps may appear in the ridge, either because the ice at that point was not loaded with debris or perhaps because the moraine had been eroded away. Such massive moraines formed while the ice front was stationary, following either advances or retreats of the front.

Most of the rocks in till are locally derived; that is, they have been carried no great distance from their source. Thus, if the surrounding countryside is chiefly limestone, the till will be mostly limestone fragments, large and small. Interspersed with them, however, may be a number of far-traveled rocks. If the latter have a distinctive composition, they may be traceable all the way back to their source, in which case they are called *indicators*. Many are known to have traveled 500 km (311 mi); none apparently can be traced more than 1200 km (746 mi). Fragments of native copper from the southern shore of Lake Superior are found as far away from their origin as southern Iowa and southern Illinois. Other more challenging examples of glacial transport are diamonds, some as large as small pebbles, found as far south as southern Ohio and Indiana and having a presumed source north of the Great Lakes. Technically the diamonds are not indicators because their actual source is not known, or if it is, it is a remarkably well-kept and presumably profitable secret.

Deposits closely associated with glaciation

There are two characteristic deposits closely associated with proximity to glaciers, be they alpine or continental. Streams draining off the front of the ice are heavily loaded with debris and build up their beds to form vast floodplains, or *outwash plains* (Fig. 12-38). Those featureless plains have been graded and regraded endlessly by ever-shifting streams. Thus, the material is fairly well sorted and, when formed, is unvegetated. Because the material is so fresh, so uniformly textured, and in the mid-continent so fine grained, it makes fertile soil. In fact, much of the best farmland there is located on glacial outwash plains. Outwash associated with alpine glaciers, however, commonly is very bouldery, and the soils that form from it are not valued as farmland.

Because the glacial mill grinds so exceeding

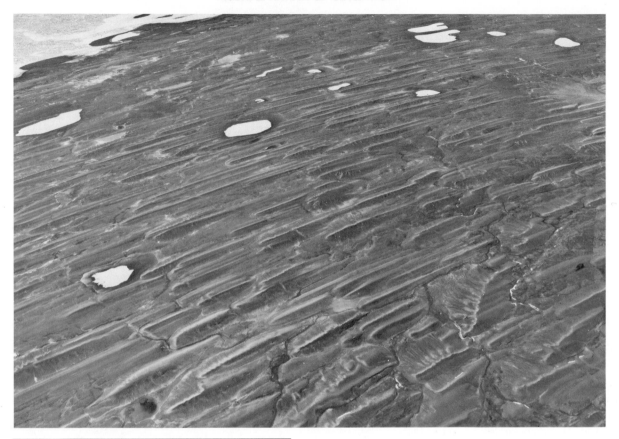

Fig. 12-35. Well-developed drumlin field, Stefansson Island, Northwest Territories, Canada. The ice flowed to the right. *National Air Photo Library, Canada*

fine, much of the surface of an outwash plain is covered with *rock flour,* a fine silt that can readily be picked up and transported by the wind. With strong winds sweeping across an open, unprotected, silt- and sand-covered plain, great clouds of dust are readily picked up and may be swept for scores of miles beyond the barren floodplain (Fig. 12-39). Deposited, the fine, wind-transported glacial debris, called loess (see Chap. 11), may blanket much of the neighboring countryside, sometimes to depths of 30 m (98 ft). Loess was broadcast over the length and breadth of the Mississippi Valley as well as across the lowlands of Central Europe and far down into the Danubian plain of Hungary. It is always thickest near its floodplain

source, and systematically thins downwind. The deposits are responsible for some exceptionally fertile soils the world over.

MULTIPLE NATURE OF GLACIATION

When the glacial hypothesis was proposed more than a century ago, people spoke of a Great Ice Age. The idea was generally held that in some mysterious way the ice advanced across the northern lands, lingered a while, and then withdrew. Today we know that the ice age was a vastly complex event, involving multiple advances and withdrawals of continental ice sheets and alpine glaciers.

A more formal term for the geologic time

Fig. 12-36. Eskers are left behind as the glacier (background) retreats. The main esker is about 20 m (65 ft) high, 150 m (492 ft) long, and is about 95 per cent glacial ice. *J.D. Ives*

during which most of the recent large-scale glaciers advanced and retreated is the *Pleistocene Epoch.* It began about 2 million years ago and ended about 10,000 years ago. The Pleistocene is not synonomous with the time of glaciations, however, because there is evidence that in some places the last period of glacial advances and retreats began before 2 million years ago in places such as Alaska and Antarctica.

A map of the United States shows the limits of tills of different ages (Fig. 12-40). How do we really know, though, that there was not more than one advance, and how do we determine how many actually took place?

Most geologists interpret the evidence available in the mid-continent to indicate that there were at least four major advances of the ice during the Pleistocene, separated by three interglacial phases when the ice withdrew, perhaps completely. Evidence for that complex succession is based in part on (1) the way in which moraines and other deposits of a later glacial stage may overlap those of an earlier advance, and (2) the degree of weathering of the glacial deposits. The original constructional pattern of older moraines and other glacial landforms will be blurred or perhaps even obliterated by subsequent erosion. Older deposits have developed a strong soil profile and younger ones a weak profile. Weathering may have progressed to depths of 3 m (10 ft) or more in older glacial deposits, and some boulders, even though they appear solid and intact, are so deeply decayed that they readily can be sliced through with a bulldozer blade. Some weathered glacial material is known by the eminently descriptive word *gumbotil.* Gumbotil was originally defined as a gray, thoroughly leached clayey soil that characteristically is sticky when wet, but hard and firm when dry. When stepped in wet in the field, it literally grabs your boots off! In places, such soils formed in interglacial periods and were subsequently buried by the till of the next glacial advance; their presence constitutes fairly firm evidence for extensive retreats of the ice for long periods of time (Fig. 12-41).

The glaciations and interglaciations of the central part of the United States generally were

named after the localities or areas in which deposits or soils were well exposed for study. The major glaciations and interglaciations are (youngest at top):

Glaciation	*Interglaciation*
Wisconsin	
	Sangamon
Illinoian	
	Yarmouth
Kansan	
	Aftonian
Nebraskan	

Names for the glaciations were taken from states, but those for the interglaciations came from Sangamon County, Illinois; Yarmouth, Iowa; and Afton Junction, Iowa.

Fig. 12-37. Major moraine systems in the mid-continent east of the Rocky Mountains. All were formed during the last major glaciation, named the Wisconsin, older moraines having been destroyed by subsequent erosion or buried by younger deposits.

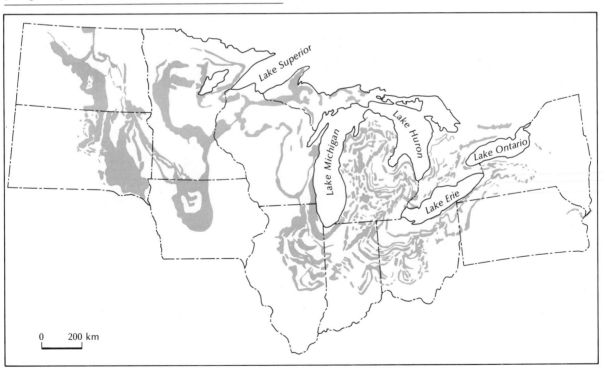

Fig. 12-38. Glacial outwash in Ekalugad Fiord, Baffin Island, Canada. The light-colored surface is recent glacial outwash; colonies of lichens have darkened the outwash surface (center) deposited about 4000 to 5000 years ago. Notice the patterned ground of the latter surface. *J.D. Ives*

Fig. 12-39. Clouds of silt rise from the floodplains of the Knik River and tributaries, Alaska. *W.C. Bradley*

It is difficult to determine exactly when the various mid-continent glaciations occurred. The Wisconsin commonly is considered to have started about 75,000 years ago and lasted until about 10,000 years ago. The only other "dated" glaciation is the Kansan, which occurred at about the same time as a 600,000-year-old volcanic ashfall deposit that resulted from an enormous volcanic explosion in Yellowstone National Park.

Alpine glaciation seems to have swung more or less in concert with the continental glaciations and interglaciations, as shown by weathering and topographic features as well as by absolute dates on the various tills. Major canyons in the high mountain ranges were glaciated, and bulky moraines commonly are found at their mouths (Fig. 12-42). In a very general way, the preservation of alpine moraines is related to the age of the till, just as it is in the mid-continent. Moraines dating from the Wisconsin glaciation are relatively well preserved, Illinoian moraines have a rather subdued topography, and older moraines have been worn down completely through the various processes of erosion. In addition to the moraines that mark the maximum extents of the ice, smaller moraines are found far up-valley, in an area adjacent to the cirques. They record slight climatic changes over the last 5000 years, enough change to have caused cirque glaciers to advance and retreat several times.

The beginning of the last period of glaciation seems to have varied from place to place.

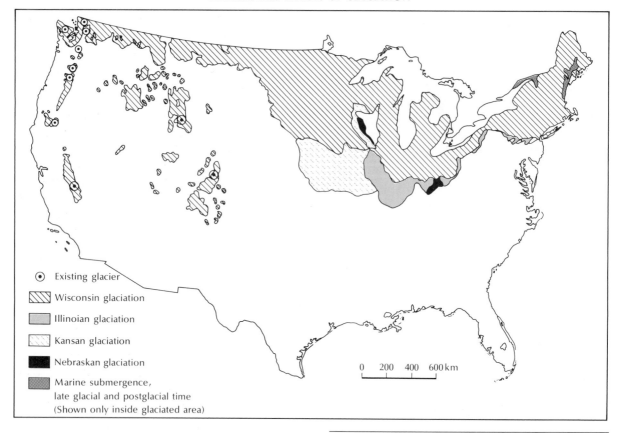

Existing glacier

Wisconsin glaciation

Illinoian glaciation

Kansan glaciation

Nebraskan glaciation

Marine submergence,
late glacial and postglacial time
(Shown only inside glaciated area)

0   200   400   600 km

Fig. 12-40. Glacial geology of the United States.

The Antarctic Ice Sheet reached a substantial size about 10 million years ago, perhaps even before that time. Extensive glaciation also took place 10 million years ago in Alaska. In contrast, the oldest glaciation recognized so far in the western United States may be about 3 million years old, or even younger. In like manner, the end of the last glaciation varies from place to place. Glaciers have all but disappeared from many of our western mountains outside of the Pacific Northwest, so that we are at present in an interglacial period. Such examples point out an important lesson for glacial geologists—climatic change does not take place in all areas at the same time. Thus, we would not expect climatic change to be synchronous on a world-wide basis. In geologic terminology, the change is *time-transgressive*, occurring in one place first, and at another place later—depending on special circumstances.

Finally, the climatic changes that produced the glaciers were sufficient to change greatly the environment in areas where glaciers did not form. One such place is the western American desert, incorporating parts of California, Nevada, and Arizona. We know, for example, that during the Wisconsin glaciation trees grew at altitudes several hundred meters below their present lower limit. The evidence comes from careful examination of the tree fragments in abandoned rat nests found where it is too hot and dry for trees to exist. The consensus is that rats do not travel far overland on a hot day; thus in those prehistoric times the trees must have gradually migrated down to the altitudes at which the nests have been found. Other evidence for climatic change is the abundant evidence that lakes of all sizes dotted the western basins at that time (see Chap. 11). Why, a canoe, which seems out of place on the top of a car traveling through that hot desert, would have been an ideal way to traverse the country during the Wisconsin glaciation.

POSTGLACIAL CLIMATIC CHANGES

Enough is now known to tell us that climate since the end of the Pleistocene has not always been the same. Most of the information comes from non-climatic sources because the length

Fig. 12-41. The kind of field evidence necessary to prove multiple glaciation. At the surface is the younger (Wisconsin) till. Since the retreat of ice some 15,000 years ago, a weak soil has formed on its surface. Beneath it is a soil that formed on older (Illinoian) till. The buried soil is deeper, more clay rich, and more weathered, representing a surface exposed to weathering and soil formation between the Illinoian and Wisconsin glaciations for much more than 15,000 years.

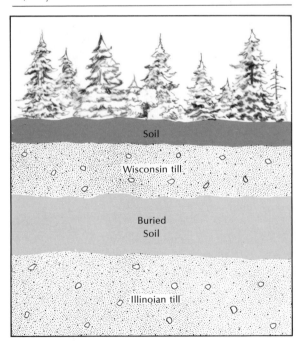

of time during which climatic observations have been made is far too short. Reliable instrumental records date only from the mid-nineteenth century; in fact, the earliest rainfall record in the modern sense was made in Padua, starting in 1725, and the earliest temperature record in Florence, in 1655. As a result, some of the scattered information seems authentic, some of it interpretive, and some of it frankly speculative. Yet we are working in the realm of human history, and historic records have survived many climatic hazards such as drought, floods, crop failure, blocking of alpine passes by long-enduring accumulations of ice and snow, and successions of unusually savage winters. Accounts of those events often were written down, others have to be inferred from such occurrences as the discovery of village sites far up a slope laid bare through the evaporation of lake waters. In an opposite sense, the original site of lake dwellings may now be concealed by a rising lake level.

A good example of a historical record that reflects changing climatic patterns is the annual flooding of the Nile. Water levels were recorded almost continuously between A.D. 641 and 1480, and flood stages between the years 1480 and 1830.

By means of such techniques as carbon-14 dating, tree-ring dating, measurement of the sizes of lichens on rocks, and historical records it is possible to determine the approximate age of mountain moraines and thus a glacier's pattern of recent oscillation (Fig. 12-43). Interpreting the evidence, earth scientists have determined that three major advances have occurred over the past 5000 years—sometimes referred to as "The Little Ice Age" (Fig. 12-44). In many places, the advance that culminated in the last several centuries was the most extensive.

For the first 50 years of the twentieth century the rate of recession for most, although not all, Northern Hemisphere glaciers was rapid. During that half-century a significant warming of the atmosphere, accompanied by shifts of marine currents and fish populations, migrations of land animals, and changes in annual temperatures and precipitation was seen. Since about 1950, however, slightly cooler and moister climates have resulted in a slowdown in the retreat of some glaciers, and in a re-advance of others. Understanding the long-range effects of those subtle changes is important because they come at a time when the world's population is expanding rapidly, and the need for increasing food production is greater than ever.

Piecing together fragmentary evidence is a task largely unfinished, although a vast store of information has been accumulated, chiefly from Europe, where a long and often turbulent record can be found. We know little, however, of what postglacial climatic changes were like in lands that are more remote, or in areas with radically different climates—such as the monsoonal tropics.

The doomed Norse colony in Greenland is an example of the impact of climatic change. The colony was founded in A.D. 984 and per-

Fig. 12-42. Moraines at the mouth of Bloody Canyon in the Sierra Nevada, California. The Illinoian moraines formed when the ice took a path to the right, as viewed downvalley, after leaving the mountain front. In contrast, the ice that laid down the Wisconsin moraines moved directly out of the canyon and deposited a lateral moraine across the upper end of the valley enclosed by the older moraines. Multiple Wisconsin moraines commonly are interpreted as the result of multiple advances and retreats. *U.S. Air Force*

Fig. 12-43. Young moraines front the Isfallsglaciären in Swedish Lapland. The numbers refer to the ages of the moraines, based on historical records for the 1910 moraine, and lichen sizes and carbon-14 dating for the older ones. *G.H. Denton*

ished around 1410. In its early history the Arctic seas were unvexed by ice, and Viking ships could make passage where today ice floes and stormy seas bar the way. The colonists raised cattle and hay, built permanent habitations, and the settlement flourished to such an extent that it had its own bishop. With a climatic change that brought the Greenland ice southward again, with the pressure of the Eskimos at their gates, with a succession of crop failures, with the rise of permafrost in the ground—so that even such shallow excavations as graves were no longer possible—and with the perils of

the ocean crossing too great for the frail vessels of that day, the colony and all its inhabitants perished.

Other examples, almost without number, might be cited of the impact of changing climates on the lives of human beings, and thus on the course of history. A powerful description of the effect of a prehistoric climatic change of the magnitude of some of those that occurred in historic time is the following passage from Palle Lauring's *The Land of Tollund Man, The Pre-history and Archaeology of Denmark:*

The change which set in altered not only living conditions but the country itself. It became windy, rainy, and foggy, and there was a fall in temperature. The change is evident in the form of a clear stratum in bogs, which indicates that the rain turned to torrents. Grey mists swept like a veil across the land, the cattle congregating miserably round the houses. The winters set in with drifting snow, frost, and yet more snow. Wondering, men advanced through cold and death-like forests, where the snow stifled every sound and where oak tree branches were weighed down by it. Cattle froze to death; wolves failed to find food; belts and sounds were overlaid with ice, rendering navigation impossible for months on end; corn would often be destroyed by frost and water, and harvest would mean waiting for the air to dry while the ears blackened and rotted. Gone were the days when young women sang as they went about clad only in corded skirts, golden-brown from the sun.

Where the present trend is leading, no person can say. Will the earth's atmosphere generally continue to warm and present ice melt, or will the air chill once more and massive glaciers start their march again?

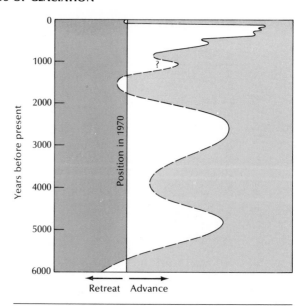

Fig. 12-44. World-wide average position of mountain glaciers during the "Little Ice Age," relative to their position in 1970.

OTHER ICE AGES

A single catastrophic deep-freeze of the earth would not fit very well with our ideas of uniformitarianism, and, when we look back through the geologic record, we do indeed find evidence for at least two periods of extensive glaciation preceding the last one. One occurred about 260 million years ago, or in the Permian Period. The evidence for it consists largely of glacial deposits known as *tillites*, which are unsorted mixtures of sand, gravel, boulders, and clay that have been consolidated. Some of the boulders show striations, and many of the tillites rest on striated rock surfaces. Other quite extensive tillites are marine in origin. In such deposits boulders and cobbles from melting sea ice appear to have dropped into much finer-grained bottom sediments, deforming their horizontal layers. The presence of a number of distinct tillite layers indicates that the Permian glaciation, like the Pleistocene, had stages and substages.

There are several puzzling things about the Permian glaciation. One is that it seems to have been limited to the Southern Hemisphere. Another is that the directions of the striations indicate that a number of Southern Hemisphere continents were attached at that time, and later drifted apart. The latter idea has been a difficult one for geologist to accept, and it will be looked into further in Chapter 19.

The oldest ice age we know of occurred about 600 million years ago at the end of the Precambrian Period, but before the extensive fossil record of the Cambrian Period. Evidence for it, too, consists of tillites, many of which were marine, indicating a widespread distribution of sea ice. In contrast to the Permian ice age, this one was not limited to one hemisphere.

Other glaciations may have occurred, but they have no records in preserved rocks. Of all the sediments that accumulate on land, glacial remnants by their very nature might be considered least likely to survive. Certainly de-

426

posits left by alpine glaciers on mountain slopes are the most vulnerable to erosion. Continental glacial deposits may not be too permanent, either.

CAUSES OF GLACIATION

Naturally, a phenomenon as challenging as an ice age has brought forth a host of attempts at explanation. No single explanation has won unanimous acclaim; each has its adherents and its detractors. At least 50 hypotheses must have been advanced over the years, but most of them contain an inconsistency that in the end is a fatal flaw. Rather than worry about pre-Pleistocene glaciations, we will concentrate on the origin of Pleistocene glaciation because more data are available on it. Among the things to be kept in mind about that particular glaciation before considering its explanation are the following:

1. The Pleistocene glaciation was a multiple rather than a single event, and at least four major advances of ice commonly are recognized in North America and Europe.
2. The glaciation appears to have been synchronous on both sides of the Atlantic, and apparently when Northern Hemisphere glaciers advanced or receded, Southern Hemisphere glaciers did the same. In other words, the entire earth seems to have responded to the same climatic pulses.
3. The advances were not of equal size, nor were the interglacial times of equal length.
4. Glacial buildup seems to have been a slow process, but the retreat was rapid and verged on rates that might be termed catastrophic. For example, it has been estimated that the North American continental ice sheet of Wisconsin age took some 20,000 years or more to attain maximum size, but it disappeared in only 7000 years.
5. Times of glacial advance appear also to have been times of lowered temperature, as demonstrated by the nature and fossil content of cores recovered from sediments at the bottom of the sea, and by evidence of lowering of the regional snowline on the

high mountains of the world. In areas such as Arctic Canada, much more precipitation is required to initiate an ice sheet than is available at present. With those observations in mind, let us turn our attention to some of the major theories of glaciation.

Terrestrial causes

The presence of widespread land areas in polar positions and the subsequent generation of glaciers and ice shelves are said to be a major prerequisite for cooling the earth and setting the stage for a glaciation. In support of that theory, Antarctica appears to have moved into a polar position at about the time of the first evidence for the last major glaciation there. High mountains also seem to be necessary, because the land surface has to intersect the regional snowline. Most of the world's major mountain ranges were elevated to their present heights during or just prior to the Pleistocene, including such lofty summits as those of the Himalayas, Andes, Caucasus, and Alps. Not only was the Pleistocene a time of unusual crustal activity, but the north-south pattern of such significant mountain ranges as those of North and South America became firmly established, and since those ranges lie athwart the general planetary circulation of the atmosphere, they may have had a significant effect on the growth and dispersal of glaciers. Neither theory explains multiple glaciation, however: only a heretic would call for mountain ranges to rise and fall in yo-yo fashion to produce the glaciations and interglaciations.

Before we go on we should mention that while one theory might explain the initiation of an ice sheet, another might better explain its continued growth. For example, in whatever way the North American ice sheet began, it could be at least partially responsible for propagating itself. The continental ice advanced on fronts thousands of kilometers long, and ice moved for hundreds of kilometers across low-lying terrain. Once the glacier reached such massive proportions, much of the snow needed to nourish it could have accumulated near its margin rather than at its center. The great temperature contrast between a glacier and its surroundings makes it operate meteorologically much like a permanent polar front, while the ice itself acts much like a rising mountain mass to alter the climate. Thus, with more rapid accumulation of snow close to the margins of the ice sheet, *ice domes* (that part of the glacier whose surface stands higher than adjacent areas) could accumulate, and from them the ice could spread out, rather than from a remote northerly center in far-off Canada.

Atmospheric causes

Other theories call upon changes in the composition of the atmosphere to explain the events of the last glaciation. One theory says that climatic variations extensive enough to trigger glacial epochs may be caused by variations in the amount of carbon dioxide (CO_2) in the atmosphere. A marked increase of CO_2 would produce a so-called *hothouse effect*, or a general rise in temperature. Energy from the sun can reach the earth's surface because it comes as visible energy—light—to which our atmosphere is transparent. Solar energy radiates from the earth back into space as heat; that is, in the infrared portion of the spectrum. But the CO_2 as well as the water vapor and ozone contained by the atmosphere is partially opaque to infrared radiation; thus solar heat is kept close to the earth. The more CO_2, the warmer the atmosphere. In fact, the warming trend that started about 1900 is considered by some to have resulted from the enormous quantities of CO_2 added to the atmosphere through the burning of coal and oil once the Industrial Revolution hit its stride. Although the theory explains how the earth could heat and cool at various times, no one has explained acceptably how the amounts of CO_2 in the atmosphere could vary enough to produce climatic changes of glacial-interglacial proportions.

Another theory involves the dust in the atmosphere. Dispersed particles can block part of the sun's rays from the earth's surface, thus cooling it down. Objections to such a theory are that major historic volcanic eruptions, such as that of Krakatoa in 1882, did not affect the weather enough to bring about a major change in the budget of glaciers or to initiate glaciers where none had previously existed. Some geologists, indeed, are turning the argument around, suggesting that increased volcanism could result from crustal stresses induced by the loading of ice on land. We see how geologic ideas can change, for what has been a popular cause at one time may well become, as more data arrive, nothing more than an effect at another time. The "dust" argument, however, is far from settled.

Astronomical causes

Astronomical explanations for the origin of glaciation make much of the fact that the eccentricity of the earth's orbit changes slowly with respect to the sun; also the inclination of the earth's axis with respect to the sun undergoes a slow change—part of the year the earth is tipped toward the sun, part of the year away from it. The Northern Hemisphere is inclined away from the sun in the winter when the earth and sun are a little closer to each other than they are in the summer. The result is that the Northern Hemisphere has slightly warmer winter temperatures than the Southern. A third variable is that the amount of the earth's inclination to the plane of its orbit may change slightly over millennia. Because the various eccentricities in the relation of the earth to the sun are not systematically linked, there will be variations over the years in the distribution of solar energy at any given place on the earth's surface but not in the total amount received from the sun.

Arguments can be marshaled both in favor of and against the theory. In favor of it is the accumulating evidence in deep-sea sediments and in marine-terrace sequences that temperature variations and glacially controlled ocean levels both seem to coincide with the temperature variations predicted by the theory. The correlation appears to apply, at least, to the last glaciation and part of the preceding interglaciation. Against the theory is the notion that Northern and Southern Hemisphere glaciation should be out of phase, when in fact they seem to have been in phase.

Another astronomical theory calls for variation in the amount of incoming solar radiation. We know that variation does occur, but does it happen with such a frequency to trigger major glaciations? No one has answers to the question and finding evidence to support the theory may be hard to prove with the fossil record.

We by no means have exhausted the hypotheses put forth, but we have presented the most current ones. Much additional work needs to be done to establish the validity of any glacial hypothesis. Nonetheless, such theories are important in provoking debate, in forcing scientists to marshal arguments pro and con, and in encouraging them to seek new evidence, both within their discipline and with other disciplines. Such work might even prevent us from initiating an irreversible climatic trend and starting a new ice age of our own making.

SUMMARY

1. Glaciers form where winter snowfall does not melt entirely, but instead builds up year by year until the mass, partly converted to ice, flows under its own weight.
2. The mechanisms of glacier movement are (1) slippage along a bedrock base and (2), the internal deformation of the ice crystals.
3. The *glacier budget* is determined by the annual accumulation and the annual loss, or *ablation* of ice. If the two are balanced the glacier, although flowing, will maintain a stationary front. If accumulation exceeds ablation, the glacier advances. In contrast, if ablation exceeds accumulation, the glacier retreats.
4. Glacial erosion is responsible for much of the world's scenic mountain terrain. *Cirques* form at the valley heads; the valley lower down is U-shaped. Also, glaciers gouge out bedrock depressions that become lakes—the legacy of glacial erosion either in mountainous terrain or in flat shield areas overrun by continental glaciers.
5. *Till* is the common sediment deposited by glaciers. It is found heaped into *moraines* around the former periphery of the glacier, or as *ground moraine* where deposited at the base of a glacier. Deposits closely associated with glaciation are glacial *outwash* along rivers that drained glaciers, and *loess*, a fine silt blown away from the glacial-age floodplains.
6. Multiple glaciation took place during the Pleistocene, with no less than four major glaciations recognized on land. The main mountain glaciations seem to coincide with the major continental glaciations, and the times of glaciations in both the Northern and Southern hemispheres may have been the same.
7. The causes of glaciation and its multiple character are not known. Many hypotheses have been put forth, but all have a major drawback in explaining the detailed story of the Pleistocene.

SELECTED REFERENCES

Andrews, J. T., 1975, Glacial systems, Duxbury Press, North Scituate, Massachusetts.

Denton, G. H., and Porter, S. C., 1970, Neoglaciation, Scientific American, vol. 222, no. 6, pp. 100–10.

Flint, R. F., 1971, Glacial and Quaternary Geology, John Wiley and Sons, New York.

LaChapelle, E. R., 1969, Field guide to snow crystals, University of Washington Press, Seattle.

Paterson, W. S. B., 1969, The physics of glaciers, Pergamon Press, New York.

Sharp, R. P., 1960, Glaciers, University of Oregon Press, Eugene.

Wright, H. E., Jr., and Frey, D. G., eds., 1965, The Quaternary of the United States, Princeton University Press, Princeton, New Jersey.

Post, A. S., and LaChapelle, E. R., 1971, Glacier ice, University of Washington Press, Seattle.

13

THE SHORE

One of the most visible and dramatic interfaces on the earth occurs where the land and the ocean confront each other. It is an area, unique both physically and biologically, that has fascinated people for thousands of years (Fig. 13-1). When an almost irresistible force meets an almost immovable object, the resulting conflict is bound to be worth observing and investigating.

WAVES

Waves can be almost hypnotically fascinating. Although one wave may look like any other, no two are ever the same. Their rhythmic beat depends not only upon the local wind for the shorter, steeper waves, but also upon distant fiercer winds that start the long, even-spaced ridges of a ground swell moving outward from a storm center half a world away.

Before we go on, a few terms relating to waves should be introduced: *Wave length* is the horizontal distance separating two equiva-

Fig. 13-1. The sea attacks a rocky coast on Mount Desert Island, Maine. *John Steenstra*

lent wave phases, such as two crests or two troughs. The *velocity* is the distance traveled by a wave in unit of time, which can be related to its other physical properties by a number of simple relationships. The *period* of a wave is the length of time required for two crests or two troughs to pass a fixed point. The *frequency* is the number of periods that occur within a set interval of time—say a minute.

As we watch the endless procession of waves, it is difficult to believe that it is the form of the wave that moves forward through the water and not the water itself. That statement may not appear to make sense at first, but watch a bottle bobbing on the surface of a bay. Waves pass under it repeatedly, but other than slowly drifting with the current the bottle holds its position remarkably well. An analogy is the rippling motion wind makes as it blows across a field of grain. Waves follow one another across the stalks of wheat, and yet the wheat does not pile up in a heap on the far side of the field. Instead, the motion in the grain results from the nodding of the individual stalks each time a wave passes through them.

As long ago as 1802 it was known that water

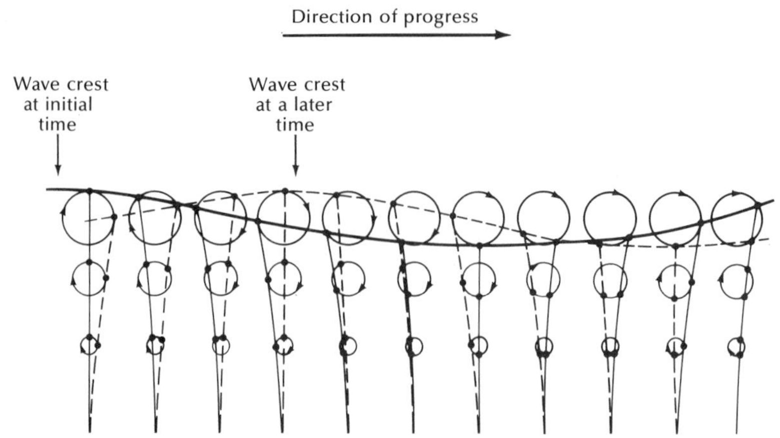

Direction of progress →

Wave crest
at initial
time

Wave crest
at a later
time

Fig. 13-2. Cross section of an ocean wave showing the paths the water particles follow. The wave profiles and positions of the water particles are shown at two moments which are one-quarter of a period apart in time. Notice that the orbit of the water particles diminishes with depth. The nearly vertical lines show how grass would be bent as a wave form passes. The stalks are vertical beneath a crest or trough.

particles within a wave do not move forward with the advancing wave itself but instead follow a circular orbit (Fig. 13-2). Detailed studies, of course, have been made since that time, but the basic principles of water motion in a *wave of oscillation* are the same as those recognized in the early nineteenth century. As the wave crest approaches, the water in the immediately preceding trough moves toward the advancing crest. Then, progressively, the particles move upward, then forward with the crest, then downward, and then seaward again in preparation for the passage of the next wave crest.

The same diagram also shows how rapidly wave motion diminishes with depth. The diameter of the circles decreases in a geometric ratio with increasing depth. For practical purposes wave motion ceases to be effective when the water depth is approximately equal to one-half the wave length.

Some of the effects of that orbital motion are familiar to every surfboarder, and others can test the effects by simply wading or swimming into the ocean a short distance from the beach. You will be conscious that the water is running strongly seaward toward the oncoming wave. As the wave surges shoreward, however, the water will sweep you strongly toward the beach. A common escape route to avoid being caught in the force of a breaking wave is to

dive to the deeper water in either the trough or beneath the next crest, where motion is less. Most of us are aware of the backward and forward pulse of the sea in the breaker zone, but not too many people realize that the motion is only part of the orbital path described by water particles within a wave.

There appear to be finite limits to the size that waves can reach. Among the largest waves whose dimensions have been established reasonably well was one that rose 34 m (112 ft) high when it was sighted off the stern of the U.S.S. *Ramapo* in 1933 during a gale in the North Pacific. Wave lengths are likely to be less than most people imagine, but they are impressively extensive at times (Fig. 13-3). One of the largest swells ever reported had a wave length of 792 m (2598 ft) with a period of 22.5 sec and a velocity of 126 km/hr (78 mph). The figures are formidable when one considers the enormous masses of water involved.

How such volumes of water are set in motion is a fair question. Almost everyone knows that the wind driving across the surface of the sea is the primary cause. Yet, how is it then, that on completely windless days a tremendous surf may belabor some exposed coast? Or that in a violent gale the wind may hammer the sea flat into a turbulent mass of dark, malevolent-looking water streaked to the horizon with foam?

Fig. 13-3. The Coast Guard cutter *Ponchartrain* wallows in
the trough of a following sea in the North Atlantic.
U.S. Coast Guard

Fig. 13-4. Stormy seas are characteristic of the oceans off Antarctica. *W.R. Curtsinger, National Science Foundation*

For the formation of large waves in deep water the following requirements must be met. First, there must be a strong wind in order to set large masses of water moving. Second, the wind must be of fairly long duration—more than just a sudden gust is needed. Third, the water must be deep, at least deep enough to round out the full circular pattern—waves 9 m (30 ft) high are not likely to grow in a water basin only 3 m (10 ft) deep. Fourth, the distance that wind friction can operate on waves—called *fetch*—is important. When waves have a long, uninterrupted run, they have an opportunity to reinforce one another. The ripples crossing a small pond are a good example. They are small on the upwind side of the pond, yet they may have grown to fair dimensions by the time they reach the downwind shore.

It is not surprising, then, that some of the largest seas are those driven before the strong westerly winds that rule south of Cape Horn. Around the margins of Antarctica an unbroken sweep of ocean encircles the earth; waves in those seas may be said to have unlimited fetch.

To return to an earlier question: how, if waves are formed by the wind, can they travel shoreward in an endless rhythmically advancing succession on a dead calm day? The answer is that such waves, to which the name of *swell* is given, may have originated in gales thousands of miles away. Waves that break on the exposed coast of Cornwall may have had their start in the far-distant reaches of the South Atlantic. Waves crashing on the west coast of

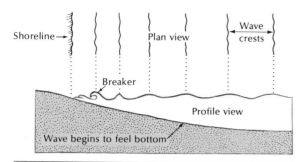

Fig. 13-5. How waves shorten, increase in height, steepen, and break as they advance into shallow water.

Fig. 13-6. Surfer riding a plunging breaker. *Aaron Chang, Surfer Magazine*

the United States may have been born in the Antarctic (Fig. 13-4), while waves that the surfers ride in Hawaii may have come from the Arctic.

A swell is made up of *long-period waves*, which outrun the more randomly distributed, *short-period waves* characteristic of a storm. Short-period waves are left behind, and the more uniformly spaced swell far outdistances the gale winds localized around some cyclonic center.

Formation of surf

The formation of surf is a complex phenomenon. The endlessly changing pattern of breaking waves—and their variations with the tide,

with wind and calm, with storm, and with the lulls between—was an inspiration to generations of painters, photographers, writers, and ordinary daydreamers. Few manifestations of the natural world are more dynamic or make us more conscious of the force of moving water than the surging mass of a strongly running surf. Indeed, without surf and breakers, most coasts of the world would lack their most picturesque element.

As waves move from deep water shoreward they begin to feel bottom when the depth of water becomes about equivalent to one-half the wavelength. As they move into shallower water near shore, their length is shortened and their height is increased relatively (Fig. 13-5). Their velocity is also reduced, and if one looks at their crests along a pier or breakwater, they do indeed seem to rise up out of the sea. In a general way, waves of oscillation break when the stillwater depth is roughly equal to 1.3 times the height of the wave. At least two major causes appear to lead to the formation of breakers. The first is a speed-up of the circular velocity of particles in a wave that has moved into shallowing water, which makes the velocity of the particles at the wave crest exceed the decreasing velocity of the wave form. The second is that in the shallow depths near shore the amount of water needed to complete the wave form is not available. That is especially conspicuous in the so-called *plunging breaker*, in which a wave curls over in a beautifully molded half cylinder, topples, and crashes with a thunderous roar (Fig. 13-6). The entrained air is violently compressed, and in its efforts to escape converts the entire roller into a froth of foam-whitened water.

Spilling breakers are those in which the crest foams over and cascades down the front of the advancing wave without actually toppling over. Such breakers ordinarily diminish in height as they move landward, and they may advance in rows simultaneously over a broad stretch of beach.

Tsunami

One of the most destructive waves on earth should be mentioned—the *tsunami* or *seismic*

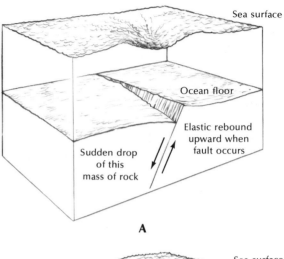

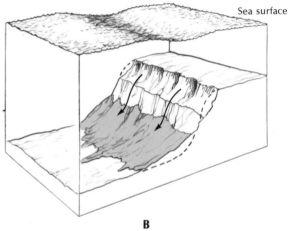

Fig. 13-7. Two common ways in which tsunami form. **A.** A sudden drop of the ocean floor along a fault causes the water surface to drop, and a wave is generated. **B.** An earthquake triggers a submarine landslide of loose sediment, which displaces water and sets up a tsunami.

sea wave. In the past it was called a tidal wave, but the term has been abandoned by earth scientists because the wave is in no way related to the tides. It is generated in response to disturbances on the ocean floor, such as fault displacement or earthquake-caused landslides (Fig. 13-7). American oceanographers chose the word *tsunami* (Japanese for harbor wave) to designate such waves. Certainly the effective destructiveness of such waves is enhanced in the progressively narrowing reaches of a bay.

The waves are barely noticeable in the open

Fig. 13-8. Houses along the Hilo waterfront were broken and moved inland by a tsunami that arrived at midnight, 23 May 1960. Automobiles were swept for whole blocks, and in some cases were piled three-high; in other places they were wrapped like limp fish around the bases of coco palms. The waves, which were not far from storm waves in dimension, moved generators and sugar-mill rolls and ripped up whole slabs of asphalt paving. Boulders from the seawall, weighing 22 metric tons or so, were transported as far as 180 m (590 ft) inland. *Y. Ichii, Honolulu Advertiser*

ocean, for they might be no more than 1 m (3 ft) high, with a wave length approaching 200 km (124 mi), and a period in the order of 15 minutes. Such waves move at the fastest natural velocities on earth, exceeding 600 km/hr (373 mph).

It is only as this oceanic ripple hits a coast that violence is unleashed (Fig. 13-8). Like most waves it too feels bottom, increases in height,

and breaks. Willard Bascom has kept track of the damage of some tsunami, and here records the effects of one generated by a colossal landslide in the Aleutian Trench. The date was April 1, 1946:

Later, over a period of years, I traveled to many Pacific shores asking about the effects of that tsunami. Remarkably often points facing into the waves and bays facing away from them were hardest hit. For example, Taiohai village at the head of a narrow south-facing bay in the Marquesas Islands four thousand miles from the earthquake epicenter was demolished. Hilo, Hawaii, only half as far from the disturbance and whose offshore topography seems precisely suited to funnel tsunamis toward the town, fared worse. There the captain of a ship standing off the port watched with astonishment as the city was destroyed by waves that passed unnoticed under the ship. Another ship, the *Brigham Victory*, was unloading lumber at Hilo when the tsunami struck. The ship survived with considerable damage but the pier and its buildings were destroyed. One hundred and seventy-three persons died and $25 million in property damage was done by the waves at Hilo that morning.

But the truly great waves of April 1 struck at Scotch Cap, Alaska, only a few hundred miles from the tsunami's source, where five men were on duty in a lighthouse that marked Unimak Pass. The lighthouse building was a substantial two-story reinforced concrete structure with its foundation thirty-two feet above mean sea level. None of the men survived to tell that story but a breaking wave over one hundred feet high must have demolished the building at about 2:40 A.M. The next day Coast Guard aircraft, investigating the loss of radio contact, were astonished to discover only a trace of the lighthouse foundation. Nearby a small block of concrete one hundred and three feet above the water had been wiped clean of the radio tower it once supported.

Today there is a tsunami warning system set up around the Pacific to help avoid such catastrophes.

Wave refraction

As waves approach land they commonly are bent, or refracted, so that they almost always parallel the shoreline closely, no matter what its configuration. Refraction takes place because of variations in the velocity of a wave as it reaches shallower water: different parts of a wave touch bottom at different times, slowing down its forward progress and changing its direction.

Wave refraction also explains the variation in the intensity with which waves attack an indented coast (Fig. 13-9). Land promontories are

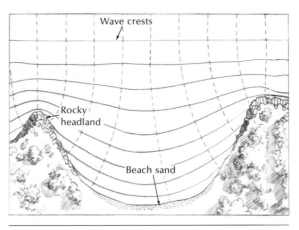

Fig. 13-9. How waves are refracted. Wave crests crowd together as they move into the shallow water around the rocky headlands; in the deeper waters of the bay, however, they are more widely spaced. Wave energy, equally distributed between the parallel dashed orthogonal lines in deep water, is concentrated on the headlands where the lines converge but is diminished in the bay, where they diverge.

beaten back because refraction focuses energy on them. In contrast, the energy is diffused through an adjacent bay, as any mariner knows. Because of the diffusion, wave energy is slight in bays, and they are commonly the site of deposition of sediments eroded from the promontories.

WAVE EROSION

It is in the shore zone, where breakers are able to bring their full force to bear against the land, that most of the erosive work of the sea is concentrated. Because there is an upper and lower limit to such attack, the work of waves might be likened to that of an enormous horizontal saw. The upper limit of their effectiveness is the maximum height that they can reach at high tide during a storm. That height will differ among coasts depending upon how boldly the land faces the sea and how strongly gales drive waves before them. Some storm-driven waves have been known to reach heights of 65 m (213 ft) or so.

The lower limit of wave erosion is much less

Fig. 13-10. Evidence of recent sea-cliff retreat in California. **(A)** (above) The cliff on this beach near Montara retreated 51 m (167 ft) between 1866 and 1971—an average rate of about 0.5 m (1.6 ft) per year. **(B)** (left) Sea-cliff retreat near El Granada destroyed this coastal road. The row of rocks was placed there in the early 1960s in a futile attempt to halt further erosion. *K.R. Lajoie, USGS*

Fig. 13-11. (Overleaf) Near-vertical sea cliffs on the Daisei Islands located off the west coast of Korea testify to the high rate of beach erosion and shoreline retreat. A stack is an isolated portion of the cliff not yet consumed by the sea. Notice the contrast in topography between the sea cliff and the hilly topography formed behind it by the subaerial processes. *Department of Defense*

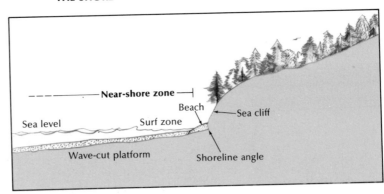

Fig. 13-12. Cross section showing characteristic landforms along a steep coast.

certainly known. Estimates vary widely, possibly reflecting the prejudices of the estimator. Observational evidence seems to indicate that most waves can generate currents capable of moving sand at depths of not much more than 12 m (39 ft)—although ripples and other signs of disturbance of the sea floor are often seen at depths of 107 m (351 ft) below sea level and even deeper. Gravel and cobbles are not moved by wave currents much below 6 m (20 ft). The lower effective limit of wave transportation and erosion is called *wave base;* obviously its depth differs on different coasts, just as the upper limit does.

Waves generally accomplish most of their erosion by abrasion, just as streams do. In times of calm, very little erosion takes place; in fact, if there is a source of sediment, deposition commonly occurs. During storms, however, when the waves are highest and most capable of carrying abrasive agents such as sand, gravel, and even cobbles, their erosive power is great. Since waves cannot erode above their maximum height, their action is restricted to a narrow zone of the shore. On a cliffed shoreline the base of the cliff is undercut, and, if the rocks are weak because of jointing or poor consolidation, landslides will occur. In such cases landslides are actually the eroding agent of the upper parts of the cliff.

The shore zone is one of the most dynamically active erosional theatres on earth. During high tide, it is submerged. When the tide is low, and especially if the tidal range is great,

then a broad expanse of the shore zone may be exposed to the air, where it can be acted upon by rain and wind. In the Arctic, the shore may be modified by shore ice, and in limestone regions the alternation of wet and dry periods also is important in erosion.

The rate of wave erosion varies from place to place. Some coasts show little change in a generation, yet in other places, erosion can be quite rapid, as testified by undercut roads, patios, and steps built down sea cliffs. Rates in excess of 1 m (3 ft) per year are not uncommon (Fig. 13-10). The rate is a function of the strength of rock making up the cliff and the violence of the waves. Cliffs along the English Channel have retreated as much as 12 to 27 m (39 to 89 ft) overnight during a single storm.

Because wave erosion has an upper and lower limit, and works horizontally, it produces landforms differing from those made by downward cutting streams and glaciers (Fig. 13-12).

The most noticeable of coastal landforms is a *sea cliff* (Fig. 13-11). Its height depends upon a variety of factors, including the power of the attacking waves, the slope of the land surface, and the resistance of its rocks. Obviously, a headland of granite is likely to be much tougher than one made of chalk. The imposing height of the chalk cliffs near Dover results from the erosional weakness of that rock as well as the fury of the sea's attack in a North Atlantic gale (Fig. 13-13).

As a sea cliff recedes before the onslaught of the waves, a planed-off rock bench is cut at its

Fig. 13-13. Vertical cliffs cut in chalk along the southern coast of England near Dover. *Aerofilms, Ltd.*

base, called a *wave-cut platform*. Sometimes it will be bare, abraded rock, interrupted, perhaps, by tide pools, and occasionally by unreduced remnants of the cliff, known as *stacks* (see Fig. 13-12). Where the shoreward portion of the platform is mantled with sand, it is called a *beach*. Seaward of the platform there may be an accumulation of wave- and current-transported material which constitutes a *marine-built terrace*. Whether or not the last feature exists depends in part upon the strength of waves and currents, especially the longshore

Fig. 13-14. Waves striking the shore in the vicinity of Solana Beach, California. The angle at which the waves meet the coast indicates that the longshore transport is toward the viewer. *R.W. Thompson*

current. Should the current be strong enough, it may sweep the sediment into the mouth of the next bay down the coast instead of accumulating it at the base of the cliff from which it came.

Beaches are the land's first line of defense against wave erosion, particularly by hurricane and other storm waves. Much of the force of storm waves is expended upon the gradual slope of the beach, and even if the beach is removed, it will usually be rebuilt by deposition in more peaceful times if a source of sediment exists. Sea cliffs do not have such renewal ability; once parts of them are eroded and removed, they are gone forever. There are protective as well as recreational reasons, then, for close observation and study of our beaches in order to preserve them.

BEACH DEPOSITION AND EQUILIBRIUM

Wave and current erosion and transport make some portions of a coast retreat and others advance, a complicated pattern of trading, as it were, along a coastline. Material that is eroded from a point or headland may be deposited by a longshore current in a nearby bay or marsh. In the long run, erosion and deposition may attain a rough equilibrium; or one may outpace the other. In Britain, for example, in the 35 years immediately preceding 1911 it is estimated that 1900 hectares (4692 acres) were lost to the sea while 14,345 hectares (35,444 acres) were gained. The latter were chiefly salt marshes, sand spits, beaches, and bars. On the other hand, the island of Heligoland in the North Sea off the German coast in A.D. 800 had

a 200-km-long (124-mi) shoreline, and was a much-fought-over piece of terrain in the Viking era. By 1900, the one-time independent duke-dom had been reduced to a strongly fortified rock only 5 km (3 mi) around, enclosed by sea walls, breakwaters, and other artificial de-fenses.

Where do beaches come from? Beach sands are a product of weathering and erosion gen-erally far inland from the ocean. The sands are brought to the coast by streams in their inex-orable trip to the sea, and there they are dumped when the stream no longer has the capacity to carry them. All the sand on any given beach, however, is not necessarily derived from the nearest stream or river. *Long-shore currents* are constantly shifting a river of sand parallel to the coast and predominantly in the same direction. Longshore currents result from the fact that, despite refractions, few waves hit the coast exactly parallel to it; most meet shore at a slight angle (Fig. 13-14). As each wave breaks, the water rushes up the beach face at an angle, but leaves it by flowing straight down the face. The result is that water par-ticles move along the beach in a zig-zag fash-ion, and along with them the sand in the surf zone. Even though on a small scale the sand particles move in a zig-zag fashion, on a larger scale the body of sand in the surf zone moves parallel to the coastline. The *sand transport,* as it is called, provides the raw material from which the waves can build beaches between storms.

An equilibrium is commonly reached along a beach that nicely balances the amount of sand available and the variations in wave energy. As we shall see, that equilibrium has many attri-butes common to stream equilibrium. Even if some of the sand is captured by submarine valleys and shunted off to deeper parts of the ocean floor, it is replaced by sand brought down to the sea by streams or by sea-cliff ero-sion, and so the balance is maintained.

Permanent loss of beaches can have devas-tating effects and, ironically enough, such loss is usually caused by our meddling with the beach equilibrium. Every dam we build up-stream accumulates behind it sand that would

otherwise reach the coast. Thus we intercept the supply, and beach erosion is increased. In the construction of beach resorts, much of the available sand may be removed to provide flat areas for the foundations of buildings.

Probably the most disruptive activity, how-ever, is the building of jetties and breakwaters for the purpose of improving harbors and creating marinas. Such interference with the longshore current and sand transport creates all kinds of havoc, resulting in damage that is extremely expensive to correct. The longshore current will deposit sand on one side of a jetty or breakwater, which acts as a dam, and sand will accumulate there until the beach is exces-sively wide. Meanwhile, on the other side, the current removes all the available sand and carries it away until there is no beach at all. In addition, whatever sand does manage to get around the end of the jetty proceeds to fill up the harbor or marina the jetty was built to protect (Fig. 13-15). In one case, at Port Hueneme, California, a jetty that extends sea-ward 915 m (3001 ft) essentially takes the sand and leads it into Hueneme Submarine Canyon.

Unfortunately we seldom recognize that our own structures are the undoing of our precious beaches. Instead we blame the ocean and at-tempt to conserve beach property by fighting the ocean rather than working with it. In addi-tion to building jetties and breakwaters to trap the sand, we also attempt to control erosion by building seawalls. They are seldom effec-tive, however, because they provide a vertical surface for the sea to batter against, rather than the gradual slope of a natural beach. The re-sult is excessive wave erosion, which not only prevents the accumulation of a new beach, but also quickly undermines the seawall itself. Commonly, too, seawalls are not built high enough to keep out the most damaging storm waves. And the water that does break over them is kept from returning to the ocean.

Another remedy often tried is a groin, a smaller version of a breakwater. The structure does indeed trap the sand on its upcurrent side, but property owners below the groin are apt to be somewhat upset when the under-nourished longshore current rapidly removes

Fig. 13-15. This breakwater at Santa Barbara, California, was constructed to protect small boats. The spit just beyond the moored boats was formed by longshore transport of sand and threatens to close off the harbor. The breakwater also cut off the supply of sand to the coast beyond. Sand is now dredged from the spit and delivered to the adjacent coast. Equilibrium is thus restored, but at the cost of a dredging operation. *John S. Shelton*

the beach. They may even erect groins on their own property. Thus groin-building has a natural tendency to proliferate and, in fact, the New Jersey shore, for one, bristles with more than 300 of them (Fig. 13-16).

Beach engineering has progressed to the point where suitable and effective structures have been designed, but usually they cost more money than the individual property owner wishes to pay, and in the absence of any clarification of legal responsibility, local governments are often unwilling to take on the added tax burden.

One of the remedies that has proved useful is an artificial beach, constructed by dumping a large quantity of sand brought in from elsewhere. Doing so is tricky, however, for if the sand grains are not the correct size, the artificial beach may disappear faster than the original beach. If there are cobbles in the im-

ported sand, as happened when the beach was renewed at Oceanside, California, the sand vanishes, leaving a very stony shore area. While stone-strewn beaches may not be suitable for recreational purposes, they do help to prevent further erosion. Off Long Branch, New Jersey, on the other hand, 459,000 m³ (600,370 yd³) of sand was dumped too far offshore. Beyond the reach of the longshore current, the sandpile remained essentially intact.

But where conditions are right, artificial beaches work quite well, and dune fields, too, are being rebuilt by adding sand and stabilizing it with a plant cover. One of the remaining problems is determining the conditions under which an artificial beach is feasible. The problem is a difficult one since measurements in a zone as active as the ocean-land interface are not easy to make. Sometimes it is a matter of finding enough sand. Usually, however, a suitable source can be found nearby—an up-coast harbor that is filling with unwanted sand, or an ancient beach a short distance offshore or even inland. But with an increasing use of the shorelines of the world, the search for sand will become of increasing economic interest. For example, a proposal once was made to bring sand from the Grand Bahama Banks to re-

Fig. 13-16. Groins along the New Jersey coast, north of Seabright. Sand transport is toward the viewer, as shown by the sand that has piled up on the far side of the groins. *John S. Shelton*

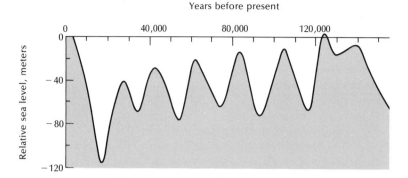

Fig. 13-17. World-wide sea-level fluctuations based on studies of marine terraces in Barbados.

plenish the eroded sand along 160 km (99 mi) of the southeast coast of Florida. The project would involve the design and construction of a special hopper dredge, as well as the payment of a suitable royalty to the government of the Bahamas.

The size and scope of beach erosion problems make it obvious that control cannot long be left to individual landowners. For reasons of economy alone, as well as the large-scale research, planning, and organization required, such operations will increasingly fall to local, state, and federal agencies. The situation may lead to much more public ownership of beach property and, we can hope, to its increasing availability for public recreational use.

SEA LEVEL FLUCTUATIONS

Along some coasts local crustal instability has caused the shore to be uplifted or depressed relative to sea level. In addition, there have been actual changes in sea level itself, and by their very nature the latter are world wide in effect. Seldom, in fact, has sea level stayed in the same position (at least over the past million years or so), but it is generally agreed that sea level has been close to its present position for only about the last 5000 years (Fig. 13-17).

Interestingly enough, the world-wide relative stability in sea level coincides roughly with the appearance of the maritime civilizations around the shores of the Mediterranean Sea and the Persian Gulf, so that the ancient harbors of the Egyptians, Persians, Phoenicians, and Minoans correspond roughly to the sea level of today. Before 5000 years ago sea level assumed a lower, but fluctuating position, and we have to look back some 120,000 years ago for the last time it was close to its present position. Looking still farther back in time (Fig. 13-18) we can see that over the last 400,000 years, sea level has seldom been as high as it has been in the past 5000 years and especially today.

Yet, sea level is far from stationary. Beginning around 1850 it started to rise, as shown by a careful comparison of tide gauges from seaports all over the world. C. A. M. King has compiled data from around the world on

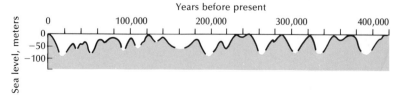

Fig. 13-18. Approximate sea-level curve for last 400,000 years, based on studies of marine terraces in New Guinea.

the annual rate of present-day rise—that for the Atlantic coast of North America is 4 to 7 mm (0.16 to 0.28 in.), Italy reports 1.7 mm (0.07 in.), Holland 1.5 mm (0.06 in.), and western North Africa 0.6 mm (0.02 in.). There are various reasons for that non-uniform rise from port to port, but nobody doubts that sea level has been on the rise.

The important question is, what accounts for the fluctuating behavior of sea level? The rise that has taken place since the mid-nineteenth century almost certainly correlates with the world-wide recession of glaciers and is a result of the return of their water to the sea. For fluctuations over the last 120,000 years (Fig. 13-17), it seems that periods of relatively low sea level correlate with major advances of continental glaciers and times of relatively high sea level with major recessions, all during Wisconsin glaciation (Chap. 12). It is difficult to make correlations further back in time, but the similarity between Figures 13-17 and 14-19 suggests a very strong glacial-interglacial influence on sea-level position. Changes brought about in such a way are called *eustatic* changes.

Sea-level variations have important ramifications. Rises in sea level over several genera-tions can wreak havoc with construction facilities and houses, especially along flat coasts where a slight rise goes a long way! The timing of the fluctuations can help our understanding of the timing of glacial periods. And a measure of the drop in sea level can tell us how much ice was stored on land. That is, we can convert the water in the 120-m (394-ft) drop in sea level to ice to estimate ice volumes, weights, and ice-cap thicknesses during the Pleistocene.

EXAMPLES OF COASTAL DEVELOPMENT

Much has been written about the development and classification of various coasts. Their formation is a complex matter, for not only do present-day processes operate on coasts, but some coastal landforms carry a legacy of the past. Three major kinds of coasts will be described below.

Emergent coasts

Along some coasts it is quite clear that the land is rising. The best evidence for such change is found in areas that were once under

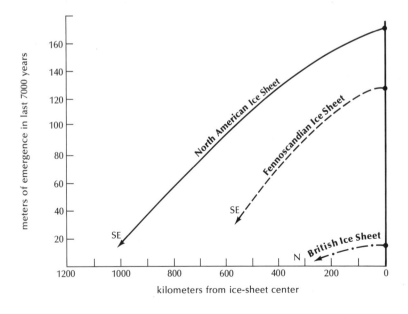

Fig. 13-19. Hypothetical cross sections across terrain occupied by former ice sheets shows how the amount of emergence (marked by shoreline features in the field) is related to the size of the ice sheet (the North American ice sheet being the largest, the British the smallest), and to the distance from the center of the sheet.

Fig. 13-20. Elevated beach deposits near Fort Severn, Canada, on the southwest shore of Hudson Bay, clearly demonstrate that the land is still rising since the disappearance of Pleistocene ice. *Dept. of Energy, Mines, and Resources, Canada*

the burden of the Pleistocene glaciers. The sheer weight of the ice pushed the earth's crust down; when the ice melted, the crust rebounded. So it is in areas that were under or near ice caps that prehistoric shoreline features are found far above present sea level. The patterns of shoreline uplift are related to the size of the ice cap; the larger the ice cap, the greater the uplift (Fig. 13-19). Emergence is still going on in those glaciated areas (Fig. 13-20). For example, the Gulf of Bothnia is rising at the rate of 0.9 cm (0.35 in.) per year, and southeastern Hudson Bay at 1.2 cm (0.5 in.) per year. It is estimated that the latter areas still have a con-

siderable amount of rebounding to do, so that we can expect more harbors to become increasingly shallow or eventually to go dry.

Other coasts far from centers of glaciation are also rising. Again, the evidence is found in old marine beaches and wave-cut platforms—collectively called marine terraces—and sea cliffs stranded far above sea level. Such landforms certainly were pushed up by tectonic movements, because, as we have mentioned,

there is no evidence that sea level has risen much above its present level for the past million years. Classic examples of marine terraces are found along the California coast (Fig. 13-21). Perhaps the most spectacular flight of terraces are those made of coral reefs on New Guinea (Fig. 13-22). It also appears that part of the New Guinea coast is the most rapidly rising one in the world—some 3 m (10 ft) in a thousand years! That means that beach

Fig. 13-21. Prominent marine terrace along the California coast, north of San Simeon. *John S. Shelton*

Fig. 13-22. Marine terraces consisting of coral reef material near Finschhafen, New Guinea. The highest terrace is about 260 m (853 ft) above sea level, and formed about 140,000 years ago; this coast, therefore, has been uplifted about that amount because the highest terrace formed when sea level was close to its present position. *Department of Defense*

deposits of the last interglaciation (about 120,000 years ago) are now more than 300 m (984 ft) above sea level.

Submerged coasts

Other coasts of the world show distinct signs of submergence. A familiar example is the coast of the Netherlands, which is in a deltaic area of rapid subsidence, and thus is sinking at a rate of nearly 10 cm (4 in.) per century. No wonder that with a total rate of sinking of around 20 cm (8 in.) per century—about half from subsidence and half from rising sea level—that, beginning with the great floods of medieval times, the Dutch has been compelled to resort to the construction of an extraordinary complex of dikes and coastal defenses upon which their national survival depends.

Typical submerged coasts are those in which the sea extends inland, sometimes for long distances, in embayments. Should the indentations have been shaped by stream erosion before their invasion by salt water, the coast is known as a *ria coast,* from the name applied to the southern shore of the Bay of Biscay. The Gulf of Maine (Fig. 13-23) and the southern coast of Brazil are other examples of ria coasts.

How a particular coast becomes embayed may be difficult to determine. Perhaps the land subsided, in which case the sea invaded or "drowned" pre-existing river valleys. Or it could be that the land remained stationary, and the postglacial rise of sea level flooded low-lying parts of it. In either case the effect is the same.

Some of the progressive changes that a ria coast is likely to undergo as a result of its modification by the waves and currents of the sea are shown in the accompanying set of diagrams (Fig. 13-24).

In the first diagram the sea has come to rest upon a landmass whose surface has been shaped by stream erosion. Former ridges now extend seaward as headlands, and the sea may reach inland as an embayment or estuary, perhaps much as Chesapeake Bay does (Fig. 13-25).

The second and third diagrams show some of the changes that might be anticipated with time. Because most of the erosional energy of the sea is concentrated on the headlands, they soon terminate in sea cliffs (Fig. 13-26). If there is a longshore current, debris from the sea cliffs or the rivers can pile up in a curved embankment called a *spit* that trails down-current from the land (Fig. 13-27). In places spits grow

Fig. 13-23. Flooded ice-sculptured valleys along the coast of Maine form a ria coast. *John S. Shelton*

almost long enough to close off a bay, and they are then called *bars* (Fig. 13-28). In other places the beach sand moving along the coast piles up in the low-energy zone behind near-shore islands and eventually a strip of sand, called a *tombolo*, connects the mainland to them. Streams entering the bay deposit sediment when their velocity is abruptly checked, build-

ing deltas out into the relatively still water. Thus, the outer exposed parts of the coast are pushed back by erosion at the same time that the innermost parts are built out by deposition.

In the later history of such a coast, as shown in the last two diagrams, the headlands may retreat inland until the coast is more or less straight, and an equilibrium might be reached

453

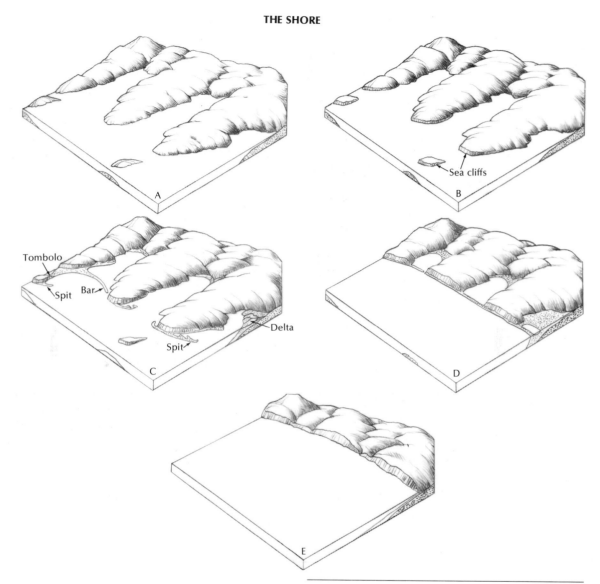

Fig. 13-24. The possible successive stages in the development of a submerged coastline.

between sea erosion and deposition of sediments from terrestrial erosion. If there is very little sediment being added, the headlands may recede as far as the innermost bay head. Then the coast will have lost its original indented character and will be cliffed along much of its length. What irregularities there are will very largely reflect the relative resistance of the rocks cropping out along the cliff face.

Plains coasts

Some coasts of the world with a low gradient and an adequate sand supply are fringed by long narrow sandbars roughly parallel to the coast but separated from it by a narrow body of water. They are generally called *barrier islands*, but may be called *barrier beaches* or

Fig. 13-25. Chesapeake Bay is a fine example of a landscape shaped mainly by subaerial processes, but later drowned by the rising sea. *NASA*

barrier spits, depending upon their form. Long chains of such barriers extend, like linked sausages, from Long Island, New York, around Florida and the Gulf of Mexico, to the southern end of Texas (Fig. 13-29). A comparable chain of sandy barriers stretches along the low coasts of the Netherlands, Friesland, and Denmark around the southeastern margin of the North Sea.

During the Civil War the sandy bars and islands along the east coast were the locale of scores of amphibious landings and minor naval operations. Gaining control of the offshore islands and the passes between them was of the utmost importance to the Union Navy and proved to be a formidable task. The shallow, unlighted passes (or openings) through the barrier islands, with their endlessly shifting channels, were a boon to the blockade runners, whose shallow-water craft were more maneuverable than the vessels of the blockading fleet.

Unpromising as such low and often unstable islands may be, on occasion they may be the sites of large and flourishing communities—including such metropolitan concentrations as Galveston and Atlantic City, as well as scores of beach and resort centers. The island itself is

Fig. 13-26. Sandspit at the mouth of a small stream, south shore of Cape Cod, Massachusetts. *W.C. Bradley*

Fig. 13-27. A bar nearly closes the opening into Bolinas lagoon, north of San Francisco, California. A tidal channel cuts through the bar and several such channels meander across the mud flats behind the bar. The position of the channel across the bar is unusual, because the longshore currents move the sand away from the viewer.
Aero Photographers

Fig. 13-28. Barrier islands fringing the outer coast of North Carolina, photographed on the Apollo 9 space mission, 1969. *NASA*

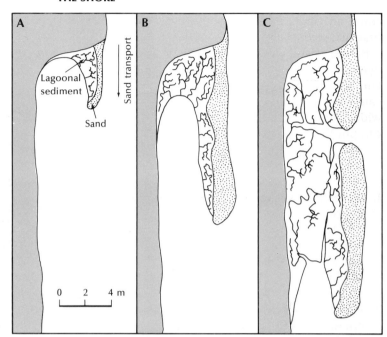

Fig. 13-29. Formation of a barrier island from a spit. In (**A**) and (**B**) the sandy spit grows and fine-grained lagoonal sediments fill in behind it. In (**C**) the spit has been breached by a tidal channel and the lower end of the spit has become a barrier island.

commonly only a few feet higher than sea level, and the most conspicuous natural objects on its surface are sand dunes, which are much overshadowed at Atlantic City by the high-density development of beach hotels. Behind the island is a broad stretch of shallow water, a *lagoon,* separating the island from the mainland.

Barrier islands appear to originate in two different ways. One is the growth and ultimate breaching of a spit (Fig. 13-29). The other, which is probably the way most barrier islands form, involves (1) the formation of a beach or dune ridges along the coast followed by (2) a slow submergence of the coast (or rise in sea level) so that the beach or dune ridges be-

Fig. 13-30. Formation of barrier islands by submergence. **A.** Active wave and wind action build up a rampart of beach and dune ridges. **B.** With a submerging coast or rising sea level, the ridges become a barrier island separating the lagoon from the ocean. Under appropriate conditions they can maintain themselves by growing upward and seaward.

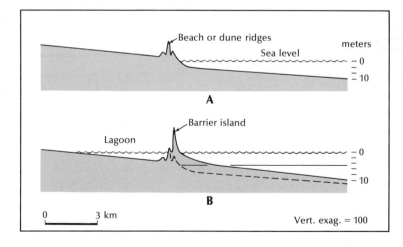

come an island and the low area behind it a brackish-water lagoon (Fig. 13-30).

Barrier islands are maintainexd by the same dynamic processes that built them initially. For example, an abundant sand supply can cause them to build seaward and increase their width, or an increasing rate of submergence can put them out of existence, below sea level.

SUMMARY

1. Waves cut away at the edges of land, and the depth of their most effective erosion is only one-half the wavelength. Water particles move in an orbital fashion as waves pass by, but the orbital motion is not completed in shallow waters; there the waves break in the surf zone.
2. *Tsunami* commonly are earthquake-generated waves that reach exceptional heights as they approach land; hence they are among the most destructive waves on earth.
3. Waves are refracted as they approach coasts, and most of their erosive power is concentrated on headlands. Bays between headlands commonly are sites of deposition of material derived from the headlands. On steep coasts, wave erosion forms a *sea cliff* and *a wave-cut platform*. Beach materials (sand and gravel) cover the platform.
4. Wave refraction produces a *longshore current* that moves beach material along the coast. In places, an equilibrium is set up between the wave energy and the amount of material moving along the coast. People interfering with the equilibrium have inadvertently caused problems by altering coastal erosion and deposition.
5. Sea level during the Pleistocene was low during times of glaciation and high during the interglaciations. At no time during the Pleistocene was it much higher than it is at present.
6. Several major types of coasts are recognized: *emergent* coasts, *submerged* coasts, and *plains* coasts.

SELECTED REFERENCES

Bascom, W., 1964, Waves and beaches, Anchor Books, Doubleday and Co., New York.

King, C. A. M., 1972, Beaches and coasts, Edward Arnold Publishers, London.

Shepard, F. P., 1973, Submarine geology, Harper and Row, Publishers, New York.

Fig. 14-1. Sampling arm of the research submersible (see Fig. 2-15) *Deepstar* plucks a sample from the ocean bottom in 183 m (600 ft) of water off Mission Beach, California. The mushroom-like objects in the background are 0.7-m (2-ft) high stalked animals called sea anemones. *U.S Navy*

14

THE SEA

Most alien of all the world's environments, and therefore most mysterious, is the sea. In the United States, a country of virtually continental extent, our interests are largely directed inward upon the affairs of the land. The time when the sea played the role in American life that it did in the time of Richard Henry Dana or Herman Melville is long gone. For most of us today the sea is an obstacle, rather than a way of life, and at best is thought of chiefly as a source of diversion along its shore. Crossing the ocean now is a matter of hours rather than the weeks and months of unremitting toil and peril it was in the day of wooden sailing ships.

Events taking place far below the surface are much less an intimate part of our lives than any seen from the deck of a small and heavily laboring vessel running before the dark seas of a mid-oceanic gale. Yet today the sea remains as almost the last great unexplored scientific frontier on earth. There is no land area, however remote, about which we do not possess more knowledge than we do of the sea floor directly off our shores.

Geologists long neglected study of the deep oceans—not because they had no interest in them, for quite to the contrary, they did. Many sedimentary rocks, for example, are marine, and in order to understand them better we need to know more about the marine environment. In the last three decades, however, technology has allowed study of the deep oceans to increase markedly, and a better understanding of those areas is emerging (Fig. 14-1). Not only does the theory of plate tectonics (see Chap. 19) rest heavily on oceanic features, but the study of deep-sea organisms sensitive to variations in temperature and in sea-water composition has revealed a fairly lucid picture of the timing of the Pleistocene glaciations and interglaciations.

Economic reasons help to explain part of our present interest in the ocean—or better, in ocean sediments. About $4 billion worth of oil and gas are extracted annually from shelf sediments, with about one-quarter of the materials coming from shelves off the United States. Although environmental problems caused by offshore drilling are numerous the solutions to

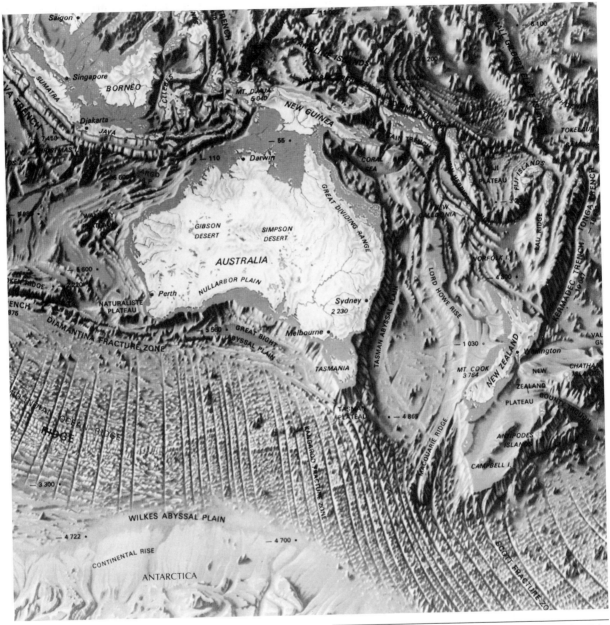

Fig. 14-2. Relief map of the rugged ocean floor in the vicinity of Australia and Antarctica. A major ocean ridge lies between the above two continents.

many of them are in sight, and the oceans' shelves probably will increase in importance as a fuel source. More industries may look to the sea floor for concentrations in ancient submerged beaches of such valuable heavy minerals as tin, gold, or diamonds; but exploration costs are high, and profitable concentrations may not be all that common. Much is made of the presence of manganese nodules in the deep sea, but the depth of the water makes mining difficult, and the manganese content is not always so high as it is in terrestrial ore bodies.

OCEAN FLOOR FEATURES

The ocean floor is far from being flat (Fig. 14-2). On a fine scale, it probably has as many irregularities as do land areas. On a large scale, however, its features are fairly regular and predictable. The origin of many oceanic features will be discussed here, but the origin of those that are tied to plate tectonics are given in Chapter 19.

Continental shelves

As we leave the beach and proceed seaward, we find a relatively shallow platform that surrounds almost all the continents and that is known as the *continental shelf*. Although it makes up only 7.5 per cent of the ocean floor, it is equal in area to 18 per cent of the earth's land area. The shelf has an average width of 80 km (50 mi), and ranges from 0 to 1500 km (0 to 932 mi) wide. It approximates a smooth, gradually sloping platform, but often has an irregular surface with depressions, hills, and terraces, showing as much as 100 m (328 ft) of relief. In formerly glaciated regions, glacial erosion and deposition give the shelf considerable relief. And, in areas of deep fiords, the deepest portion of the shelf might even be the landward part. The outer edge of the shelf is called the *shelf break,* and is characterized by a sudden steepening of the slope. The depth of the shelf break ranges from 20 to 550 m (66 to 1804 ft) and averages 133 m (436 ft).

Most continental shelves are composed of sediments and sedimentary rocks that average 2 km (1 mi) thick, and many shelves are given form by natural dams (Fig. 14-3). Common around the Pacific Ocean are tectonic dams, formed by long blocks of rocks that have been broken or folded and pushed upward. The basins thus formed behind the dams are then filled by sediments from the continents. Sometimes the dam rises above the surface of the sea and exists as islands, such as the Farallons near San Francisco and the Channel Islands off southern California. At other places, chiefly the east coast of the United States, such dams are very ancient and sediments have not only filled the basins behind them but have covered the dams as well. In such cases, and where there are no dams, the shelf sediments form a great wedge, oftentimes thickening in their seaward direction.

Off the coast in the western Gulf of Mexico, the dams are giant domes of salt, called *diapirs,* which have pushed up through overlying sedimentary rocks. In tropical waters, dams are often formed by outlying reefs that grew millions of years ago and that are composed of ancient coral and marine microscopic plants, or algae as they are called, and of more recent marine corals. Shelves off the eastern Gulf coast and southeastern United States have that type of dam.

During the Pleistocene, when much of the ocean's water was withdrawn from the basins to lie in great ice sheets on the land, extensive parts of the continental shelves must have been exposed to subaerial and glacial erosion. In places the erosion exposed the tops of the dams, revealing igneous and metamorphic rocks which have not been covered to this day. Climate during the ice age did not remain cold for the entire period, but fluctuated between colder and warmer times. Those fluctuations, of course, were reflected in changing levels of the sea. When temperatures remained constant for any length of time, the waves cut cliffs and terraces in the exposed continental shelves. Such wave-cut features are now submerged, and some of them have since been

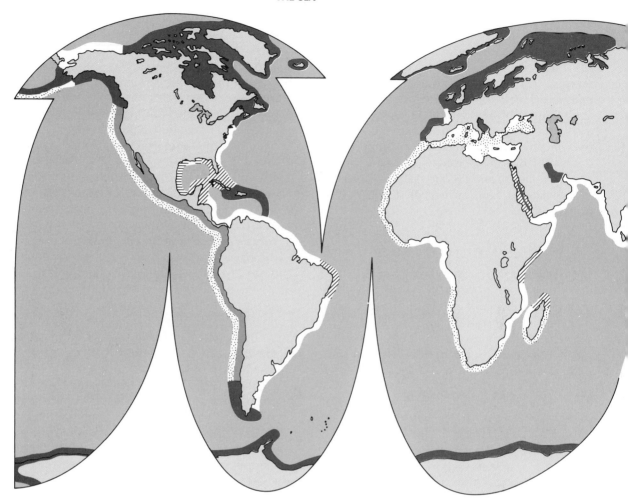

Fig. 14-3. Character of the continental shelves, by origin. Sediment deposited before formation of the dams is shaded gray; that deposited later is unshaded. The lines through the sediments indicate the orientation of bedding planes.

buried by younger sedimentation. Others, however, still persist and can be detected by echo-sounders. They provide us with clues to the earth's recent history. The fluctuating sea level also resulted in a complex pattern of sediment distribution on the shelf. In fact, in time some of the ice-age sand and gravel beach deposits might become major sources of building materials as more readily accessible deposits on land are used up.

Continental slopes

The *continental slopes* extend downward from the shelf break to the deep-ocean floor. It should not be thought that the gradients of the slopes and shelves are very great, a conclusion that might be drawn from looking at typical drawings of cross sections of ocean basins, because such drawings have high vertical exaggeration. The average slope of the shelves is 0°07', while that of the continental

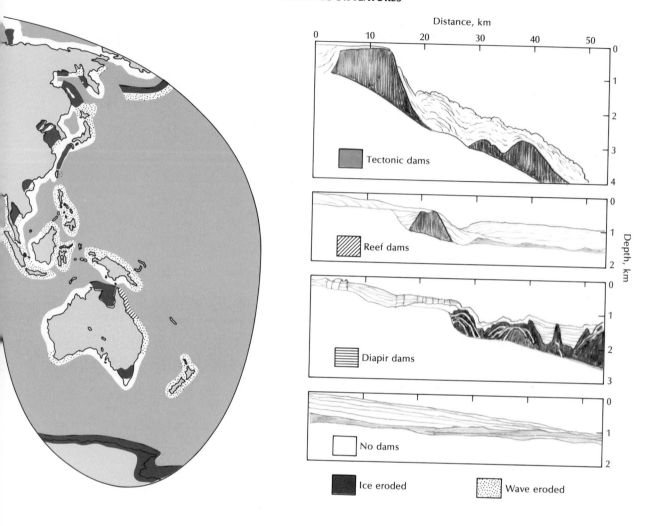

slopes is 4°17' for the first 1825 m (3986 ft). Like the shelves, the slopes are not gradually descending, flat expanses, but have basins, hills, valleys, and canyons. In general, the slopes are a quite rugged transition zone, about 16 to 32 km (10 to 20 mi) wide, which connects the two main levels of the earth's surface—sea level, or close to it, and ocean bottom, 3660 m (12,000 ft) below sea level.

Submarine canyons

One of the most characteristic features of the continental slopes is the numerous valleys, or canyons, that have been cut into them. Some are caused by landslides and other gravity mass movements in the unstable materials of the slopes, and others follow breaks in the underlying rocks. Those valleys are among the most impressive features of the ocean floor and the most difficult to explain.

Submarine canyons are found throughout the world, but it is very difficult to pick out one as typical because so few have been studied in detail. Probably the best known is La Jolla Submarine Canyon and Fan Valley, with its major tributary, Scripps Canyon. Those canyons could hardly escape intensive investigation, located as they are just off the coast of La Jolla, California, and the site of Scripps Institution of Oceanography, as well as being within a few miles of the Naval Undersea Research and Development Center. The canyons are of interest also as the location of the first dives by a submersible in the United States when the *Trieste* explored there in 1960. In addition, they head very close to shore, making their shallower portions down to 60 m (197 ft) easily accessible to small boats and scuba-diving scientists. Deeper portions have been studied with Jacques Cousteau's *Diving Saucer*.

From the beach, a flat, gently sloping, sand-covered terrace extends seaward. About 213 m (699 ft) from shore at a depth of 12 m (39 ft), the bottom suddenly drops away in the precipitous 24-m (79-ft) headwall of La Jolla Canyon (Fig. 14-4). At the base of the cliff is the wide, bowl-shaped head of the canyon. Seaward, the valley widens out on each side and then narrows again until it is a rock-walled gorge whose steep sides are covered with a lush growth of marine plants and animals. The bedrock of the canyon floor shows through its normal sandy covering in some places; in others there are thick mats of seaweed and other organic material. In places progress down the canyon is by giant steps or terraces caused by slumping in the loose bottom sediments.

Partway down, Scripps Canyon joins La Jolla Canyon, and a glance through the submarine gloom reveals that its walls have been so undercut at their bases that they actually overhang as much as 6 m (20 ft). Gradually the height of the walls decreases, and the valley widens until it is a little more than 1 km (0.6 mi) wide with an entrenched channel wandering across it. The walls are composed of semi-consolidated clay rather than hard rock, and there are natural levees on either side of the

Fig. 14-4. Overhanging canyon wall cut in solid rock at a depth of 40 m (131 ft) in Scripps Canyon off the coast of California. *U.S. Navy*

valley. We have entered the La Jolla Fan Valley which is cut not in the La Jolla Terrace as was the canyon, but in its own fan-like deposits. Eventually its depth below its surroundings decreases until it merges with the sea floor. Such submarine fans, built by sediment channeled down the canyon, are not uncommon, and if a number of canyons or other submarine valleys are spilling sediment out onto the ocean floor, a marine equivalent of an alluvial plain is formed. That sediment-laden boundary of coalescent fans between the continental slope and the deep-ocean floor is called the *continental rise*.

Larger canyons exist in other parts of the world. One off the California coast—known as the Monterey Canyon—rivals the Grand Canyon of the Colorado River in its vertical dimen-

sion (Fig. 14-5). The Hudson Canyon is the most famous off North America's east coast. It heads off Long Island, New York, trends southeastward in a fairly straight line, and is about 915 m (3001 ft) deep. For world-record length, we look to the deep canyon off the mouth of the Congo River, for it extends 240 km (149 mi) offshore.

The origin of submarine canyons has been the subject of debate by geologists for decades, and it is not necessary here to go into the many theories that have been proposed, only to be discarded. Much work remains to be done before the solution can be found, but current thought seems to favor subaerial erosion (see Chap. 13) as a starting point for many canyons. The theory is reasonable in view of the great lowering of sea level during Pleistocene glaciations. Yet subaerial erosion could not carve out canyons as continuous features onto the deep-sea floor. If a canyon were to result solely from such action, sea level would have to drop to the level of the floor of the canyon. Past glaciations have removed some water from the oceans, but hardly enough to expose the canyons to the realm of subaerial erosion.

The consensus, then, is that in order to understand the origin of submarine canyons, we need to understand better the submarine processes of erosion in such regions. We know that strong bottom currents exist throughout the ocean, and they have been observed in canyons as well. Such currents can transport sand-size sediment. Marine organisms no doubt play their part in the mechanical and chemical disintegration of canyon walls, some of which are hard bedrock. Density currents of heavily laden water triggered by the slumping of oversteepened and unconsolidated material near the heads of canyons are another force commonly called upon as an erosional agent. They are called *turbidity currents*, and their role in the erosion of submarine canyons is accepted by some but not by others.

Turbidity currents, as the name implies, are clouded or muddy streams of moving water. They were first described scientifically in 1840 by a Swiss engineer, Forel. He noticed that the muddy waters of the Rhône disappeared where the river flowed into the clear waters of Lake Geneva, and reasoned, quite correctly, that the colder water of the Rhône laden with glacial

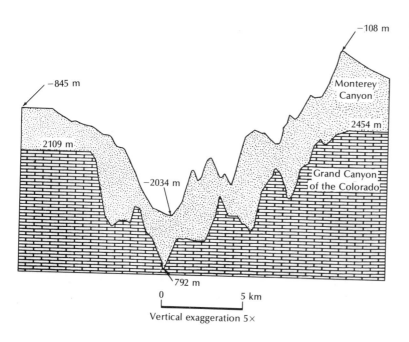

Fig. 14-5. Transverse profile across Monterey Canyon off the California coast, compared to one of the Grand Canyon of the Colorado River.

−845 m

−108 m

Monterey Canyon

2109 m

2454 m

−2034 m

Grand Canyon of the Colorado

792 m

0   5 km

Vertical exaggeration 5×

silt was heavier than the water of the lake. Therefore, the river sank through the lake water and flowed as an undercurrent along the bottom.

Much the same thing happens where the muddy Colorado River flows into Lake Mead, whose water is backed up for several kilometers behind Hoover Dam. A muddy current makes its way along the bottom of the lake to deposit sediment all the way to the powerhouse intake towers. The surprising effectiveness of such a current apparently was not anticipated at the time the dam was designed. The working assumption made then was that much of the river sediment would be deposited in more or less conventional way in a delta at the upper end of the lake.

Two major problems are present: (1) the abrasive effect of the sediment particles on the turbines, and (2) sediment transport across the entire length of the lake, which might seriously shorten its life expectancy since it will fill more rapidly than planned. A model of the lake was prepared in order to study the circulation of the silt-laden bottom current,

and photographs show how the turbidity current glides along the reservoir floor below the less dense and clearer water of the upper levels of the lake.

That turbidity currents may exist in the sea was suggested strongly on the afternoon of November 18, 1929, when a severe earthquake shook the Grand Banks off the Newfoundland coast. Apparently the shock was violent enough to trigger what may have been a submarine landslide or slump, which was soon converted into a roiled cloud of suspended sediment that swirled away down the continental slope. Such an event normally would go unnoticed were it not for the unique circumstance that the path followed by the earthquake-induced current was directly across one of the greatest concentrations of submarine cables in the world. It is one of the communication networks linking North America with Europe, and in that nexus Newfoundland is the easternmost land station in the New World.

One after another the cables were broken that afternoon and night, and, because the

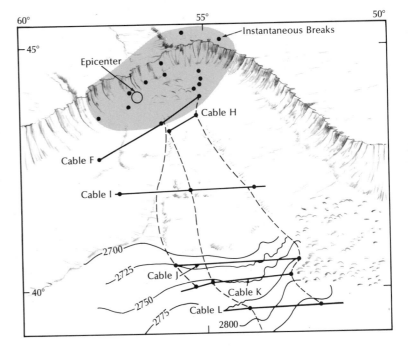

Fig. 14-6. The setting of the Grand Banks earthquake of 1929, showing the area in which the cables broke simultaneously, and the positions of parts of other cables that broke as the turbidity current progressed downslope. The dark circles are points of breakage, and the dashed lines are paths that parts of the turbidity current might have taken. The numbers on the contour lines in the lower part of the diagram represent water depth in fathoms (1 fathom = 1.8 m or 6 ft).

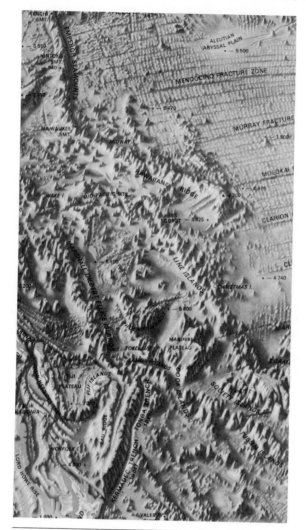

Fig. 14-7. Features of the ocean bottom in the western Pacific. Deep trenches are clustered along the Asiatic margin of the ocean.

estimated that, during the early part of its life, the current had a velocity of about 20 m/sec (66 ft/sec) or more, but that the velocity gradually decreased with time until the last cable, located about 600 km (373 mi) from the epicenter, broke some 13 hours after the earthquake.

Whether the cable-breaking pulse traveling across the sea floor was a turbidity current or not probably never will be known. About all we do know from that isolated episode, and a few other kindred occurrences, is that occasionally currents are set in motion in the depths of the sea that for short periods are capable of doing an immense amount of work and of shifting great volumes of sediment.

THE ABYSS, OR DEEP OCEAN

At the bottom of the continental slope in the depths of the sea is the *abyss*. Its floor stretches for unimaginable distances from continental margin to continental margin. Some parts of the floor of the abyss are a dead-level plain and those so-called abyssal plains leave an unwaveringly smooth line on a fathometer kilometer after kilometer. Some appear to be sediment covered, as is a broad expanse in the northwest Atlantic Ocean south of Newfoundland. Others, such as one in the Indian Ocean, seem to be the nearly level surfaces of vast subocean floods of basalt, perhaps comparable to the Deccan lava plateau on land in nearby India. Hills known as *abyssal hills* commonly rise above the plain but remain well below sea level.

Other parts of the sea floor are more varied, and include such diverse forms as broad rises and plateaus that may have quite gentle slopes. Ridges are narrower features with steeper marginal slopes and narrow crests. Many of them are submarine mountain ranges. Outstanding among them is the Mid-Atlantic Ridge, which stretches down the full length of the ocean and is practically equidistant from the continents on either side. Where the ridge pierces the surface of the sea its lofty peaks are the foundations of islands such as Ascension, the

instant of time as well as the point of rupture could be determined electrically, the pattern of their failure was determined (Fig. 14-6). It turned out that the cables broke progressively downslope, and when they were fished up for splicing, many of them were found to be chaotically snarled and jumbled, as though they had been thrust aside by a giant hand.

From the distance separating the individual breaks and from the time of failure it was

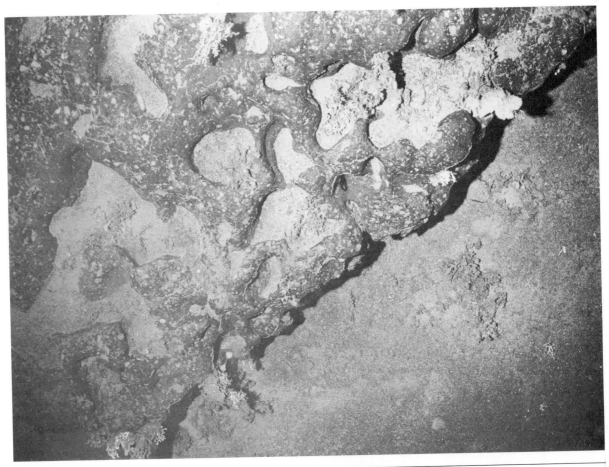

Fig. 14-8. Pitted lava with attached coral and bryozoan colonies forming the surface of the Kelvin Seamount, western North Atlantic, at a depth of 903 m (2962 ft).
Bedford Institute of Oceanography

Azores, Iceland, St. Helena, and Tristan da Cunha. Highest of the latter is Pico in the Azores, which towers 2300 m (7544 ft) above sea level and stands on a base that rises 6100 m (20,008 ft) above the ocean floor.

A curious feature of the Mid-Atlantic Ridge is the median valley, an essentially continuous central depression, or rift, with a nearly level floor and steep sides that follow the crest. The deepest part of the rift lies some 1000–1300 m (3280–4264 ft) below the adjacent ocean floor. In some ways the trough resembles the rift valleys of eastern Africa.

Almost equally curious are the great *trenches* countersunk below the general level of the ocean floor. Like many mountain ranges, they are long, narrow features (Fig. 14-7). The more prominent ones are in the Pacific, and in general they are clustered along the Asiatic margin. Except for the trenches off the coast of Chile and Peru, such submarine wrinkles are associated with the chains of volcanic islands that festoon the sea between the trenches and the adjacent mainland, following roughly the same pattern as the island arc.

No deeper spots on earth exist than those

found in the trenches. The Mariana Trench off the Mariana Islands at 10,910 m (35,785 ft) and the Tonga Trench north of the Tonga Islands at 10,770 m (35,326 ft) are the two deepest known at present. It is interesting to note that adventurers, in their efforts to get as far from sea level as possible have always gone up—to the top of Mount Everest—when in fact the points farthest from sea level are down, hidden from view.

The recognition of the extent and pattern of the deep, elongate trenches on the ocean floor is an outgrowth of World War II, especially in the Pacific. With naval vessels continually crossing its normally unfrequented waters, an enormous mass of fathometer records were accumulated. When the records were compiled, the great system of trenches paralleling the Mariana Islands—one of the world's immense submarine mountain chains—became clearly apparent. The work of surveying those

hidden features has continued to the present, and slowly our knowledge of their shape, if not their origin, is developing.

Another discovery was that not all oceanic volcanoes are associated with trenches, island arcs, and ridges. Most, in fact, are not high enough to rise above the surface of the water. They are known as *seamounts* and have the familiar shape of land volcanoes of the intermediate type, with steep sides and a small summit area (Fig. 14-7). Generally they stand about 1 km (0.6 mi) above the ocean floor. The number of such submerged volcanoes is almost unbelievable. In 1964 the oceanographer Henry Menard estimated that there are 10,000 in the Pacific Basin alone, which makes them one of the ocean floor's most prominent features. Although some are isolated cones, many others occur in clusters and linear arrangements. All so far investigated are of basaltic composition (Fig. 14-8).

Fig. 14-9. Perspective diagram of Bikini Atoll and Sylvania Guyot adjoining it on the west.

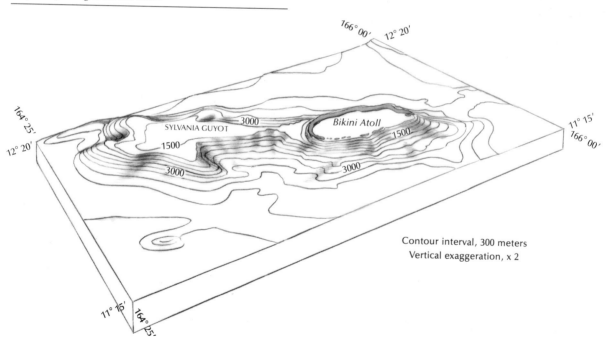

Contour interval, 300 meters
Vertical exaggeration, x 2

Much more puzzling than seamounts are features called *guyots* (Fig. 14-9). They, too, are volcanic mountains, but their summits form a nearly level submarine plateau. Their strange name honors Arnold Guyot (1807–1884), a Swiss geographer and associate of Louis Agassiz, who came to the United States more than a century ago to teach at Princeton University. Guyots commonly rise to within 910–1520 m (2985–4986 ft) of sea level. Their surfaces typically consist of barren, planed-off volcanic rock, but many are covered with rounded boulders, which may have been shaped when the platform stood closer to sea level. Some guyots have been photographed, and rock samples have been dredged from them. Fossil coral belonging to the Cretaceous Period (more than 70 million years ago) has been found attached to some of the samples. In one guyot in the Pacific near Bikini fossils of *Globigerina*, a minute, single-celled, surface-dwelling organism that lived during the Eocene Epoch (about 50 million years ago) had sifted down into cracks in the truncated summit.

How were the guyots' summits planed off to such remarkable uniformity, and how did they reach the great depths at which we find them? If we argue that they were beveled by wave attack—which seems the most logical solution —then we are caught in the dilemma of believing either (1) that sea level has risen in the not-very-remote geologic past, or (2) that the guyots were carried down to their present depth by regional subsidence. Although the latter explanation is the more likely one, it raises many questions—such as the likelihood of great expanses of the ocean floor sinking and the mechanics of transferring huge quantities of material at subcrustal depths. Plate tectonics (see Chap. 19) can be used to explain the location of guyots far below sea level.

Before we leave the subject of submarine volcanoes, we should point out their enormous size. We commonly conceive of terrestrial volcanoes as large features, but they are dwarfed by the size of many of their submarine cousins (Fig. 14-10).

CORAL ATOLLS

The origin of coral *atolls*, ring-shaped islands made up of the skeletons of marine animals, has been another subject of intense discussion among geologists (Fig. 14-11). Their appeal reaches back across the centuries, since the first western European seafarers made their

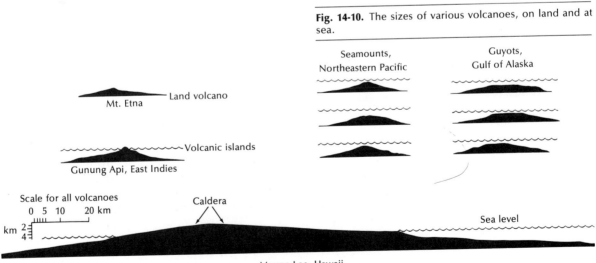

Fig. 14-10. The sizes of various volcanoes, on land and at sea.

Land volcano
Mt. Etna

Volcanic islands
Gunung Api, East Indies

Seamounts, Northeastern Pacific

Guyots, Gulf of Alaska

Scale for all volcanoes
0 5 10 20 km
km 2
4

Caldera

Sea level

Mauna Loa, Hawaii

Fig. 14-11. Barrier reef encircling a volcanic island, an intermediate stage in the formation of an atoll. *Office of Naval Research*

way into the distant reaches of the tropical seas. That is the world made famous by Herman Melville, by Robert Louis Stevenson, by the stirring events arising out of the contentious voyage of H.M.S. *Bounty,* and, in our time, by the naval conflict that surged back and forth across the vast domain of the Pacific in the years 1942–45. For those who are interested, the modern world of those far-distant islands is described in a most effective way in the novels and essays of James Michener.

Probably relatively few people realize that coral reefs once excited the scientific world as greatly as they fired the thoughts of the more romantically inclined literary figures of the nineteenth century. Few environments provide a setting more unlike that of the Western world nor a way of life more different from ours than that of the sea-roving Polynesians.

Early in the nineteenth century the more expansionist powers fitted out a number of exploring expeditions, and in many cases the expeditions included a naturalist among the ship's company. The status of such persons in

that nautical world was never high, but we can be grateful today that a few dedicated men blazed the scientific trail for us. Charles Darwin (1809–82) will always stand chief among them. As a young man of twenty-two he was assigned as a naturalist to H.M.S. *Beagle*, a 10-gun brig of about only 230 metric tons, scarcely 30 m (98 ft) long, and carrying 74 persons. The ship circumnavigated the world, and Darwin's zoological observations on that voyage, when sifted through, compiled, and analyzed, contributed to the theory of evolution put forth in the *Origin of Species*.

Few people are aware that Darwin's scientific training had been in geology as much as in any other branch of science, and that on the voyage of the *Beagle* he made great numbers of observations on rocks, volcanic features, and fossils. Among the many wonders he beheld, few aroused his interest more than the fairyland world below the sea created by the corals. Though he actually saw only a few reefs, Darwin had the insight to recognize that there were three major kinds: (1) fringing reefs, (2) barrier reefs, and (3) atolls; and that the three were related to each other in a logical and gradational sequence.

Darwin believed that the succession from one reef type to another could be achieved by the upgrowth of coral from a sinking foundation, such as a subsiding volcano. So long as the rate of coral growth was more rapid than the rate of sinking, Darwin argued that the progression would be from a fringing reef through the barrier stage, and, with the disappearance through subsidence of the central island, only a reef-enclosed lagoon, or atoll, would survive (Fig. 14-12).

Fringing and barrier reefs are not so difficult to explain—even such imposing ones as the Great Barrier Reef of Australia—as are the atolls. Some atolls are very large—Kwajelein in the Marshall Islands of the South Pacific is 120 km (75 mi) long and averages as much as 24 km (15 mi) across. Most are far smaller. Nearly all true atolls are low—few have a freeboard of much more than 3 m (10 ft). The reef ring consists of solid coral and over it the sea

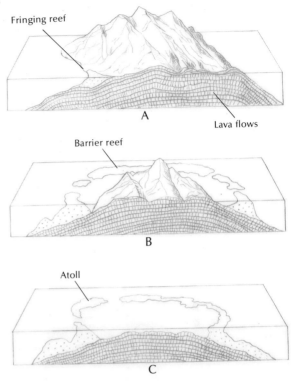

Fig. 14-12. Succession of reef types around a sinking volcanic island, from a fringing reef to a barrier reef to an atoll.

sweeps freely at high tide, and with telling effect during storms. The rest of the time the reef may be just barely awash. In the ring are the reef islands—in a sense they are like sand dunes—which are composed of thousands upon thousands of small fragments of ground-up coral, algae, pieces of shell, and the like. On these are rooted the pandanus and the coco palm—lifegivers of the Polynesian world.

In that two-dimensional world one is eternally aware of the encircling sea. No place on earth could be more vulnerable to the unbridled savagery of a hurricane, whose impact on such a defenseless world is vividly described by Nordhoff and Hall in their novel *Hurricane*. During one of those gales the sea rises with the fall of barometric pressure, and storm-driven waves can strip off the patiently accumulated reef islands like so much bark off

a tree—together with all the people and their coco palms, taro patches, boats, villages, and even their graves.

Research since Darwin's time generally has upheld his ideas on the formation of atolls. If, for example, the reef foundation was sinking as the reef built up, then atolls should be underlain by hundreds of meters of coralline material, all of which grew in shallow waters. Darwin realized that drilling a hole through a reef would be the surest test of his theory, as he suggested in 1881 in a letter to Alexander Agassiz, an American oceanographer and the son of Louis Agassiz.

Several atolls were drilled prior to World War II, but the results regarding their origin were not conclusive. Fortunately, as part of the environmental studies made in the Marshall Islands in connection with atom-bomb testing, the U.S. Navy drilled a series of deep holes. A total of five were put down on Bikini in 1947, three shallow and two deep, the latter to 404 and 767 m (1325 and 2516 ft). The drill holes remained in coralline material all the way. Three deep holes drilled in 1951–52 on nearby Eniwetok were more decisive. The two deepest, 1266 and 1389 m (4152 and 4556 ft), went completely through a 1200-m (3936-ft) cap of shallow-water reef limestone and bottomed in a typical mid-Pacific volcanic rock, basalt. The age of the fossils in the basal limestone is Eocene, which meant that Eniwetok atoll is at the top of a coralline accumulation that has grown upward during the last 60,000,000 years of earth history. The rate of subsidence does not appear to have been constant, but ranged between perhaps 15 m (49 ft) and 51 m (167 ft) per million years, slowing down in the later, compared with the earlier, history of the reef.

Thus some of the events ushering in the nuclear age, a little more than a century after Darwin saw his first atoll, serve to vindicate his altogether remarkable insight. There can be little doubt that at least two of the great atolls of the western Pacific in the Caroline and Marshall Islands resulted from the upgrowth of reef organisms on a slowly sinking foundation.

Another complicating factor in atoll forma-

tion, or for that matter in the formation of reefs in general, is that of changing sea level with glaciations and interglaciations. During glacial times, with sea level at least 100 m (328 ft) below the present level, the reef would be exposed to weathering and battering by waves. With the advent of an interglaciation, however, sea level rises and reefs build upward. Run that cycle many times, as has happened over at least the past million years, and the formation of the circular islands becomes quite complicated, especially in the top 100 m (328 ft) or so.

Many other perplexing features have come to our notice as increasing numbers of oceanographic vessels cross the seas of the world, and as the technology of marine surveying improves. Great escarpments, several of them more than 1600 km (994 mi) long, interrupt the sea floor in places. Four of them, separated from one another by roughly equal spacing, strike almost due west from the coasts of California and Mexico into the Pacific, at least to the longitude of Hawaii. Their relationship to geologic structures on land is difficult to decipher. Rather than being a continuation of the prominent land fracture, the San Andreas fault, the two off the California coast intersect the fault very nearly at right angles. One of the escarpments off the coast of Mexico possibly may be a seaward continuation of the line of Mexican volcanoes aligned along the nineteenth parallel. That volcanic system includes not only the lofty cones of Popocatépetl and Ixtaccihuatl outside Mexico City but also the recently active volcanoes of Parícutin and of the island San Benedicto off the coast.

OCEAN SEDIMENTS

A great variety of sediments blanket the ocean floor. Most people are familiar with the sand and gravel of beaches. Seaward of the beach, silt and clay mantle the continental shelf. What of the material that covers the deep floor of the sea?

Many of the descriptions of sediment found on the floor of the abyss are taken from the reports of the H.M.S. *Challenger* expedition of

a century ago (Fig. 14-13). The *Challenger* left England in December 1872, making an epic oceanographic study of the Atlantic, Pacific, and Southern oceans before returning to England in May 1876. Six scientists gathered data to better understand the physical, chemical, and biological properties of both the ocean water and the sediment that blankets the ocean floor. Today we know that the pattern of the sea floor, although it may resemble the one worked out by the men of the *Challenger* in broad outlines (Fig. 14-14), is more complex in detail.

In brief, the *Challenger* demonstrated that generally on the deep sea floor between the lower part of the continental slope and a depth of around 4 km (2.5 mi) the most abundant sediment is *ooze*. That wonderfully descriptive

Fig. 14-13. H.M.S. *Challenger* passes by an iceberg during its voyage.

Fig. 14-14 A. Distribution of sediment types on the sea floor. **B.** Sediment thickness. **C.** Rate of growth of sediment thickness.

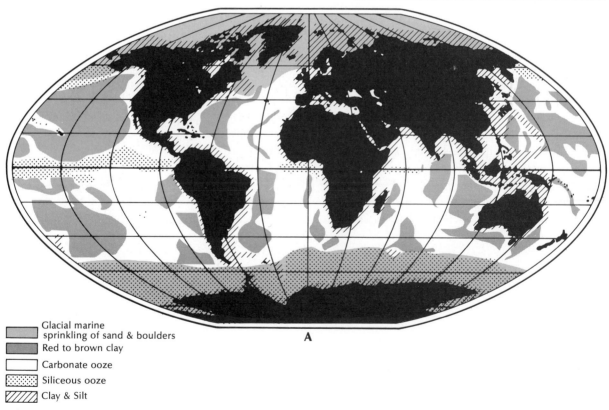

Glacial marine
sprinkling of sand & boulders
Red to brown clay
Carbonate ooze
Siliceous ooze
Clay & Silt

A

476

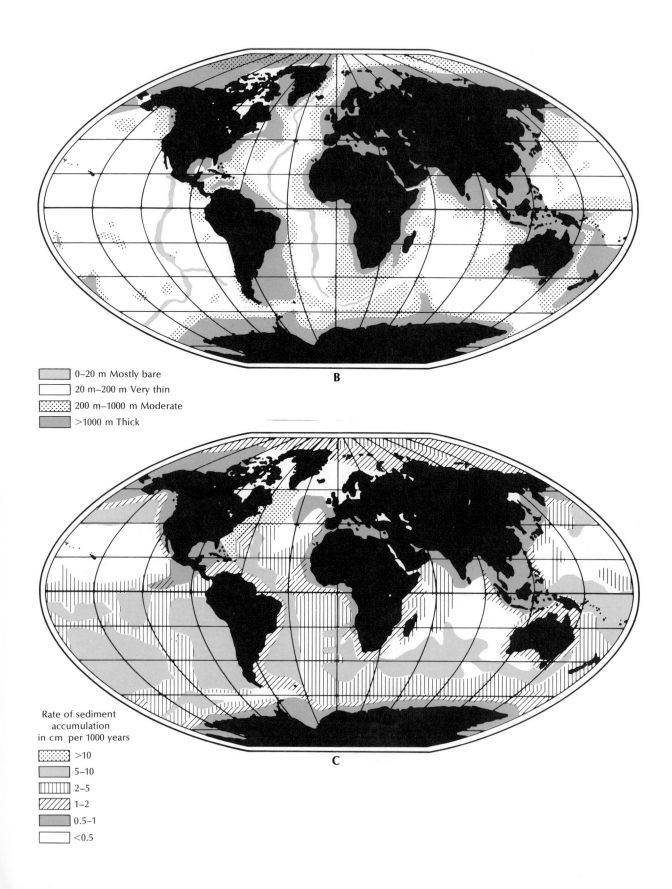

0–20 m Mostly bare
20 m–200 m Very thin
200 m–1000 m Moderate
>1000 m Thick

B

Rate of sediment
accumulation
in cm per 1000 years

>10
5–10
2–5
1–2
0.5–1
<0.5

C

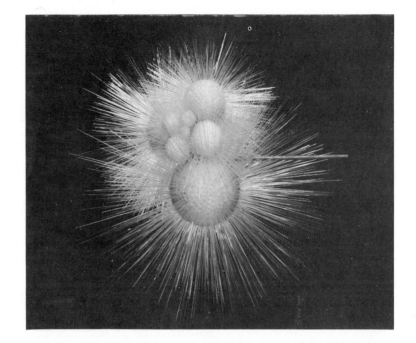

Fig. 14-15 A. (above) Glass model of the Foraminifer *Globergina* as it appears floating in the ocean waters. Once the shell is shed, the spines are lost, and as a fossil it looks similar to the other Foraminifera in **(B)**. *American Museum of Natural History* **B.** (below) Middle Eocene ooze (45 million years old) from a Deep Sea Drilling Project core in the western Indian Ocean. Most of the fossils are radiolaria, but a few are Foraminifera (F) and sponge spicules (S).
Scripps Institute of Oceanography

term gives an impression of what organic flour is like. Photographs of ooze, as well as verbal descriptions by bathyscaphe divers, suggest that it looks like an ivory-colored blanket. The slightest disturbance sends a dust-like cloud swirling up through the dark water, making us realize what an impalpable powder it must be. Ooze does not form in place on the sea floor, but accumulates as the result of a gentle, unceasing "snowfall" of the remains of ornate, microscopic, free-floating surface organisms. When they die, their minute remains sift down from the sunlit surface to the lightless floor of the sea. It is only because the terrestrial supply of sediment is so slight that such organic deposits can build up.

Not all oozes are the same, however; their composition varies systematically across the ocean floor. Much of the abyssal plain beneath tropical and warm-temperate seas, in depths of less than about 4000 or 4500 m (13,120 or 14,760 ft), is drifted over with a carpet of microscopic shells of Foraminifera, a single-celled creature that secretes calcite (Fig. 14-15A). Like so many single-celled creatures, they do not die, but reproduce themselves by division. That is, one organism divides to make two, each of which grows a new shell, while the original shell, now vacated, is cast off to sink to the bottom of the sea. As the shell drifts downward the temperature and chemical conditions of the water are such that the carbonate dissolves, generally below 4000–4500 m (13,120–14,760 ft).

In other parts of the ocean, *siliceous oozes,* which consist mainly of *radiolaria* (the coarse-meshed objects in Fig. 14-15B) and *diatom* remains, predominate. They are minute animals and plants, respectively, that secrete a silica cell. The siliceous oozes are found in areas of either high productivity of such organisms combined with low productivity of the carbonate forms, or in places where the water depths are so great that the carbonate forms dissolve. Although siliceous forms also dissolve in sea water, their rate of dissolution is less than that of carbonate forms—thus, they can survive as fossils where carbonate forms cannot.

Siliceous oozes consisting mainly of diatoms are abundant in a broad band in the colder seas marginal to both the Arctic and Antarctic. Those cold-water-inhabiting plants swarm in the frigid seas in uncountable multitudes, and in a sense compose the true pastures of the sea. How curious that the largest creatures on earth, the Arctic whales, should depend for sustenance on the humble diatom, among the smallest of living things.

Like their floral counterparts on land, diatoms flourish during part of the year, reach a climax, and then decline. Often when they die, they appear to perish in hordes. The water in protected inlets, such as fiords, then becomes green and soupy looking, and oars and boats immersed in such a flood of organic matter grow green and slimy and acquire a distinctly fishy fragrance.

A unique sedimentary accumulation is the monotonous soft carpet of red to brown clay that floors the deeper parts of the oceans. In terms of area it is the most widespread of all the deposits of the earth. Curiously, it is almost wholly inorganic, save for such exotic souvenirs of marine life as the ear bones of whales and the teeth of sharks—both relatively insoluble in sea water. The fact that some of the sharks' teeth are those of species long dead, and known to us only as fossils, is evidence of the extraordinarily slow rate at which the clay accumulates.

The clay is extremely fine grained, and much of it comes from land. It was transported to the sea by rivers, waves, or the wind and then carried far and wide by ocean currents. Gradually it sifted down, particle by particle, through the lightless depths to accumulate as an impalpable flour on the bottom of the sea. Volcanic and cosmic dust also contribute to the carpet of clay, but they probably only make up a small proportion of the whole.

An unusual deposit in some places on the deep ocean floor is large nodules—sometimes up to the size of cannon balls (Fig. 14-16)—of manganese dioxide (MnO_2). Superficially they resemble cobbles or boulders, which they are not, because they grew by accretion of ion upon ion on a suitable substrate, such as a

Fig. 14-16. Manganese nodules of cannon-ball size at a depth of 5417 m (17,768 ft) in the southwest Pacific. *Smithsonian Oceanographic Sorting Center*

shark's tooth. The nodules may also contain appreciable amounts of iron, cobalt, nickel, and zinc, and are so abundant in some places that, fanciful as it may sound, dredging them to the surface has been suggested as a possible means of obtaining those relatively scarce metals. One estimate is that 2 billion tons of nodules lie on the ocean floor—clearly the largest mineral deposit on earth. These minerals are chemical precipitates, and the nodules grow at an excessively slow rate. Some estimates of the rate of nodule growth are 2–4 mm (0.08–0.16 in.) per million years.

As one might expect, the thickness of sediments varies from place to place on the ocean floor (see Fig. 14-14B), as does the rate of deposition (see Fig. 14-14C). One reason the thickness varies is that the ocean floor varies in age from very young over the oceanic ridges to progressively older proceeding away from the ridges (see Chap. 19). The rate of deposition varies with proximity to continents, erosion rate on the continents, and the rate of biological productivity of the organic oozes. Add to that the prevailing ocean currents and the solution of the carbonate or silicate remains as they settle to the bottom, and one can see that the rate is a difficult one to calculate.

In places, relatively coarse-grained sediments are found on the abyssal plains. For example, it is not uncommon to find broad, fan-like expanses of sand spreading out across the ocean floor from the continental margins. Such sands are found in regions where ooze normally would be found, and often contain the remains of organisms which, to the best of our knowledge, live in shallow waters near the margins of the ocean basin. It is likely that many such sediments were carried to their resting places by turbidity currents moving off the continental slope or down submarine canyons.

Perhaps more anomalous are the presence of land-derived boulders far from the continental shores (Fig. 14-17). No ocean current could have moved rocks of such size by lateral transport into their present position. Rather, they seem to have fallen straight down to the ocean floor, after having been carried far offshore by icebergs (Fig. 14-18). Such stones are found concentrated around glaciated areas and areas of pack ice, and the resulting sediment is described as *glacial marine* (see Fig. 14-14A).

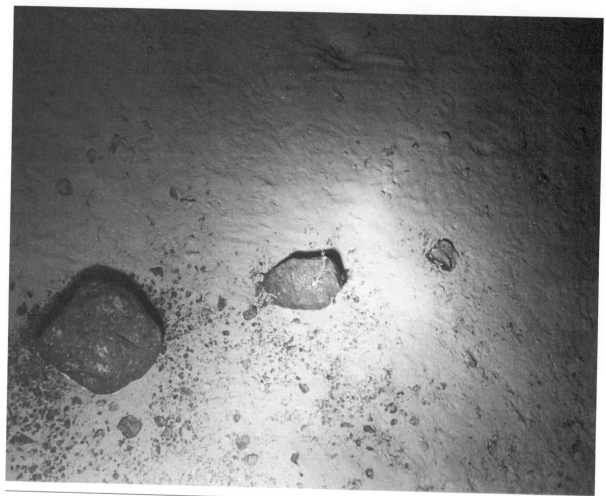

Fig. 14-17. Boulders on the sea floor at a depth of 3562 m (11,683 ft) in the Drake Passage between South America and Antarctica. *Smithsonian Oceanographic Sorting Center*

Fig. 14-18. Iceberg with boulders carried piggyback fashion. As the iceberg rotates or melts, the boulders and sands rain down on the deep sea floor thousands of meters below.

GLACIATIONS AND INTERGLACIATIONS— THE STORY FROM THE OCEAN DEEP

Recent intensive study of numerous cores from the floor of the sea has caused scientists to alter their ideas on the numbers and timing of Pleistocene glaciations. How was their thinking changed? A sample core is a historical record of events in earth history, with the youngest events recorded near the top of the core and the older ones at the base. Further, many scientists think that the deep oceans were repositories of sediment throughout the Pleistocene and even farther back in time, and that erosion in those areas is virtually non-existent. Thus the record preserved by the sea is nearly complete.

What do ocean sediments tell us about glacial periods, however? For one thing, some cores show glacial-marine deposits interbedded with normal marine deposits. In other cores, the Foraminifera population changes in number, in species, and in chemical composition with depth. Marine life such as the latter seems to be sensitive to either the changes in ocean water temperatures or water composition that accompany climatic change and the stockpiling of water on land as ice.

When the data are carefully analyzed, one can see a procession of alternating warm and cold intervals that somehow must be related to glaciation and interglaciations on land (Fig. 14-19). The revelation of more glacial periods than had hitherto been recognized on land was fairly revolutionary. From Figure 14-19, for example, one can postulate six major glaciations in little more than the past 400,000 years. How are the six to be correlated to the traditional four recognized in the land record? Nobody has come up with the answer acceptable to the majority of earth scientists. The other interesting aspect of the curve in Figure 14-19 is its sawtooth shape, which can be interpreted to mean that ice caps take a long time to build up —say some 50,000 years or more—but that their

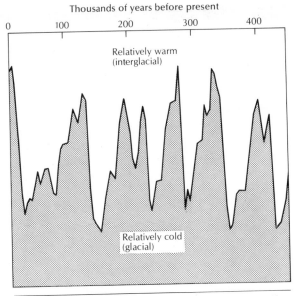

Fig. 14-19. Chemical data on fossil shells of Foraminifera can be used to help date the time of glacials and interglacials in the past. Here is presented the interpretation of such data from a core brought up from the floor of the Caribbean Sea.

disappearance by melting verges on the catastrophic—it is all over in as little as 10,000 years!

Much work still needs to be done. We have to learn to better read the deep-sea sediment record. We also need to think through the correlation of the oceanic record with the glacial record on land; that will require more thorough field work, but field work is challenging, and often a major reason why earth scientists have chosen their profession. Also, from studying the relationship between warm and cold intervals and periods of glaciation as shown in Figure 14-19, we might well be able to predict the timing of future glaciations, the climatic conditions signaling their onset, or the catastrophic warming conditions that accompany their disappearance.

SUMMARY

1. The ocean floor consists of several major topographic features. The *continental shelf* forms the relatively shallow, gently sloping platform that is found off the coast of most continents. The shelf gives way seaward to the *continental slope,* which extends down to the ocean floor. Submarine canyons, some larger than any on land, are cut into both the shelf and slope in places. Major features of the deep ocean floor are ridges and the deep trenches, both of which form prominent topographic lineations and have their origin in plate tectonics (see Chap. 19).

2. Volcanoes are a prominent feature of the ocean. Some rise to the surface as islands, others are submerged and are called *seamounts,* and some submerged ones have flat tops and are called *guyots.* The flat tops of the guyots are generally ascribed to erosion when the sea water was shallower than at present. Plate tectonics (see Chap. 19) can be used to explain their present position far below sea level. *Atolls* consist mainly of the buildup of coral limestone deposits on slowly subsiding volcanoes.

3. Deep ocean sediment consists mainly of the fine-grained *oozes* of calcareous or siliceous composition (the remains of minute animals and plants) and clay. In places nodules of ore minerals are present. In other places, gravel clasts rest upon the fine-grained ocean bed far from shore. They probably were carried far offshore in icebergs and deposited when the ice melted.

4. The chemical and biological properties of the oozes vary with depth in ocean cores, and the changes give important clues as to the number of glaciations and interglaciations in the Pleistocene. In short, there are more such periods recorded in the ocean sediments than in the record of terrestrial glacial tills.

SELECTED REFERENCES

Heezen, B. C., and C. D. Hollister, 1971, The face of the deep, Oxford University Press, New York.

Menard, H. W., 1964, Marine geology of the Pacific, McGraw-Hill Book Co., New York.

The Ocean, a Scientific American book (1969), W. H. Freeman and Co., San Francisco.

Thurman, H. V., 1975, Introductory oceanography, C. E. Merrill Publishing Co., Columbus, Ohio.

Fig. 15-1. Sinkholes southeast of Winslow, Arizona, were formed mainly through the removal by solution of parts of the Kaibab Limestone. *John S. Shelton*

15

GROUND WATER

In Xanadu did Kubla Khan
A stately pleasure-dome decree
Where Alph, the sacred river, ran
Through caverns measureless to man
Down to a sunless sea.

Coleridge's verse quoted above reflects the re-markable image that many people have of the nature of water within the earth. Some are prone to speak glibly of underground rivers flowing for miles beneath the parched surface of some of the world's most absolute deserts, and to many of us springs are nearly as mysterious as they were to people long ago.

Springs played a leading role in Greek and Roman mythology. An example is the spring of Arethusa, which appears on the island of Sicily in the ancient harbor of Syracuse. The river god Alpheus, in pursuit of the wood nymph Arethusa, flowed as an underground river all the way from Greece to Sicily, where he finally caught the elusive spirit and changed her into a spring.

Comparable beliefs were held in those early days. They were certainly picturesque and more colorful than the prosaic opinions we now hold. Generally there were two leading schools of thought. One held that springs drew their water from the sea—how the salt was eliminated and how the water was elevated to the great heights it reached in mountain springs remained unanswered questions. The

other belief was that springs and streams had their own origin within subterranean caverns, large enough perhaps to have atmospheres of their own from which water condensed as a sort of rain to feed them. Aristotle (384–322 B.C.) was content with that idea since rainfall was obviously inadequate:

The air surrounding the earth is turned into water by the cold of the heavens and falls as rain, [so] the air which penetrates and passes into the crust of the earth also becomes transformed into water owing to the cold which it encounters there. The water coming out of the earth unites with the rain water to produce rivers. The rainfall alone is quite insufficient to supply the rivers of the world with water. The ocean into which the rivers run does not overflow because, while some of the water is evaporated, the rest of it changes back into the air or into one of the other elements.

Seneca (3 B.C.–A.D. 65) gave the seeming death blow to any concept so preposterous as one that water in the ground had anything to do with rain:

Rainfall cannot possibly be the source of springs because it penetrates only a few feet into the Earth whereas springs are fed from deep down. . . . As a diligent digger among my vines I can affirm my observation that no rain is ever so heavy as to wet the ground to a depth of more than ten feet.

It was not until the mid-seventeenth century that two Frenchmen, Pierre Perrault (1608–80)

485

and Edme Mariotte (1620–84) demonstrated that there was a relationship between rainfall and the discharge of springs. The approximate relationship of evaporation from the sea to the amount of rainfall and runoff was worked out by the astronomer Edmund Halley—known to us for the comet named after him. According to the geologist Oscar Meinzer (1876–1948):

Perrault made measurements of the rainfall during three years, and he roughly estimated the area of the drainage basin of the Seine River above a point in Burgundy and of the run-off from this same basin. Thus he computed that the quantity of water that fell on the basin as rain or snow was about six times the quantity discharged by the river. Crude as was his work, he nevertheless demonstrated the fallacy of the age-old assumption of the inadequacy of the rainfall to account for the discharge of springs and streams. Perrault also exposed water and other liquids to evaporation and made observations on the relative amount of water thus lost. He also made investigations of capillarity, established the approximate limits of capillarity in sand, and showed that water absorbed by capillarity cannot form accumulations of free water at higher levels.

Marriotte computed the discharge of the Seine at Paris by measuring its width, depth, and velocity at approximately its mean stage, making the velocity measurements by the float method. He essentially verified Perrault's results. In his publications, which appeared after his death in 1684, he defended vigorously the infiltration theory and created much of the modern thought on the subject . . . he maintained that the water derived from rain and snow penetrates into the pores of the earth and accumulates in wells; that this water percolates downward till it reaches impermeable rock and thence percolates laterally; and that it is sufficient in quantity to supply the springs. He demonstrated that the rain water penetrates into the earth, and used for this purpose the cellar of the Paris Observatory, the percolation through the cover of which compared with the amount of rainfall. He also showed that the flow of springs increases in rainy weather and diminishes in time of drought, and explained that the more constant springs are supplied from the larger underground reservoirs.

In ancient times a considerable use was made of water from the ground. Wells were the center of village life for centuries and still are over much of the world. Not only were wells the focal point of village life, they were absolutely essential to the survival of a walled city or castle. It would be a foolhardy baron who attempted to hold off a siege without an intramural source of water.

The demand for water created by large concentrations of people in urban centers has resulted in a more extensive development of underground sources of water than most people realize. Almost everyone knows of the heroic measures the Romans took to conduct water to their cities by building imposing, valley-spanning aqueducts. Oddly enough, the Romans knew little about water in the ground. They placed their chief dependence on springs and streams with the result that they went prodigious distances to the Apennines for water when they had a perfectly adequate supply almost directly underfoot had they dug for it.

But other ancient people did dig for water, and their underground pursuit led to the construction of remarkable burrow-like excavations. The chief example are the *kanats* of ancient Persia, now Iran. They center largely around Teheran, and for the most part are dug in the gravels of the great apron of alluvial fans at the base of the Elburz Mountains. The kanats are long passages that serve as collection galleries in the porous gravels of the fan. Some are 24 km (15 mi) long, and individual tunnels may be as much as 152 m (499 ft) below the surface. In the old days they were truly multipurpose structures because they served as a source of drinking water and as a means of sewage disposal. In general a kanat follows a water-bearing layer of sand or gravel within the fan, and every few hundred yards is connected with the surface by a shaft sunk during construction.

In Egypt a remarkable water-collecting tunnel system was built around 500 B.C. Constructed in the Nubian sandstone, it gathered water which was probably introduced into the rock as seepage from the Nile. All told, the tunnel system has a length of 160 km (99 mi) or so, although no one can say with certainty because it has almost entirely caved in. Water still escapes from the tunnel entrance, which was

once throught to be a spring. Actually, that extraordinary system was so important that the temple of Ammon was dedicated in its honor.

Not only did men of long ago drive tunnels to intercept water in the ground, but they drilled wells to surprising depths. An outstanding achievement among dug wells was one at Orvieto, in Italy, which was sunk to a depth of 61 m (200 ft) in 1540. Two spiral staircases lined the walls, one above the other, with one being used by descending, the other by climbing, water-bearing donkeys.

Wells were drilled to great depths, as much as 1500 m (4920 ft) in China, for example. Deep wells drilled at Artois in France in the twelfth century and at Modena in the Po Valley of Italy flowed water at the surface, exciting great interest since they were the first true artesian wells of medieval times.

In this chapter we will discuss the origin and occurrence of ground water, the ways to prospect for it, the problems of contamination in underground reservoirs, and features such as hot springs and geysers, which are a ground-water phenomenon as well as an energy resource.

ORIGIN OF GROUND WATER

Nearly all the water in the ground comes from precipitation that has soaked into the earth. Additionally, some water is included with marine sediments when they are deposited, and some reaches the upper levels of the crust when it is carried there by igneous intrusions, volcanoes, and hot springs. In practical terms, however, the latter sources provide only a minor part of the total budget of usable water in the ground.

Many things happen to water that falls as rain or snow, as we saw in the discussion of the hydrologic cycle (see Fig. 10-2). For the United States, earth scientists Luna Leopold and Gordon Wolman have estimated that water is divided up as follows: An average of 76 cm (30 in.) fall as rain each year. Of that amount, approximately 53 cm (21 in.) are returned directly to the atmosphere by evaporation and transpiration. Only 23 cm (9 in.) run off in

streams directly to the sea, and of the total runoff nearly 40 per cent escapes by the Mississippi river—an impressive fraction of the continental supply.

Where does ground water come from, then, if the budget balances as closely as the above figures? The answer is that although the amount of water entering the ground by infiltration is slight—perhaps as little as 0.25 mm (0.01 in.) per year in some places, more in others—with the passage of many millennia, great quantities of water slowly accumulate in the ground. It is that vast reservoir built up gradually over thousands of years that we draw on today—unfortunately in some areas more rapidly than it is replenished. Where the latter is so, for example in the Tucson-Phoenix area in Arizona and parts of the Great Valley in California, the ground water can be considered as a non-renewable resource that will be depleted at some time in the future.

The total amount of ground water in the world has been estimated at some 8.4 million km³ (1.8 million mi³). A glance back at Figure 10-2 indicates that the amount is more than that in the rivers and lakes at any one time, but less than that in the oceans or glaciers. It is that great readily available abundance that makes ground water so important to the development of agriculture, communities, and industries.

OCCURRENCE AND MOVEMENT OF GROUND WATER

Probably many of us have only seen ground water standing in shallow wells—very often green, scummy, and unappetizing. If we were to determine its altitude and then to compare it with the level at which water stands in nearby wells, we would quickly discover that in many regions the water surface is at the same level. The surface at which water stands in wells is called the *water table*. All the voids, or openings, in rocks below the water table are filled with water, or are saturated. Above the water table, in the zone of *aeration*, the pore spaces in the ground may range from completely dry

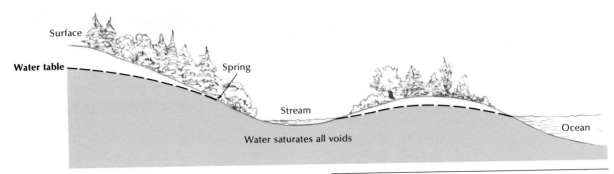

Fig. 15-2. Position of the water table in a humid region.

to partially full. In many places, the water moves downward through the zone of aeration to the water table. Such movement, however, is quite slow.

Actually, the water table is rarely dead level. Instead it is more likely to be a blurred replica of the ground surface (Fig. 15-2), rising under hills and sinking under valleys. It intersects the surface at lakes and streams and also at springs. Sometimes it adds water to a stream, especially if the stream is at a low level (Fig. 15-3). Such a situation is common in humid regions, and the stream is called an *effluent stream*. If the stream flows above the water table and thus adds to the supply of water in the ground, it is an *influent stream*. The latter are common in arid regions. The position of the water table related to an effluent stream is more-or-less stable. That beneath an influent stream, however, is apt to fluctuate. It will intersect the surface when the rivers are flowing but will drop below the surface if they run dry.

A closer look at the zone of aeration shows that it is actually made up of three zones: (1) the *zone of soil moisture*, (2) the *intermediate zone*, and (3) the zone encompassing the *capillary fringe*.

The zone of soil moisture is the portion of the profile most familiar to us. It is the ground layer that becomes wet after a rain or a lawn watering. When completely saturated it may become a quagmire; at other times it may be bone dry and dusty.

Commonly there is a lower margin to the surface zone of soil moisture. It may be only a few centimeters down or it may be several meters.

Fig. 15-3. Position of the water table with respect to influent and effluent streams. The former are common in desert regions, and the latter in humid regions. Arrows depict the flow of water beneath the water table.

Influent stream

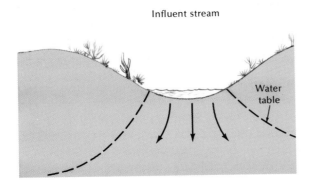

Effluent stream

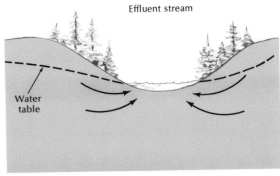

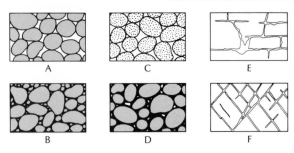

Fig. 15-4. Various degrees of rock porosity. **A.** Highly porous well-sorted sandstone. **B.** Poorly sorted sandstone of low porosity. **C.** Highly porous well-sorted pebble conglomerate made up of porous sandstone pebbles. **D.** Well-sorted sandstone of low porosity (cementing material fills much of the void space). **E.** Limestone rendered porous through solution. Some of the voids could eventually grow into large underground caverns. **F.** Dense igneous rock rendered porous by joints.

When we dig into the ground it generally becomes drier, until most of the soil moisture seems to have disappeared—as Seneca believed it did. Typically, though, in that intermediate belt the water percolates slowly downward through openings until it reaches the water table. How well or how rapidly it percolates depends largely on the *porosity* and *permeability* of the ground. What those two terms mean will be described a few paragraphs ahead.

Extending a short distance upward from the water table is the capillary fringe, thread-like extensions of water that have migrated upward in the minute passageways between the individual soil grains. The movement is achieved in about the same way that kerosene climbs the wick of a kerosene-burning lamp, or water rises in the confines of a narrow glass tube in a chemical laboratory. The fringe is usually less than about 4 m (13 ft) thick.

Contrary to what many people think, the zone of high-water content under the water table does not continue indefinitely downward. In other words, drilling a well to greater depths will not necessarily increase the flow of water. With increasing depth the pore spaces in the rocks close up, and their water-bearing capacity diminishes until they may become completely dry. In deep mines the upper levels may require constant pumping to keep them from flooding, while the lower levels may be so dry that water has to be brought down for use in drilling.

Porosity

Porosity is of the greatest importance in controlling the movement of water in the ground. We are familiar with the general meaning of the word when we think of a porous substance as one that contains many holes. Actually, porosity is expressed as the percentage of the total volume of the rock that is occupied by openings. If half the volume of a rock is taken up by openings, the material has a porosity of 50 per cent; if only one quarter, then 25 per cent, and so on.

Many factors determine the porosity of a rock. In clastic sedimentary rocks the packing arrangement is important, but so too is the degree of sorting, the porosity of the clasts, and the amount of cementing material in the void spaces. Limestone itself may be quite dense, but masses of it are quite porous because dissolution of the rock can result in large cavities and thus in high porosity. The porosity of igneous and metamorphic rocks is largely determined by joint frequency, because the rock itself is so dense (1 per cent, or less, porosity).

It is important to note that grain size does not influence porosity in clastic sediments. BB's or basketballs, if packed in the same manner, would have identical porosities. In fact, relatively fine-grained materials, such as silt, may have higher porosities than such seemingly open material as gravel. Among the most highly porous materials are newly deposited muds, such as those of the Mississippi Delta. They may actually reach the incredible value of 80 to 90 per cent porosity—they contain so much water that the individual particles scarcely touch one another. Quicksand is another example. The porosity of most materials, however, is less than 15 to 20 per cent. Various degrees of rock porosity are shown in Figure 15-4.

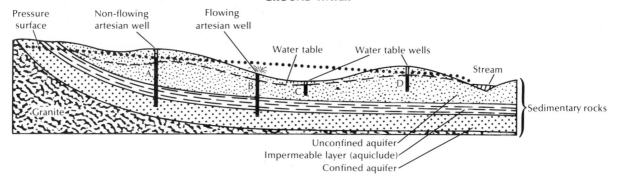

Fig. 15-5. The two main kinds of aquifers and the geologic setting necessary for artesian wells and an aquiclude.

Permeability

Permeability is a measure of the capability of a rock to transmit a liquid through it. Therefore the actual size of the openings in a rock is much more important than the percentage of open space. For example, a silt or clay may have a higher porosity than a gravel, but since the void spaces are so small the permeability is less. Large void spaces obviously are more permeable than small ones. Also important are connections between the openings. If the pore spaces in a rock do not join, the water will not flow, no matter how large those spaces may be.

Aquifers

Not all rocks are equally permeable, nor do they all have equal water-holding capacities. A layer such as a permeable, highly porous sandstone not only may be able to hold much more water than its enclosing rocks, but it also may provide a route along which ground water moves with relative freedom. Such a favorable layer that readily yields water to a well is called an *aquifer.* In contrast, a layer that is too impermeable or too tight to accept water, such as one high in silt and clay, is appropriately called an *aquiclude*.

There are two common kinds of aquifers (Fig. 15-5). One is an *unconfined aquifer,* which may be nothing more than a surficial layer of permeable sand or gravel. The other is a *con-fined aquifer,* a layer of permeable sandstone enclosed between layers of impermeable rock. Typically, a sandstone aquifer may crop out in a band paralleling a mountain front, dip below the adjacent plain, and flatten as it extends farther away from the mountain. Such an aquifer, called the Dakota Sandstone, dips east of the Rocky Mountains and under the Great Plains portion of the Dakotas and Colorado (Fig. 15-6). The first wells were drilled into that sandstone in the 1880s, and since then it has yielded a prodigious quantity of water.

Water wells

Wells are constructed to tap such underground supplies of water as those described. In an unconfined aquifer the water level in the well lies at the water table (see Fig. 15-5, wells C and D). The water level varies, however, due to precipitation patterns as well as to pumping practices. Those who have lived on a ranch dependent on a well for irrigation water, for example, are fully conscious of the fact that when the well is pumped the water level in it drops. A short time after pumping ceases, the water rises, although not always to the same level, should the amount of water removed be exceptionally large.

How much does a single well affect the water table of an entire district? Does the water level in adjacent wells rise or fall in concert? The answers to such questions have been established through observation in many localities

over many years. If a well is pumped heavily, and water is taken out of the ground faster than it can be replenished, then the water table is pulled down in the form of an inverted cone centering on the well, known as a *cone of depression*. Obviously the water level in nearby wells will be affected more drastically than in more distant ones. Studies show that the effect of an individual well may be felt by others over distances of as much as 0.4 km (0.25 mi). Hard pumping in many wells causes the rims of individual cones to overlap until the water level of an entire basin is lowered.

Some wells that tap confined aquifers flow at the surface of the ground (Fig. 15-7). They are called *artesian wells* from the Roman province of Artesium, now Artois, in southern France. To almost everyone, the term artesian means a well that flows freely. Yet in practice the word has a more restricted use, and is now applied to a well in which the water is under pressure because a confined aquifer has been penetrated.

Whether or not water reaches the top of the ground depends on the relationship of the *pressure surface* and the shape of the terrain (see Fig. 15-5). The pressure surface is the level to which water rises in a confined or unconfined aquifer. Theoretically, in a confined system it would be equal to the highest point in the aquifer. However, the pressure surface does not coincide with that point because of

Fig. 15-6. Diagram of the Black Hills and the surrounding plains. Water that falls in the mountains enters the Dakota Sandstone, a major water supply, and slowly travels along that confined aquifer to great depths beneath the dry plains. (Adapted, by permission, from Fig. 32.39 in *Physical Geography*, 3rd ed., by Arthur N. Strahler. Copyright © 1969 by John Wiley & Sons, Inc.)

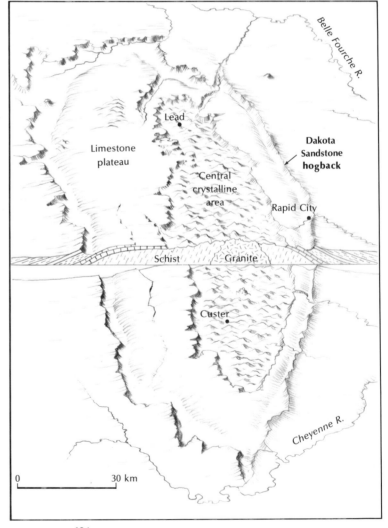

491

the frictional loss of energy that takes place as the water moves through the aquifer; hence the pressure surface lies at a lower level and declines in altitude away from the recharge area.

Artesian wells can be flowing or nonflowing. They flow where the pressure surface is above ground (Fig. 15-5, well B). In contrast, if the pressure surface is below ground, water must be pumped (Fig. 15-5, well A). The pressure in a confined aquifer, however, may cause the water to rise to a higher level than that of the water table in an adjacent, unconfined aquifer.

Should a large number of wells tap an artesian reservoir, the pressure will drop, and the flow will ultimately diminish. Such is the case with the Dakota Sandstone and the other aquifers associated with it. Where 40 to 70 years ago, water poured out of the ground under pressure high enough in some places to operate waterwheels, today, after the drilling of about 10,000 wells, pressure has dropped to the point where many must be pumped, and in flowing wells the yield has greatly diminished.

It might be imagined that under ideal conditions the amount of water withdrawn from an aquifer would be replenished quickly. But replenishment takes place slowly, whereas pumping goes on at a rapid pace. The result is that the water table is lowered and wells must be drilled to greater depths at greater cost.

The problem of overdraft is especially critical in the Southwest, where ground water is critical to the billion-dollar agricultural industry as well as to other industries. In parts of California and Arizona, for example, maximum water-table decline approaches or exceeds 5 m (16 ft) per year. Obviously the drawing of water supplies cannot go on forever, and other sources of water are being sought. In California, for example, an aqueduct has been constructed to transfer surface water from the northern part of the state, where supplies are more plentiful, to the southern counties.

Coastal areas have their own peculiar problems when it comes to pumping overdrafts. Ground water beneath the land is fresh but that beneath the ocean is salty. Fresh water, being

Fig. 15-7. Artesian well flowing in Montana. *J.R. Stacy, USGS*

less dense than salt water, rests on the latter, and the contact between the two dips beneath the land. If fresh ground water is removed at a rate greater than its replenishment, the salty ground water encroaches on the land.

A case in point is an area in Long Island, New York, where some residents began to get more salt water in their drinking water than they had bargained for. A good freshwater aquifer underlies the island, and before extensive ground-water development took place, salty ground water did not enter the aquifer (Fig. 15-8). In earlier days, much of the fresh ground water was returned to the aquifer via cesspools. But as population and the number of cesspools increased, contamination became

such a problem that sewage was dumped into the sea. By that time water was being removed from the system more quickly than it was replenished, and the interface between fresh water and salt water encroached on the island. Deep wells that once tapped fresh water had become brackish.

If the situation is allowed to continue, a very important fresh-water aquifer will be destroyed. What can be done? Perhaps one of the simplest solutions would be to reclaim the sewage and to pump water back underground. In that way the fresh water–salt water interface would be forced seaward, away from the domestic water wells.

Extensive withdrawal of ground water can have other deleterious side effects—for example, the land may subside. Sinking occurs because unconsolidated sediments compact when ground water contained in the sedimentary pores is removed. In some areas the amounts of subsidence are large indeed: up to 2 m (6.5 ft) in the Houston-Galveston area, 1 m (3 ft) at Las Vegas, and 8 m (26 ft) in the southern portion of the Great Valley of California. The areas involved measure from hundreds to thousands of square kilometers.

Such subsidence creates enormous problems. Coastal areas suffer from the encroachment of the sea, and levees have to be built to keep the waters back. Perhaps a larger problem is that water-transport systems that require gravity flow, such as canals and sewers, can be disrupted by the changes in gradient that accompany subsidence. Californians are especially aware of the problem because the very expensive California Aqueduct, which brings water to the southern part of the state, crosses subsiding areas. And, as if that were not enough, compacting sediments can exert enough stress on the well casings to cause their rupture. One way to counteract such sinking is to recharge the aquifer with water.

CONTAMINATING GROUND WATER

Pollutants are a by-product of our society, and so far we have tried almost everything to remove them from sight. For example, many pollutants have been dumped or piped into our streams and oceans. Because so many areas rely heavily on ground water, we must guard against contaminating our aquifers. Recharge areas for aquifers especially should be kept from possible contamination.

A common practice today is to dispose of refuse in sanitary landfills (garbage dumps). What makes the operation "sanitary" is that periodically the refuse is bulldozed over and buried beneath compacted earth. Such landfills can become unsanitary, however, if the water draining through them picks up undesirable materials in solution and comes in contact with the water supply, and so contaminates the latter. Hence, communities should locate their dumps only after the geology and the ground-

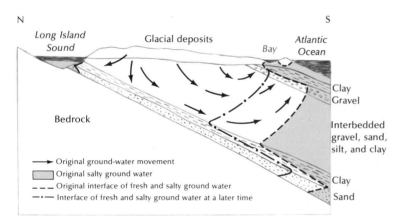

Fig. 15-8. Ground-water conditions on Long Island, New York.

N   S

Long Island Sound   Glacial deposits   Bay   Atlantic Ocean

Bedrock

Clay
Gravel

Interbedded gravel, sand, silt, and clay

Clay
Sand

→ Original ground-water movement
▢ Original salty ground water
- - - Original interface of fresh and salty ground water
-·- Interface of fresh and salty ground water at a later time

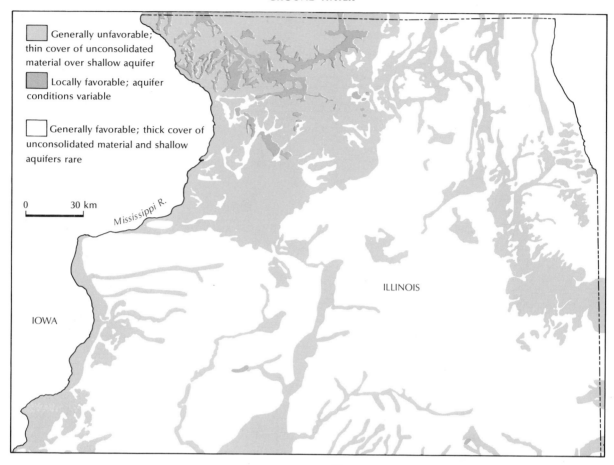

Generally unfavorable; thin cover of unconsolidated material over shallow aquifer

Locally favorable; aquifer conditions variable

Generally favorable; thick cover of unconsolidated material and shallow aquifers rare

0   30 km

Mississippi R.

ILLINOIS

IOWA

Fig. 15-9. Map of northern Illinois indicating the chances of success in locating a suitable site for sanitary landfill.

water conditions of an area have been thoroughly studied. Where sufficient information on the subsurface is available, regional maps have been prepared to aid planners in siting landfills (Fig. 15-9).

In places it is feasible to dispose of liquid waste by pumping it to great depths. Obviously the geology of an area must be studied with great care in order to avoid any possibility of contaminating present or future aquifers. In one case (see Chap. 17), deep pumping of radioactive wastes temporarily made Denver, Colorado, a seismically active region.

PROSPECTING FOR GROUND WATER

"Prospecting" for water is not so strange as it may sound, for ground water is essentially a mineable resource. Radiometric dating has shown that thousands of years may be required to accumulate water in underground reservoirs. Furthermore, in places the permanent lowering of water tables with continued pumping indicates that the annual rainfall is not sufficient to replenish the amount of water removed. Yet the demand for ground water increases, and the need to prospect for new

water supplies is becoming more and more evident to urban planners, irrigation scientists, hydrologists, and others.

The search for ground water has produced its colorful prospectors, although they may not be so picturesque as the grizzled old gold prospector with his burro and pick and gold pan. It has long been customary to use "water witching" to determine the location of a new well. The water witch, or dowser, walks back and forth over the land, holding two ends of a forked stick, or divining rod, keeping the stick horizontal (Fig. 15-10). When the stick, through some magical power, dips sharply downward of its own accord, the dowser announces that this is the place to dig. More often than not the patently unscientific method brings in a good well, at least often enough in humid regions to perpetuate the belief in water witching, even among those who should know better. One reason it is successful is that in many places in the world there is a plentiful supply of ground water no matter where a well happens to be drilled. Another is that some water witches are very observant. They have noticed that the water table is closer to the surface in valleys than on hilltops, or that certain types of vegetation are indicators of water. Even though scientists condemn the practice, it is estimated that about 25,000 water witches are at work in the United States.

Much more accurate methods, however, are now used to find the vast quantities of water required by people everywhere. The developing countries especially—many of which are in arid regions where water must be found before agricultural production can be increased— require reliable and accurate prospecting.

There are several direct methods of searching for water. Mapping the rock units in an area will show potential aquifers, as will the records of existing wells. Geophysical methods used in petroleum exploration can also be used to help locate aquifers (for example, the seismic methods described in Chap. 17).

Prospecting for ground water in sedimentary rocks is not too difficult because the positions of the rocks at depth are predictable. Prospecting in the glaciated part of the mid-continent, however, is much more difficult. In places the glacial tills of low permeability are interlayered with permeable stream deposits consisting of sands and gravels. The latter are good ground-water sources, but they vary in thickness over short distances and do not form extensive blanket deposits as some older sandstones do (e.g., the Dakota Sandstone in Fig. 15-6). It is very difficult to predict where they will occur.

Fig. 15-10. Seventeenth-century water witch at work. From Pierre de Vallemont La physique occulte, Paris, 1663. *Rare Book Division, New York Public Library*

495

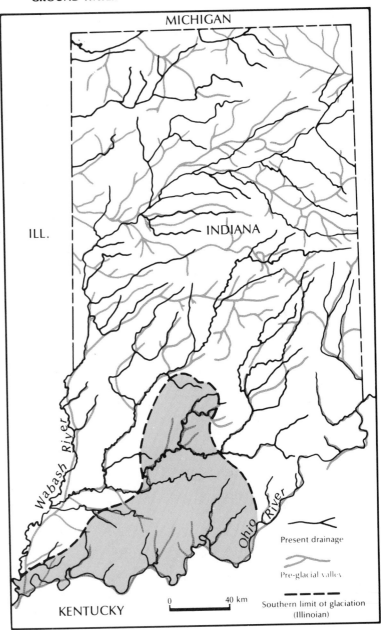

Fig. 15-11. Map of Indiana showing pre-glacial valleys, the southern glacial boundary, and the present drainage system. Because glacial deposits cover much of the state, extensive drilling is required to locate the aquifers contained in the buried valleys.

In places the gravels of pre-glacial valleys buried beneath the fills are sought as ground-water reservoirs (Fig. 15-11).

Once aquifers are located we can call upon computers to help us decide how to develop the resource most efficiently. Computer models for an entire ground-water system can simulate the flow of the water, the conditions at each well, the amount of water being removed, and the actual water-table level at any time. They can also predict the effects of future withdrawals and help to plan the distribution of wells and the timing of water removals. In some areas efficient management

may require the collection and storage of rain-fall runoff in huge ground-water reservoirs to prevent excessive loss through surface evaporation. In others it may be economically feasible to recharge a ground-water system by pumping into it water brought from a distant source.

GEOLOGIC ROLE OF GROUND WATER

Water in the ground does work of geologic significance comparable in many ways to the more visible achievements of rivers, glaciers, lakes, and the sea on the earth's surface.

A cementing agent

Among the more significant accomplishments of ground water is that of providing the means by which the various natural cements, such as calcite ($CaCO_3$), silica (SiO_2), and iron oxide (Fe_2O_3), are introduced into the pore spaces of unconsolidated sediments. Such cements are reasonably soluble substances and may be dissolved from rock or soil layers by water when it starts its journey underground. Later, when the saturation is sufficiently high, and temperature and pressure relationships are right, those substances may be precipitated out of solution. Gradually, as they are deposited on the surface of individual grains, much as scale is deposited in a kettle or hot-water heater, they bind the grains together. In that way, pore spaces are drastically reduced in size until finally they may become sealed off almost completely, with an accompanying fall-off in permeability.

Underground caverns

Ground water plays a unique role in areas underlain by rocks that are readily soluble in water. Limestone, marble, gypsum, salt, and other evaporite deposits are examples of such rocks, and when they dissolve and slowly waste away, large hidden caves are formed. The Carlsbad Caverns in New Mexico, Mammoth Cave in Kentucky, and the caves decorated by Stone Age peoples near Lascaux in southern France are renowned. And there are scores of others in many parts of the world. Their dark, silent recesses have intrigued explorers since the earliest days, and even today there are few countries or states without active speleological groups within their borders.

The origin of limestone caves has long been debated and is far from settled. The crux of the debate is whether the solution responsible for the removal of thousands upon thousands of cubic meters of solid rock in some of the larger caverns occurred above or below the water table. Two factors give rise to the argument. On the one hand, in sections of caverns above the water table the leading present-day process appears to be deposition rather than solution. At least deposition is the process responsible for making stalactites and stalagmites. On the other hand, water below the water table is often saturated with lime already; it cannot pick up any more and thus solution stops. A continuing supply of circulating water with a low lime content is called for, seemingly an unlikely situation in a region underlain dominantly by limestone.

A further complicating factor is that many caverns include deposits of clay, silt, and even gravel, which has led some geologists to conclude that those caves were eroded, at least in part, by subterranean streams. Such rivers are fairly common in limestone terrains such as those in Indiana and Kentucky.

A theory that appears to be applicable to the Carlsbad Caverns is that caves were formed by solution at a time when the water table stood higher than it does now. As a result of canyon-cutting by nearby streams, the water table was lowered and passageways made by solution along joints and bedding planes were opened to the air. It was possible then for such distinctive features of the cave world as stalactites, stalagmites, columns, and ribbons and sheets of travertine ($CaCO_3$) to grow.

Few geologic phenomena arouse more curiosity than the strange, in fact, eerie patterns made by dripping water in the timeless darkness of caverns underground. Most familiar of

Fig. 15-12 A. (above) Stalactites hanging from the roof of a cave, Lehman Caves, Nevada. *Hal Roth* **B.** (opposite) Stalagmites growing upward from the floor of a cave, Carlsbad Caverns National Park, New Mexico. The stalagmite in the center has joined a stalactite to form a pillar. *National Park Service*

those to visitors to the great number of national, state, and privately controlled caves are the icicle-like pendants of *travertine* hanging down from the cave roof (Fig. 15-12A). They are *stalactites,* and they normally form where dripping water seeps from the rocks above the cave. When the water reaches the air of the cave some of the carbon dioxide contained in solution escapes, and calcium carbonate ($CaCO_3$) is precipitated. Also, if some of the water evaporates, a residue of calcium carbonate is deposited. Because the drop of water that hangs suspended momentarily from the

cave roof is likely always to be about the same size, the tiny ring of travertine it leaves nearly always will have the same diameter. Gradually, a series of rings pile up to form a long pendant, customarily with a narrow tube extending through its full length. Seldom, though, is such perfection actually achieved or long maintained. The tube may become plugged, the amount of water may vary, or new holes may break out along the sides rather than at the tip. All such changes lead to the great variation in form that stalactites show.

Stalagmites (Fig. 15-12B) are deposits built

Fig. 15-13. Karst topography near Kwei-Lin, Kwangsi Province, China, creates a surrealistic landscape. A probable explanation for the strange towers is that the limestone terrain may have been lowered by erosion, leaving the former caves exposed as valleys and the solid rock between the caves as steep hills. *H.E. Malde, USGS*

upward from the cave floor, and characteristically they grow below stalactites. When a drop of water falls from the rim of a stalactite it may lose some of its carbon dioxide, or its dissolved lime may become concentrated through evaporation, with the result that more lime is deposited. Thus a counterpart accumulation of lime gradually builds up from the floor of the cave to oppose the stalactite growing downward from the roof. Stalagmites, unlike stalactites, do not have a central tube, and, because they are built up by the water that spatters over their surface, they usually are thicker and more diversified in shape. With two such structures growing toward one another, if the initial distance separating them is not too great, and if an adequate rate of growth is maintained, stalactites and stalagmites eventually meet and fuse to form a pillar.

Other cave deposits may take on a wide variety of shapes—fluted, columnar, or sheetlike masses—that are often enhanced by indirect lighting effects in the more frequently visited caves.

Karst

The landscape that may develop in a region underlain by limestone differs in a multitude of ways from one characteristic of less soluble rocks. Parts of China display such a landscape well (Fig. 15-13). But perhaps the best-known region of that type is the Karst, the portion of Yugoslavia bordering the Adriatic, the Dalmatian Coast. It is one of the picturesque coasts in the world, with the sea penetrating far inland in long, fiord-like inlets. They are separated by barren, whitish limestone ridges and islands that contrast vividly with the wine-dark waters of the sea—to use the 2700-year-old imagery of Homer. *Karst*, in fact, is the name given to similar topography formed by the solution of rocks everywhere.

Dalmatia is one of the historic coasts of Europe. The once-forested slopes of the now barren hills of what was then known as Illyricum provided timbers for the galleys of Rome, and later for the wide-ranging vessels of the Venetian Maritime Republic. Today it is a harsh, stony land, and it is difficult to visualize the widespread forests and soils that once mantled its slopes before destruction by overcutting, overgrazing, and active erosion.

The region has one of the heavier rainfalls of Europe, yet it is strikingly devoid of surface streams. Limestone—if joints and other fractures abound—is so permeable that rainwater

Fig. 15-14. Sinkhole topography in Indiana. *John S. Shelton*

sinks rapidly into the ground. A stream will flow for short distances, disappear underground, and then reappear several kilometers away as a river emerging full-born from a giant spring. Such a limestone terrain is pocked with large numbers of closed depressions, some large, some small. Commonly the depressions are floored with clay, and that thin accumulation of reddish soil is likely to be all that is available for agriculture. In Yugoslavia the larger depressions may be several kilometers across—large enough at any rate to shelter a village and its surrounding patchwork of fields. The origin of the large depressions is uncertain. While they are partly caused by removal of material by solution, they also appear to be the result of the folding and faulting of the underlying limestone.

Smaller, closed depressions in Yugoslavia and elsewhere are almost certainly caused by solution. Some of them extend downward into the earth through near-vertical shafts, which commonly lead to deep caverns. In North America such solution pits are called *sinkholes,* and some may hold small lakes if they are floored with a layer of clay (Fig. 15-14). Should the clay seal be broken, then the lake will drain away through solution channels into the underlying limestone.

Sometimes sinkholes serve as natural wells. Their steep sides extend downward for scores or even hundreds of meters, until they intercept the water table which stands as a pool of water, somber and green, at the bottom. Renowned examples of such formation are the *cenotes* of Yucatan. Mexico's Yucatan Peninsula is a nearly level limestone plain, devoid of surface streams because the rainwater sinks almost immediately into the ground. When the peninsula was the site of the Mayan Empire, the dense agglomerations of people at such cities as Chichén-Itzá were dependent upon so slender a supply of water as the dank fluid at the bottom of a limestone sink. No wonder that, to preserve that tenuous link with survival, a maiden burdened down with bangles and ornaments was ceremoniously hurled into the cenote each year in order to assure a continued supply.

Sinkhole formation is still an active process, as some unfortunate landowners have discovered. It is a vexing geologic hazard common to many areas in Florida and in other parts of the country where it causes damage to homes and building foundations (Fig. 15-15) and makes the maintenance of stable road beds for highways very difficult. Research is now being directed at finding ways to locate areas that are prone to sinkhole formation, so that future damage can be avoided. One method of study is aerial photography with remote-sensing devices. Data thus obtained may reveal thermal and vegetation patterns that could indicate the presence of caverns subject to collapse.

Geysers and hot springs

By far the most spectacular manifestation of ground water is its appearance at the surface in the form of geysers and hot springs. Certainly they are the leading attraction of Yellowstone National Park, and it is a rare household that does not include a member who has seen Old Faithful run through its repertoire. Yellowstone is not the only geyser area in the world; in fact, the extensive one of Iceland gives its name to this sort of aqueous outburst, since all are named for a large Icelandic spring, *geysir* (Fig. 15-16). Another large and touristically attractive geyser region is the Rotorua region of North Island in New Zealand. It is currently being developed as a source of thermal power.

Although the actual process that goes on in an erupting geyser is something of a mystery, enough is known of the physical laws operating that a plausible explanation can be offered. Incidentally, its general elements are much the same as the one advanced by the German chemist Robert Wilhelm Bunsen, whose burner is known to every student of chemistry.

A generally held view is that ground water percolating downward in a geyser area comes in contact at depth with a source of heat. That source may be cooling volcanic rocks, or steam, or other gases given off by magma. Even though the water at the bottom of a tube may be heated to 100°C (212°F) it does not boil

Travertine deposits at Mooney Falls, Havasu Canyon, Arizona. Carbonate-rich stream water fed by springs precipitates travertine along the course of the stream.
Edwin E. Larson

(Above) Green Pool, West Thumb area, Yellowstone National Park. Heated ground water, along with volcanic emanations, rises to the surface, forming geysers and pools. *William C. Bradley* (Below) Seep cave in Jurassic Navajo Sandstone, Glen Canyon, Utah. A spring seeping out along a break in massive sandstone strata can bring about undermining (sapping). *Edwin E. Larson*

Fig. 15-15. Sinkhole in limestone terrain, Florida, formed recently by the collapse of a cavern roof. Such events are rather common; 1000 sinkholes have formed in Shelby County, Alabama, in the last decade alone. *Wide World*

Fig. 15-16. Geyser erupting in Iceland close to the location of the Great Geysir. *Iceland National Tourist Office*

because the boiling point is raised with an increase in pressure. We are more familiar with the opposite effect—the lowering of the boiling temperature in the thin air of high mountain tops to the point where potatoes, for example, do not cook through.

Thus the temperature at the bottom of a column of water may rise above the boiling point at normal atmospheric pressure. However, nothing is likely to happen until all the water is heated to the top of the column, perhaps to the point where it begins to surge, or spill over the rim. Should enough water drain off, then the pressure throughout the column is reduced, with the result that superheated water near the bottom flashes into steam. That action is enough to propel the whole column of water upward, and since a similar pressure reduction and near-instantaneous conversion to steam occurs throughout its length, a mixture of hot water and steam is hurled skyward—in Old Faithful for about 46 m (150 ft). It is the details of how the subterranean geyser reservoir is filled after being blown clear, and how some geysers achieve their remarkable periodicity that have become the center of much of the debate.

The castellated rims, platforms, and particolored deposits surrounding the geysers and hot springs of Yellowstone are especially interesting features to park visitors. In general, there are two kinds of hot-water deposits.

503

Fig. 15-17. Travertine terraces at Mammoth Hot Springs, Yellowstone National Park. Photographed by W. H. Jackson in 1864. *The Metropolitan Museum of Art, Rogers Fund, 1974*

Those deposited directly from mineral-rich geyser water often are composed of silica—supplied in part from the underlying volcanic rock—and are called *siliceous sinter.* They are likely to be grayish colored and to consist of amorphous silica, very much like opal. Limy deposits, made by calcareous algae that can survive in the temperatures of hot springs and pools, are called *travertine* (Fig. 15-17).

Hot springs are more widely distributed over the face of the earth than geysers. There are more than one thousand in the United States, and most of them are located in the montane parts of the Far West. Fundamentally, hot springs are a consequence of ground water coming into contact with a source of heat in the earth's crust. Typically, the source may be volcanic rocks that have not yet lost all of their initial heat. Or it may be *juvenile* water; that is, water freed for the first time by cooling igneous bodies at depth.

Geothermal energy

Naturally occurring areas of hot water and steam are known as geothermal regions, and people have long cast a speculative eye on them and asked, "Are they of any use?" Although the development of such areas is still in its infancy, the answer is "Yes." Such areas are being sought out, for they have a high potential for driving turbines to produce electricity.

The first commercial geothermal venture was in Italy in 1904. Since that time, producing areas have been sited in Japan, the Kamchatka Peninsula of the Soviet Union, North Island of New Zealand, in Mexico, and in Iceland. Such an energy source is particularly appealing to nations like Japan where oil and coal are almost nonexistent, hydroelectric power is limited, and about 70 per cent of its fuels must be imported. In Iceland volcanic steam is used to heat buildings, and steam pipes placed in the fields warm the soil so that crops that ordinarily would not survive in that severe climate can be grown.

In the United States the first and the only successful large-scale exploitation of geothermal energy is at The Geysers in northern California, only 145 km (90 mi) north of San Francisco. In spite of its name, the area has no geysers, but steam rises from hot springs, wells, and fumaroles—vents from which gases

and vapor are emitted. Steam from wells several thousand meters deep is piped to turbines that in 1975 produced 502,000 kilowatts of electricity. Continued expansion of the facility calls for the production of 908,000 kilowatts by 1978.

The producing geothermal areas, as well as those with potential for development, all are in areas of recent volcanism. Magma or hot masses of rock at depth are the primary sources of heat, which is transferred to the surface by rising hot water and steam. Of the water involved in such a system, fully 95 per cent originates at the surface and is circulating; the rest originates at the source of heat in the form of volcanic steam.

Geothermal energy, although locally very important, will not have a great impact on the total energy picture. In 1971, for example, geothermal energy accounted for only 0.08 per cent of the total world electrical capacity. It is estimated that the figure might rise to as much as 10 per cent in the future.

Yet geothermal energy, like most resources, is not without its problems. The hot waters carry materials in solution which, if abundant, can precipitate solids that clog pipes. Or, they might pollute the surrounding surface and ground-water supplies. Thermal pollution or surface subsidence accompanying water withdrawal might also create problems. Still another problem is how to assess the lifetime of a facility. If the energy is used more rapidly than it is produced, it will eventually be depleted—too familiar a story about most natural resources.

SUMMARY

1. Rainwater percolates beneath the surface of the ground, eventually filling the pores to become *ground water*. The latter is an extremely valuable source of water.

2. The top of the ground-water zone is the *water table*, and its configuration generally parallels that of the land surface. Where the water table intersects the land surface, water flows; springs are a common example.

3. *Permeability* of rock is the main property controlling the rate at which ground water flows from one point to another. Rock units that are permeable and hold sufficient quantities of water are called *aquifers*. Porous sandstone beds make some of the best aquifers.

4. Before aquifers are extensively used, they should be studied to avoid certain problems: too-rapid lowering of the water table, land subsidence accompanying excessive withdrawal, encroachment of salty ground water in coastal areas, and pollution of ground water.

5. The geologic role of ground water ranges from the cementation of sand deposits into sandstone to the formation of underground caverns and the stalactites and stalagmites for which many are so famous, to the formation of the peculiar topography called *karst*.

6. Geysers and hot springs also are ground-water phenomena, and in places their energy is harnessed for geothermal power.

SELECTED REFERENCES

Davis, S. N., and De Wiest, R. J. M., 1966, Hydrogeology, John Wiley and Sons, New York.

Jennings, J. N., 1971, Karst, Australian National University Press, Canberra.

Leopold, L. B., 1974, Water, a primer, W. H. Freeman and Co., San Francisco.

Meinzer, O. E., 1939, Ground water in the United States; a summary, U. S. Geological Survey, Water-Supply Paper 836-D.

Muffler, L. J., and White, D. E., 1972, Geothermal energy, The Science Teacher, vol. 39, pp. 40–43.

Poland, J. F., and Davis, G. H., 1969, Land subsidence due to withdrawal of fuels, *in* Reviews in Engineering Geology, D. J. Varnes and G. Kersch, eds.: Geological Society of America, Boulder, Colorado.

Todd, D. K. 1959, Ground water hydrology, John Wiley and Sons, New York.

Fig. 16-1. Sedimentary strata deformed on a large scale; Murdafil, Iran. *Aerofilms, Ltd.*

16

DEFORMATION OF ROCKS AND MOUNTAIN BUILDING

When the crust of the earth is subjected to deformational stress it buckles and breaks, and the rocks become permanently deformed. In places, originally horizontal strata are tilted and folded (Fig. 16-1) and in others the rocks are cracked and often offset along faults (Fig. 16-2). The phase of geology that deals with the deformation of the earth, especially that recorded in ancient rocks, is called *structural geology*.

Geologists are now aware that the rocks in the outer part of the earth are currently being bent, broken, uplifted, and depressed at countless places in response to forces acting deep beneath the surface. The fact that the earth is in dynamic readjustment is brought home to us forcibly every time there is a large earthquake—a tangible indication that stresses in the earth have built to the point at which rocks fracture and suddenly shift.

The rates of deformation are slow, however, so that geologists who have been making observations for only a short moment of earth history, see today what appear to be only minor modifications in the rocks at the earth's surface. Yet, given long periods of geologic time, those nearly imperceptible small-scale changes can accumulate, eventually to bring about deformation on a grand scale. Surface rocks can be broken and separated along faults by as much as tens or even hundreds of kilometers; sedimentary rocks can be crenulated into great and small folds that resemble those of a gigantic rumpled tablecloth; mountain chains of spectacular proportions, like the Himalayas, can be pushed up where none before existed; and continents can be rent asunder and ocean basins formed.

The realization that forces working inside the earth can bring about such drastic changes was a long time dawning. Leonardo da Vinci (1452–1519), the Renaissance genius, observed fossil shells preserved in the rocks of the Tuscan hills of Italy and precociously reached the conclusion that the Apennine Mountains had once been covered by the sea—like the mythical lost city of Atlantis. The same arguments were made later by Nicolaus Steno (1630–87) in Florence, who made the additional observation that since sedimentary rocks were deposited essentially horizontally, those that deviated in

Fig. 16-2. San Andreas Fault, California, a fracture along which considerable movement has occurred. The view is south along Elkhorn scarp; the Carrizo Plain is to the left. The dark sharply defined pattern left of the fault is vegetation along a fence line. *R.E. Wallace, USGS*

angle from horizontal must have been tilted by some force after the sediments had been laid down (Fig. 16-3).

The idea that deformation of rocks is accomplished by the accumulation of small deformational increments over long periods of geologic time is a uniformitarian view. Charles Lyell, as

you recall, was the champion of that idea in the early nineteenth century, and in fact he provided one of the examples that demonstrated the painful slowness of the process. Because he was able to visit many lands, he acquired a world outlook conspicuously lacking in many of his Victorian contemporaries. And few spots

are more favorable than the shores of the Mediterranean he visited in providing an understanding of the rates at which deformation proceeds. Buildings had been constructed along the shores of that nearly tideless sea for millennia and the more durable ones were made of stone, which had withstood the ravages of time and weather in the relatively dry climate remarkably well. Those built close to the shore served as unusually sensitive recorders of changes in the sea level.

Lyell became aware that all three of the surviving columns of the Temple of Jupiter Serapis, not far from Naples (Fig. 16-4), had lines encircling them about 7 m (23 ft) above sea level. Below the line each of the columns is riddled with holes bored by shallow-water marine shellfish through a band almost 3 m (10 ft) wide. He concluded that the uppermost line on each column represented a former stand of the sea level. And the holes drilled by the sea-inhabiting mollusks further supported

Fig. 16-3. These tilted strata near Ship Rock, New Mexico were once nearly horizontal. *John S. Shelton*

Fig. 16-4. Temple of Jupiter Serapis, near Naples, Italy, showing the zone of marine shellfish borings made in the columns during submergence.

his belief that the temple was once partially submerged to that level.

The historical record bears out Lyell's conclusion. Apparently the temple was built around the second century B.C., and, although it was constructed on land, it must have started to subside shortly thereafter, because a new floor was built on a fill of about 45 cm (17.5 in.) covering the original mosaic floor, and false bases were built around the columns. At an unknown later date a fall of volcanic ash buried the court of the temple to a depth of about 3 m (10 ft). Continued subsidence of the land eventually allowed the sea to invade the entire structure. The 3-m (10-ft) band on each of the columns where they are pierced by borings of marine organisms represents the depth of water between sea level and what was then the mud- and ash-covered floor of the Mediterranean.

When the land started to rise again is not known. In medieval times the sea extended inland to the bluff behind the temple. Certainly the rise was well under way by A.D. 1503, for

in that year, according to Lyell, Ferdinand and Isabella of Spain granted to the University of Puzzuoli "a portion of land, 'where the sea is drying up' (*che va seccando el mare*)." Also, according to Lyell, the main rise of the land occurred at the time of the historically destructive eruption of Monte Nuovo in A.D. 1638. Then "the sea abandoned a considerable part of the shore, so that fish were taken by the inhabitants; and, among other things Falconi mentions that he saw two springs *in the newly discovered ruins*" (Lyell).

That example, based on observations he made in 1828, demonstrated to Lyell that much of the deformation of the earth goes on relatively slowly, and is a far cry from the catastrophic explanations most of his contemporaries relied upon to account for the more striking elements of the landscape. His book, *The Principles of Geology* (1830), is the first genuine textbook of geology, and it exerted a profound influence on the Victorian world of letters. Tennyson, with many other literary lights of the period, was excited by Lyell's discoveries, as the poet's emotion-charged passages indicate:

There rolls the deep where grew the tree
  O earth, what changes hast thou seen!
  There where the long street roars hath been
The stillness of the central sea.
The hills are shadows, and they flow
  From form to form, and nothing stands;
  They melt like mist, the solid lands,
Like clouds they shape themselves and go.

In Memoriam, 1850

The vertical fall and rise of the land surface Lyell documented near Naples is minor, however, when compared to the subsidence now affecting much of the Netherlands. Everyone knows that the coast of Holland is protected by dikes and that much of the country, including the cities of Rotterdam and Amsterdam, is below sea level. Few realize how unremitting and laborious a struggle must be waged to hold back the sea, because the coastal part of Holland is sinking at the rate of about 21 cm (8 in.) per century. That makes the

achievement of the Dutch all the more remarkable.

The hills along the coast near Los Angeles, California, provide yet another illustration of uplift of a segment of land from the sea.

A person approaching the Palos Verdes Hills, the bold headland that partially shields the harbor of San Pedro in southern California, is impressed by the promontory's seaward slope, which rises above the sea like a cyclopean stairway (see Fig. 13-21). Wave-cut terraces separated from one another by steep cliffs rise in thirteen steps to 396 m (1299 ft) above the sea. They are evidence that the coast in that area has been elevated vertically—nearly 402 m (1319 ft)—so recently geologically that fossil seashells preserved on the flat terraces are identical to ones living today in the nearby ocean.

Another even more impressive set of uplifted wave-cut platforms exist marginal to the coast of Peru. Some of them are 16 to 24 km (10 to 15 mi) wide and are littered with marine shells appearing as fresh as if they had lived in the sea only yesterday.

And as a final example, high on the flanks of the world's loftiest mountain, Everest, are found water-laid rocks containing marine fossils that lived in the sea 60 million years ago.

Rates of movement

What do we actually know about the rate at which the earth's crust is being deformed? The answer is "not a great deal," because accurate surveying records go back only about two centuries, and in many cases the best-surveyed areas are among the more stable parts of the earth. The Coast and Geodetic Survey ran levels in 1906, 1924, and 1944 across Cajon Pass between Victorville and San Fernando in the Transverse Ranges of southern California. The measurements show that the area around the pass where it crosses the San Gabriel–San Bernadino Mountains rose as a gentle arch during those 38 years by about 20 cm (8 in.) or at the rate of about 53 cm (21 in.) per century. As James Gilluly of the U.S. Geological

Fault

Fig. 16-5. Steeply tilted strata adjacent to a fault in Dinosaur National Monument, Utah. The Green River is in the foreground. *Philip Hyde*

Survey points out, though the rates may appear modest indeed, with sufficient time—and time is a quantity not lacking in geology—such a rate yields an uplift of 122 m (400 ft) in 25,000 years for the San Gabriel–San Bernadino Mountains. Mount Everest, at a comparable rate, could have reached its present height in 2 million years. Deformation of the earth need not be a uniform, continuous process, but can occur in abrupt jumps. For example, in 1899, accompanying a severe earthquake in the area centering around Yakutat Bay, Alaska, a large portion of the earth's surface was uplifted as much as

14.4 m (47 ft). And during the 1964 Alaskan earthquake, a maximum uplift of 11.9 m (39 ft) was recorded in the region of Patton Bay.

These examples of the rate of earth movement have all been in a vertical direction, either the land sinking or rising, or, in the case of the Temple of Jupiter Serapis, doing both. Horizontal shifts have been recorded, too. In 1906, roads, fences, and so forth were offset along the San Andreas Fault by as much as 6.4 m (21 ft) almost instantaneously. Repeated surveys also show a gradual creep of points on one side of the fault past corresponding ones

on the other side without any discernible break of the ground surface. Recent geodetic and geophysical measurements along the San Andreas Fault indicate a steady shift of about 2 cm (0.79 in.) per year of one fault block laterally past the other. And as we shall see in Chapter 19, the slow drift of continents across the face of the earth and the opening of ocean basins proceeds at rates of about the same magnitude.

ANCIENT DEFORMATION RECORDED IN ROCKS

One step in the study of an area's structural geology is to describe the geometry and complexity of the folds, fractures, and faults (Fig. 16-5). When the structural features are small,

they can easily be investigated by direct observation. Some structures are so large, however, that they can only be studied piecemeal, and then their evaluation in its entirety requires the making and studying of geologic maps.

The kinds of rocks that best lend themselves to structural study are sediments. Most strata were deposited horizontally, and any deviation from that position or offsets in the continuity of beds indicate deformation. Moreover, strata that possess distinctive rock properties can be traced easily for long distances, enabling study of deformation over wide areas.

If a sedimentary layer is tilted at an angle to horizontal it is said to be a *dipping* bed: the amount of *dip* is the angle between the bed and the horizontal surface as determined by a leveling instrument.

Fig. 16-6. Strike and dip of a tilted bed. The strike is represented by the line of intersection of the sea surface with the dipping bed. The dip is perpendicular to the strike and represents the path a marble would take as it rolls down the bed. *William Estavillo*

The line of intersection made by the dipping stratum and the horizontal surface is called the *strike* of the bed. The dip is measured in a plane that is perpendicular to the strike.

Perhaps the best way to visualize the dip and strike of an inclined stratum is to imagine a single dipping bed of sandstone that projects out of the calm sea (Fig. 16-6). The intersection of the sea surface (a horizontal surface) with the bed is the strike, and the dip is the amount of tip of the bed in a plane perpendicular to the sea-level trace, measured downward from the horizontal. A round pebble placed on the surface of the tilted bed would roll down the dip direction of the bed: the line of dip thus possesses a directional quality.

The two ends of the strike line point in opposite compass directions. That is, if one end of the line points toward the northeast, the other end points southwesterly, and those directions can be measured by means of a compass. By convention we only record the end of the strike line that makes an acute angle with the true north direction. A strike line measured and denoted by N 30 E, would be one in which one end points to a position 30° east of north, and, of course, the other end 30° west of south. On a map the geologist would draw a short segment with that orientation, as shown in Figure 16-7.

The dip can be represented on a map by a short line drawn in the direction of dip, at right angles to the strike line and attached to it. Consider the dip of such a bed to be 40° and the dip direction toward the southeast. In a field note-

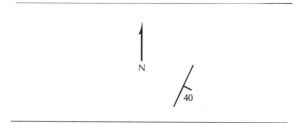

Fig. 16-8. Map symbol for a bed striking N 30 E and dipping 40° toward the southeast.

book the dip and the strike direction are written as follows: N30E40SE, and the map representation is that shown in Figure 16-8.

Vertically dipping beds are denoted on the map thusly: ―+―. Horizontal beds, of course, lack both a dip and a definable strike and field geologists use the symbol ⊕ to denote that situation on a map. If a bed is rotated past 90°, that is, is *overturned,* the geologist measures the dip and strike as before but denotes on the map with the symbol ⌐┐ that the bed is overturned.

By measuring the dips and strikes of sedimentary strata and plotting them on a map it is generally possible to determine the large-scale fold geometry and offsets or abrupt changes in the geometry that might indicate faulting.

Folds

Some structural processes result in the warping or folding of the rocks near the earth's surface. In sedimentary rocks the fold structures generally are more easily determinable, and they will appear as small to large-scale wave-like crenulations in the strata (Fig. 16-9).

The folds resembling wave crests are called *anticlines*—from the Greek, to be inclined against itself, and those resembling wave troughs are called *synclines*—also from the Greek, to lean together (Figs. 16-9 and 16-10). The sides of such folds are called *limbs,* or *flanks,* and a line drawn along the points of maximum curvature of each bed is termed the

Fig. 16-7. Map symbol for the strike of a dipping bed, in this case N 30 E. Notice that the two ends of the line point in opposite compass directions.

Fig. 16-9. Large-scale crenulations in strata; anticlinal and synclinal folds of the Grande Chartreuse north of Grenoble in the French Alps. *Swissair-Photo*

axis. It is apparent that in open and some asymmetrical anticlines the limbs dip away from the axis, whereas in the same kinds of synclines they dip toward it (Fig. 16-11). The axis of most anticlines would correspond to the ridge line of a roof, for example, just as the axis of most synclines would correspond to the keel of a ship.

A plane connecting the axial lines in succes-

sive beds in a fold is called the *axial plane.* It can be flat or curved, and vertical to flat lying (Fig. 16-12), depending on the straightness of the axis and the difference in dip angle between the two flanks. Small-scale recumbent folds are shown in Figure 16-13. Large-scale recumbent anticlinal folds, largely containing deformed metamorphic and igneous rocks in their centers, have been called *nappes* and are

Fig. 16-10. Anticlines and synclines near the Sullivan River in the southern Rockies of British Columbia.
Geological Survey of Canada

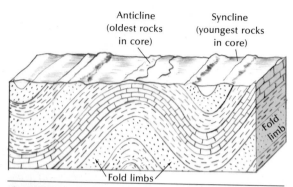

Anticline
(oldest rocks
in core)

Syncline
(youngest rocks
in core)

Fold
limb

Fold limbs

Fig. 16-11. An open symmetrical anticline and syncline. As a result of erosion, as shown, the oldest rocks will be found in the core of an anticline and the youngest in the core of a syncline.

Fig. 16-12. Different types of anticlinal folds, showing in each case the relation of the axial plane and the axis to the fold geometry.

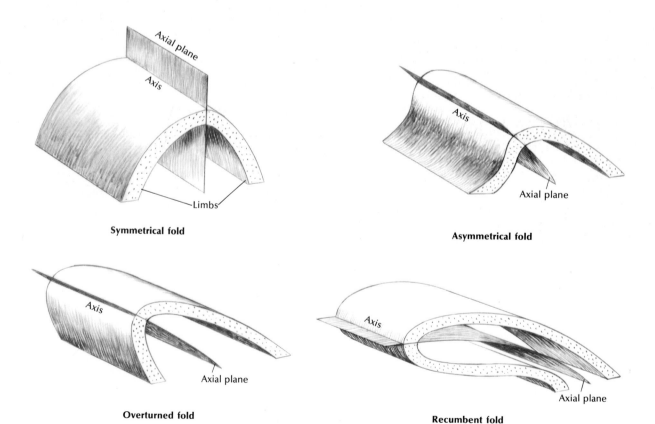

Axial plane

Axis

Limbs

Symmetrical fold

Axis

Axial plane

Asymmetrical fold

Axis

Axial plane

Overturned fold

Axis

Axial plane

Recumbent fold

Fig. 16-13. Recumbent folds in slate folded about 350 million years ago. *Geological Survey of Canada*

Fig. 16-14. Symmetrical folds with horizontal axes (**A**) before erosion and (**B**) after erosion. Notice how the resistant layers form parallel ridges after erosion.

Fig. 16-15. Symmetrical plunging folds (**A**) before erosion and (**B**) after erosion. Notice how the more resistant layers form series of arcuate, V–shaped ridges after erosion.

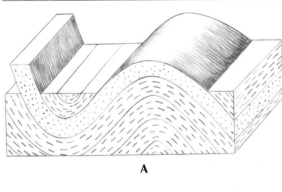

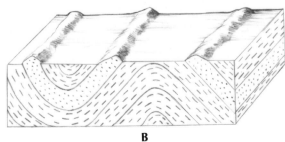

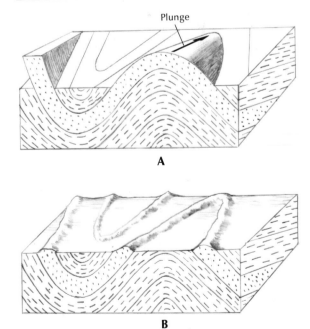

conspicuously developed in the higher reaches of the Alps.

Deep mines and oil wells show that in deformed sedimentary rocks the geometry of a fold deep underground may be very different from its geometry at the surface. Such structures are called *disharmonic folds*. Trying to second-guess what a fold at the surface will do at depth is the sort of gamble that makes the work of a geologist such an intellectually stimulating challenge.

Folds in which the axes are horizontal are said to be *non-plunging,* and they could be represented by a piece of tin roofing placed on a table (Fig. 16-14). In most cases, the axes dip at an angle to the horizontal and the folds plunge into the ground (Figs. 16-15 and 16-16). Many folds plunge in one direction at one end of the fold and in the other direction at the other end. They are said to be *doubly-plunging folds* (Figs. 16-17 and 16-18).

Careful study of the preceding diagrams and photographs of folds should bring out the following relationships more clearly than many paragraphs would:

1. Anticlines plunge in the direction that their sides converge (come together).
2. Synclines plunge in the directions that their sides diverge (spread apart).

Fig. 16-16. Plunging anticline in sediments along the eastern flank of the Front Range, near Fort Collins, Colorado. Notice the similarity of the eroded pattern of this fold to the diagram in Figure 16-15B. *Edwin E. Larson*

3. Thus in anticlines it can be seen that the rocks become progressively older toward the axis; in synclines, they are younger.
4. The dip measured on the axis also gives a measure of the amount the fold plunges. Note also that the strike of the beds where they cross the axis is at right angles to the axis in both the anticline and the syncline.

Anticlines that lack a well-defined elongation and that plunge from a point nearly equally in all directions are called *domes* (Fig. 16-19), and the corresponding synclinal structures are called *basins*.

One other fold to be described is the *monocline* (from the Greek word for "one inclination"), a one-limbed structure with horizontal strata on either side of steeper-dipping strata (Fig. 16-20).

Joints

Nearly all rocks visible on the earth's surface are cut by cracks and fractures (Fig. 16-21). They are so commonplace that few people grant them more than casual notice. A minor number of the cracks, or *joints*, as they are called, are the result of cracking of thin sheets of igneous rocks as they crystallize and cool. When igneous rocks are broken into a close-spaced set of prisms, the pattern is called *columnar jointing*, as we learned in the study of igneous rocks in Chapter 4.

Most joints are a result of deformation of the rocks, although the exact details of the relationship are not at all clear. Many occur in closely spaced subparallel alignment—called a *joint set*—and not too uncommonly, three such sets are developed that are mutually perpendicular (Fig. 16-22). Jointing is less likely in laminated or foliated rocks than it is in the texturally massive varieties.

FAULTS

Faults are fractures along which significant movement has occurred. In some cases, the offset amounts to a centimeter or less; in

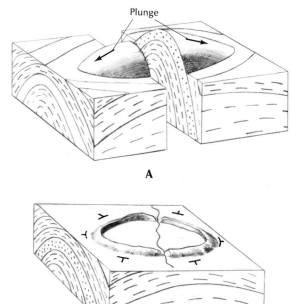

A

B

Fig. 16-17. Doubly–plunging anticline (**A**) before erosion and (**B**) after erosion. Dip and strike symbols show attitudes of the layers.

others, it may amount to as much as 200 km (124 mi).

Because fault movement involves the sliding of one block past another, many fault surfaces are smoothly polished and grooved (Fig. 16-23). These features, called *slickensides*, provide one clue to the direction of fault slippage, but they record only the most recent event. Previous events may have involved slippage in totally different directions, but the slickensides associated with them in general will have been obliterated by the most recent slippage.

Often too, the rocks adjacent to a fault will have been pulverized or ground to bits forming a clayey, soft material called *fault gouge*. In some instances the rocks in the fault zone may be broken and sheared, creating a *fault-zone breccia*.

If fault movement occurs close to the earth's surface, the surface itself may be broken and

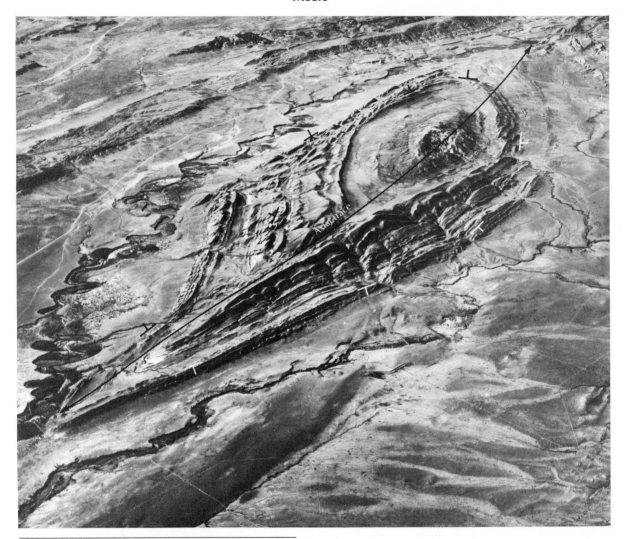

Fig. 16-18. Little Maverick Dome, northwest of Riverton, Wyoming. A doubly–plunging anticline, or elongate dome, showing the trace of the doubly–plunging axis and appropriate dip-and-strike symbols. *John S. Shelton*

offset. The resulting low linear cliff is called a *fault scarp* (Fig. 16-24). Quite a number of such low, fault-induced cliffs have appeared suddenly in widely scattered parts of the world where earthquakes are especially prevalent. Such historically recent scarps are especially common in Japan, New Zealand, and India, and in the contiguous United States in California and Nevada.

The tremendous eastern wall of the Sierra Nevada in the Mount Whitney sector also is often cited as an example of a fault scarp (Fig. 16-25). Although it is true that the eastern front of the range is the result of fault action, the escarpment is not actually an exposed fault plane. Steep as the mountain front appears when viewed full face, when looked at from the side the average inclination of most of the

Fig. 16-19. (opposite, above) Upheaval Dome, Canyonlands National Park, Utah; an almost perfect example of a dome structure.
Eros Data Center, USGS

Fig. 16-20. (opposite, below) Looking south along the monocline that forms the west flank of Raplee Anticline, Colorado Plateau, southern Utah. The upland beds to the left are part of Monument Upwarp. In the foreground is the San Juan River. *Don L. Baars*

ridges can be seen to be about 25°, while exposed faults within the range have average dips of 60 or 70°. It is apparent that the face of the range is a fault-controlled erosional landform. The high part of the Sierra Nevada has been uplifted along faults, but weathering and erosion have reduced the original steep fault slope from nearly 70° back to about 25°.

Since fault planes have all the geometrical attributes of a dipping stratum, we can describe their orientation in space by means of strike-and-dip notation. For example, a particular fault plane might trend N 45 W and dip 52° toward the southwest.

Fig. 16-21. Closely spaced joints in nearly horizontal sandstone strata near Mexican Hat, Colorado Plateau, Arizona.
Edwin E. Larson

Fig. 16-22. Three sets of joints nearly at right angles (two are perpendicular to the rock face, and the third is parallel to it), in granitic rocks on the east face of Mount McAdie in the Sierra Nevada, California. *Tom Ross*

Fig. 16-23. Slickensided fault surface at Plum Ridge, just north of Klamath Falls, Oregon. *G.K. Gilbert, USGS*

Fig. 16-24. Looking northwest along trace of Red Canyon fault scarp, which was produced suddenly during 1959 Hebgen Lake (Montana) earthquake; Grayling Creek is in the foreground. Notice the disrupted road. *I.J. Witkind, USGS*

Fig. 16-25. The Sierra Nevada, California, showing the accordant summit of the range and the steep eastern face. Mount Whitney, the highest peak within the contiguous United States, is in the left background. Although faulting has caused uplift of the mountain block, the steep fault surface has been modified and flattened by erosion. *Roland von Huene*

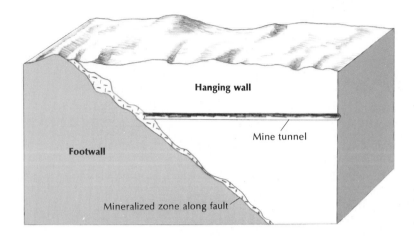

Hanging wall

Mine tunnel

Footwall

Mineralized zone along fault

Fig. 16-26. Relation of a hanging wall and foot wall along a fault.

Fig. 16-27. Two possible displacements on a fault leading to identical offset of dipping stratum. **A.** Unfaulted dipping bed, with incipient fault plane shown. **B.** and **C.** Two types of fault motion that could produce offset. **D.** Faulted dipping bed as observed in field.

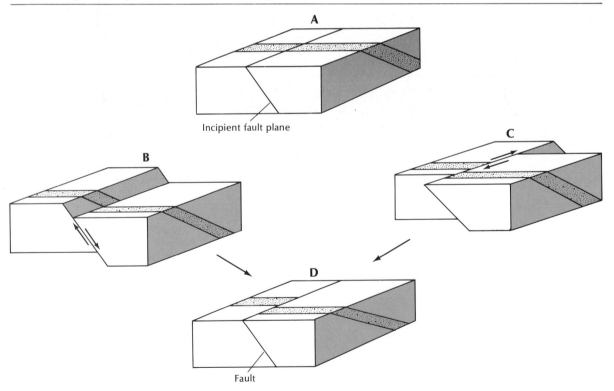

A

Incipient fault plane

B

C

D

Fault

Many faults, because they provide avenues for circulation of underground fluids, have been mineralized by ore-bearing solutions. Miners have long been aware of that fact, and numerous tunnels and shafts have been dug to reach the riches trapped in the fault zones. A miner working a mineralized fault zone where the fault dips at a moderate angle literally stands on the block on one side of the fault, with the face of the other block hanging above his head. The block beneath his feet is called the *foot wall* and that above his head the *hanging wall* (Fig. 16-26). In the description of faults that follows we will use those two common terms.

Apparent relative movement

In most cases it is almost impossible to tell the exact direction of movement of the two blocks on either side of a fault from observation of the

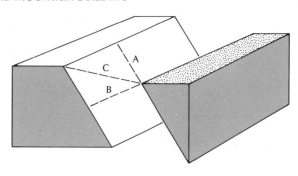

Fig. 16-28. The types of slip on a fault plane: A. Dip slip; B. Strike slip; and C. Oblique slip.

Fig. 16-29. Types of faults. **A.** Initial block of stratified rock, **B.** Normal fault, **C.** Reverse fault, and **D.** Strike-slip fault.

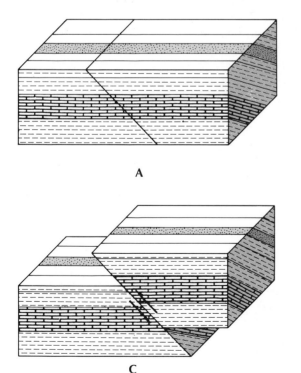

A

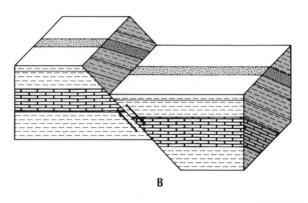

B

C

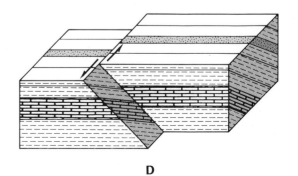

D

rocks alone. In Figure 16-27, for example, the final pattern (16-27D) which is all we see, could have resulted from (1) slippage of block A upward, in relation to block B—as seen in Figure 16-27B, (2) the horizontal slippage of block A relatively past block B (Fig. 16-27C), or slippage at an oblique angle such that some vertical and some horizontal displacement was involved. In most cases, then, we can determine only an apparent relative movement.

Known directions of relative displacement—normal, reverse, strike-slip, and overthrust faults

In those cases where a line, as for example a fold axis, or volume of small dimensions is cut by a fault, the exact nature of the relative displacement can be determined.

The term used to denote the actual relative displacement of the once adjacent points, measured in the plane of the fault, is the *slip*. Figure 16-28 illustrates the three kinds of slip. *Dip slip* (A) is movement on the fault parallel to the dip; *strike slip* (B) is a measure of displacement parallel to the fault strike; and *oblique slip* (C) is movement at an angle to the dip and strike of the fault. A fault in which the displacement is primarily dip slip and in which the foot wall has moved up relative to the hanging wall is called a *normal fault* (Fig. 16-29B). If the displacement is primarily dip slip but the foot wall has moved down relative to the hanging wall, the fault is called a *reverse*, or *thrust*, fault (Fig. 16-29C). Depending on the dip angle of the fault, it can be described as a low- or high-angle fault: if less than 45°, *low angle;* if more

than 45°, *high angle*. If the dip angle on the thrust fault plane is less than 10°, it is called an *overthrust fault*. If the slip is primarily parallel to the strike of the fault it is called a *strike-slip fault* (Fig. 16-29D).

High-angle normal and reverse faults Many faults show displacements that are primarily dip slip in nature. Along some of the larger faults of that type, displacements up to 2 km (1 mi) or more have occurred.

If a block rises relatively above the blocks on either side of it the fault-bounded upland block is called a *horst* (from the German word meaning, among other things, a ridge). Relatively down-dropped linear blocks between horsts are called *graben*, after the German word for "trough or trench" (Fig. 16-30). A renowned example is the Rhine graben, which is followed by the Rhine River from Basel to Mainz. There the valley is linear and trench-like with walls that are fault scarps marginal to highlands—the Schwarzwald to the east in Germany and the Vosges to the west in France.

An impressive set of graben is visible as the nearly continuous line of fault-bordered troughs that extend the length of Africa and part of the Middle East, from Mozambique to the Dead Sea. Those great down-dropped segments of the earth's crust are far too large to be referred to as trenches, and for that reason the frequently used term *rift valley* is appropriate.

One of the better-known rift valleys holds Lake Tanganyika, whose length is 672 km (418 mi) and whose width ranges from 32 to 64 km (20 to 40 mi). The water surface is 771 m (2530 ft) above sea level, but the bottom is

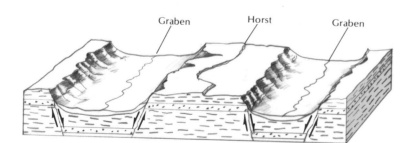

Graben   Horst   Graben

Fig. 16-30. Structural and physiographic relations of a horst and adjacent graben, a feature found in many parts of the world.

Fig. 16-31. The trace of the right-lateral San Andreas Fault, Carrizo Plain, California, *A.M. Basset*

506 m (1660 ft) below. Lake Albert Nyanza, possibly the most impressive, is set in a trough whose eastern wall is 305 to 457 m (1000 to 1499 ft) high and whose mountainous western slope rises to 2439 m (8000 ft).

Strike-slip faults In most strike-slip faults, which primarily involve displacement parallel to a fault strike, the dip of the fault plane is nearly vertical. If you place two bricks side by side on a table and move them backward and forward, you will have reconstructed the movement of a typical strike-slip fault.

Most of the world's faults that show the greatest amount of movement, measurable in

places up to scores of kilometers, are strike slip in nature, as, for example, the San Andreas Fault.

Since the 1906 California earthquake, the San Andreas Fault and others like it have been studied in increasing detail, and such faults have been recognized in parts of the world as distant from one another as California, Canada, Scotland, Switzerland, the Philippines, Japan, and New Zealand. With such a wide range of occurrences in so many lands, it is not surprising that a widely varied terminology has been applied to strike-slip faults. British geologists are likely to call them *transcurrent faults*, or *wrench faults*. In the United States commonly

they are called *lateral faults;* thus we can speak of *left-lateral* faults for those in which the side opposite a person facing the fault has apparently been displaced to the left, and *right-lateral* if the movement apparently has been to the right (Fig. 16-31).

From the statements made thus far, it should be apparent that the San Andreas is one of the world's more renowned examples of a strike-slip fault. The map (Fig. 16-32), showing the trace of the fault, demonstrates its great length, roughly 960 km (597 mi) from where it cuts the coast north of San Francisco until its several branches disappear beneath the waters of the Gulf of California. The map also shows that it is not a single fracture at its southern end but rather is a complex system of faults.

Theoretically, in a fault system as great as the San Andreas, it would seem reasonable that the distance rocks are offset on either side of the fault would be easy to determine. Actually, it

Fig. 16-32. Map of California showing the trace of the San Andreas Fault. Notice how the fault splays and splits in its southern portion.

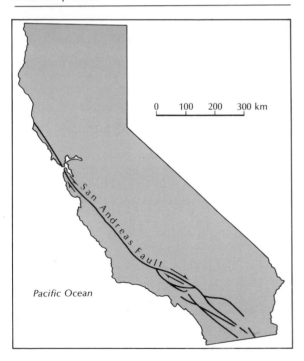

is very difficult. Part of the problem arises from the fact that the strike of the rocks cut by the San Andreas is nearly the same as that of the fault itself, so that sharply defined intersections are lacking. Also, ancient displacements along the fault are concealed beneath more recent deposits, which in a sense have covered over old scars. A tentative estimate by the geologists M. L. Hill and T. W. Dibblee would place the amount of movement along the fault complex at 192 km (119 mi) in the last 60 million years and possibly 560 km (348 mi) since 120 million years ago—roughly the age of the oldest rocks marginal to the fault. Another well-documented example of a strike-slip fault along which considerable lateral movement has occurred is the Great Glen Fault in Scotland. Almost everyone who has ever looked at a map of Scotland is likely to wonder about the nearly continuous vale that extends across the middle all the way from the North Sea to the Atlantic Ocean. The valley is also the site of Loch Ness, home of the legendary sea serpent. Fortunately, the rocks and structures on either side of the Great Glen Fault can be matched up with some confidence, and according to the British geologist W. Q. Kennedy, they indicate a strike-slip movement of about 104 km (65 mi), with the northern part of the Scottish Highlands being displaced relatively southwestward.

Overthrust faults Geologists struggling to interpret the perplexing relationships of the Northwest Highlands of Scotland early in the nineteenth century—around the time of the War of 1812—were puzzled to find sandstone, shale, and limestone interbedded in what appeared to be a normal sequence with gneiss and schist. Even such a noted person as Charles Lyell accepted it as a conformable succession of seemingly related rocks. Fortunately for Lyell's reputation, no one in that day was really certain of the true nature of metamorphic rocks.

However, with continued field work throughout the nineteenth century, the complacent state of mind brought about by that miscon-

Fig. 16-33. McDermitt Lake and Mounts Gould, Pollack, and Allen in Glacier National Park, Montana. Bailey Willis demonstrated the existence of large low-angle overthrust sheets in this region. *Bailey Willis, USGS*

ception was increasingly disturbed by (1) the awareness that metamorphic rocks are not normally interbedded with sedimentary rocks, (2) the fact that some of the contacts separating different rock units are not normal depositional surfaces but are faults, and (3) the discovery of rocks containing fossils of earlier geologic periods which were found resting on rocks containing fossils known to have lived later in geologic time.

In a region as renowned for disputation as Scotland, it is no wonder that the "Highland Controversy" is one of the more celebrated intellectual conflicts in the history of geology. By 1861, the view gained acceptance that perhaps those aberrant geologic relationships were the result of the gliding of great sheets of rock over one another in a fashion similar to what had been found in 1849 to be part of the internal structure of the Alps. As a result of nearly a century of effort, the existence of large-scale, almost horizontal displacement of many miles' extent was demonstrated to the satisfaction of most. There in the Scottish High-

lands the presence of at least three rock slices was established, and they were shown to have moved from the southeast toward the northwest carrying regionally metamorphosed schists over younger unmetamorphosed sedimentary rocks.

Such faults, as described earlier, are called *overthrust faults,* and in typical examples the displacement may be very large: 16–24 km (10–15 mi) are not uncommon—and in some cases may amount to 50–60 km (31–37 mi).

In the United States the pioneer work in establishing the existence of thrust faults was done around 1900, in large part through the efforts of Bailey Willis, who for many years was a professor of geology at Stanford University. Willis had been doing field work in Glacier National Park (Fig. 16-33). He showed that the eastern wall of the Rocky Mountains in the park area consists of sedimentary rocks that are more than 500 million years old resting on rocks whose age ranges from 60 to 130 million years. The differences in those unlike strata are further accentuated by the fact that the older sedimentary rocks are much more resistant to erosion and form the castellated ridges and steep cliffs scored by deep canyons that make the park landscape so renowned. The younger rocks, however, are less resistant and are responsible for the gently rolling, undulating landscape of the high plains of Montana to the east. The explanation Willis came up with was that the older sediments slid over the top of the younger sequence along a large overthrust fault (Fig. 16-34).

Fig. 16-34. Cross section of an overthrust fault.

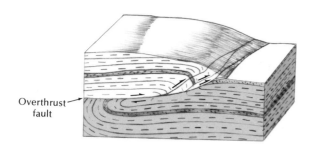

Overthrust fault

One of the remarkable features of the great thrust fault of Glacier Park—to which the especially appropriate name of Lewis Overthrust was given to commemorate the pioneer exploratory efforts of Meriwether Lewis in 1803–06—is the outlying peak, Chief Mountain (Fig. 16-35). That mountain is a remnant of the hanging wall of the fault, isolated by erosion, and thus in a sense is a "mountain without roots." Such an erosionally isolated fragment of a thrust sheet is called a *klippe* (from the German word for cliff; pl., *klippen*). Conversely, erosion may succeed in cutting through the overthrust plate of such a fault and exposing the rocks beneath (Fig. 16-36). The place that grants us such a revealing insight into the nature of the reaches below the thrust plane is called a *fenster,* which is the German word for window.

Many thrust faults are not regular geometric planes, but have complexly curved surfaces; in places the dip may be 10° or less, in others it may steepen to 45° or so (see, for instance, Fig. 16-36). Some even have been folded or bent after formation by a continuation of deformational processes. Whether or not they have been folded, they appear to have very gentle dips near the leading edge of what may once have resembled a tongue-like lobe, and steeper dips in the root zone to the rear.

Having established the geometry of those great over-riding rock plates which have produced such large-scale displacement—somewhat like the telescoping of an old-fashioned collapsible drinking cup—geologists are now ready to attack the central problem of determining the mechanism responsible for their existence. Were they shoved from behind? Are they the result of gravitational gliding of great masses of rock down gentle slopes? Neither possibility can be advocated with certainty. Recent investigations, however, have shown that if the pore spaces within the rocks are filled with water under abnormally high pressure, so that a buoyant effect results, less force than had formerly been thought necessary is required to overcome friction and set such a mass in motion. Large overthrust sheets,

Fig. 16-35. Looking northwest toward Chief Mountain with Cable Mountain directly to its left, Glacier National Park, Montana. Chief Mountain, composed of sediments over 500 million years old, sits on top of rocks ranging from 60 to 130 million years in age (forming the lower, gentler slopes) and represents a remnant of an overthrust sheet that moved for many kilometers over the Lewis Overthrust fault. *Doug Erskine, Glacier National Park*

then, may be able to slide downhill on relatively gentle slopes.

Many overthrust sheets are found flanking some of the larger complex mountain chains of the world (Fig. 16-37). It appears probable that as mountain building progressed, the central part of the range was uplifted to some degree, thereby providing the slope necessary for overthrust slippage—and the blocks of sediments slid, like large single landslide blocks, downslope for many kilometers or even tens of kilometers. Quite commonly, the

(Above) The Virgin anticline, Utah, looking North. *John S. Shelton*. (Below) Landsat image of the Appalachians in Pennsylvania, showing the Susquehanna cutting through folded sediments resulting from a mountain-building episode affecting the Appalachian Geosyncline at the end of the Paleozoic Era. *USGS*

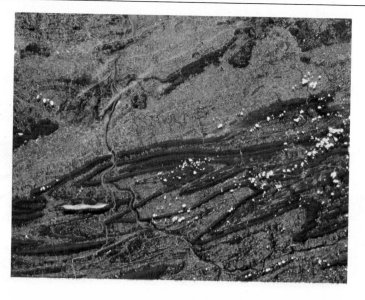

(Above) Machu Piccu, in the Peruvian Andes, is the remnants of an Inca stronghold on the Urubamba River. The rock spires and towers were eroded from a Late Cretaceous-Tertiary pluton intruded into Precambrian and Paleozoic rocks. (Below) Mont Blanc, French Alps, viewed from Talèfre. Spectacular scenery carved by rivers and glaciers in a geologically recently developed alpine mountain range. *Both photos, Jeffrey House*

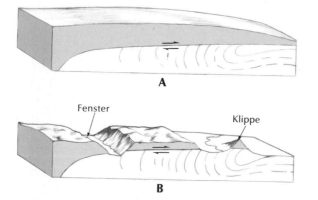

A

Fenster   Klippe

B

Fig. 16-36. Diagram showing the existing situation along the Lewis Overthrust fault (**A**) soon after movement along it ceased and (**B**) after erosion has brought about partial dissection. The erosional remnant isolated from the thrust sheet is termed a klippe and the window eroded through the sheet, a fenster.

Fig. 16-37. The Keystone Overthrust fault in the Spring Mountains west of Las Vegas, Nevada, separates dark-colored Cambrian limestone above from light-colored Jurassic sandstone below. *John S. Shelton*

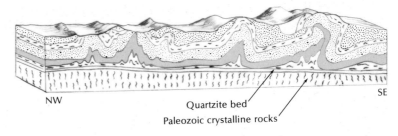

Fig. 16-38. Décollement in the Jura Mountains, western Europe. The lowest, nearly vertical structure consists of Paleozoic crystalline rocks (mostly metamorphics). Directly above is a thin bed of flat-lying Triassic quartzite, then a structurally weak unit consisting largely of layers of shale and salt along which most of the sliding has occurred.

rocks in the upper plate (hanging wall), especially if they are composed of relatively weak sediments, will be folded complexly and cut by numerous smaller thrust faults. Such an overthrust sheet, composed primarily of rumpled sediments, is called a *décollement*, and is common in the Jura Mountains marginal to the Alps (Fig. 16-38) and in parts of the Appalachian Mountains in the eastern United States.

MOUNTAINS AND MOUNTAIN BUILDING

The description of structural features such as folds, joints, and faults, is only a prelude to a discussion of mountains and mountain-building processes.

Of all the landforms and structural features on the earth's surface, none, surely, is closer to the heart of geologists than mountains (Fig. 16-39). The greatest variety of rocks is visible in their valleys and on their ridges and peaks. In the long, linear mountain belts, metamorphic and plutonic rocks commonly are found in close association in the axial portions and outward from that, sedimentary rocks are broken, overthrust, and complexly folded. Many of the more dramatic aspects of erosion are concentrated in mountains. Landslides and other kinds of mass movement have their maximum development; streams are more powerful because their gradients are steeper; frost-riving efficiently subdues the higher alpine summits; lastly, it is in high mountains that glaciers add the final touch to make montane scenery a source of joy and inspiration to many dwellers of the lowlands.

That mountains can be a place of solace and of beauty is a relatively new point of view.

Mountains were greatly feared by travelers in medieval times, and rightly so. Roads crossing them were few, and almost all were rough and tedious. Inns were incredibly crude by our standards and distances between them in terms of travel time were great. Dangers from landslides, cold, snow, and armed highwaymen were very real. Certainly few people would have been deranged enough to climb a mountain just for the sake of climbing it.

By the eighteenth century a change set in. Not only had climbers started Alpine ascents, but there was an awakening intellectual curiosity as to the nature of the mountain world. One of the leaders in that emerging inquiry was a Swiss, Horace Benedict de Saussure (1740–99) who, in addition to making the first ascent of Mont Blanc, wrote a four-volume work, *Voyages dans les Alpes*, which contains, among descriptions of birds, flowers and trees, a first account of the complex structure of the rocks of those famous mountains.

Mountains have been explored scientifically through the years since the mid-eighteenth century, until today we know vastly more than our predecessors about the rocks and structures found within them. Yet we still do not understand clearly how mountains are formed, and what processes operate within the earth's crust to raise some of them—such as the Andes and Himalayas—to their imposing heights.

Every mountain and mountain range is unique, from whatever standpoint it is judged. Therefore, like many other natural phenomena, mountains are difficult to divide into classifications that have real meaning in the natural world. We are prone to assign to rigid, pigeonhole categories, features that may have

Fig. 16-39. Mt. Gardner, Ellsworth Mountains, Antarctica.

had more than one kind of origin or that tend to merge with one another rather than having sharp boundaries. The abbreviated classification below serves to differentiate the more distinctive types. Yet it is not so arbitrary as to be inflexible.

Volcanic mountains are built up of an accumulation of volcanic eruptives, such as ash, pumice, bombs, and lava flows.

Fault-block mountains owe their elevation to differential movement along faults, so that some parts of the crust are raised and others are lowered relative to one another.

Oceanic ridges and rises form a nearly continuous elongate mountain system that extends throughout the world's ocean basins.

Folded and complex mountains generally consist of igneous and strongly deformed sedimentary and metamorphic rocks. Normally they occur in great elongate belts and require, on the average, at least several hundred million years to form.

Volcanic mountains

Some of the world's best-loved and most scenic peaks are volcanoes. Among the more familiar are Fujiyama, Mount Rainier, Mount Etna, Mauna Loa, and lofty Andean summits such as El Misti and Aconcagua. Mauna Loa, which, counting the submerged as well as the visible part, rises about 9150 m (30,012 ft) from a base 145 km (90 mi) in diameter on the sea floor, is the world's largest isolated mountain mass.

Volcanoes constitute a fairly high percentage of the world's mountains. Add to them the extinct or dormant centers, such as the Cascade volcanoes of the western United States (Fig. 16-40), and the total number is large. Could the waters of the sea be rolled away, we should be doubly impressed because the peaks of many of the volcanic islands would loom far above the surrounding abyssal plain, and we

should also see for the first time the hidden slopes of numerous submarine volcanoes.

However, volcanic mountains differ fundamentally from all the others in our classification since they are accumulations of material piled up on the earth's surface. In Chapter 5 we discussed the great diversity of form that volcanic mountains may show; there is no need to repeat the description here. The chief point to be emphasized is that volcanoes are heaps of pyroclastic material or of lava or both. Although they may be grouped in clusters, or even in chains as in the Cascades, they do not form the long and nearly continuous ridges so characteristic of folded or complex mountains, such as the Alps or the Himalayas. Characteristically, volcanoes rise as conical or dome-like mountains above their surroundings.

Fault-block mountains

We have already discussed fault-block mountains to some degree in connection with the description of normal and reverse faulting. They are conspicuously developed in the Basin and Range province of the United States from Colorado westward to California and from Idaho southward to Arizona.

To the geologists who accompanied the early exploring expeditions, those mountains were a challenge. Obviously their structure was wholly unlike that of the central Appalachians, with nearly continuous ridges broken at long intervals by water gaps such as those of the Cumberland, Potomac, and Susquehanna rivers. By the 1840s the outlines of the geologic structure of the Pennsylvania portion of the Appalachians had been deciphered; it was shown that the internal arrangement of the mountains was a succession of synclines and anticlines whose crests and troughs much resembled a train of waves.

Although men looked for a similar geometry in the Great Basin, its discovery eluded the nineteenth-century geologists who were attached to the various railway surveying parties or to military expeditions. Instead of being rela-

Fig. 16-40. A chain of volcanic mountains form the Cascade Range in California, Oregon, and Washington. Mount Jefferson is closest, and in the background, from left to right, are Mounts St. Helens, Rainier, Hood, and Adams. *John S. Shelton*

Fig. 16-41. Looking toward the east side of the Sierra Nevada and the floor of Owens Valley, California. The mountain block has been relatively uplifted along a high-angle frontal fault that runs approximately along the base of the escarpment. *Roland von Huene*

tively simple like the central Appalachians, the internal structure of many of the Great Basin ranges is wonderfully complicated, with intricate patterns of thrust faults, folded stratified rocks, and complex hierarchies of igneous intrusions. Certainly there was no wave-like progression of rhythmically folded strata.

It remained for one of the United States' greatest geologists, Grove Karl Gilbert (1843–1918), to solve the riddle of the origin of the Great Basin ranges. It was a difficult feat indeed, for scientifically, next to nothing was known of the remote and arid Southwest. Not even an adequate map existed to show the location and extent of many of the desert

ranges. As a young man of twenty-eight, Gilbert accompanied an exploring party of the U.S. Army Corps of Engineers. The difficulties imposed by the hostile terrain were severe enough, but Gilbert, as a civilian, also had to contend with the arbitrary decisions governing the movements and route of a militarily oriented topographic surveying department. The accomplishment of so much in so short a time, while he was still an untried and largely self-taught geologist, earned him a place among the pioneers of scientific exploration.

In simplified terms, Gilbert saw that the unusual topographic form of the generally straight-margined mountains, separated from

one another by broad, gravel-floored basins, was most plausibly explained by the presence of one or more faults along their margins, faults by which they had been uplifted. Such features, bounded at one or both lateral margins by large, high-angle normal or reverse faults are called *fault-block mountains* (Figs. 16-41 and 16-42). For the time (1872) his was a bold generalization, but investigations of those and other mountains of the Far West, as well as of comparable ones around the globe in recent years since then, support Gilbert's prescience in making a sound generalization from the limited data available to him.

As often is true about scientific discoveries, later work demonstrates that an originally simple concept becomes complicated as more and more information comes to light. No one as yet has devised a completely satisfactory explanation for the fault-bordered mountain ranges in the Basin and Range province. It appears that geologists generally agree on the nature of the boundary faults—they are relatively straight along their strike; their dip is steep, perhaps 60 to 70° and they resemble normal faults more than other types. There is evidence that the fault surface of some mountain-front faults decreases in dip downward and, several kilometers beneath the

surface, incline at angles of only 40 to 50°, or less (Fig. 16-42).

Clearly, vertical movements in the earth's crust have been responsible for the origin of fault bordered mountains and troughs. There is no evidence of compression or of shortening of the earth's crust, the process that commonly is believed necessary, at least in part, for the formation of structures such as anticlines, synclines, and thrust faults. However, there is evidence of extension in the Basin and Range country that has affected it across its entire east–west width—about 700 km (435 mi). If one were to construct a model of the numerous horsts and graben, and to slide the blocks back along their inclined fault planes to their original unbroken condition, there would be a reduction in the east–west dimension; that is, at right angles to the fault trends, equivalent to about 64 km (40 mi). That amount of extension has occurred in the region during the last 25 million years.

In addition to east–west stretching there is abundant geologic evidence that the area in general has been uplifted during the same time period, particularly during the last 10 million years or so. No one really knows what could bring about both the uplift and stretching of a large segment of continental crust. It has been

---

Fig. 16-42. Typical fault block mountains of the Basin and Range province, western United States. Many of the range-border faults decrease in dip downward, as shown. Arrows at the base of the block indicate directions of extension required to form these mountains.

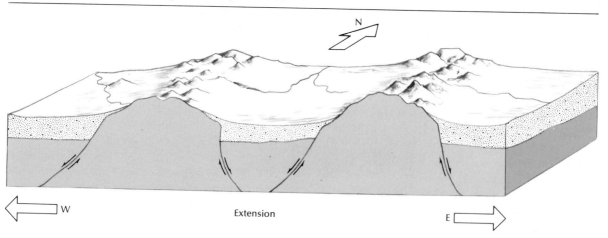

suggested that the East Pacific Rise, a zone where two plates are pulling apart (diverging), does not die out where it butts into the North American plate near the southern tip of Baja California, but continues northward beneath the continent (Fig. 16-43). The extension and uplift of the basins and ranges, then, are thought to represent cracking and rifting of the crustal blocks above the land-locked rise system. But a drawback to the theory is the extremely wide zone of extension. Most earth scientists think that the zone of stretching associated with a diverging plate boundary should be concentrated in a narrow band right along its crest—as in the rift systems in the Red Sea area and eastern Africa.

Moreover, many earth scientists do not believe that the East Pacific Rise extends beneath the continent at all. In regard to the latter, they feel either that the rise ceases to exist where it abuts Baja California or has been offset northward along the San Andreas Fault system. They point to the small segment of a zone of divergence found in offshore Oregon and Washington as the other end of the severed rise system.

There is one bit of geophysical data that probably has some bearing on the origin of the Basin and Range province. The entire western part of the United States, including all of the areas that have been structurally and volcanically active during the last 30 million years, is underlain by an anomalous upper mantle that extends 100 km (62 mi) or so downward, below the base of the crust. The mantle is anomalous because it is less dense than it should be. Some geophysicists therefore have speculated that in some way the upper mantle has changed chemically to produce a relatively less-dense (and accordingly more-voluminous) material. In response to the increase in volume in the mantle, the crust above has been uplifted and stretched. What might cause such chemical change? With our present state of knowledge, all we can do is to shrug our shoulders resignedly and shake our heads.

Fault-block mountains larger than those in the Basin and Range province (commonly rising to elevations over 3800 m; 12,464 ft) are found

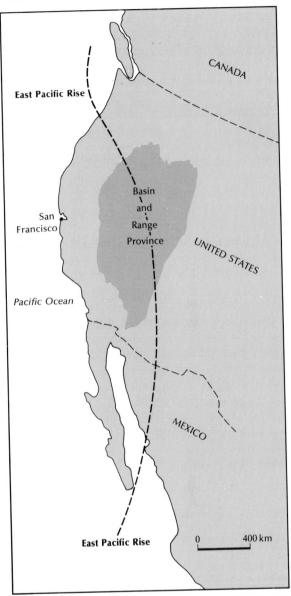

Fig. 16-43. Western North America, showing the general area of Basin and Range province and the possible location at depth, of the East Pacific Rise divergence zone.

in New Mexico, Colorado, and Wyoming. In fact, most of the large, generally north-south to northwest-southeast trending ranges in those three states, which include the Sangre de Cristo, Front, Gore, Sawatch, Owl Creek, Wind

River, and Big Horn ranges, owe their existence, in part, to block faulting that has occurred along one or both sides of the mountain blocks. In many cases the mountain block has been raised so high that erosion has stripped off most of the overlying Paleozoic and Mesozoic strata to expose the Precambrian crystalline rocks beneath—rocks that resided at least 7625 m (25,010 ft) below sea level only about 70 million years ago. Oil-well drilling and field studies have shown that most of the faults are almost vertical at depth but, as they are traced upward, they flare outward into high-angle thrust faults. In some instances the thrust fault flares outward to such an extent that it becomes a low-angle thrust, or overthrust fault. It is almost as if the mountain block expanded as it rose along the frontal faults, much in the way a cork expands as it is pulled from a wine bottle.

Whatever the cause of those mountains, it has pushed the rocks of the upper mantle and the overlying crust upward for distances as great as 13 km (8 mi). Vertical movements of crustal blocks of such large magnitudes are hard to rationalize within any existing theory of mountain building. In fact, the existence of such mountains so far from the active plate margins has proved to be a particularly sharp thorn in the side of the plate-tectonic model.

Oceanic ridges and rises

The world's longest mountain chain lies mostly beneath the ocean's surface. As described in Chapter 2, the ridge-and-rise system stretches almost unbroken nearly 64,000 km (39,770 mi) throughout the world's ocean basins. The range, basically unlike any on the continents, appears to be caused by a broad, linear rise of the ocean floor. It is cut by abundant normal faults that parallel the ridge axis and commonly it is offset, for distances up to 1000 km (621 mi) and more, along fracture zones that cross it nearly at right angles.

The ridge system is the site of many shallow-focus earthquakes that occur directly beneath the ridge axis and along parts of the cross-cutting fracture zones. Volcanic activity is common along the ridge axis, and many earth scientists feel that the elevation of the ridge is due in part to the accumulation of basaltic lavas and the injection of multitudes of dikes at shallow depths. Some believe, however, that the elevation of the linear welt is caused by a volume increase resulting from serpentinization of olivine-rich rocks along a narrow zone beneath the sea floor.

It has become apparent that oceanic ridges and rises are located astride the zone where two plates of lithosphere are being pulled apart —the zones of divergence. Such oceanic mountain chains, then, are closely associated in some way to the origin and movement of lithospheric plates. We will leave further discussion of the subcrustal process responsible for such features to the chapter on continental drift and plate tectonics (Chap. 19).

Folded and complex mountains

When geologists think of mountain-building processes they usually have in mind those that produce *alpine chains*, the complex linear mountain belts found on all of the continents (Fig. 16-44). Because of their complexity, no two alpine chains are exactly alike, either in their features or their history of development. Much, if not most, of the thought on the origin of mountains has been given to consideration of how alpine chains came to be.

Alpine chains are essentially continuous, extending for several hundred up to 1600 km (994 mi) or more in length, and contain most of the following features:

1. A core of structurally complex, dynamo-thermally metamorphosed rocks (generally sedimentary or sedimentary and volcanic in origin) which are locally intruded by and intimately associated with granitic plutonic rocks, particularly batholiths.
2. Elongate belts of thick sections of relatively unmetamorphosed sedimentary rocks, usually on one but in places on both sides of the core, which have been complexly folded, overturned, and thrust faulted. Dé-

Fig. 16-44. Looking north along the sheer west front of the Sentinel Range, Antarctica. Mount Tyree (top center) rises 4965 m (16,285 ft). Deformed quartzites of Paleozoic age underlie the west escarpment and most of the visible mountains. *USGS*

collement structures are common. The degree of folding and faulting dies out with distance from the mountain core until, at a sufficiently great distance, the rocks are essentially flat lying. The thickness of the sedimentary section decreases progressively from the core region outward.

Most alpine mountains, just after they are formed, apparently rise high above sea level to form linear topographic prominences, such as the Alps and Himalayas. Even after erosion has essentially reduced them to a low-relief feature once again, if the main geologic elements just outlined are recognizable, it is still possible to describe the feature as an alpine mountain chain. The Appalachians in eastern North America certainly do not have the surface grandeur of many mountain ranges. Yet they possess all of the features just enumerated in a nearly classical manner and can be spoken of as an alpine chain, even though the last mountain-building activity there occurred about 225 million years ago (Figs. 16-45 and 16-46).

Commonly, but not always, alpine chains are formed where thick sections of sedimentary rocks have accumulated. That relationship was commented on in 1859 by James Hall (1811–98), a young and energetic member of the pioneering New York Geological Survey, at a time when much of upstate New York was still a frontier. He found that Paleozoic strata underlying the Mississippi Valley states, such as Iowa and Illinois, were much thinner than those of identical age underlying the Allegheny Plateau of New York and Pennsylvania. From that observation, as well as from the knowledge he acquired through reading, he reasoned that many of the world's mountains had somehow been formed in regions with abnormal thicknesses of sedimentary rocks.

A contemporary of Hall's, James Dwight Dana (1813–95), had done his stint of wandering as a geologist attached to the vessels of the U.S. Exploring Expedition (1838–42). It was a round-the-world investigation that has disappeared into unmerited oblivion in the scientific memory of our country but deserves to stand with the much better-known voyages of the *Beagle* and the *Challenger*. Dana was greatly impressed by Hall's discovery of the abnormal thickness of sedimentary rocks exposed in the Appalachians, and, recognizing the rather unusual circumstances under which such sediments accumulated, coined the name *geosynclinal*, now shortened to *geosyncline*, for the immense trough in which they were laid down (Fig. 16-47).

Fig. 16-45. Main structural features of the Appalachian Mountains.

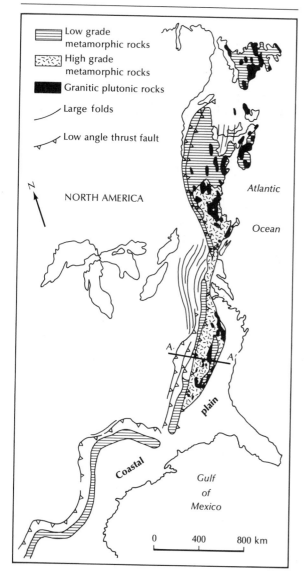

545

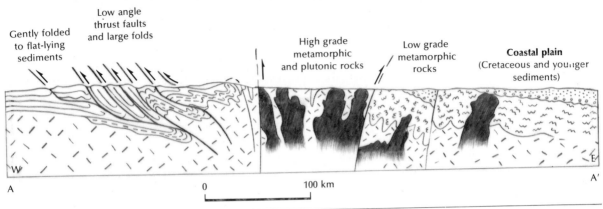

Fig. 16-46. Cross section A-A' (see Fig. 16-45) across the Appalachian Mountain belt. In the axial portion are high-grade metamorphic and plutonic rocks; outward from the axis the degree of metamorphism and the intensity of deformation decreases.

Although the two men agreed on the physical relationships of the rocks and structures in the Appalachians, they differed strongly on how they got that way. Hall believed that the earth's crust was bowed down as a consequence of load imposed upon it by a localized accumulation of sediment. The more sediment that was deposited in such a wedge, the deeper the crust would be bowed downward. Obviously, Hall reasoned, such a situation made for a rather uncertain equilibrium. The comparatively weak material of the sediments could be crumpled readily and thrown into contorted folds by inward movement of the vise-like margins of the trough. Fracturing of the infolded sediments could also provide avenues through which magma could invade shallower zones in the earth's crust, or possibly even reach the surface.

Dana, on the other hand, did not believe that the weight of such accumulated sediments was capable by itself of bending the crust downward. Without the knowledge that seismology has given us today of varying densities of the material in the earth's crust and mantle, he recognized that only with great difficulty could relatively light, water-soaked, and unconsolidated or recently consolidated sedi-

ments displace subcrustal material whose density was higher. Dana believed that the earth was slowly contracting through time and that the contraction of the interior (very likely, he believed, through loss of heat) caused the formation of a geosyncline and eventually led to the development of folds and faults in the thickened geosynclinal sedimentary section, much like the wrinkles on the skin of a dried apple. Ultimately a mountain chain like the Appalachians would be formed.

In the century since then, both points of view have continued to find their adherents, but as we have learned more and more about moun-

Fig. 16-47. Hypothetical cross section through the Appalachian geosyncline as it filled with sediments, showing the thickening of strata toward the axial portion. The maximum thickness of sediments amounted to about 12,200 m (40,016 ft).

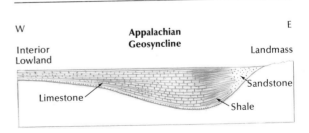

tains and the physical properties of the earth, we have been forced to search for more sophisticated theories. Plate tectonics, for example, can more fully explain all of the features of alpine mountains and the other geologic features of the earth as well.

In order to become more familiar with the way an alpine mountain chain is developed, let us look in more detail at the Appalachian system. Erosion during the last 225 million years since the last stages of its development has been both a blessing and a bane. Because of erosional stripping we can now look deep into the innards of the range to see, more or less, what transpired far below the surface. Unfortunately, the upper levels of rocks, which undoubtedly were different from those below, have been stripped away, removing the record of the processes that were operating near the surface in the geosyncline. There is general agreement, however, that the Appalachians progressed through an evolutionary sequence somewhat as follows:

1. A lens of sediment perhaps 9150 to 12,200 m (30,012 to 40,016 ft) thick, 3200 km (1988 mi) long, and 480 km (298 mi) broad, accumulated slowly during the later part of Precambrian time and throughout the Paleozoic Era—roughly from around 650 million to 225 million years ago. From all the evidence, the source of most of the sediment was east of the present-day Appalachians. Sometimes the source land stood relatively high, and streams emptying into the geosynclinal sea carried coarse material such as boulders, gravel, and sand. At other times the source land may have stood lower. One certainty from the record of the geosynclinal strata is that considerable volcanic activity was concentrated in the region. The volcanoes probably made their appearance as volcanic island arcs that grew from time to time in the geosynclinal trough. Some grew sufficiently high to project above the sea that covered the geosyncline, only to stagnate and to be planed off by subsequent erosion, their debris added to the continuous rain of

sediments filling the trough. The evidence of the physical and paleontological record is that the water was generally less than 2000 m (6560 ft) deep all during the 425 million years the trough was in existence. During part of its history, the time in which the coal beds of West Virginia and Pennsylvania were being laid down, some of the geosyncline was a broad marshland above sea level. At other times it may have been an expansive delta, perhaps resembling the Mississippi Delta of today. The trough continued to downwarp during its development and the sediments continued to be carried in to fill it.

As time passed, much more happened than a continuous but well-nigh imperceptible sinking of the geosynclinal floor. From time to time the rocks were squeezed and deformed. The disturbed intervals are recorded by unconformities in the rocks, and by structures such as anticlines, synclines, and faults. There were occasional volcanic episodes, and a succession of alternating incursions and retreats of the sea.

2. Sporadically, during the development of the range, rocks in some parts of the deeper central portion of the basin were heated to temperatures high enough to bring about regional metamorphism, and slates and schists were formed from the sedimentary and volcanic rocks. In some portions of the geosyncline temperatures became high enough for melting to occur, and bodies of granitic magma were formed that worked their way upward to fairly shallow levels in the crust.

3. Finally, the long depositional episode ended at the close of the Paleozoic Era, about 225 million years ago. Then, the thousands of meters of accumulated strata were further metamorphosed and thrown into folds, or were broken by great, low-angle thrust faults—the latter chiefly in the southern part of the range.

Undoubtedly at that time a mountain range of considerable stature must have

stood in bold relief to the lower-lying countryside around it.

4. Omitting some minor episodes of deformation and volcanism, the mountains eventually were worn away, and in their place was created an erosional surface that had been beveled across the rocks and structures of the former chain.

Essentially the same general sequence of events has occurred in other parts of the world, at different times, as for example in western North America during the Paleozoic and Mesozoic eras. Commonly, the rates of development in the other ranges were different, or, because of local geologic situations, the number of steps varied. Each alpine range, then, is the product of a unique set of events.

Modern counterparts of ancient geosynclines and plate tectonism One might ask where in the world today are geosynclines being developed and filled with sediments and alpine mountains being uplifted? The problem, of course, is in recognizing—mainly from features exposed at the surface—that mountain building is going on at depth.

What are the outward signs of metamorphism at depth? How does one recognize that magmas of batholithic proportions have formed and are slowly rising? A lack of volcanic activity is no sure sign since batholiths vent at the surface only after they have reached rather shallow levels. Conversely, the presence of volcanism may not be a valid indicator since we know of many volcanoes and large plutons that have developed in response to processes other than alpine mountain building.

Will an alpine chain be formed wherever sediments accumulate to great thicknesses? The answer is a tentative, no. In the region of the Gulf Coast of the United States, near the delta of the Mississippi River, sediments have piled up to a thickness of 12,000–15,000 m (39,360–49,200 ft)—a thickness equivalent to that found in many alpine geosynclines of the past. Yet there are few geologists foolhardy enough to say that the north side of the Gulf of Mexico is the site of a developing alpine chain. The great thickness of the sediments seems related mostly to the dumping of large amounts of clastic debris in the restricted Gulf waters by one of the world's largest rivers—and probably has little to do with mountain-building processes. Moreover, sedimentation in the region has not been associated with volcanism, seismic activity, or any of the other aspects of mountain building we might expect. Deep drilling in search of oil in the area has shown that the sediments are still largely unaltered metamorphically and structurally.

In other parts of the world—Australia and the USSR for example—there are found old sequences of essentially unaltered and undeformed sedimentary rock as thick as those on the Gulf Coast. Certainly from the point of view of thickness alone, the deposits are geosynclinal in character; yet mountain building has not occurred in those areas.

Theory of alpine mountain building Several theories have been advanced to explain the development of alpine mountains, but most have proved restricted in scope and inadequate in meeting the test of time. Probably the best explanation of alpine features is the plate tectonics model, which proposes that the likely place to expect mountain building is along the boundary where two plates converge (Fig. 16-48). Such a place is characterized by active andesitic volcanism, seismic activity down to 700 km (435 mi) in many arcs, and linear troughs deep enough to accumulate great thicknesses of sediments derived from the erosion of nearby continents and from volcanic islands. Such areas, of course, correspond to the island archipelagoes that border the Pacific Ocean today. Sediments that fill the geosynclines can be carried down to great depths in the active zone of convergence, where they can be heated, dynamothermally metamorphosed, melted to produce batholithic magmas, and folded and faulted. A diagrammatic

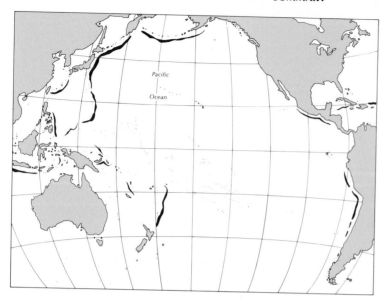

Fig. 16-48. Deep ocean trenches, shown in black, demarcate zones of convergence. Notice that most border the Pacific Ocean basin.

illustration of how the process might unfold with time is given in Figure 16-49.

One island-arc segment, that of Japan, has had a long and complex geologic history. The rocks exposed in the islands are complexly deformed, show signs of having been dynamo-thermally metamorphosed, contain numerous granitic stocks and batholiths interspersed with the metamorphic rocks, and include considerable thicknesses of sediments derived both from the Asian mainland and from the eruption and erosion of andesitic volcanoes. Many geologists point, therefore, to Japan as a site of active mountain building through the process of plate convergence. But only time will tell whether the extrapolation is sagacious.

SUMMARY

1. Rocks in the outer part of the earth are currently being bent, broken, uplifted, and down-dropped in response to the action of internal forces.
2. Evidence that deformation occurred in the geologic past is seen in *jointed, folded,* and *faulted* rocks. Folding produces *synclines* and *anticlines*, which may be *symmetric, asymmetric,* or *overturned. Faults* are classified on the basis of known directions of relative displacement (*normal fault, reverse* or *thrust fault, strike-slip fault,* and *overthrust fault*).
3. There are many kinds of mountains: volcanic, fault-block, oceanic ridges and rises, folded and complex. Most difficult to understand are the generally linear *alpine chains*, such as the Appalachian Mountains. They are complex in structure and require up to hundreds of millions of years to form. Most contain a core of dynamothermally metamorphic and associated batholithic rocks flanked on one or both sides by sedimentary rocks. The degree of deformation in the strata progressively lessens from the core outward.
4. Most geologists believe that mountains are formed at converging plate boundaries, but modern counterparts of the ancient geosynclines are difficult to identify. They are possibly in part the ocean deeps and associated island-arc systems.

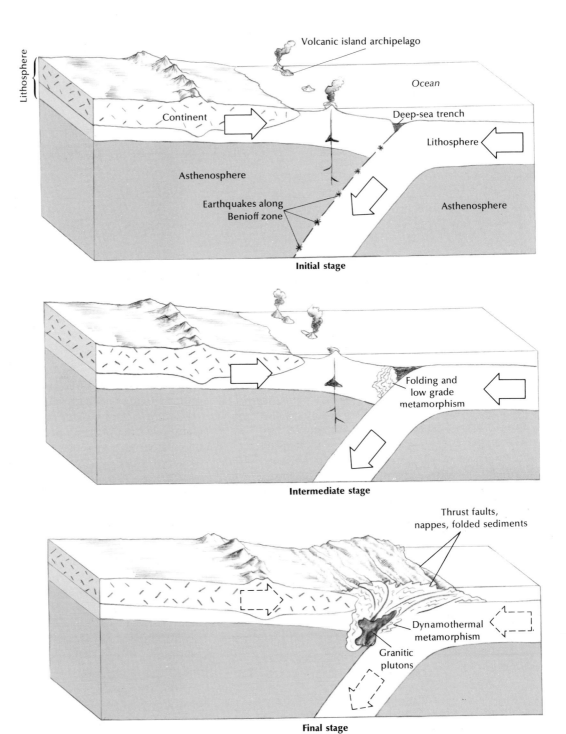

Lithosphere

Volcanic island archipelago

Ocean

Continent

Deep-sea trench

Lithosphere

Asthenosphere

Asthenosphere

Earthquakes along
Benioff zone

Initial stage

Folding and
low grade
metamorphism

Intermediate stage

Thrust faults,
nappes, folded sediments

Dynamothermal
metamorphism

Granitic
plutons

Final stage

Fig. 16-49. Postulated sequential development of an alpine mountain system along a zone of convergence, according to the model of plate tectonics.

SELECTED REFERENCES

Billings, M. P., 1960, Diastrophism and mountain building, Geological Society of America Bulletin, vol. 71, pp. 363–98.

———, 1972, Structural geology, 3rd ed., Prentice-Hall, Englewood Cliffs, New Jersey.

Clark, S. P., Jr., 1971, Geologic structures, Chap. 2, pp. 9–25 in Structure of the earth, Prentice-Hall, Englewood Cliffs, New Jersey.

Collett, L. W., 1927, repr. 1974, The structure of the Alps, R. E. Krieger Pub. Co., Huntington, N.Y.

Curtis, B. F., ed., 1975, Cenozoic history of the southern Rocky Mountains, Geological Society of America, Memoir 144.

de Sitter, L. U., 1964, Structural geology 2nd ed., McGraw-Hill Book Co., New York.

Dewey, J. F., and Bird, J. M., 1970, Mountain belts and the new global tectonics, Journal of Geophysical Research, vol. 75, pp. 262–65.

Eardley, A. J., 1951, Structural geology of North America, Harper and Bros., New York.

Gilluly, James, 1949, The distribution of mountain building in geologic time, Geological Society of America Bulletin, vol. 60, pp. 561–90.

———, 1970, Crustal deformation of the western United States, in The megatectonics of continents and oceans, H. Johnson and B. C. Smith, eds., Rutgers University Press, New Brunswick, New Jersey, pp. 47–73.

Griggs, D. T., 1939, A theory of mountain building, American Journal of Science, vol. 237, pp. 611–50.

Hill, M. L., 1947, Classification of faults, American Association of Petroleum Geologists Bulletin, vol. 31, pp. 1669–73.

——— and Dibblee, T. W., Jr., 1953, San Andreas, Garlock, and Big Pine faults, California, Geological Society of America Bulletin, vol. 64, pp. 443–58.

Hsu, K. J., 1958, Isostasy and a theory for the origin of geosynclines. American Journal of Science, vol. 256, pp. 305–27.

Kay, M., 1951, North American geosynclines, Geological Society of America, Memoir 48.

Kelley, V. C., 1955, Regional tectonics of the Colorado Plateau and relationship to the origin and distribution of uranium, University of New Mexico, Geology Publication no. 5, Albuquerque.

Kennedy, G. C., 1959, The origin of continents, mountain ranges, and ocean basins, American Scientist, vol. 47, pp. 491–504.

King, P. B., 1959, The evolution of North America, Princeton University Press, Princeton, New Jersey.

Murray, G. E., 1961, Geology of the Atlantic and Gulf coastal province of North America, Harper and Brothers, New York.

Oakeshott, G. B., 1966, San Andreas fault: geologic and earthquake history, 1966, Mineral Information Service, October, pp. 159–65, California Division of Mines and Geology, Sacramento.

Rubey, W. W., and Hubbert, M. K., 1959, Role of fluid pressure in mechanics of overthrust faulting, Geological Society of America Bulletin, vol. 70, pp. 167–206.

Spencer, E. W., 1969, Introduction to the structure of the earth, McGraw-Hill Book Co., New York.

Umbgrove, J. H. F., 1947, The pulse of the earth, Martinus Nijhoff, The Hague, Netherlands.

17

EARTHQUAKES AND THE EARTH'S INTERIOR

Many of recorded history's major catastrophes have been related to earthquake activity. It has been estimated that during the last 4000 years about 15 million people died as a result of earthquakes. Each year more than a million earth tremors occur around the world and of those, about 50 are intense enough to cause significant loss of property and lives. About 10 each year can be considered major events capable, in the right circumstances, of bringing about untold death and destruction within a moment's time.

The most sinister aspect of an earthquake is that it strikes with little or no warning. At one moment the landscape is serene; at the next it gyrates and dances in a frenzy of motion.

EARTHQUAKE CASE HISTORIES

Perhaps the best way to become acquainted with the way in which earthquakes affect the earth and human beings is to recount a few of the numerous case histories. As yet, the United States has been fortunate in that there has been relatively small loss of life and property resulting from earthquake activity. Yet, in the last 200 years there have occurred several tremors of sufficient intensity to have caused immense damage had they struck in an urbanized community.

United States

San Francisco In the spring of 1906 San Francisco was the queen city of the California coast and gateway to the Orient. Her business districts hummed by day and the Barbary Coast by night. She had no reason to know that disaster was but a tick of the clock away. Early in the morning of April 18, 1906, the schooner *John A. Campbell* was running on a southeast course with Point Reyes in northern California due east 233 km (145 mi). With no warning the vessel shuddered suddenly with almost the same sensation as if she had run aground. The startled crew could scarcely believe their senses because the chart showed a depth of 2400 fathoms (14,389 ft) at their position.

Although the crew had no way of knowing it,

Fig. 17-1. The San Andreas Fault is one of the major faults in the United States, running nearly two-thirds of the length of California. Almost without exception the trace of the fault, here shown in the Carrizo Plains, is easily discernible throughout its length. *William A. Garnett*

Fig. 17-2. Earthquake damage in San Francisco, California, April 18, 1906. Although the quake caused much damage, the fire that subsequently swept unchecked through the city caused most of the property loss.
California Historical Society

they were only a few among the many whose daily routine was disturbed, or even ended forever, by the events set in motion at 5:12 A.M. throughout a region covering about 975 thousand km² (376 thousand mi²) surrounding San Francisco. That disaster of more than seventy years ago is still a most instructive example today because it devastated an essentially modern city. The population of the United States is becoming increasingly urbanized, and many of the problems faced at that time by the people

of San Francisco are exactly the same ones that might confront Civil Defense agencies today, with the additional burden of immense traffic jams, unimaginable a generation ago.

San Francisco in 1906 in some ways resembled the city of today, but in others it was very different. The cobblestone streets along the waterfront were clangorous with the movement of iron-tired wheels of horse-drawn drays. The pungent atmosphere of the waterfront was compounded in large part by the

aromatic by-products of the multitude of horse-powered vehicles, as well as the fragrance of roasting coffee and the beery blasts that emanated from the dark, block-long sawdust-floored saloons. The San Francisco waterfront was an infinitely more picturesque sight then than now, with the delicate tracery of the upper yards and rigging of the wind-driven Cape Horners rising about the pier sheds, and the waters of the bay whitened by the paddle wheels of ferries and Sacramento River boats.

Much of that colorful world, including the "Barbary Coast," was obliterated in a series of violent shocks in the early morning hours when most people were asleep (Fig. 17-2). Had the earthquake struck later when people were up and about, the casualty toll would have been much larger. How many people died will never be known, but the number may have been as high as 700. Many transient residents in such structures as sailors' boarding houses simply vanished, and, since even the sketchy pre-1906 records disappeared in flames, many of the former permanent inhabitants could not be traced. In several of the investigations made after the event, almost all the destruction was attributed to the fire that followed the tremor (Fig. 17-3). Fire was indeed the leading destroyer, but earthquake damage was not negligible, perhaps averaging as much as 25 per cent.

Newer buildings in the San Francisco of 1906 looked much like those in the older downtown sections of American cities today. By 1906,

Fig. 17-3. Sacramento Street, San Francisco, just after the 1906 earthquake. Notice how the brick fronts of the buildings spilled across the streets. The people are watching one of the great fires resulting from the quake. *Arnold Genthe, California Palace of the Legion of Honor*

riveted steel frames were coming into wide use for taller structures. Exterior walls were more commonly faced with masonry than they are today, and reinforced brick was widely used for smaller commercial buildings. In that day the Victorian influence prevailed, and most buildings were crusted with ornamentation and gingerbread—real earthquake hazards. San Francisco was unusual in one regard, and that to its sorrow, in that it was one of the larger cities of the world constructed principally of wood. Typically, the residential section consisted of block-long rows of wooden multistoried houses or apartments crowded next to one another on narrow lots. The fire was essentially unstoppable once it reached those structures.

The occurrence of the San Francisco earthquake emphasized that the kind of ground on which buildings are constructed is critical in determining the extent and nature of structural damage. Those founded on solid rock showed slight damage when compared with virtually identical structures built on waterlogged or unconsolidated ground. Damage was especially severe, for example, in downtown San Francisco, where an area about 20 blocks square was built on ground reclaimed from the bay after the Gold Rush of 1849. There, on a sludgy foundation made up of sunken ships, water-soaked refuse, bottles and bodies, all buried under poorly consolidated mud and silt, the most severe damage was concentrated.

Fires from a variety of causes broke out at many points almost immediately after the strongest shocks. At first there was little awareness that fire was the true enemy. People either gawked at the blazes or attacked them on a piecemeal basis—and not too successfully, because the alarming discovery was made almost at once that there was no water for the fire hoses. Most of the water lines beneath the streets had ruptured at innumerable points, and the water flow dropped to a pathetic trickle.

Fire started near the waterfront and swept inland across the broken city. Should you ever visit San Francisco, try to visualize the swath, eighteen blocks deep, swept by fire from the Embarcadero at the waterfront inland to Van Ness Avenue, the first wide street where a fire line could be held. Elsewhere vain attempts were made to check the advancing flames by dynamiting whole rows of buildings to keep the fire from leaping from roof to roof as a crown fire does in the forest. Unfortunately, the dynamiters had little expertise, and as often as not the explosions hurled burning material far and wide, commonly resulting in new fires in buildings as far as a block away.

Among those who witnessed the spectacle of the San Francisco earthquake and fire were a youngster, John Ohrwall, and his parents. His father was so shocked by the catastrophe that he packed up all he could salvage and moved the family about 160 km (99 mi) away, to just south of Hollister; far enough, he thought, to be free from the specter of subsequent quakes. As fate would have it, he traveled southward, paralleling the trace of the San Andreas Fault, and finally built his new home right next to the fault. John Ohrwall eventually went into the wine business and became superintendent of the Almaden Winery. Unbeknownst to him, one of the winery buildings was constructed exactly on the fault line; now, each year, interested scientists flock to the Almaden Winery, not only to sample the excellent product, but to view and measure the cracking and splitting of the walls resulting from continued earth movement. Fate has not been so cruel as it might have been, for along that section of the San Andreas, the fault walls do not stick for long periods of time. As a result, that section of the fault is relatively free from strong earthquakes—at least for the present.

From technological and scientific points of view, the San Francisco quake proved to be a particularly significant event. Immediately following the quake the California State Earthquake Commission, appointed by the governor but supported financially by the Carnegie Institution, conducted an exhaustive investigation. Hundreds of people were interviewed and evidence was collected from every damaged area.

Sudden slippage along the San Andreas Fault was the primary cause for the earthquake, and that fact was demonstrated beyond any reasonable doubt (Fig. 17-4). Not only was faulting established as the mechanism, it was shown unequivocally that the displacement was strike-slip. Before 1906 the importance of that type of fault movement was only dimly appreciated. But there for all to see was a nearly continuous trail of furrowed ground, fractured barns, and other features that stretched for hundreds of kilometers, with all the evidence showing that the western side of the San Andreas Fault moved horizontally northward with respect to the eastern side, which moved southward. The fault has a length of perhaps 966 km (600 mi) on land; during the 1906 quake, slippage occurred along about one-half of its length—from Point Arena north of San Francisco to San Juan Bautista to the southeast.

The maximum offset of 6.4 m (21 ft) was near Tomales Bay north of the city; elsewhere in the neighborhood of San Francisco it held rather consistently at around 4.6 m (15 ft). Most observers noted that the maximum displacement was in rather soft ground which appeared to have "lurched," thereby amplifying the offset. A separation of about 5 m (16 ft) appears to be closer to the true maximum offset.

Part of the study authorized by the California Earthquake Commission was a remeasurement by topographic survey stations in and around San Francisco. On the basis of the results the geologist Harry Fielding Reid formulated a theory for the generation of earthquakes. He realized that there are blocks of the earth's crust which slide past one another along faults, in response to large-scale forces. As long as the blocks on either side of a fault are free to slide, slippage will be continuous and smooth. If, however, the walls on opposite sides of the fault become "locked" together because of irregularities along the fault surface, slippage will proceed in fits and starts. Long time-spans during which no movement occurs would be followed by short moments of accelerated motion. Reid thought that during the times when the walls were locked the rocks adjacent to the fault would be progressively bent and would accumulate *elastic strain energy*. When the crust failed and rapid movement occurred, the strained rocks would snap back (rebound) —much like a springboard does after a swimmer has dived into a pool—and an earthquake would be spawned. The larger the amount of strain energy released, the larger the quake. Reid's explanation, which is diagrammed in Figure 17-5, is called the *elastic-rebound* theory, and today is still the accepted model for the generation of a quake.

New Madrid, Missouri, 1811–12 Although the San Francisco earthquake caused a great deal of damage it was not the most severe quake (from the standpoint of the area over which it was felt) to jolt the U.S. mainland in the last 200 years. The one that claims that distinction occurred in the unlikely location of New Madrid, Missouri, and took place about eight years after the Louisiana Purchase.

The New Madrid quake occurred at the time of the War of 1812, when the region was still barely opened to settlement, and life on the Mississippi was harsh. The tremor was felt over an area of around 2,500,000 km² (965,250 mi²)— from the Canadian border to the Gulf of Mexico and from the Rocky Mountains to the Atlantic Ocean (Fig. 17-6). It originated in the central part of the Mississippi Valley, very near Cairo, Illinois, where the Mississippi and Ohio rivers unite. A more unlikely spot would be hard to select—in a vast lowland plain, remote from any actively growing mountain range. By far the largest number of historic earthquakes in the United States have been concentrated in the mountainous parts of the Far West.

Although aftershocks occurred for a year following, the main earthquake effects resulted from three major shocks, beginning very early in the morning of December 16 and continuing into December 18, and again on January 23 and February 7, 1812. Were shocks of a similar magnitude to occur today, most of the cities of the central Mississippi Valley would be flattened, and newspapers, news magazines,

Fig. 17-4. Linear valley eroded along the trace of the San Andreas Fault, Mecca Hills, Riverside County, California. *USGS*

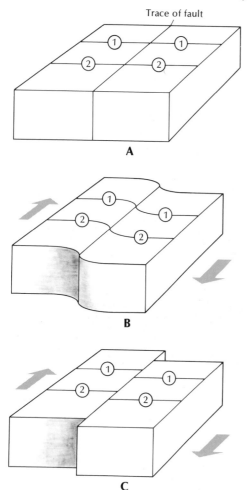

Trace of fault

A

B

C

Fig. 17-5. Three stages in the movement of two blocks of the earth's crust past one another, as postulated in Reids's theory of elastic rebound. **A.** Undeformed blocks with imaginary straight lines (1,2) extending across the trace of the fault; **B.** The two blocks move as indicated by the arrows but the fault is "locked," resulting in bent lines (1,2) near the fault trace; **C.** The blocks continue to move, leading to sudden slippage along the fault, and an earthquake. The lines are again straight but offset along the fault trace.

newsreels, and TV programs would carry photographs of little else.

According to the American seismologist Myron L. Fuller, whose investigation a century later resulted in a detailed picture of the extent to which the landscape of the Mississippi Valley was altered by the cataclysm:

The ground rose and fell as earth waves, like the long, low swell of the sea, passed across its surface, tilting the trees until their branches interlocked and opening the soil in deep cracks as the surface was bent. Landslides swept down the steeper bluffs and hillsides; considerable areas were uplifted, and still larger areas sunk and became covered with water emerging from below through fissures or little "craterlets" or accumulating from the obstruction of the surface drainage. On the Mississippi great waves were created, which overwhelmed many boats and washed others high upon the shore, the return current breaking off thousands of trees and carrying them out into the river, sand bars and points of islands gave way, and whole islands disappeared.

Fortunately, the history of that earthquake was better documented than might have been expected since there were a number of very capable observers in the region at the time, including the renowned naturalist John James Audubon. The region was visited by Sir Charles Lyell at the time of his American tour in 1846, and he left a wealth of observations made when the evidence was still fresh.

There were three truly noteworthy geologic effects of the earthquake: (1) the appearance of low cliffs cutting across country—very possibly fault scarps—some of which produced waterfalls around 2 m (6.5 ft) high where they intersected the Mississippi River; (2) the elevation of long, low, arch-like ridges or domes, along which former swamp areas were uplifted perhaps 3 to 6 m (10 to 20 ft). One of the ridges was 24 km (15 mi) long; (3) the sudden appearance of the so-called "sunken ground," which is a broadly depressed part of the Mississippi floodplain extending along the river 240 km (149 mi), and also the site of two very large lakes that came into being—St. Francis and Reelfoot. The latter, now a bird sanctuary, is an imposing sight; the gray trunks of cypress trees drowned more than a century ago stand in somber dark waters, which in places are as much as 6 m (20 ft) deep.

The New Madrid quake was unusual in that (1) it consisted of three main shocks of extraordinary magnitude occurring over a period of nearly two months, and (2) it occurred in an area normally not thought of as earthquake

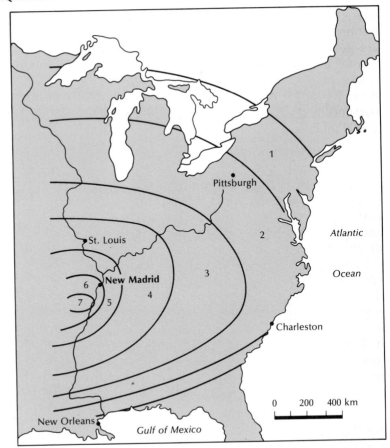

Fig. 17-6. The New Madrid earthquake: generalized zones of equal intensity east of the Mississippi. Intensity progressively increases from moderate in Zone 1 to total damage in Zone 7. The shock would have been felt over an area of about 2,500,000 km² (965,250 mi²).

prone. At a recent meeting of earth scientists, it was pointed out that small earthquakes are still being generated near New Madrid, which implies that the stress system responsible for the 1811–12 quake is still active. The real reason for the large quakes is unknown. Some scientists have suggested that they were related to loading along the southern margin of North America as a result of sedimentation by the Mississippi River during the last few million years.

Japan

The islands of Japan are even more beset by earthquakes than is the West Coast of the United States. Since 1700, approximately

200,000 people have died from quakes in Japan; since 1900 there have been more than 25 earthquakes equal in intensity to the San Francisco quake of 1906.

About 100 million people live on the islands, which together make up a land area about equal to that of California. Of necessity people are crowded into restricted urban areas, a factor that surely has increased the probability of earthquake damage. Many of the tremors occur just offshore, commonly along the landward side of the deep trench that skirts the islands. Some of the earthquakes are associated with rapid changes in submarine topography, either as a result of offset along faults or slumping of loosely consolidated sediments that have accumulated in bays or other near-

560

shore areas. Not uncommonly, after such sea-bottom changes, tsunami (seismic sea waves) are generated and cause great devastation in coastal population centers. In 1869 an earthquake in the central Pacific sent a wall of water 34 m (111 ft) high against the Japanese coast north of Tokyo, resulting in a death toll of 27,000 people.

China

China too has been beset by earthquakes. As recently as 1976, a large quake struck in northeastern China producing extensive damage, and, although reports are only fragmentary, a death toll of hundreds of thousands.

Undoubtedly the largest loss of life from earthquake activity ever recorded occurred during the 1556 tremor in China's Shansi region. At least one million people perished as a result of the quake, mostly from the collapse of buildings, and from the famine and pestilence that followed as the countryside lay in ruin.

Europe

Southern Europe is not free from earthquakes, particularly around the Mediterranean Sea. Italy, for example, has been struck by six large quakes since 1780 which have caused a total death toll of 150,000 people.

One of the most renowned and best known of the earth tremors that have struck southern Europe occurred in the ancient city of Lisbon, Portugal, on the banks of the Tagus River, on All Saints' Day, November 1, 1755 (Fig. 17-7). There were three major shocks, at 9:40 A.M., 10:00 A.M., and at noon. Many of the people who were at Mass perished in the collapse of the multitude of medieval churches whose spires had been so notable a feature of Lisbon's skyline.

With buildings made of loosely bonded masonry, huddled along narrow, irregular streets, no wonder most of them collapsed in one thunderous crash, perhaps not greatly different from the succinct description given by Voltaire in *Candide:*

When they had recovered a little, they walked towards Lisbon, hoping to get something to eat after their ordeal, as Candide still had some money left; but they had scarcely entered the city when suddenly the earth shook violently under their feet. The sea rose boiling in the harbour and broke up all the craft anchored there; the city burst into flames, and ashes covered the streets and squares; the houses came crashing down, roofs piling up on foundations, and even the foundations were smashed to pieces. Thirty thousand inhabitants of both sexes and all ages were crushed to death under the ruins.

A startling phenomenon to the unhappy survivors at Lisbon, as well as to observers at other places along the Portuguese coast, was the sudden appearance of a great wave (seismic sea wave) set in motion by a submarine shock off the coast.

In Lisbon, the wave reached a height of about 6 m (20 ft) as it swept up the Tagus River. Ships and boats were smashed together and sank, and much of the wave's fury was concentrated on a newly constructed marble pier, the *Cais de Pedra,* whose level surface was black with people fleeing from the burning city. Their drowning gave rise to the legend that the pier and all the people on it had been engulfed in a great fissure that opened and closed over them so that they were never seen again. In fact, Lyell believed on the basis of a visit to Lisbon that such a burial had occurred. As H. F. Reid, the geologist who came to prominence following the 1906 San Francisco quake, pointed out, it simply was not true. The quay probably broke up under the earthquake shock and the attack of sea waves, and in his words, "the story of the engulfing of the Lisbon quay must be ascribed to the love of the marvelous and the mysterious, which has not yet entirely disappeared from the accounts of earthquakes."

The distance to which effects of the Lisbon earthquake were felt throughout Europe was one of the remarkable features of the shock. It was strong throughout the Spanish peninsula and well into France, and also caused severe damage in Morocco (although there is some

Fig. 17-7. Contemporary engraving depicting the destruction wrought by the Lisbon earthquake, 1755. *Prints Division, New York Public Library*

feeling that the destruction there was due to an entirely separate shock of close proximity).

The Lisbon earthquake had a profound effect on European thinking. It led Rousseau to point out in his *tout est bien* philosophy that earthquakes are all for the best; it is the evils of civilization that are bad. If we were not cooped up in cities, earthquakes would not kill us. Voltaire struck a devastating blow of ridicule at such unthinking optimism. As we have seen, Candide and his insufferable companion, Dr. Pangloss, were caught in that earthquake. For Dr. Pangloss it proved to be a time to test his belief that every horror encountered was a necessary event in the pre-established harmony of the universe.

Spiritually, the Lisbon disaster was a challenge to the theologians of the time. Why did one of the most devout and justly renowned cities of the world suffer such a fate at the hands of a loving God—so that the innocent should perish in such multitudes along with the guilty? According to T. D. Kendrick in his excellent study, *The Lisbon Earthquake*, which is concerned primarily with the related themes of eighteenth-century earthquake theology and the end of optimism, a widely held belief with regard to the earthquake was the following:

The reason why God had overthrown Lisbon was not only because He intended to shock the whole of Christendom into a state of penitent obedience to Him by the staggering destruction of such a celebrated and wealthy city, one that was perhaps, thanks to its maritime trade, the best-known city in Europe; but also because Portugal was a kingdom under the special and principal care of Heaven, so that according to the rules of the divine discipline, the Portuguese for their own good and as a result of the heavenly priority that was their due, were singled out for the honour of being the first to be punished and those who were punished most severely.

Politically, the Lisbon earthquake was a prophetic event because it provided the setting for the playing out of the role of a modern dictatorship; the dictator being the Marquez de Pombal. Although he was ultimately to seize all the trappings of power, to reduce the king to ineffectiveness, to expel the Jesuits, and to execute his opponents, during the disaster Pombal unquestionably preserved order where chaos would have prevailed and never faltered in his determination to rebuild a greater Lisbon. The modernization of the city, the cutting of boulevards, and the restoration of its ancient glories—if not its liberties—were essentially

Pombal's doing. In the tradition of most dictatorships the contributions of others were written out of the official history, and to Pombal were attributed almost superhuman virtues and accomplishments.

South America

The west coast of South America is also subject to violent earth tremors. Charles Darwin, during his globe-encircling voyage aboard the H.M.S. *Beagle,* was present in Chile in 1835 at about the time that the region near Concepción was struck by an earthquake of moderate severity. Unaccustomed to such activity, Darwin wrote an account of the event, even noting that some parts of the earth had been uplifted by as much as 3 m (10 ft) after the quake.

On May 21, 1960, a particularly devastating series of earth tremors shook Chile. The first of the tremors occurred on Saturday and caused widespread damage in and around Concepción; they continued all the next day (Fig. 17-8). At 2:45 P.M., Sunday, an unusually strong shock was felt, and many persons throughout southern Chile left their homes and stood about in the streets. Fortunately, they were still standing there at 3:15 P.M. when the main shock rocked the entire region.

Pierre St. Amand, a well-known seismologist, in his discussion of the Chile quake, provided a lucid account of the events associated with the first large shock.

The motion of the ground during the main shock was as if one were at sea in a small boat in a heavy swell. The ground rose and fell slowly with a smooth, rolling motion, smaller oscillations being superimposed on larger ones. In Concepción, cars and trucks parked by the side of the road rolled to and fro over a distance of half a meter when they bobbed up and down in response to the movement of the ground. The tops of the trees waved and tossed as in a tempest. Some already damaged buildings fell. The earthquake itself was silent; not a sound came from the earth. The period of vibration was of the order to ten to twenty seconds or more. The shaking lasted fully three and a half minutes and was followed for the next hour by other

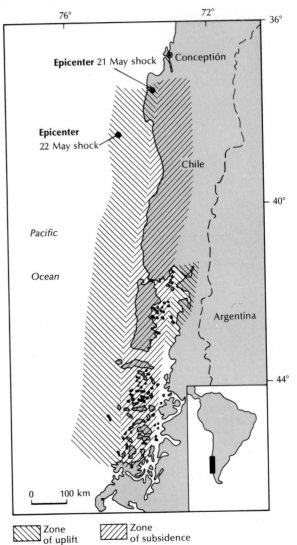

Fig. 17-8. Epicenters of the Chilean earthquake (May 21 and 22, 1960) and the area affected by uplift and subsidence accompanying the earthquake. Incredibly large forces changed the elevation of a huge area of the earth's crust.

shocks, all having a slow, rolling motion. . . . In the Region of the Lakes . . . the movement began smoothly and continued for some two minutes, just as in other localities, when suddenly, a loud subterranean noise was heard followed by a sharp jarring motion and a more rapid, less regular vibration of the earth. Similar reports were obtained at

other points to the east of the Lakes, and it seems from these that another earthquake took place here . . . while the ground was still shaking from the first shock.

Aftershocks continued to be felt for months. St. Amand recorded 119 shocks, from the main one in May 1960, until June 1961. While none was so severe as the main quake, 32 of the aftershocks were very strong. Because the active area was so large, 160 km (99 mi) wide and almost 1600 km (994 mi) long, and therefore included a wide variety of land features and surface conditions, the Chilean earthquake showed almost all the effects that such motions can have, with one notable exception. Even though a thorough search was made, both on the ground and from the air, no places were found that showed large offsets of the surface along a fault trace.

Other effects on the land surface, however, were numerous. Landsliding was one of the most common; the earthquake in many places triggered the sudden movement of unstable material. Cracks in the ground were caused by the settling of fill and the subsidence of areas underlain by soil that had liquefied. Liquefaction also caused huge earth flows, probably because of the presence of quick clays. In the harbor of Puerto Montt a ship was caught in a current of sand and mud that flowed into the bay; thereby it was given the unique distinction of being the first ship to go aground in a landslide. The ship was later converted into a small but unusual hotel.

In the area of the most intense shaking, some trees were snapped off, while others were uprooted and fell. In some cases the broken branches formed a circular pile of debris on the ground around the trunk.

Flooding by sea water resulted from drops in the level of the land. Some areas, however, were left 1.5 to 2 m (5 to 6.5 ft) above sea level. The rivers, too, produced flooding. In places the severe shaking compacted the poorly consolidated or unconsolidated material of the river banks, and the banks were lowered considerably. The motion of the earth actually shook water out of the ground, so that the rivers were unusually full. Almost continuous rain added to the problem.

Two days after the main shock, the volcano Puyehue in the Andes east of Concepción started an eruption that continued for several weeks. Steam and ash issued from a fissure about 300 m (984 ft) long and from several smaller openings. The last stage of the eruption was characterized by a number of flows of viscous lava. St. Amand wrote:

The local newspapers, in an unparalleled burst of enthusiasm, reported that 12 volcanoes had exploded and that two new ones had been formed. Lava was reported to be flowing down the sides of several of the volcanoes, and towns were said to have been buried. Because of the bad weather and poor visibility in the central valley, the inhabitants of the valley towns all believed that the volcanoes were, indeed, erupting and were concerned by the situation for a period of several weeks. Newspapers and news magazines all over the world repeated and enlarged on these stories.

The Chilean Region of the Lakes lies inland from the coast but is still well within the area affected by the earthquakes. The lakes exhibited a feature known as a *seiche*. The motion of the quake set up in each lake a wave which oscillated back and forth in the lake basin. The water sloshed, much like the familiar wave in a bathtub. Seiches in the Chilean lakes were small compared to that in Hebgen Lake, Montana, during the 1959 earthquake there. Hebgen Lake was an artificial body of water held in place by a dam. The seismologist John Hodgson wrote:

An eyewitness standing in the moonlight on Hebgen Dam and looking down its sloping face could not see the surface of the water, so far had it receded. Then with a roar it returned, climbing up the face of the dam until it overflowed the top, and poured over it for a matter of minutes. Then the water receded again, to become invisible in the moonlit night. The fluctuation was repeated over and over, with a period of about seventeen minutes; only the first four oscillations poured water over the top of the dam, but appreciable motion was still noted after eleven hours.

Fig. 17-9. Adobe buildings in Huaraz, Peru, destroyed by the 1970 earthquake. The epicenter of main shock was about 150 km (93 mi) west of Huaraz. *USGS*

A similar withdrawal of the water was noticed in the Chilean quake in the ocean itself. Shortly after the main shock, the coastal waters withdrew in a jumbled and confused state well below the lowest low-tide line. The inhabitants knew from experience what to expect and fled from their homes to nearby hills to wait and watch for the tsunami. They had from 15 to 30 minutes to wait and to complete their evacuation, whereupon the sea returned in a mighty wave that reached 6 m (20 ft) in height at some places and extended as much as 3 km (2 mi) inland. The waves continued for the rest of the afternoon, generally diminishing in height, although the highest wave is reported to have been the third or fourth rather than the first.

Leveling surveys run sometime after the 1960 quake indicated that two strips parallel to the coast, each about 1500 km (932 mi) long and 150 km (93 mi) wide, had changed in elevation; one went up while the other went down (Fig. 17-8). The values varied locally, but it appears that, on the average, uplift amounted to about 1 m (3 ft) and subsidence about 1.5 m (5 ft).

Although the Chilean quake of 1960 was of greater magnitude, the one that struck parts of Peru on May 31, 1970, caused ten times the destruction and loss of life. The quake was centered about 25 km (15.5 mi) off the coast and caused extensive damage on land. About 30,000 people lost their lives, largely as a result of collapse of buildings (Fig. 17-9). The quake also set in motion a chain of events that led to a catastrophic rock avalanche and the demise of thousands more in the towns of Yungay and Ranrahirca (see Chap. 9).

OCCURRENCE OF EARTHQUAKES

Earthquakes occur essentially all over the globe. Their distribution, however, is not haphazard; they tend to occur in linear belts or zones of seismic activity—zones which earth scientists now believe mark the boundaries of rigid plates (see Chap. 19). Earth tremors are generated at the plate boundaries where the plates converge, diverge, or slide past one another (Fig. 17-10). Away from the plate edges, earthquake activity is generally absent. The San Andreas Fault in California marks one such plate boundary between the Pacific Plate on the west and the North American Plate on the east.

It is noteworthy that a belt of seismic activity

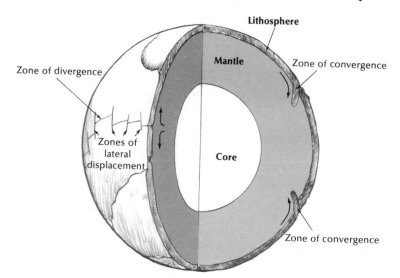

Fig. 17-10. The earth, showing the zones where plates converge, diverge, and slide laterally.

(mostly related to a zone of convergence) almost completely girdles the Pacific Ocean basin and is essentially coincident with the zone of active andesitic volcanism which also rings the Pacific (see Chap. 5). Most earth scientists believe that both phenomena—seismicity and volcanism—are effects of large-scale forces within the earth pushing the plates together.

Numerous and often large quakes occur in an east-west swath that cuts across southern Europe and Asia. Those quakes also apparently mark a zone of convergence of two main plates and perhaps a number of smaller ones.

Several earthquake belts occur within the ocean basins. Those that are coincident with the oceanic ridges and rises, such as the Mid-Atlantic Ridge or the East Pacific Rise, mark zones of divergence; that is, zones where the plates are moving apart. Volcanic activity occurs along such zones as well but the lavas tend to be basaltic rather than andesitic. One land-locked zone of divergence marked by both earthquake and volcanic activity runs along the east side of Africa. It is marked topographically by a nearly continuous line of elongate basins—the African Rift valleys.

Earthquakes also occur in zones where two plates slide laterally, one past the other. Such boundaries are relatively abundant in the ocean basins and are represented by the fracture zones that cross the oceanic ridges and rises.

Rarely do earthquakes originate at the earth's surface; generally, the fault movement starts at some depth below ground. The point of initiation of rupture or slippage is called the *focus*, whereas the point on the earth's surface vertically above the focus is designated the *epicenter* (Fig. 17-11). Once a rupture starts it propagates along the fault at a speed of several kilometers per second. The total length of the tear associated with the 1906 San Francisco earthquake was about 432 km (268 mi). Most quakes, however, originate from much smaller ruptures.

The focus of an earthquake can be as deep as 700 km (435 mi). At greater depths it seems likely that rocks are so plastic that they cannot accumulate the strain energy necessary to generate a seismic event. Arbitrarily, quakes that originate at depths of less than 70 km (43 mi) are called *shallow focus*; those originating between 70 and 300 km (43 and 186 mi), *intermediate focus*; and those starting between 300 and 700 km, *deep focus*. Most earthquakes taking place today fall into the category of shallow focus.

Along the plane of rupture, strain energy is converted and released in two different energy forms—heat and seismic wave motion. About half of the released energy goes into heating the rocks directly adjacent to the fault zone. The seismic vibrations, on the other hand, travel away from the source in all directions, much as ripples radiate from the point where a pebble is dropped into a quiet pond. As the waves travel outward, the energy they carry is distributed over a progressively larger perimeter, and the intensity of wave motions diminishes.

Some seismic waves travel through the interior of the earth and are called *body waves;* others travel at and close to the surface of the earth and accordingly are called *surface waves.* It is the motion of surface waves that we feel during an earthquake. Such waves can be set in motion either directly by fault movement at or close to the surface or, indirectly, by body waves that emanate from the focus and strike the ground surface near the epicenter directly above. The violence and complexity of the shaking at the earth's surface is related to the magnitude of the released strain energy, the proximity of the focus, the duration of the quake, and the physical nature of the surface materials. Unconsolidated mud is prone to the development of much larger waves than are crystalline igneous or metamorphic rocks. We shall return to the discussion of body waves later in the chapter, when we explore what those waves tell us about the interior of the earth.

PEOPLE AND EARTHQUAKES

What can be done to soften the destructive effect of earthquakes on people and their trappings? Certainly, it is inconceivable that urban areas will be moved to less-seismic regions. Managua, Nicaragua has been destroyed at least four times by earthquakes, the latest as recent as 1971; yet its residents continue to rebuild and to try to forget the past. California is hit by hundreds of quakes each year, and still its population continues to burgeon (Fig.

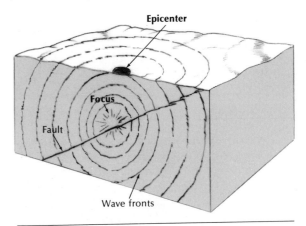

Fig. 17-11. Relation of focus to the epicenter of an earthquake. The focus is the site of initial movement along the fault; the epicenter is a spot on the earth's surface directly above the focus. Also shown are wave fronts of seismic waves radiating out from the focus.

17-12). And there is still the possibility, even in the seismically quiet regions of the world, that a rare but nonetheless destructive quake might occur, as for example that in New Madrid, Missouri.

At present, there are two approaches to the problem of possible earthquakes. One is concerned with establishing an awareness of the effects of quakes and with the enforcement of building codes; the other is concerned with the prediction and control of the earthquakes themselves.

It has become apparent that through careful selection of building sites and major facilities and the establishment of minimum construction requirements, the effects of earthquakes can be lessened.

In major earthquakes, buildings constructed in essentially the same way will be most severely damaged if they stand on filled, unconsolidated, or water-soaked ground, and least damaged if they are founded on solid rock. Selection of an actual building site is a factor over which we can exercise some control, yet, it is surprising how little thought has been given to the problem in most regions of high earthquake hazard. Fortunately, as we become

Fig. 17-12. Looking south along trace of San Andreas Fault, from Mussel Rock on the west side of San Francisco Peninsula, across Daly City, toward San Andreas Lake. During the 1906 earthquake, the line of lateral displacement ran from Mussel Rock for almost 160 km (99 mi) into the distance. A quake of similar magnitude today would cause untold loss of property and life in the now-urbanized region. *R.E. Wallace, USGS*

more aware of the physical world around us, the matter of choosing more stable building sites is receiving increased attention.

The element over which we have the most direct control is the type of construction used in the buildings we inhabit. It is probably most important to construct a building that will vibrate as a single unit. It helps, too, if it is not too rigidly structured. Fortunately, most wooden-framed houses fit those requirements, especially if the mudsills are bolted to a concrete foundation, and the studs are strengthened by angle-bracing at the corners. Also important is the use of reinforcing rods, run the length of brick chimneys and tied to a concrete slab at the chimney base. Among the more memorable sights after the 1933 Long Beach earthquake were the hundreds of brick chimneys snapped off at the roof line.

Among the more vulnerable buildings are those with roofs extending over a comparatively broad span, such as supermarkets, bowling alleys, and auditoriums. If the roof of such a structure is arched, it exerts an outward thrust against the bearing walls, which may be hammered down, as it were, until they collapse completely.

It is necessary to consider the geologic structure of an area when determining the placement of major transportation routes, aqueducts, pipelines, water-storage facilities, power plants, and the like. Geologists now know the major fault traces in most areas and can generally tell which ones are active at present or are likely to have been active recently. Most municipalities construct their vital industries with that knowledge in mind. For example, in the selection of possible building sites for nuclear power plants, those that would otherwise meet specifications may fail to qualify because of the presence of faults that appear to have been recently active.

Regardless of how carefully we select building sites and how well we build, a major earthquake in a populated area will cause property damage and loss of life (Fig. 17-13). If only we could predict earthquakes. Population centers could be evacuated, thereby reducing the loss

Fig. 17-13. The San Fernando earthquake of 9 February 1971 caused the collapse of this overpass in Los Angeles County, California. *USGS*

of life, and at least some preparations could be made, thereby reducing property damage.

Such desires may seem fanciful and beyond attainment. Yet earth scientists, particularly in those countries where earthquakes are prevalent and resources are available (the United States, Japan, the USSR, and China) are trying to reach those goals. *Earthquake prediction* has assumed a high priority in all of science and is another example of how science can be used to benefit people. In China today, earthquakes have been designated the "number-one naturally occurring enemy of the people," and work in the field of earth science has been directed singularly toward reliable prediction.

One type of general long-range seismic forecasting is based on the aerial distribution and intensity of earthquakes that have occurred in the past. From such analysis, seismic-risk maps can be generated which give an inkling of the likelihood of earthquakes of a certain size for any geographic area (Fig. 17-14).

How would we go about predicting an earthquake? Are there any telltale signs? One line of approach to earthquake prediction involves determining how much strain has accumulated along a locked fault plane and at what rate. Earth scientists look for changes such as tilt or strain in the relative and actual positions of points on the earth's surface—either by geodetic or other measurement. Or they look for changes in the physical properties of the rocks along the fault, which are affected by strain buildup. Such changes may be revealed by measuring the local magnetic field, the electrical conductivity of the rocks, or the velocities of seismic body waves (Fig. 17-15).

Scientists also set up sensitive seismic instruments along a fault zone and "listen" to the crackling of the earth as strain builds and the fault prepares to slip. As the time of slippage approaches, the number of micro-earthquakes per observational interval generally increases. It may be an increase in micro-activity that accounts for the common observation that dogs and horses become increasingly nervous and jittery just before a large quake.

All of our techniques are still rather crude, and earth scientists have been unable to predict earthquakes precisely. Many seismologists believe that changes in the speeds of body waves, which apparently reflect cracking of the rock adjacent to the fault and their subsequent infilling by ground water, will become our best means of prediction. But the method has had only mixed success to date. For example, seismologists recently predicted on the basis of changes in body-wave velocities that in the region of Hollister, California, an earthquake of a certain magnitude would occur within a prescribed three-month period. Hollister is continuously shaken by small quakes, and residents, when they heard the prediction, stated that it was just like predicting rain in Seattle, Washington, during the winter. Much to the chagrin of the predictors, no quake at all occurred during the specified time period. Shortly after they acknowledged failure, a quake of approximately the predicted magnitude occurred.

If and when we acquire the ability to predict routinely an earthquake of a certain magnitude, another problem will arise—relating to how citizens in the affected area respond to the prediction. For example, what if a potentially disastrous earthquake was predicted for the vicinity of San Francisco on or about a certain date? Many would panic and flee helter-skelter away from the danger area. Accidents and loss of life could occur and who would be responsible? What if the alarm was a false one, and after a week or so people returned to find their houses ransacked, pets lost or dead, crucial jobs interrupted? Who would shoulder the blame? What if, after they returned to the city, an earthquake did hit, causing much destruction? What would be the reaction of the surviving citizens? What would be the effect on the economy of a city such as Los Angeles if it were predicted that a large earthquake would occur definitely sometime within the next two years?

Those and other equally sticky questions have led some scientists to conclude that the only way to deal with earthquakes is to control them. But is there any evidence that we can influence their behavior? The answer appears to be "yes," and the approach to controlling them centers around a seemingly unlikely substance —water. About 15 years ago a deep well was drilled into the Precambrian rocks underlying the Rocky Mountain Arsenal near Denver, Colorado. The purpose of the well was to dispose of radioactive waste fluids. Beginning in 1962 and continuing for three years thereafter many thousands of gallons of waste fluids were pumped down the well. Shortly after pumping began, earthquakes began to be felt in the Denver area. Officials at the arsenal denied any connection between the pumping and the earthquakes, but a Denver geologist, David M. Evans, noted that the frequency of quakes was directly related to both the amount of wastewater pumped in during any period and the pressures used to pump it in (Fig. 17-16). In 1965, the pumping was stopped and the earthquake activity slowly died away.

Most earth scientists have concluded that the

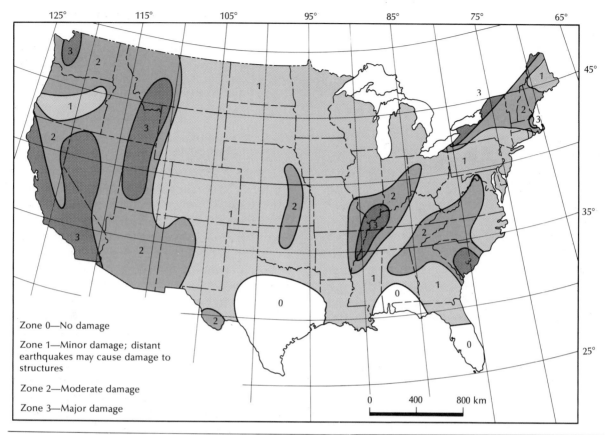

Zone 0—No damage

Zone 1—Minor damage; distant earthquakes may cause damage to structures

Zone 2—Moderate damage

Zone 3—Major damage

Fig. 17-14. Seismic-risk map of the United States, based on the known distribution and extent of damaging earthquakes, evidence of strain release, and the study of major geologic structures and provinces believed to be associated with earthquakes. From the standpoint of seismic risk, Louisiana, Alabama, Florida, and southern Texas are the least hazardous places to live in the United States.

Denver earthquakes represented the release of tectonic strain stored in the Precambrian rocks adjacent to the fault. Before the pumping the walls were locked, and no faulting or quakes were recorded. By forcing wastewater down the hole, water pressure at depth was generated, a pressure sufficient to force the walls of the fault apart and thereby effectively lubricate it. The fault "unlocked" and slippage, in response to the accumulated tectonic strain, occurred.

Such incidents have been documented elsewhere, leading geologists to speculate that pumped water might be used for the control of fault movement. Let us take the San Andreas Fault as an example. Along a segment of the fault, three equally spaced wells would be drilled down along the fault plane to a depth of 4600–6100 m (15,093–20,014 ft). Ground water would be pumped out of the two wells at either end of the fault segment, drying up those parts of the fault and effectively locking them. Then enough water would be pumped down the center well to lubricate the fault and permit movement. Supposedly only a small-magnitude quake would result, and afterward

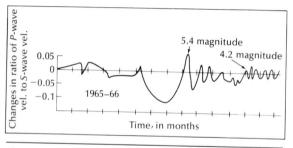

Fig. 17-15. Two earthquakes in the Garm district, U.S.S.R. (1956–66), were preceded by drops in the ratio of *P*-wave to *S*-wave velocity. (See p. 576 for definitions of *P*- and *S*-waves.)

that section of fault would be temporarily free from strain. The process of drying out and lubricating would be repeated up and down the fault until stored strain energy had been released along its entire length.

There are many unknowns surrounding such an approach to earthquake control. Even if it would eventually work, large earthquakes could be triggered during the early stages of the strain release program. At the moment no federal, state, or local agency feels ready to accept

Fig. 17-16. Correlation between quantity of wastewater pumped into a deep well and the number of earthquakes near Denver, Colorado.

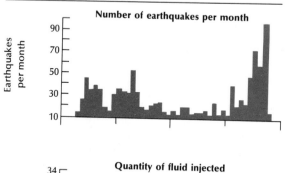

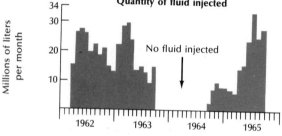

the responsibility of risking this method of control.

The same could be said about earthquakes as is said of the weather: Everyone talks about them but no one does anything about them. But perhaps in the near future that will all change.

MEASURING THE HEARTBEAT OF THE EARTH—SEISMOLOGY

There is another aspect of the earthquake story, one that is rather undramatic and has to do with the systematic study of earthquakes and seismic waves. Through it we can hope to learn a great deal about the nature of earthquakes around the world and about the material inside the earth through which body waves pass.

Earthquakes, simply because they are dramatic natural events, have invited speculation as to their origin even from the earliest times. Greek philosophers concocted many theories to explain them, explanations that today read like mythology, but not half so interesting as the adventures of the far-wandering Odysseus. For example, Aristotle, who lived in the fourth century B.C., and whose interpretation of natural phenomena once had the force of dogma, believed that earthquakes resulted from the escape of air trapped within the bowels of the earth.

The Lisbon earthquake in 1755 prompted the first serious effort to study earthquakes scientifically. By 1700, from observations of swinging chandeliers, John Mitchell, an English physicist, surmised that earthquakes produced wave motion in the rocky crust of the earth. In 1859, Robert Mallet (1810–81), an English scientist who had been thinking about earthquakes in theoretical terms, had an opportunity to study the results of a real one when he visited Italy and saw the damage done to hill towns in the Apennines east of Naples by a destructive quake in 1858. Although he believed incorrectly that earthquakes were explosive in origin and were related to volcanic activity, he laid the foundations of observational *seismology*.

The word seismology is a common one today; it comes from the Greek *seismos,* for earthquake. By observing such things as the direction in which buildings and monuments fell and the nature of cracks in the ground, he worked out a rough method for determining the source of an earthquake. Mallet was among the first to try an experimental approach in earthquake study by setting off explosions and measuring the travel times of the resulting waves. He also started an earthquake catalog and through it achieved some understanding of the geographic distribution of earthquakes.

The seismograph

The founder of modern seismology was an English mining engineer, John Milne, who went to Japan in 1875 as one of a group of men who wanted to bring the technology and educational systems of the West to the Japanese people. By 1880, he had sufficient stimulus, from a nearly continuous exposure to earthquakes, to emerge as the first full-fledged seismologist. Through his efforts the seismograph, the instrument used to record the vibrations set up by an earthquake, was transformed from a scientific curiosity into a precision instrument.

The perfection of the instrument is the result of continued effort over the many years since the rudimentary one was contrived by L. Palmieri in Italy in 1855. The problem that had to be solved initially was how to measure the vibrations from an earthquake at the earth's surface when the instrument itself is attached to the surface and is shaking. The solution was a mixture of simplicity and elegance. A large mass was suspended from a thin filament (a pendulum), thereby virtually isolating the object from the earth's surface—it was in essence floating in space (Fig. 17-17). When earthquake waves set the ground into motion at the recording site, the mass, because of its large inertia, tended to remain at rest while the ground beneath it and the point of filament suspension shimmied and shook. A stylus, or pen point,

was attached to the mass so that it could trace the motion of the ground as it moved to and fro beneath the pen.

The first primitive seismographs consisted of only a single, vertically suspended pendulum, and they scribed their record in smoothed sand or on a stationary piece of smoked paper (Fig. 17-17). Since the stylus traced its path back and forth past the same spot on the ground, each earth tremor produced a bewildering array of lines. But the problem was overcome when the recording paper was attached to a plate which moved beneath the pen at a known rate; thus it could not write over its previous trace (Fig. 17-18). In most modern seismographs, the pen records its wiggles on a drum which turns at precisely four revolutions per hour. As the drum revolves it slowly spirals ahead; it is in essence a spiraling clock. The recording pen also makes a tic mark on the record after each minute, so that the onset of any motion to the hour, minute, and second can easily be read off. For ease of comparison of their records, all stations around the world have agreed to use the same time base; when the paper on their various drums is changed each day, it is reset and synchronized with Greenwich Civil Time.

A somewhat more complicated problem was associated with the early seismographs. During a quake the ground moves in a complicated manner, in three dimensions; that is, the motion at any one moment can be simultaneously east-west, north-south, and up-down. By means of a single pendulum suspended vertically it was impossible to obtain a readable record of the three-dimensional movement. In the late 1800s, Milne, who had long struggled with the problem, hit upon the idea of using three separate seismographs at each station, each one capable of recording one of the dimensions—E-W, N-S, and vertical (Fig. 17-19). In order to measure the horizontal movements he mounted the pendula in horizontal positions, but at right angles to each other, and for measuring vertical movement, he retained the pendulum suspension in a vertical position. The breakthrough established the study of earthquakes as a quantitative field

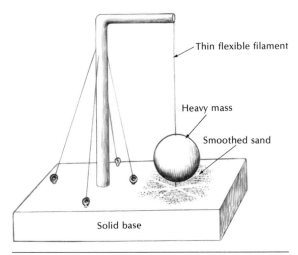

Fig. 17-17. Simple model of a seismograph in which a heavy mass is suspended from an arm on a thin wire. During an earthquake, the base (solidly attached to the earth), moves as does the arm, but the mass, because of its inertia, is relatively unaffected by shaking.

Fig. 17-18. In a refinement of the seismograph design, a rotating drum was placed beneath the stylus. After Fig. 23-14A (p. 394) from THE EARTH SCIENCES, 2nd Edition, by Arthur N. Strahler. Copyright © 1963, 1971 by Arthur N. Strahler. Reprinted by permission of Harper & Row, Publishers, Inc.

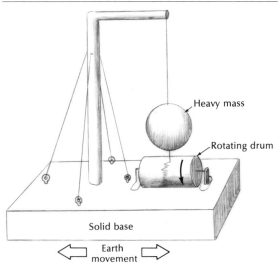

of science. By 1882, Milne-Shaw seismographs had been installed at 50 cooperating stations around the world. Scientists throughout the world were then able to measure and compare records of the most complicated shakings of the earth. It is no wonder that Milne is considered the founder of seismology.

To be sure, other problems, such as amplification of signals and damping of vibrations set up in the seismograph by a quake, had to be overcome as the seismographs of the nineteenth century gave way to the modern instruments. Several of the recent improvements in design are a result of research in electronics and experimental physics. For instance, in one seismograph today, relative motion between the pendulum and earth is sensed, amplified, and recorded electromagnetically.

Today's precision seismographs are relatively easy to use, maintain, and standardize. Most are installed at permanent stations, but portable models are becoming increasingly popular. The latter are set up to record quakes at some strategic location for a few days, weeks, or months, and then are moved elsewhere. As you may recall, portable seismographs were taken to the moon on several of the Apollo missions and have been relaying to the earth recordings of "moonquakes" ever since. Some seismographs are built specifically to monitor the intensely strong vibrations created close to the epicenter; others are capable of measuring incredibly weak seismic rustlings. Most seismographs can be variably tuned in order to pick up long or short seismic waves, as desired. Some special instruments are capable of recording the exceedingly long wavelength seismic vibrations associated with large quakes, which require about an hour for the passage of one wavelength.

The number of operating seismographs stands at more than 1000 today. Seismograph stations are not uniformly distributed around the world: countries that have the most earthquakes generally also operate the largest number of seismographs. Since the early 1960s about 120 of the stations have agreed to operate according to a specified procedure and to

transmit their records to a central collection agency at Golden, Colorado, maintained by the Earthquake Information Center of the U.S. Geological Survey. Most theoretical seismologists today use the records and do not themselves operate seismograph stations. The copying and distribution center for the world-wide network of stations is at Boulder, Colorado, and functions as a part of the National Oceanic and Atmospheric Agency (NOAA).

Seismology has come a long way in the last 100 years. It can truly be said that we are capable of listening closely to the heartbeat of our planet. As a result, we have begun to learn not only a great deal about earthquakes but about the interior of the earth as well.

Waves and wiggles Most of the time the pens on a set of seismographs are relatively quiet. They record only low-level background vibrations that arise from a variety of causes: electronic instrumental noise, traffic vibrations near the site, wind gusts, and even the pounding of heavy surf on a distant coastline (Fig. 17-20).

From time to time, however, the seismographs will be shaken by waves emanating from an earthquake somewhere in the world, and the pens will record the onset, duration, and nature of the incoming wave motion (Fig. 17-21). For a quake that is relatively close by, the instrument generally will record three major wave pulses arriving in succession at the station. The first to arrive, generally a relatively low-amplitude wave, is called the *primary*, or *P wave*. The second pulse, termed the *secondary*, or *S wave*, is distinguished by a sudden onset and is normally of greater amplitude than the *P* wave. And the *S* wave is followed by a wave train of higher amplitude and longer wavelength called the *long*, or *L wave*. Depending on where the quake is located in relation to the recording station, each of the three seismographs (E-W, N-S, and vertical) will show some variation in character in each of the three wave trains.

The *P* and *S* waves travel through the earth and accordingly both are called body waves,

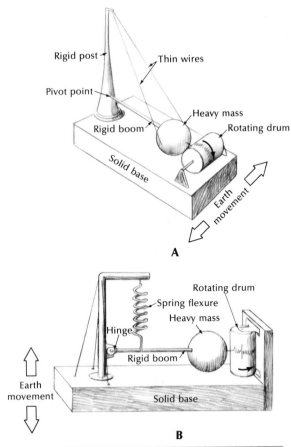

Fig. 17-19. Basic design of modern seismographs. A. Horizontal-type pendulum seismograph for measurement of N-S or E-W motions. B. Hinged pendulum seismograph for measurement of vertical motion. After Figs. 23-15 and 23-14B (p. 401) from THE EARTH SCIENCES, 2nd Edition, by Arthur N. Strahler. Copyright © 1963, 1971 by Arthur N. Strahler. Reprinted by permission of Harper & Row, Publishers, Inc.

whereas the *L* waves follow paths in the outer layers of the earth and are called surface waves.

The nature of motion for each type of wave has been studied from a theoretical standpoint and in the laboratory. It has been found that *P* waves travel through a solid by means of an accordion-like, push-pull sort of oscillation (Fig. 17-22). As the wave travels, the rock at any one place is at one moment squeezed together and at the next pulled apart. That type of compressional wave motion is familiar to us all since it is characteristic of sound waves as they move through the air and into our ears. Sound, as we hear it, is simply the recording of the

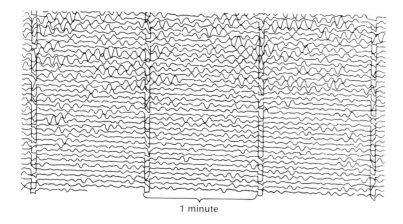

Fig. 17-20. Enlargement of seismograph record showing low-level background vibrations that arise from a variety of causes. One-minute interval marks are transcribed on the record for precise determination of the onset of seismic signals.

to-and-fro motion of our ear drums in response to the push-pull waves moving through air. Sound travels in a similar way (but more quickly) in water. By means of sonar (the beaming and receiving of sound signals underwater), a ship's personnel can locate nearby ships, pinpoint obstacles, and gain a knowledge of the topography of the ocean bottom over which the ship is traveling.

Compressional sound waves travel fastest in a solid. That is because the atoms are tightly bonded to each other, causing a high degree of elasticity. And anything that will increase the degree of elasticity will cause the *P* waves to be transmitted faster. For instance, when a rock specimen is squeezed in a hydraulic press, the

atoms in the rock are pushed closer together, thereby increasing its elasticity and resulting in an increase in the speed with which *P* waves travel through it. Therefore, other things being equal, deep inside the earth where pressures are extremely high, *P* waves travel more quickly than they do near the earth's surface. Increasing the temperature of a rock specimen, on the other hand, has a tendency to *decrease* elasticity and will cause a corresponding reduction in the transmission speed of *P* waves. Obviously, at any point inside the earth the elasticity will be the combined result of rock type, confining pressure, and temperature.

The *S* wave always travels more slowly than the *P* wave, and the reader would be correct

Fig. 17-21. Actual seismogram of north-south motions of two aftershocks recorded in Fairbanks, Alaska, a few days following a major earthquake in 1964. The onset of *P* and *S* waves are indicated for the first event. *S* waves and *L* waves of the first shock obscure the onset of the *P* wave of the second event. The foci of the events were about 800 km (497 mi) south of the seismograph station.

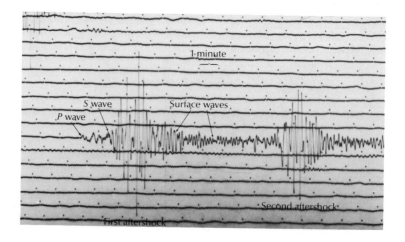

in assuming that the difference in arrival time is due to a difference in the manner in which an S wave travels. It has been shown that in a solid, another type of oscillation is possible besides push-pull motion; seismic energy also can be carried by a shaking motion of the material tranverse to the direction of passage of the wave (Fig. 17-23). In that mode, the rock particle at any one point moves back and forth sideways while the S wave travels on in a snake-like fashion. It is much like the motion made by shaking a piece of a rope; the waves travel in S-shaped patterns as the material slides from side to side. Rocks exhibit a lower degree of elasticity to such motion; therefore, the S wave is slower than the P wave. But, of course, an increase in confining pressure will cause an increase in the elasticity of the rock, and the S wave will increase in speed. At depth in the earth, the S wave is expected, therefore, to move more quickly than near the surface. And an increase in temperature will cause a decrease in elasticity and a slowing of the wave. Fluids and gases lack the type of elasticity that will permit the passage of S waves. Therefore such waves cannot travel through water or air.

The L waves, the slowest of the three, travel in a rather complex way close to the surface of the earth. Because their passage is restricted to the outer levels of the earth, their speed is relatively constant.

All earthquakes generate all three types of waves. How they appear on the record at any one seismograph station will depend on the size and nature of the rupture at the point of origin, the distance from the focus to the seismograph station, the depth of focus of the quake, and the kind of material through which the waves travel before reaching the station.

We can gain insight into how waves travel by plotting the arrival time of the waves generated by a single earthquake at stations located at increasing distances from the quake epicenter (Fig. 17-24).

If many stations are used, it is possible to construct a travel–time curve (Fig. 17-25). As you can see, the plot for each wave is decidedly

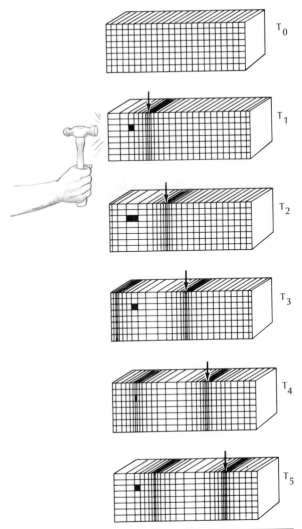

Fig. 17-22. Successive stages in the small-scale deformation of a rock by a P wave. As the sequence progresses, a zone of maximum compression (marked by the arrows) passes through the solid. As wave after wave passes through the block, the darkened square (representing any small volume of the material) shakes back and forth, undergoing alternate compression and expansion.

different. The L-wave plot is almost a straight line, whereas those of the P and S waves are curved. Of the two body waves, the plot of the P-wave arrivals bends more sharply than does that of the S waves. Moreover, from about 10,000 km or 6214 mi (equal to about 102° of

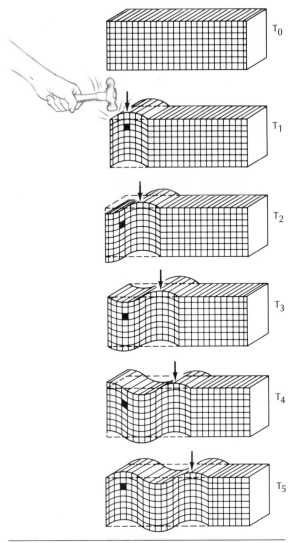

T_0

T_1

T_2

T_3

T_4

T_5

Fig. 17-23. Successive stages in the small-scale deformation of a rock by an *S* wave. As the sequence progresses, the *S*-wave crest (marked by the arrows) passes through the block, and the material shakes sideways. Any particular volume (darkened block) undergoes cyclic changes in shape in response to the passage of waves.

earth circumference), to about 16,000 km or 9942 mi (equal to about 143°) from the epicenter, no *P* or *S* waves are directly recorded. It is as if seismographs in that portion of the globe were in the shadow of some seismic obstacle. That zone is called the *shadow zone*.

At arc distances beyond 143°, seismic waves are again received, but their travel time is much longer than expected, as if they had been slowed during their passage through the central part of the earth.

Both *P* and *S* waves start out together from an earthquake much as in a footrace. However, a *P* wave is about twice as fast as an *S* wave, and the farther the two waves race, the wider becomes the gap separating the wave fronts (Figs. 17-24 and 17-25). The inequality in speed between *P* and *S* waves is indeed fortunate, for it provides us with a means of determining the distance from a seismograph station to the earthquake focus. For example, in the seismograph record in Figure 17-26, the *S* wave arrived about 8 minutes after the *P* wave. The travel-time curve (Fig. 17-25) indicates that the separation in arrival times is equivalent to about 8800 km (5468 mi). Now, seismologists cannot tell from which direction the quake came, so they cannot pinpoint the epicenter. They can only say on the basis of one set of travel–time data that the epicenter lies at a distance of about 8800 km from the station. A circle of that radius drawn around the seismograph station on a map will contain all of the possible points of origin (Fig. 17-27). If the same quake is recorded at another station, however, it is possible to narrow down the quake origin. Seismologists at the second station can similarly determine the distance to the epicenter and draw on a map a circle of appropriate radius with their station at the center. The circle will intersect that drawn around the first station at either one or two points. If information is available from a third seismograph station, a single intersection will be determined and thereby the earthquake epicenter can be located (Fig. 17-27).

In practice, three circles will not usually cross at a single well-defined point. For one thing, many earthquakes do not start at a point but commence nearly simultaneously along a fault line. The 1906 San Francisco quake, for example, originated along a fracture hundreds of kilometers long. For another, some quakes do

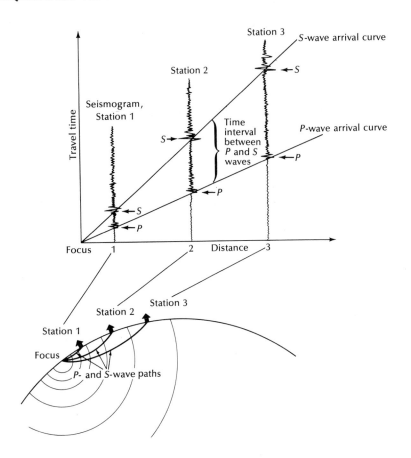

Fig. 17-24. Increase in the arrival time of *P* and *S* waves with distance from the epicenter. Notice that the time interval between the arrivals also increases. As shown, the *P* wave is faster than the *S* wave.

not originate near the surface of the earth but at depths of as much as 700 km (435 mi). The travel-time curves given in Figure 17-25 are fine for shallow-focus quakes, but different curves must be used for deep-focus quakes. The existence of the latter was first suggested by H. H. Turner, a seismologist, in 1922, and during the following decade their reality was widely accepted only after debate and discussion. Part of the evidence for their existence is the different sort of seismogram they write compared to that of shallow-focus earthquakes. For one thing, the *L* waves are either lacking or ambiguous. Also, the *P* waves arrive sooner than they would have had they started near the surface. After all, they have the advantage of a considerable head start in their race toward a seismograph station. As we have already

noted, deep-focus quakes commonly originate in zones of convergence, where two plates come together. Most earthquakes originate around the periphery of the Pacific Ocean but some come from beneath the Himalayas, the Mediterranean Sea, Indonesia, and the Caribbean. The nature of rupture at a depth of several hundred kilometers is not exactly clear since at those depths it might be expected that the rocks are so hot that they would tend to flow. Yet the quakes do occur, and the amount of energy released during any one event is just about the same as that from a near-surface quake. Today the accurate location of the focus of most quakes is accomplished by the use of high-speed computers. The records from as many stations as possible are used in the calculation. After the focus is roughly de-

termined by use of the travel-time curves, as outlined above, the position of the focus is continuously moved (figuratively) in small steps laterally and vertically, and the travel times recalculated to all stations. The best estimate of the position of the focus is accomplished when the calculations lead to the best fit of calculated travel times to all stations involved, with the actual travel times measured directly from the records.

Fig. 17-25. Travel-time curves for earthquakes originating at depth of less than 100 km (62 mi). Notice how the *P* and *S* wave curves bend, indicating an increase in speed with distance.

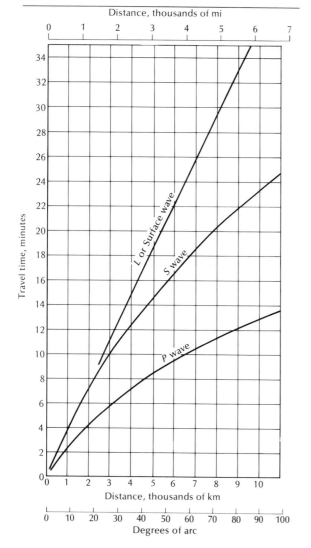

Intensity versus magnitude

As soon as the earth starts to rupture, *P* and *S* waves begin their outward journey, carrying the vibrational energy initiated by the seismic disturbances in all directions. The size of the waves (their amplitude—which is measured from the crest of one wave to the trough of another and divided by 2) at any one point varies with the size of the event. Small ruptures produce small waves, whereas large ruptures produce large oscillations. It is possible to obtain some idea of the magnitude of an earthquake at its source by measuring the amplitudes of *P* and *S* waves as they are recorded on a seismograph. Of course, the amplitude diminishes as the waves travel away from the disturbance and as they carry the vibrational energy into an ever-enlarging volume. Therefore, it is necessary to make a correction that takes into account the distance of the recording station from the seismic event.

A scale of earthquake magnitudes based on that line of reasoning was first devised by the seismologist Charles Francis Richter in 1932 and bears his name. It enables seismologists at a distant recording station to get a pretty good idea of the amount of strain energy released during an earthquake (Table 17-1). The scale is roughly exponential, so that a difference of only one division corresponds to a difference in source energy of about 30 times. During a magnitude 4 event, for example, the energy released is not just twice that released during a magnitude 2 event, but about 900 times as great. The largest earthquakes ever recorded have measured about 8.9 on the Richter scale: they correspond to the energy produced by detonation of about 100 million metric tons of TNT. In contrast, a magnitude 1 quake would be equivalent to the energy released by the detonation of less than half a kilogram (1 lb) of TNT. You might wonder why no quake larger than 8.9 has ever been recorded. It appears that there is a limit to how much strain energy can be stored before the rocks fail and slippage occurs.

Each year, many hundreds of thousands of magnitude 1 and 2 quakes occur; only about

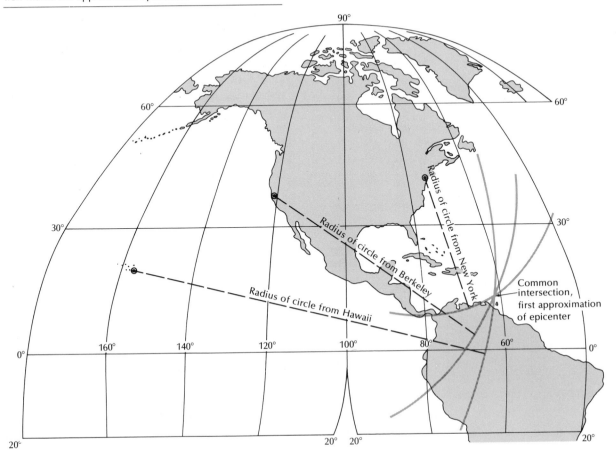

Fig. 17-26. Portion of a seismogram from Berkeley, California, recording an earthquake on September 20, 1968. Both the *P* and *S* waves are detectable, with the *S* wave arriving almost exactly 8 minutes after the *P* wave.

Fig. 17-27. From the time lag between *P* and *S* waves recorded at three separate stations, the approximate epicenter of an earthquake can be determined. The distance to the event is calculated and a circle of that radius is drawn around each of the three stations. The common intersection marks the approximate epicenter.

Table 17-1. Earthquake intensities, magnitudes, statistics, and energies

Characteristic effects of shallow shocks in populated areas	Approximate Richter magnitude	Number of earthquakes per year	Energy (Approx. equiv., metric tons of TNT)
Damage nearly total	≥8.0	0.1–0.2	>5,668,750
Great damage	≥7.4	4	907,000
Serious damage, rails bent	7.0–7.3	15	181,400
Considerable damage to buildings	6.2–6.9	100	27,210
Slight damage to buildings	5.5–6.1	500	907
Felt by all	4.9–5.4	1,400	136
Felt by many	4.3–4.8	4,800	27
Felt by some	3.5–4.2	30,000	1.8
Not felt but recorded	2.0–3.4	800,000	6×10^{-3} to 0.8

Source: Data from B. Gutenberg.

one larger than magnitude 8 occurs every five to ten years (Table 17-1). Any quake larger than magnitude 6 can be considered a major event and capable of producing considerable destruction.

The extent of damage done by an earthquake depends (as we have previously learned) on a variety of factors, such as proximity of the focus, duration, amount of energy released (Richter magnitude), type of rock underlying the site, strictness of building codes and the type of structures, and density of population. The amount of destruction for any particular quake will vary as those factors vary. An earthquake scale—based on the damage wrought by a quake—is called an *intensity scale,* and it ranges from a minimum number, corresponding to a barely discernible event, to a maximum number corresponding to total collapse of buildings and great loss of life (Table 17-1). The Modified Mercalli Scale is perhaps the most commonly used intensity scale today. It was formulated by the Italian seismologist Giuseppe Mercalli (1850–1913) in 1902 and modified in 1931 by American seismologists Harry O. Wood and Frank Neumann.

Seismic waves and the earth's interior

Our concrete knowledge of the interior of the earth is restricted to observations of rocks once buried deep below the surface, to cuttings from drill holes thousands of meters deep, and to rock fragments ripped from the walls of volcanic conduits and carried to the surface by eruptions of lava.

Fortunately, there are means by which we can make realistic inferences about the inaccessible parts of the earth. Of all the means, seismology has proved the most useful. From the analysis of countless earthquake records, seismologists and other earth scientists have put together a model of the structure and composition of the earth. Of course, we know that it is only a first approximation and that it will be modified somewhat as time passes and as new data become available. Most earth scientists feel so confident about the general features of the model, however, that they would expect few surprises even if at this moment we could drill a hole to the center of the earth.

Seismology has been instrumental in helping us from our quandary, largely because body waves—both *P* and *S* waves—radiate outward

in all directions from an earthquake. Many traverse the deeper parts of the earth as they make their way toward distant seismograph stations, and in the wiggles that they write on the seismogram, they carry information about the materials through which they pass. Before trying to decipher the results, let us first look at some of the properties of body waves.

Body waves, like all traveling waves, are subject to bending (refraction) and reflection. In a uniform material under uniform temperature and pressure, a seismic body wave will travel at a constant velocity—which depends on the density and elastic properties of the material. If the wave passes into a material of different density or elastic properties, its velocity generally will change, and refraction of the wave will generally occur (Fig. 17-28).

For example, you are aware that light rays, as they pass from the air into the glassy material of a pair of eyeglasses and back again, are refracted, or bent, in such a way that the rays are focused upon the retina of the eye.

Reflection of waves will occur when they encounter a sudden velocity change (discontinuity) in the material through which they are passing. As an example, light rays are reflected when they encounter a highly polished surface. Reflection surfaces generally are not totally reflective, particularly to seismic waves. Some of the wave energy passes through the discontinuity and into the material beyond, where it continues to propagate (Fig. 17-29).

If the rocks through which a wave is traveling increase in temperature, there will be a reduction in speed; an increase in confining pressure will cause the opposite behavior. In a fluid, in which all rigidity has been lost, the S waves cannot propagate at all, and the P waves are greatly slowed.

As we mentioned earlier, the waves that arrive directly at progressively more distant stations travel at increasingly greater speeds. That can mean only that those waves go deeper into the earth and move through materials that permit progressively faster wave propagation. The implication is that the earth is not uniform in its physical properties at depth. Seismol-

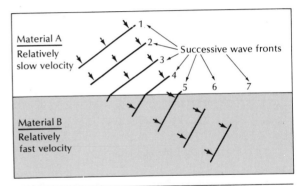

Fig. 17-28. Refraction of a light wave as it moves from one material (A) to another (B). The wave travels more quickly in (B) than in (A); therefore when it strikes the interface and is partly in both materials, it is bent, or refracted.

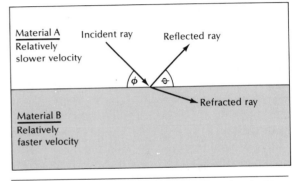

Fig. 17-29. Refraction and partial reflection of an incident light wave as it encounters a surface of discontinuity (sharp break) between one material (A) and another (B). The angle at which the ray is reflected from the interface (θ) equals the angle of impingement (ϕ) of the incident ray.

ogists now know that the earth is made up of concentric layers of materials. Figure 17-30 shows in a simple way how seismic waves behave as they pass through a layered body.

By studying the details of travel time and distance data collected from countless earthquake records from all over the globe, seismologists have been able to put together a consistent picture of how body-wave velocities vary with depth (Fig. 17-31).

The results are intriguing, for they show that

sharp velocity discontinuities exist within the earth; that *S* waves cannot be propagated between 2900 and 5000 km (1802 and 3107 mi), and that the increase in *P*- and *S*-wave velocities in the outer 700 km (435 mi) of the earth is far from uniform.

In general, the velocities of both *P* and *S* waves increase down to 2900 km (1802 mi). However, *S*-wave velocities are lower than expected between about 100 and 250 km (62 and 155 mi), which has led to the designation of the region as the *low-velocity zone*.

At 2900 km (1802 mi) the *P*-wave velocity drops drastically to a value close to its value near the earth's surface, and the *S* wave disappears entirely. Such behavior is exactly what would be expected if at 2900 km the rock changed from solid above to fluid below. The change in rock properties appears to be very sharp and has been designated the *core-mantle boundary*. The solid mantle is thought to consist of iron- and magnesium-silicate minerals, and the liquid core principally of iron. Within the core and at about 5100 km (3169 mi) the *P* wave velocity appears to increase abruptly, and there is even a hint of an *S* wave. Seismologists think that is caused by a change from a fluid *outer core* to a solid *inner core*.

The existence of the sharp velocity discontinuity at the core-mantle boundary also provides an explanation of the so-called "seismic shadow zone" mentioned earlier (Fig. 17-32). Earthquake waves can move through the mantle to a depth of 2900 km (1802 mi), but waves that probe any deeper strike the core-mantle boundary, and part of the energy is sharply reflected back to the surface. But part

moves into the core and is greatly refracted as it does so. Because of the extreme refraction at the core-mantle boundary, no direct *P* waves reach the earth's surface between 102 and 143° arc distance from the epicenter. Only *P* waves can traverse the fluid part of the core but they do so at reduced velocity, which accounts for the increased time necessary for waves to travel through the central part of the earth.

One other major seismic discontinuity has been found, and it lies near the outer margin of the earth. It is the *Mohorovičić discontinuity*, named for Andrija Mohorovičić (1857–1936), the Yugoslavian seismologist who first suggested its presence in 1909. The name has since been shortened to the *Moho* or *M discontinuity*, which has enabled many a seismologist to continue discussion at a cocktail party rather than retire to tongue-tied oblivion. The Moho separates rocks above in which *P* waves have velocities of 6 to 7 km/sec (3.7 to 4.3 mi/sec) from those below which have *P* wave velocities of about 8 km/sec (5 mi/sec). By tradition rocks above that seismic-velocity break are called the *crust* and those below it, the *mantle*. Under the continents the crust generally varies from 30 to 60 km (19 to 37 mi) in thickness, whereas under the ocean basins it is about 6 to 8 km (4 to 5 mi) thick (Fig. 17-33).

Recently, on the basis of its seismic properties, as well as its mechanical strength, the outer 100 km (62 mi) of the earth have been designated the *lithosphere*, the layer extending from 100 to 250 km (62 to 155 mi) the *asthenosphere* (Fig. 17-34), and the layer below 250 km as the *mesosphere*. The lithosphere (meaning rock sphere) therefore includes both the crust

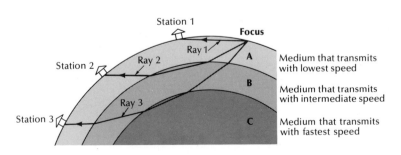

Fig. 17-30. Transmission of three earthquake waves in a hypothetical three-layered earth. Ray 1 travels only in layer A, and is the slowest. Ray 2 travels both in layers A and B, moving more quickly through B, the intermediate-speed layer. Ray 3 travels through all three layers, and moves with the highest average velocity through layer C. At each place that a wave encounters a velocity discontinuity, it bends.

and the upper part of the mantle. It is cooler, more brittle, and stronger than the asthenosphere. The large plates of the earth, which appear to slide about to produce the effects described within the plate-tectonic model, consist entirely of lithospheric material.

The asthenosphere (meaning zone of weakness) corresponds to the low-velocity zone, mentioned previously, in which S waves slow down greatly. There, it is thought that the effects of increasing temperature predominate over the effects of increasing confining pressure. Thus, the rocks are less rigid. The presence of a relatively weak zone is thought by many to be absolutely necessary to the mobility of the lithospheric plates above. In fact, many scientists have concluded that temperatures are so high in the asthenosphere that the rock is partially (up to 10 per cent) melted, and that the molten part is the source of much of the iron- and magnesium-rich lava that erupts from volcanoes at the earth's surface.

The mesosphere (middle sphere) represents the mantle zone beneath the asthenosphere where the pressure again becomes great enough to cause the rock material to be strong.

Seismologists have also been able to tell us how the inside of the earth varies in density. Making such determinations involves combining (1) data from laboratory experiments, (2) theoretical considerations that relate density and body-wave velocities, and (3) a knowledge of the general distribution of mass in the earth (see Chap. 2). The result, as shown in Figure 17-35, clearly indicates that density increases with depth and does so in a fashion that parallels the steps in the body-wave velocity distribution. The greatest jump in density occurs at the core-mantle boundary. At the center of the earth, rocks apparently possess specific gravities as high as 12 or 13; that is, they are four to five times as heavy as are most rocks at the surface.

What kinds of rocks could account for the variations in body-wave velocities and density? The elements making up the rocks must be those that are seemingly abundant within our solar system. Such a conclusion is based on our

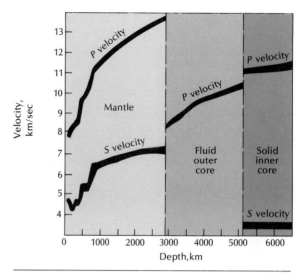

Fig. 17-31. How P and S wave velocities change with depth. The variation in speeds is represented by the variable thickness of the lines. Notice the lack of S waves and the reduced velocity of P waves in the fluid outer core.

Fig. 17-32. The shadow zone for an earthquake originating at the North Pole. As shown in the cutout, P and S waves travel normally until they intersect the liquid core. Then the P waves slow down abruptly and bend toward the center of earth (accounting for the shadow zone), and the S waves disappear entirely. Beyond 143° of arc from the focus, P waves are again received directly.

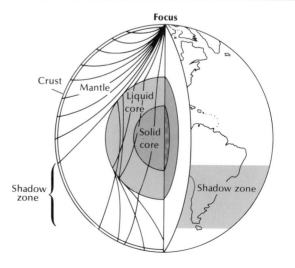

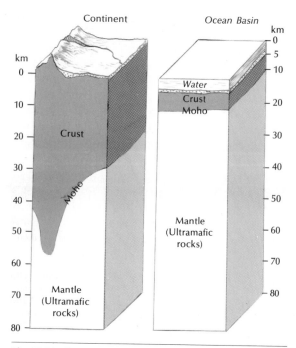

Fig. 17-33. Exaggerated schematic view of the relationships between crust and mantle beneath the continents and ocean basins. The continental crust, which has a relatively low density (2.7 g/cm^3), floats isostatically in a high-density mantle and projects downward as much as 60 km (37 mi) beneath some high mountains. The oceanic crust, with a density of about 3 g/cm^3, is relatively thin. After Fig. 23-27 (p. 401) from THE EARTH SCIENCES, 2nd Edition, by Arthur N. Strahler. Copyright © 1963, 1971 by Arthur N. Strahler. Reprinted by permission of Harper & Row, Publishers, Inc.

knowledge of the elemental abundances in nearby stars (particularly the sun) and in meteorites that fall to the earth's surface. In its passage around the sun the earth encounters a nearly continuous barrage of meteoritic debris. Most of the meteorites are small and are consumed as they pass through our oxygen-rich atmosphere. But some are large enough to resist total destruction and they fall far and wide over the earth. Most of those that are recovered can be placed into two main groups: the *iron meteorites* and the *stony meteorites*. Iron meteorites consist primarily of iron-nickel alloys whereas stony meteorites consist of silicate minerals and of lesser amounts of iron-nickel alloys and sulfides. Stony meteorites have a number of fascinating properties, but the one most pertinent to our discussion is that the relative abundances of non-volatile elements—such as magnesium, silicon, aluminum, calcium, and iron—in them are about the same as they are in the sun and other stars. Earth scientists have concluded, therefore, that the relative abundances of those elements in an average stony meteorite is a good representation of the average composition of the planetary bodies in our solar system, including the earth (Table 17-2).

The relatively dense core of the earth, which accounts for about 30 per cent of its mass, is

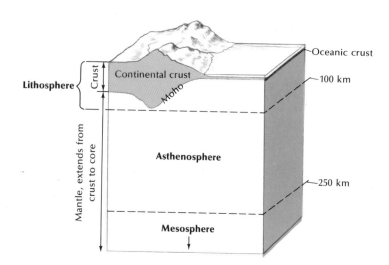

Fig. 17-34. Threefold division of outer part of the earth based on its mechanical properties. The lithosphere, which includes both the crust and the upper part of the mantle is strong and relatively brittle; the asthenosphere is relatively weak and capable of flowing; and the mesosphere is strong.

Table 17.2. Average composition of stony meteorites and hypothetical composition of the earth

Element	Stony meteorites, weight per cent	Earth, weight per cent
O	33.24	29.5
Fe	27.24	34.6
Si	17.10	15.2
Mg	14.29	12.7
S	1.93	1.93
Ni	1.64	2.39
Ca	1.27	1.13
Al	1.22	1.09
Na	0.64	0.57
Cr	0.29	0.26
Mn	0.25	0.22
P	0.11	0.10
Co	0.09	0.13
K	0.08	0.07
Ti	0.06	0.05

thought to be most probably composed of iron with smaller amounts of silicon, nickel, and sulfur. The mantle, on the other hand, accounting for about 67 per cent of the earth's mass, appears to be made of silicates—mostly those of iron and magnesium with only small quantities of aluminum, calcium, and sodium. In the outer part of the mantle the rock is made up chiefly of olivine with lesser amounts of magnetite and pyroxene. That conclusion is confirmed by analysis of the nodules and clasts found in basaltic lavas and kimberlites which were ripped from the mantle and carried to the surface during volcanic activity.

In the outer 700 km (435 mi) of the mantle there is some variation in the physical state of the iron-magnesium silicates, apparently in response to variations of temperature and pressure (Fig. 17-36). We have already noted that in the asthenosphere temperatures appear to be high enough and confining pressure low enough so that partial melting of the rocks may take place. Below 250 km (155 mi), as confining pressure becomes dominant the material becomes solid and, in the next 450 km (280 mi), undergoes at least two changes in physical

state into more-dense crystalline forms of the iron-magnesium silicates. What appears to happen, and this has been verified by laboratory experiments in which pressures equivalent to those as far down as 600 km (373 mi) are generated, is that as pressure continues to increase downward, the less-dense crystalline substances, which were stable near the outer part of the mantle, reach a point where they can no longer withstand the pressure. Then they collapse to more-compact, denser forms. One such collapse (*change of state*) occurs at a depth just above 400 km (249 mi) and another between 650 and 700 km (404 and 435 mi). Each change of state is associated with an increase in the body-wave velocity. Below 700 km no further changes are evident, and the material of the mantle undergoes continued compaction and progressive increase in body-wave velocities down to the core-mantle boundary. Evidently, below 700 km the rocks are in the densest state possible.

The earth's crust, as we know from direct observation, consists of a diverse suite of rocks composed of the lighter silicate minerals rich

Fig. 17-35. Curves depicting the estimated density of the earth. Notice the sharp increase at the mantle-core boundary. The density of rocks at the earth's center is thought to be about 4 to 5 times that of surface rocks.

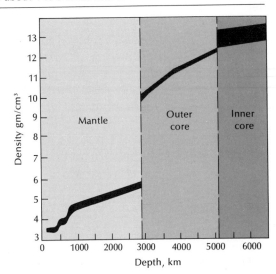

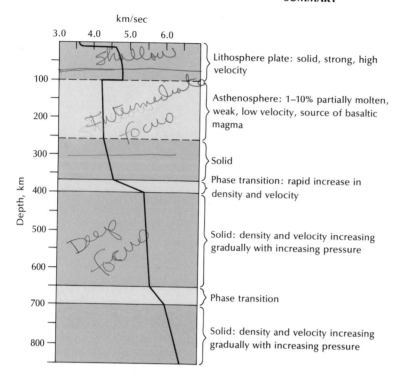

Fig. 17-36. Estimated nature of the outer part of the earth based on variations of the *P*-wave velocity with depth. The velocity break at 100 km (62 mi) represents the boundary between the lithosphere and asthenosphere. Increases in velocity near 400 km (249 mi) and 700 km (435 mi) apparently represent changes of state.

in calcium, sodium, aluminum, and potassium. In the ocean basins those silicates are composed mostly of plagioclase, pyroxene, and olivine. In contrast, the continents are composed of a complex of the relatively light granitic, metamorphic, and sedimentary rocks.

Temperature and pressure also vary with depth. Our best guess as to how they vary is given in Figure 17-37.

In many ways the realm beneath our feet is as mysterious as interstellar space. It is true that we have come a long way toward understanding, but the goal of full knowledge is still far distant. Just as it has been so up to now, the squiggles recorded on the seismograph will undoubtedly prove to be the principal way by which we progress toward that goal.

SUMMARY

1. The San Francisco earthquake, 1906, resulted from slippage along the San Andreas Fault, which amounted to about 5 m (16 ft). Fire subsequent to the quake caused a major portion of the city's destruction. In trying to explain the earthquake, H. F. Reid postulated the *elastic-rebound theory*.

   One of the largest quakes ever felt in the United States centered in New Madrid, Missouri, in 1811–12. The death toll and property damage was low because of a low population density at that time.

2. Japan, China, southern Europe, and South America are also earthquake prone and have experienced a number of severe and destructive quakes.

3. *Shallow-focus* quakes are defined as those down to 70 km (43 mi); *intermediate-focus* quakes, between 70 km and 300 km (43 and 186 mi); and *deep-focus* quakes between 300 and 700 km (186 and 435 mi). Earthquakes occur in linear belts; many scientists feel that such belts delineate the edges of tectonic plates, which are converging, diverging, and sliding laterally. Shallow-, intermediate-, and deep-focus earthquakes are associated with convergent zones, whereas shallow-focus quakes are asso-

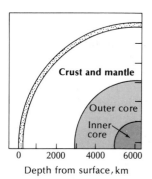 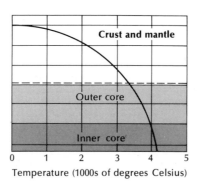 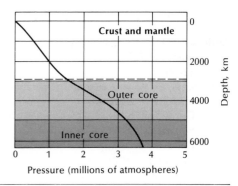

Fig. 17-37. Variation in temperature and pressure with depth inside the earth. Notice that initially the temperature increases rapidly but quickly tapers off. One atmosphere of pressure equals about 1 kg/cm² (14.7 lb/in.²), the air pressure at sea level.

---

ciated only with zones of divergence and lateral translation.

4. Amelioration of earthquake damage is being approached in two ways: through *earthquake prediction* and *earthquake control*. At present, prediction appears more feasible than control.

5. The first waves to arrive on a seismograph are *P waves*, then *S waves*, and finally *L waves*. P waves are *compressional* waves and can travel through solid or liquid. S waves are *transverse* waves that can only travel through a solid. L waves are the slowest and travel at or close to the earth's surface. Since P waves travel faster than S waves, it is possible to estimate the distance to a seismic disturbance by the separation in the arrival times of the two latter waves.

6. The energy released during an earthquake is measured on the *Richter magnitude scale*. The largest magnitude ever recorded is about 8.9. Nearly a million quakes occur yearly; of them only about ten are above magnitude 7. A quake of magnitude 8 or larger occurs every 5 to 10 years. Earthquake damage depends on several factors—magnitude, proximity, duration of quake, type of ground, and building construction.

7. Seismic data indicate that the earth is a layered body. The *crust* extends down to the *Moho discontinuity*, which occurs as deep as 70 km (43 mi) beneath the continents and down to 8 km (5 mi) beneath the oceans. Below the crust, the *mantle* is composed of ferromagnesian silicates, and extends down to 2900 km (1802 mi). Below that is the *core*, which is liquid in character down to 5100 km (3169 mi), and solid down to the center of the earth.

8. On the basis of seismic properties, the outer 100 km (62 mi) of the earth has been designated the *lithosphere* and the layer between 100 and 250 km (62 and 156 mi) the *asthenosphere*. The asthenosphere, corresponding to the *low-velocity zone*, is relatively weak and is thought to be necessary to the mobility of the lithospheric plates above.

9. Seismic speeds and laboratory data indicate that two *changes of state* occur in the outer part of the mantle in response to increased pressure. One occurs at about 400 km (248 mi) and the other at about 650 to 700 km (404 to 435 mi). Each is associated with an increase in body-wave speed.

SELECTED REFERENCES

Anderson, D. L., 1962, The plastic layer of the earth's mantle, Scientific American, vol. 207, pp. 52–59.

Baranzangi, M., and Dorman, J., 1969, World seismicity maps compiled from ESSA, Coast and Geodetic Survey, Epicenter Data, 1961–67, Seismological Society of America Bulletin 59, pp. 309–80.

Båth, M., 1973, Introduction to seismology, John Wiley and Sons, New York.

Bronson, W., 1959, The earth shook, the sky burned, Doubleday and Co., Garden City, New York.

Bullen, K. E., 1954, Seismology, Methuen and Co., London.

Byerly, P., 1954, Seismology, Prentice-Hall, Englewood Cliffs, New Jersey.

Davison, C., 1936, Great earthquakes, Thomas Murby and Co., London.

Eiby, G. A., 1957, Earthquakes, Frederick Muller, London.

Evans, D. M., 1966, Man-made earthquakes in Denver, Geotimes, May–June, pp. 11–18.

Fuller, M. L., 1914, The New Madrid earthquake, U.S. Geological Survey, Bulletin 494.

Gutenberg, B., and Richter, C. F., 1949, Seismicity of the earth, Princeton University Press, Princeton, New Jersey.

Heacock, J. G., ed., 1971, The structure and physical properties of the earth's crust, Geophysical Monograph 14, American Geophysical Union, Washington, D.C.

Hodgson, John H., 1964, Earthquakes and earth structure, Prentice-Hall, Englewood Cliffs, New Jersey.

Howell, B. F., 1959, Introduction to geophysics, McGraw-Hill Book Co., New York.

Iocopi, R., 1964, Earthquake country, a Sunset Book, Lane Book Company, Menlo Park, California.

Kendrick, T. D., 1956, The Lisbon earthquake, Methuen and Co., London.

Lawson, A. C., and others, 1908. The California earthquake of April 18, 1906, Report of the State Earthquake Commission, Carnegie Institute of Washington.

Leet, L. D., 1948, Causes of catastrophe, Whittlesey House, McGraw-Hill Book Co., New York.

Lovering, J. F., 1958, The nature of the Mohorovičić discontinuity, Trans. American Geophysical Union, vol. 39, pp. 947–55.

Oakeshott, G. B., and others, 1955, Earthquakes in Kern County, California, during 1952, California State Division of Mines, Bulletin 171.

Plafker, G., Ericksen, G. E., and Concha, J. F., 1971, Geological aspects of the May 31, 1970 Peru earthquake, Bulletin of the Seismological Society of America, vol. 61, pp. 543–78.

—— and Savage, J. C., 1970, Mechanisms of the Chilean earthquake of May 21 and May 22, 1960, Geological Society of America, vol. 81, pp. 1001–1030.

Poldervaart, A., and others, 1955, Crust of the earth, Geological Society of America, Special Paper 62.

Reid, H. F., 1914, The Lisbon earthquake of November 1, 1755, Seismological Society of America Bulletin, vol. 4, pp. 53–80.

Richter, C. F., 1958, Elementary seismology, W. H. Freeman and Co., San Francisco.

Robertson, E. C., ed., 1972, The nature of the solid earth, McGraw-Hill Book Co., New York.

Saint Amand, P., 1961, Los Terremotos de Mayos— Chile, 1960, Technical Article 14, U.S. Naval Ordinance Test Station, China Lake, California.

Shepard, F. P., 1933, Depth changes in Sagami Bay during the great Japanese earthquake. Journal of Geology, vol. 41, pp. 527–36.

Sutherland, M., 1959, The damndest finest ruins, Coward-McCann Book Co., New York.

Verhoogen, J., 1956, Temperatures within the earth, in Physics and chemistry of the earth, vol. 1, Pergamon Press, New York.

Witkind, J. S., 1962, The night the earth shook; a guide to the Madison River Canyon earthquakes area. Department of Agriculture, U.S. Forest Service Misc. Publication 907.

Wyllie, P. J., 1971, The dynamic earth: textbook in geosciences, John Wiley and Sons, New York.

——, 1975, The earth's mantle, Scientific American, vol. 232, no. 3, pp. 50–63.

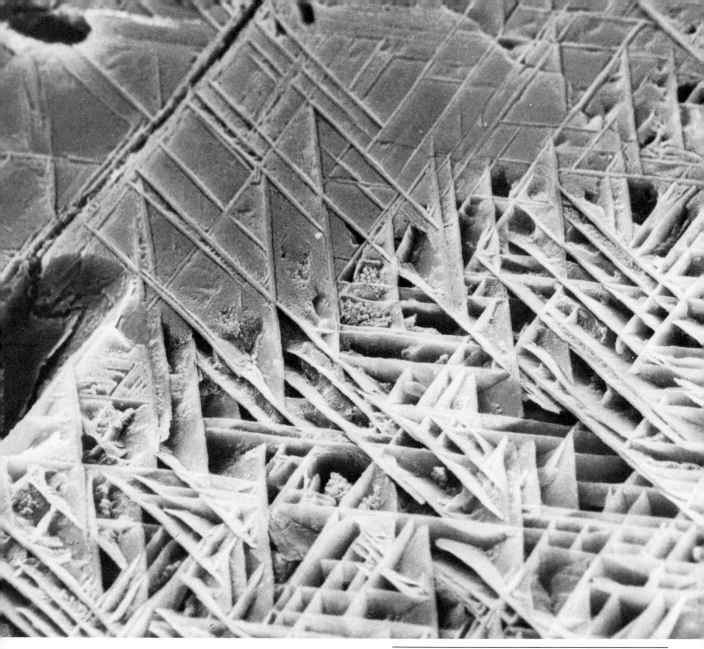

Fig. 18-1. Magnetite photographed under the scanning electron microscope. The grain shown was etched in hydrochloric acid to bring out relief. The magnetite zones tend to etch out, whereas the iron-titanium oxide stands out in a thin-walled structure. The magnification is about 1225 times. *Rick Hoblitt and Edwin E. Larson*

18

THE MAGNETIC FIELD OF THE EARTH

The ancient Greeks divided all matter on the earth into four categories: fire, water, air, and earth. They were aware that the last of those, earth, had several recognizable forms, among them a material called *lodestone,* which could attract other pieces of lodestone or iron to itself (Fig. 18-2). Today, we know the material lodestone as magnetite (Fe_3O_4), one of the common oxides of iron found at the earth's surface. The attractive force of magnetite was so mysterious that the Greek philosopher Thales of Meletus (640–546 B.C.) finally attributed a soul to the material to explain its strange properties.

By the early part of the Middle Ages, it was found that if a piece of lodestone or a piece of iron that had been rubbed on it was suspended so that it could freely turn, it always came to rest in the same position. It was not until around A.D. 1100, however, that a magnetic compass was invented which could be used by travelers on land and sea for finding their way. The inventor, whose name has been lost from the historical record, was probably Chinese in origin. Another century, apparently, had to pass before Europeans became aware of the design and use of a compass (Fig. 18-3).

Many strange and curious properties were attributed throughout the ancient times and Middle Ages to the magnet. It was supposed to give comfort and to increase one's grace; it could cure hemorrhages and toothaches; and it was considered useful in bringing about reconciliation between husbands and wives.

In 1269 Petrus Perigrinus de Maricourt, a French crusader, not only provided a detailed description of a compass, but fashioned a sphere out of lodestone and investigated the nature of the magnetic field surrounding the sphere by means of a small magnetic compass needle. Although his study was not very sophisticated, it was nonetheless the first scientific investigation of any magnetic material.

A much more detailed investigation of magnetic properties was carried out by William Gilbert (1540–1603), an English physician and physicist, who published his results in 1600 in a book called *De Magnete.* He found that magnetic materials possess poles, and that *unlike poles* attract whereas *like poles* repel each other. He also noted that a compass needle comes to rest in only one orientation at any one spot on the earth's surface (Fig.

Fig. 18-2. Lodestone (magnetite) is strongly magnetic. The magnetic field (lines of force) surrounding the samples shown here are made evident by a sprinkling of iron filings. *Jan Robertson*

18-4). He was aware that the earth possesses a magnetic field, and even went so far as to propose that the entire earth acts like a gigantic spherical magnet.

In the nineteenth century, physicists began to explore the mysteries of electricity, a phenomenon thought to be totally different from magnetism. It came as a shock, therefore, when in 1819 the Danish physicist Hans Christian Oersted (1777–1851) announced that a wire carrying an electric current would cause the deflection of a compass needle. Oersted's discovery was but the beginning of a brilliant series of studies by physicists André Ampère (1775–1836), James Maxwell (1831–79), and Michael Faraday (1791–1867) which showed that electricity and magnetism are but two sides of the same coin. It was their work that clarified the relationship between the two phenomena.

Fig. 18-3. An early magnetic compass was made by placing a fragment of lodestone on a tiny wooden raft. Such an assemblage will rotate until its magnetic field aligns with that of the earth. Stars mark the poles of the lodestone. Original drawing from Athanasius Kircher, *Magnes*, 1643. *New York Public Library*

During the twentieth century, physicists and geophysicists have continued to investigate the nature of magnetic materials and all of the phenomena related to the earth's magnetic field. They have begun to see how the earth's field is generated and have become aware that it has fluctuated and varied, both on a geologically short-term and long-term basis.

THE EARTH'S MAGNETIC FIELD

The magnetic field of the earth is so familar to us that we rarely give it a thought. We verify its

Fig. 18-4. William Gilbert's diagram of the earth with small magnets showing inclination of field. From William Gilbert, *De Magnete*, 1600. *New York Public Library*

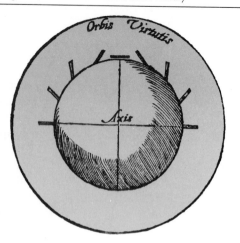

existence, however, each time we use a magnetic compass to orient ourselves. The field is an invisible force field that everywhere permeates the earth and extends outward from the earth's surface for hundreds of thousands of kilometers; even the moon 384,000 km (238,618 mi) away, is, as the astronauts have found, not beyond its effects.

The best way to visualize the main elements of the field is to place a bar magnet, which has two magnetic poles, at the center of a circle representing a cross section of the earth (Fig. 18-5). Invisible lines of force, the number being determined by the strength of the magnet, converge at one magnetic pole and diverge at the other magnetic pole. The use of the north and south magnetic pole designation is by convention of physics only. Each line of force is continuous, so that one emerging from the one magnetic pole will bend through space and eventually re-enter the magnet at the other pole, as shown. It is apparent from the illustration that the field lines are essentially parallel to the "earth's surface" at the equator but that they increase in angle of intersection with the surface toward the north. At a spot above the earth's north magnetic pole, the field lines point vertically downward. A similar angular relationship is shown by the lines that emerge south of the equator. The lines are more concentrated near the poles than they are near the equator; the result is that the field is about twice as strong at the poles as it is at the equator. Since the earth is three-dimensional, lines of force like those shown in cross section would be found completely encircling the earth's spheroid.

In order to map the magnetic field lines just described we can use a small magnetic compass, moving it about from place to place on the line circumscribing the bar magnet. At each place the compass needle will be pulled into parallel alignment with the field lines at that point, and a unique orientation of the compass will result (Fig. 18-5). The north end of the needle will always point toward the north, but the angle that the needle makes with the line

representing the earth's surface will vary at each location. One can actually "see" the field lines by sprinkling iron filings or finely crushed magnetite on the paper. The tiny fragments will be pulled into crude, curved lines that approximate those of Figure 18-6.

The simple *dipolar field* we have just studied is, as Gilbert realized back in 1600, very similar to the main dipolar field of the earth. It is no wonder he believed that a large bar magnet existed inside the earth.

The earth's dipolar field is generated in some way inside the metallic iron core. The processes that lead to its generation lie hidden so far below the surface that we can only speculate on what they might be. Certainly there is no deeply buried permanent magnet since the central regions of the earth are much too hot for any known, likely material to remain magnetic. However, as you recall, the outer core most likely is composed primarily of liquid iron. Being metallic it can conduct electric currents and being fluid it can flow. Geophysicists now believe that in response to local heating the liquid core flows convectively while the earth, including the core, spins on its axis.

If, for any number of reasons, a weak magnetic field is initially present, electric currents will be created during the flow, which will in turn create new magnetic fields in a complex interaction of electromagnetic coupling.

Once formed the new magnetic field will be amplified and constrained to lie more or less along the axis of rotation of the earth; it is therefore dipolar. The mechanism can be crudely modeled in the laboratory, and has been termed a *self-exciting dynamo* (Fig. 18-7). Since the dynamo action is controlled in part by the rate and scale of movement in the liquid iron core, the field will vary in intensity and to some degree in direction as changes in the fluid motions in the core occur. Thus, although the field appears relatively stable from day to day, over hundreds of years it will demonstrate a degree of variability. In fact, the field can vary greatly in intensity, even dropping to zero. When it regenerates it may even be oriented so that the magnetic poles are reversed.

The north and south poles today do not coincide with the rotational or geographic poles: the north magnetic pole lies at 71°N latitude, 96°W longitude. This appears to be a momen-

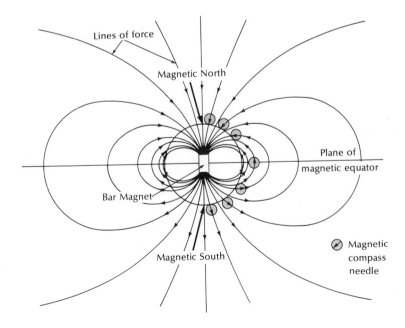

Fig. 18-5. Field lines, around a bar magnet, shown in cross section through the bar's axis. Arrows on the field lines indicate the direction of the magnetic field; the magnetic compass needle is aligned parallel to them. Notice that the inclination of the compass needle is horizontal at the magnetic equator and vertical at the magnetic poles.

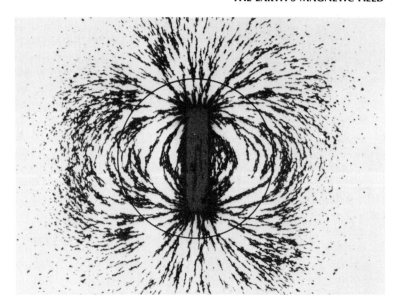

Fig. 18-6. Iron filings sprinkled over a bar magnet enable us to observe a magnetic field similar to that of the earth. Note the similarity between this figure and Figure 18-5. *Jan Robertson*

tary situation. In time, the magnetic poles will slowly shift in geographic position such that, over 5000 years or so, their average positions will match those of the rotational poles.

The latter is very significant, for if we can determine the average positions of the magnetic poles for an ancient geologic time—through a study of the rocks themselves, as we shall see—then we can also determine the positions of the geographic poles. Such information is of particular importance to geophysicists who study how the continents have moved and shifted with respect to the geographic poles. The way in which they determine the position of the magnetic poles during any particular geologic time is by measuring

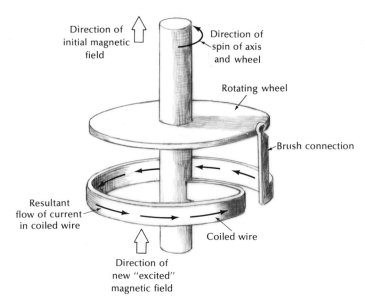

Direction of initial magnetic field

Direction of spin of axis and wheel

Rotating wheel

Brush connection

Resultant flow of current in coiled wire

Coiled wire

Direction of new "excited" magnetic field

Fig. 18-7. Model of a simple self-exciting dynamo. If the dynamo is to function, the wheel must be turning and an initial magnetic field must be present. The resultant electric currents produce a new "excited" magnetic field, which is sufficient to keep the dynamo functioning even if the initial magnetic field is removed.

the directions of magnetism contained in the rocks of that age.

ROCKS, MAGNETISM, AND PALEOMAGNETISM

Essentially all rocks contain some minerals that can become permanently magnetized. Magnetite (Fe_3O_4) is the mineral chiefly responsible for most of the magnetism in rocks, which can be measured by means of a sensitive instrument called a *magnetometer*. Magnetite will retain its permanent magnetization up to a critical point—580°C (1076°F)—known as its *Curie temperature*. The latter takes its name from the noted French chemist Pierre Curie. Above 580°C the mineral will be essentially nonmagnetic. Thus if a lava solidifies and cools, it will remain unmagnetized until the temperature drops to 580°C, the Curie temperature of magnetite. As the temperature continues to decline, the lava will acquire a permanent magnetization—called *remanent magnetization*—parallel to the earth's magnetic field lines at that locality and at that time. The remanent magnetization will be figuratively frozen into the lava, so that regardless of later changes in the earth's field, the magnetization in the lava will record the initial field direction acquired during cooling. The area of geology that deals with fossil magnetization in ancient rocks is called *paleomagnetism*. Since the field directions for the earth's dipolar field are unique for any latitude, as we learned earlier in the chapter, it is possible to determine where the north magnetic pole (and rotational pole) was at the time a rock crystallized and cooled. That information can be used in reconstructing the drift paths of the continents through time.

Rocks can acquire a permanent magnetization in ways other than cooling of their magnetic minerals. For example, it is possible for small magnetic clasts that accumulate during sedimentation to orient themselves like tiny compass needles, as they settle through the waters in a lake or ocean. Thus sedimentary rocks, too, can acquire a remanent magnetization at the time they form.

How stable is remanent magnetization? Rocks vary, but many of the more stable kinds can retain the record of the original field direction for many hundreds of millions and even billions of years. In the Beartooth Range of Montana are basaltic dikes containing magnetization that has been frozen in for more than 2.5 billion years. Its presence provides one proof that the earth possessed a magnetic field—and therefore the liquid iron core necessary to generate the field—as long ago as that time.

REVERSALS OF THE EARTH'S MAGNETIC FIELD

From a study of the permanent magnetization in rocks, it is now known that episodically during the earth's history the magnetic field has been oppositely oriented—that is, the magnetic poles have shifted by 180°. Such a field is said to be reversely oriented, and the rocks formed at those times will record *reverse remanent magnetic directions*. The last major event of reversed orientation of the field lasted until about 690,000 years ago. Since then, the field has been dominantly normal (parallel to today's field) except for minor fluctuations. Through radiometric dating of lava samples from around the world it has been possible to establish the sequence of polarity changes, at least for the last few million years (Fig. 18-8). The sequence is called the *magnetic polarity time-sequence* and, as Figure 18-8 shows, the earth's field has switched polarities about every 200,000–300,000 years. The longest polarity interval has lasted nearly 700,000 years; the shortest only about 10,000 years.

The transition from a normal polarity to a reverse polarity, or vice versa, does not take place instantaneously. Rather it takes about 1000–5000 years—a short geologic time interval. As the field changes over, it remains at a very low intensity. No one knows how the low magnetic field affects animals and plants during changes of polarity, but experiments on small organisms have indicated that normal life cycles could be greatly disrupted. In fact, some geologists have suggested that some of the ex-

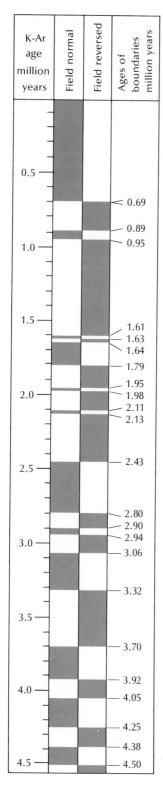

K-Ar age million years	Field normal	Field reversed	Ages of boundaries million years
			0.69
			0.89
			0.95
			1.61
			1.63
			1.64
			1.79
			1.95
			1.98
			2.11
			2.13
			2.43
			2.80
			2.90
			2.94
			3.06
			3.32
			3.70
			3.92
			4.05
			4.25
			4.38
			4.50

tinctions recorded in the fossil record are directly related to changeovers in the polarity of the earth's magnetic field. Further experiments will be necessary, however, to fully evaluate those suggestions.

It is interesting to note that in the recent past the earth's magnetic field has been steadily decreasing in intensity each year. It is possible that the field of today is headed for another changeover in the relatively near future. At the present rate of decay, however, another 2000 years would be required for a complete change. The decrease, of course, may only be part of a short-term fluctuation, completely unrelated to a reversing of the earth's polarity.

During the last 10 to 15 years, oceanographic vessels have been taking core samples of the upper sedimentary layers that cover the ocean floors. Those sediments were deposited during the last several million years, and they contain a record in their remanent magnetization of magnetic polarity changes just as the rocks on land do. It has been possible to correlate the reversals of magnetization recorded in the seafloor cores with the magnetic polarity time sequence and thereby to provide a method for dating them throughout their lengths (Fig. 18-9). In that way, rates of sedimentation have been established for different parts of the ocean basins, and fossils contained in the cores have been accurately dated.

OCEAN-FLOOR MAGNETIC ANOMALIES

With the advent of oceanographic studies many ships towed magnetometers behind them to measure the earth's magnetic field as

Fig. 18-8. Magnetic polarity time sequence back to about 4.6 million years ago based on potassium-argon dating of lava flows with known paleomagnetic directions. Notice that the field has bounced back and forth between normal and reverse polarity. Individual polarity intervals are as short as 10,000 and as long as 690,000 years.

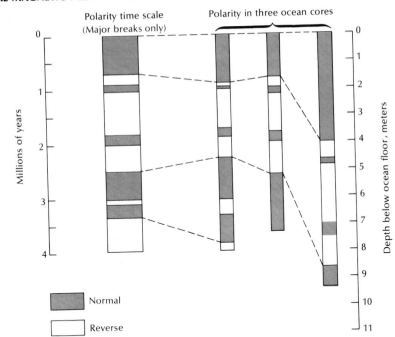

Fig. 18-9. Magnetism in deep-sea sediments. As the field changes from normal to reverse and back again, the accumulating sediments acquire corresponding magnetic directions. Hence, core samples will contain a record of the changes, which can be correlated with the magnetic polarity time sequence.

they plowed through the seas. After data had been recorded for a few years, it became apparent that the intensity of the field varied in a systematic but oscillatory way. In some places it was slightly stronger than the average, and in others it was slightly less intense. When plotted on a map the anomalously high and low regions were found to be oriented in alternating linear belts paralleling the axes of the oceanic ridges. From oceanographic vessels, detailed magnetic surveys were then made to see how the pattern varied across the oceanic ridges and rises. The results of one such survey over the Mid-Atlantic Ridge south of Iceland is shown in Figure 18-10. Not only were the linear magnetic anomalies verified, but the pattern on both sides of the ridge was found to be bilaterally symmetrical, outward from a higher-than-average linear anomaly that ran along the ridge axis. Whatever the pattern on one side of the ridge, its mirror image was recognizable on the other side.

It was several years before the full significance of the magnetic belts and their bilateral symmetry was realized. In 1963, two English geophysicists, F. J. Vine and P. M. Matthews, provided a plausible explanation. Parts of the ocean floor beneath the sedimentary layers, they concluded, must be magnetized in a normal direction—that is, parallel to the field of today; whereas other parts must be reversely magnetized. The higher-than-average field intensities supposedly correspond to the normally magnetized stripes, and the lower-than-average anomalies correspond to the reversely magnetized stripes (Fig. 18-11). The explanation was based on the idea that the fossil magnetism that paralleled the present-day field (*normal direction*) would add to the total intensity of the field, but that which was oppositely directed (*reversely magnetized*) would detract a little from it. The bilateral symmetry could only mean that for each normal or reversely magnetized stripe on one side of the ridge, there was a nearly identical counterpart on the other side. That is, paired anomaly stripes existed, parallel to and one on either side of the ridge axis.

When the widths of the magnetic stripes on one side of a ridge axis were plotted against

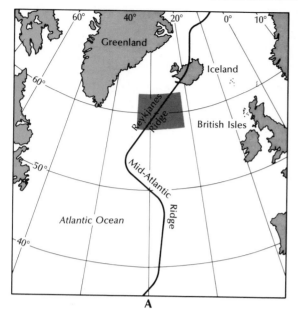

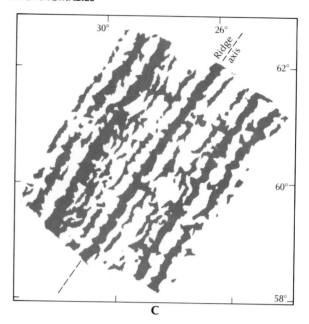

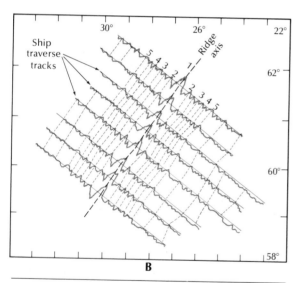

Fig. 18-10. Magnetic anomalies mapped in the area of Reykjanes Ridge, south of Iceland (**A**). In (**B**) is shown the relative variation in the earth's magnetic field determined along selected ship traverses across the ridge, and in (**C**) the inferred areal distribution of the variations. The black areas represent higher-than-average field density and the white areas, lower-than-average. There is a bilateral symmetry to the anomalies, such that those on one side of the ridge are mirrored on the other—as indicated by the numbers in (**B**).

the land-based magnetic polarity time sequence they matched in relative position nearly exactly, right up to the present (Fig. 18-12). The youngest subsediment materials occurred along the ridge axes and became progressively older outward from it in both directions. At the time, the discovery was an exciting one, for it indicated that new material was being added continually to the crust along the ridge axis, and that subsequently it was being split and moved laterally away in the manner of two outward-moving conveyor belts. Dating of the magnetic stripes provided strong proof indeed that the oceanic ridges and rises are zones of divergence and that the ocean basins are young features continually in the process of being formed.

Since the magnetic polarity time sequence was well dated it was possible to date the oceanic linear magnetic anomalies and to provide rates of spreading along the zones of divergence. They proved to be about 2–6 cm (0.8–2.3 in.) per year.

In the last few years, extensive deep drilling in the oceans has verified Vine's and Matthews's theory. The character of linear mag-

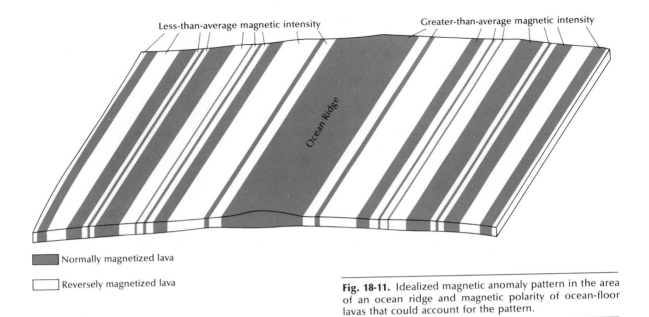

Less-than-average magnetic intensity

Greater-than-average magnetic intensity

Ocean Ridge

Normally magnetized lava

Reversely magnetized lava

Fig. 18-11. Idealized magnetic anomaly pattern in the area of an ocean ridge and magnetic polarity of ocean-floor lavas that could account for the pattern.

Fig. 18-12. Match of the linear magnetic anomaly pattern with the magnetic polarity time sequence, from which it is inferred that the ocean floor is spreading apart. Apparently, lava flows and dikes are added to the region of the ridge axis, where they cool and become magnetized paral-

lel to the field direction at that time. As divergence continues, the field changes polarity, new material is added, and older material is carried away from the axis in both directions.

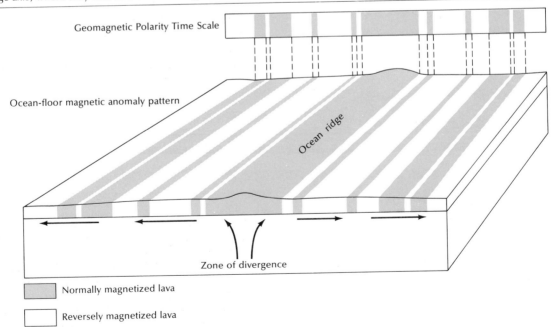

Geomagnetic Polarity Time Scale

Ocean-floor magnetic anomaly pattern

Ocean ridge

Zone of divergence

Normally magnetized lava

Reversely magnetized lava

netic anomalies has provided key insights into how ocean basins form as well as enabling precise tracking of the movements of lithospheric plates and the attached continental masses.

Certainly, the knowledge from magnetic studies concerning plate motions and zones of divergence has become fundamentally significant in the development of the plate-tectonic theory.

SUMMARY

1. The properties of a *magnetic field* were first investigated by William Gilbert in 1600. Magnetic materials possess *poles* which if alike repel each other and if unlike, attract each other. Later experiments demonstrated that electricity and magnetism are intrinsically related.

2. The earth's magnetic field is *dipolar*. The field lines parallel the earth's surface at the equator and are perpendicular to the surface at the magnetic poles. At intermediate latitudes the field lines are inclined to the surface; the higher the latitude, the steeper the angle.

3. The earth's magnetic field is generated inside the liquid iron core in the manner of a *self-exciting dynamo* in a complex interaction of electromagnetic coupling.

4. Essentially all rocks contain magnetic miner-

als (most commonly *magnetite*) and can acquire a *permanent magnetization* (*remanent magnetization*) parallel to the earth's field at the time they form; for example, lavas become magnetized as they cool below 580°C (1076°F), (the *Curie temperature* of their magnetic minerals), and sediments can acquire a remanent magnetization through alignment of magnetic clasts as they settle through water.

5. Periodically the earth's magnetic field has been reversely oriented. Radiometric dating of lava samples has established a succession of dated reversals (*magnetic polarity time-sequence*). The reversals recorded in sea-floor cores can be correlated with that sequence, thereby providing a method of dating them throughout their lengths and of establishing sedimentation rates.

6. Magnetometer surveys in the oceans reveal a magnetic anomaly pattern composed of stripes of higher-than-average field intensities alternating with stripes of lower-than-average intensity. The stripes parallel the oceanic ridge axes. Higher-than-average stripes correspond to normally magnetized ocean crust, and the lower-than-normal stripes to reversely magnetized crust. The pattern can be correlated with the magnetic polarity time sequence, showing that the oceanic ridges and rises are zones of divergence and providing a means of determining the rate of sea-floor spreading.

SELECTED REFERENCES

Cox, A., 1969, Geomagnetic reversals, Science, vol. 163, p. 237.

Doell, R. R., Cox, A., and Dalrymple, G. B., 1964, Reversals of the earth's magnetic field, Science, vol. 144, pp. 1537–43.

Heirtzler, J. R., Dickson, G. O., Herron, E. M., Pitman, W. C., and Le Pichon, X., 1963, Marine magnetic anomalies, geomagnetic field reversals, and motions of the ocean floor and continents, Journal of Geophysical Research, vol. 73, p. 2119.

———— Le Pichon, X., and Bacon, J.G., 1966, Magnetic anomalies over the Reykjanes ridge, Deep-Sea Research, vol. 13; pp. 427-44.

Irving, E., 1964, Paleomagnetism, John Wiley and Sons, New York.

Nagata, T., 1961, Rock magnetism, Marizen, Tokyo.

Parasnis, D. S., 1961, Magnetism, Harper and Brothers, New York.

Phillips, O. M., 1968, The heart of the earth, Freeman, Cooper, and Co., San Francisco.

Takeuchi, H., Uyeda, S., and Kanamori, H., 1970, Debate about the earth, rev. ed., Freeman, Cooper, and Co., San Francisco.

Fig. 19-1. Continental drift in action. The coastlines of the Arabian Peninsula and the Somali Republic in northeast Africa fit so well that there is little doubt that at one time the two regions were joined and are now moving apart from one another. *NASA*

CONTINENTAL DRIFT AND PLATE TECTONICS

CONTINENTAL DRIFT

The idea that one great land mass may have once existed on the earth, a land mass that subsequently broke into drifting continental-size fragments, is certainly not a new one. However, most of the earliest recorded suggestions of continental drift were casual remarks made after a cursory examination of a world map and had little substance to back them up.

As early as 1620, Sir Francis Bacon in his book *Novanum Organum* commented that the west coast of Africa and the east coast of South America were so strikingly similar that it could not be by accident alone. Soon after, in 1658, the Frenchman R. P. François Placet also remarked that the similarity of those coastlines suggested that the continents had been joined together at one time, and he conjectured that their separation may have occurred during the biblical flood. About 200 years later, Antonio Snider-Pelligrini wrote in *La Création et Ses Mystères Dévoilés* that the continents border-

ing the Atlantic must have once been contiguous; he based his conclusion on slightly more substantial evidence—the similarity in fossil plants in the coal beds of North America and Europe. He even went so far as to present a reconstruction of the supercontinent before its breakup. By 1908, two scientists, B. Taylor and H. B. Baker, had reached the same conclusion from their studies of the distribution of mountain chains around the world.

But all of those precursory suggestions were made in such an offhand manner that they evoked little support and even less interest in the subject of continental drift. Furthermore, such ideas ran contrary to the dogma of geology, which by 1900 was developing into a full-fledged science. Most geologists believed the continents to be permanent and fixed and regarded the weakly supported suggestion that continents could slip and slide about on the earth's surface as pure geologic fantasy. Little did they realize that a storm of controversy regarding that very issue would erupt not too far

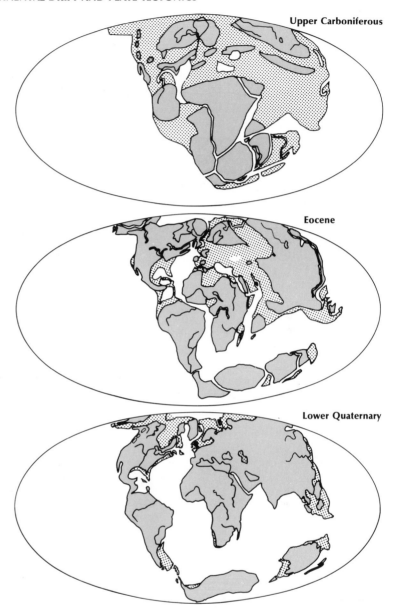

Upper Carboniferous

Eocene

Lower Quaternary

Fig. 19-2. Breaking up of the world's continents according to Wegener. He called the original giant continent Pangaea. The white areas are deep ocean, the darkened areas land, and the stippled areas shallowly submerged continental shelf.

in the future and would last for more than 50 years.

Alfred Wegener in 1910 was an instructor of the newly developing field of meteorology in a school in Germany. During the course of his work that year he too was struck by the similarity of the shapes of the coastlines of South America and Africa, just as were others before

him. Yet it was only a fleeting idea, and in the crush of his duties Wegener pushed the thought from his mind. Then, in 1911, he came across an article on the similarity of fossils on the two continents that rekindled his interest. Unlike the earlier advocates of continental drift, Wegener became obsessed with the idea and was tireless in his efforts to uncover any

evidence, from whatever source, that might indicate the former continuity of the land masses around the Atlantic Ocean. He finally published his speculations with supporting evidence in 1915. Wegener was convinced that at one time all of the continents of the world were part of one giant continental mass that he called *Pangaea* (Fig. 19-2). At some time during the Jurassic period, Pangaea split along many cracks, and the continental fragments began to drift apart. The response to the publication initially was mild, but by 1924, a full-fledged furor had developed.

R. T. Chamberlin, an American earth scientist, wrote soon after: "Can we call geology a science when there exists such differences of opinion on fundamental matters as to make it possible for such a theory as this to run wild?" Even in 1944, the American geologist Bailey Willis maintained that the theory as propounded by Wegener was nothing but a fairy tale and should be minimized because of its deleterious effect on students.

Wegener, in his synthesis, had attempted to bring together as many pieces of corroborating data as he could. He included evidence from physical geology, geodesy, geophysics, paleontology, zoology, and paleoclimatology. He was not an expert in any of those fields and chose, unfortunately, some poor and even inaccurate examples to bolster his argument. Seizing upon his errors, the opposition had a merry time ridiculing Wegener's hypothesis. Leading geologists ranted and railed against the evils of the drift theory with an almost religious fervor. Although most prefaced their remarks with pleas for open-mindedness on the subject, it was plain to see that their own minds were closed as tightly as the jaws on a bear trap. Yet in their own eyes they saw themselves as guardians of the truth, defending their science against a whimsical wave of idiotic speculation.

One of the main points the opposition made was that there appeared to be no viable mechanism for bringing about continental movement. Wegener had outlined two possible mechanisms—one related to gravitational attraction between the continents, the moon, and the sun, and the other related to the centrifugal pull on the spinning earth—but almost everyone agreed that the forces produced in both instances would be woefully inadequate to bring about drift. Since a workable mechanism was lacking, most earth scientists felt that the hypothesis of continental drift could be discounted.

Wegener stood up well to the fusillade of criticism and continued to look for more and better evidence to support his idea right up to the time of his death at the age of fifty. In 1930, while on his fourth expedition in the exploration of Greenland, he set out from the northernmost base in central Greenland bound for the west coast, and was never seen again.

Although most earth scientists were outspokenly against Wegener's hypothesis, a few farsighted individuals became enamored with the idea, and they too tried to accumulate data that would support it. Perhaps his most capable disciple was the South African geologist Alexander du Toit. He was aware that the fate of the theory would depend largely on the quality of the data that was used as evidence. So he pioneered geologic studies in both Africa and Brazil to ferret out undeniable proofs of drift. In the course of his work, du Toit came to believe that not one but two supercontinents existed before drift took place. He named them *Gondwana* and *Laurasia*. The former supposedly consisted of all the continents now in the Southern Hemisphere, and the latter of North America, Greenland, Europe, and Asia. An oceanic waterway he called the *Tethys Sea* separated the two land masses.

The great Scottish geologist Arthur Holmes (1890–1965) also was fired by Wegener's idea and came to accept it completely. Holmes in 1927 suggested that perhaps some sort of thermal convection in the earth's mantle was responsible for the movement of continents (Fig. 19-3). So at last, the advocates of the drift theory could point to a mechanism that could be powerful enough to move continents. No

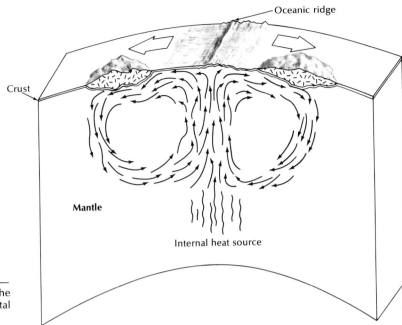

Fig. 19-3. Hypothetical currents in the mantle and their relation to continental drift.

Fig. 19-4. Fit of South America and Africa at the 183-m (100-fathom) isobath as proposed by Wegener. An isobath is a line of equal topographic elevation below sea level. Although there are gaps between the continental margins, the fit is relatively good.

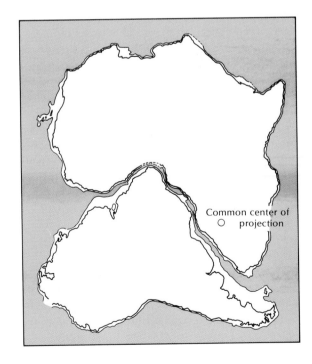

one knows even today whether the mantle can flow convectively or not; the majority of earth scientists believe that it can, however.

As the controversy continued, the number of adherents to Wegener's hypothesis slowly increased. In time many influential geologists, such as Felix Vening-Meinesz (1877–1966), S. Warren Carey, and Sir Edward Bullard, were listed among its supporters and the theory became impossible to ignore. What was some of the evidence that led more and more geologists to accept the theory?

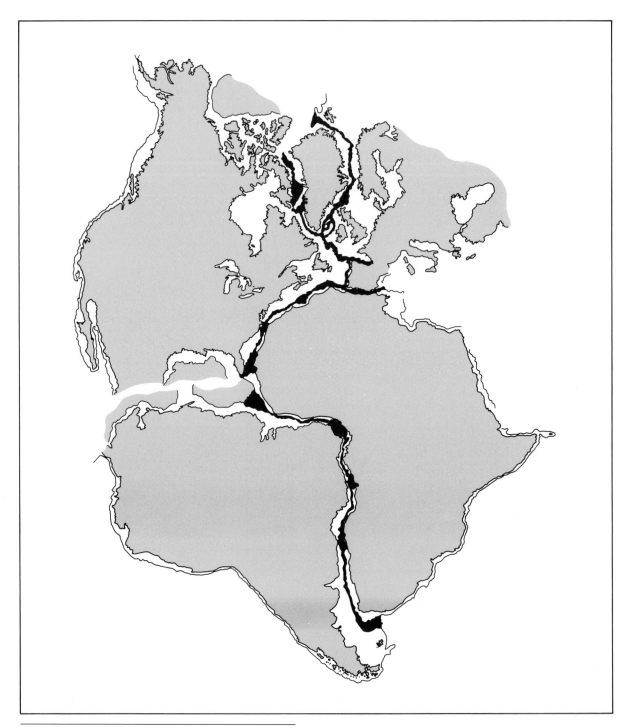

Fig. 19-5. Computer fit of all continents bordering the Atlantic. The thickness of the black line represents the amount of overlap.

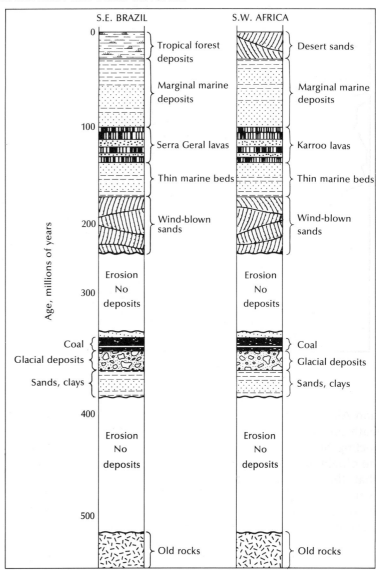

Fig. 19-6. Generalized columnar section of rocks from southeast Brazil and southwestern Africa. Notice the great similarity in sections up to about 100 million years ago.

Evidence of continental drift

Fit of continental coastlines One of the strongest pieces of evidence was how coastlines fit together. The earliest reconstructions, however, were made on maps, which, as two-dimensional representations of a three-dimensional globe, distort earth features to some degree. So the first "fits of continents" were not so exact as hoped for (Fig. 19-4). In fact,

as the opposition pointed out, the continents did not appear to fit together very well at all. Some of the poor fit, of course, resulted from the land masses being matched at the coastlines. Later, earth scientists realized that shoreline processes can greatly modify the coastal margins, and conceived of the idea of fitting the continents at a line *below* sea level, a line which more truly represented the shape of the continent. In 1955, Carey fitted South America

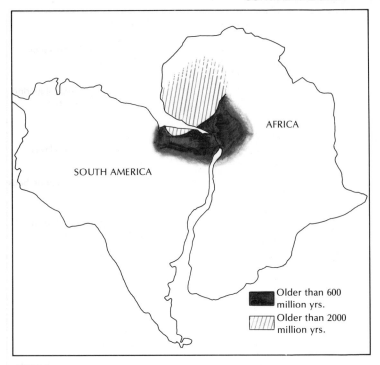

Older than 600 million yrs.

Older than 2000 million yrs.

Fig. 19-7. Matching the boundary between rocks of different ages in Africa and South America. The age boundary shown extends from vicinity of Accra, Ghana, to that of São Luiz, Brazil.

and Africa together at the 2000-m (6560-ft) iso-bath and used very careful methods for pro-jecting his reconstruction onto a map. The fit he obtained was so excellent that he remarked that the argument that South America and Africa do not closely fit together should never be used against the drift theory again. In 1965, Bullard and two associates showed that by us-ing a computer program they could obtain an even better fit of the continents around the Atlantic Ocean. They chose the 1000-m (3280-ft) isobath as their depth-of-fit line, and were amazed at how small the gaps and overlaps between the fitted continents were (Fig. 19-5).

Stratigraphic and structural similarities If the continents seemed to dovetail, did geologically significant features also match across the con-tinental boundaries? Such an idea might be termed the "torn newspaper" concept. If a page of newsprint is ripped into several pieces and reassembled, then not only should the pieces fit back together, but the lines of print should continue across the torn boundaries as well.

One type of geologic feature that could pos-sibly be matched up is the truncated rock record preserved on two or more continents. In Figure 19-6 are represented two generalized stratigraphic sections, one representing the rock record preserved in southeast Brazil and the other in southwest Africa. Both should be closely similar if the conjectured fit of the con-tinents has any merit—and they are nearly exact in detail. Such data cannot be used to prove beyond a doubt that drift has occurred, but it is telling evidence.

In the late 1960s, Patrick Mason Hurley, an American geochronologist, found that in the vicinity of Ghana in West Africa there is a dis-tinct boundary between rocks of two different ages: one more than 2000 million years and the other about 600 million years. The boundary trends southwestward to the west coastline of the continent. Hurley established a collabora-tive effort between the geochronology labora-

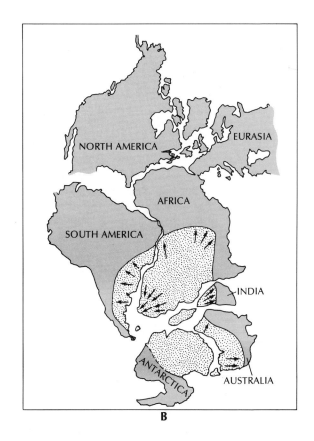

Fig. 19-8. Distribution of late Paleozoic glacial deposits as seen (**A**) on a present-day map and (**B**) on a map in which continents have been reassembled according to Wegener. Directions of glacial flow are indicated by arrows. In the present-day map the glaciers appear to originate in the oceans; in the reconstruction glacial directions form reasonable patterns.

tories at the Massachusetts Institute of Technology and the University of São Paulo, Brazil, to see if the same boundary could be found in South America, on the east coast of Brazil. Indeed, a boundary between 2000- and 600-million-year-old rocks was found, which appeared to be an extension of the African trend when the two continents were fitted together at their coastlines (Fig. 19-7). Hurley had been convinced of the reality of continental drift.

Segments of mountain chains in Europe, Greenland, and North America all are about the same age. One segment runs through Scandinavia, Scotland, and Northern Ireland; one through eastern Greenland; and one through eastern North America. Are they all parts of different ranges that developed almost simultaneously, or are they parts of one extensive range that has since been split apart during subsequent continental drift? If the continents are reassembled in the manner suggested by Bullard, then most of the mountain segments fit together in a nearly continuous chain. Again, the evidence supports the idea of drift but by itself is not conclusive.

Paleoclimatology All the continents in the Southern Hemisphere were affected by glaciation during the Pennsylvanian and Permian periods as shown in Figure 19-8A. But rather mystifying was the fact that many glacial deposits

and polished and scoured bedrock surfaces are located close to the equator, where the likelihood of continental glaciation is well-nigh impossible under any circumstances. Moreover, from the alignment of glacial grooves and the orientation of shatter marks left on the bedrock, it was possible to determine the direction of flow of the ice that created them: more often than not, the indications were that the ice had flowed from the sea toward land. That, too, was curious since geologists know of no glaciers today or during the recent ice ages that were formed in the oceans and which flowed onto the land. The situation was just the opposite.

If one reconstructs Pangaea (or Gondwana), however, it is apparent that all of the glaciated regions fit together into one large area unbroken by seaways (Fig. 19-8B). Then the flow-direction data fit together in an understandable manner.

Such evidence, which was also presented by Wegener, is not by itself an ironclad proof of the reality of drift, but when it is added to evidence already presented, the case for drift is measurably strengthened.

From a study of the rocks on the different continents, it is possible to obtain an idea of the climates that existed when the rocks were laid down. The distribution of the mapped paleoclimates make a disjointed, checkerboard pattern on the continents as they now stand. When the continents are pieced back together, however, the patterns appear continuous and reasonably explicable.

Fossils If all the continents once had been one, animals and plants that lived there should have been able to migrate freely across the land, restricted only by the availability of natural habitats. Fossils found on the various continents today, then, would show a similarity in ancestry dating back to the time when the continents were contiguous. Wegener and his followers attempted to point out such similarities in fossils. They noted, for example, that 64 per cent of Carboniferous and 34 per cent of Triassic reptile groups are the same for all southern continents. But the data available on the subject in the early 1900s was sketchy, to say the most. Even in 1943 when George Gaylord Simpson, one of the world's leading paleontologists, reviewed the situation he concluded that the paleontologic literature was too full of mistaken identities, unsubstantiated claims, internal inconsistencies, and worthless conclusions to be of any use in determining whether drift had occurred or not.

As more and better paleontologic data have been presented, Simpson's objections have begun to fade. Similarities in fossil forms have been shown to exist between the continents. As recently as 1969, fossil land-restricted amphibia and reptiles of Triassic age were discovered on the isolated frozen continent of Antarctica—exactly like Triassic fossils found in Africa, Madagascar, and Australia.

Paleomagnetism In the 1940s and 1950s paleomagnetism was a growing branch of the field of geophysics. One of the leaders in that field was the renowned English geophysicist Stanley K. Runcorn. He and his associates measured the fossil magnetism in a great many rock samples, and, in the manner outlined in Chapter 18, calculated from the data the positions of the north pole for rocks that ranged in age from late Precambrian to Recent. When they plotted the poles for European rocks on a present-day map it was found that they fell on a curved path that ran from south of the equator to the present north pole. The younger the pole the closer toward the present-day geographic pole it plotted. The plot, called an apparent polar wander curve, seemed to indicate that, with regard to Europe, either the north pole had moved or wandered or that the European continent had shifted and rotated in position with time. Today, for a variety of reasons, most paleomagnetists favor the latter explanation.

The significant thing is that when Runcorn plotted the north poles for rocks collected from North America and covering the same time span, he obtained a similarly shaped apparent polar wander curve, but it was dis-

placed about 40° toward the west from the European curve (Fig. 19-9). The similarity in shape between the two curves seemed to indicate that both Europe and North America had undergone similar changes in latitudes and rotation relative to the pole. The displacement between the two curves Runcorn attributed to continental drift. When North America was "moved" back to adjacency with Europe, the two curves became nearly coincident over most of their paths from the Precambrian through the Triassic. From that point in time onward, the paths diverged. Runcorn concluded that drift must have been initiated in the Triassic and that it had been going on ever since.

Runcorn published his results in 1962, at about the time when the theory of continental drift was still unacceptable to most earth scientists. The paleomagnetic evidence appeared so incontestable that it swayed many scientists who had been fence-sitting up to that time. Additional paleomagnetic studies of rocks of all ages from all of the continents have completely confirmed the first results.

Fig. 19-9. Two apparent polar wander curves from the Precambrian Era to the present, one based on data from North American rocks, and one on data from European rocks. The paths are similar in shape but not coincident in position, suggesting that North America and Europe have been drifting since the Triassic Period.

Linear magnetic anomalies We have already discussed the significance of oceanic magnetic anomalies (Fig. 19-10) in Chapter 18, but we should reiterate that their existence not only has confirmed that the ocean basins were cracking open along giant fractures but provided insight into the way that the continental masses actually drifted apart (Fig. 19-11). By identifying and correlating the anomalies in various parts of the ocean basins it was possible in many cases to determine when continental fragmentation began and what drift paths the continents took. The name given to the phenomenon of continental drift by opening of ocean basins is *sea-floor spreading*.

Deep drilling in the oceans If the oceanic ridges and rises are actually zones of divergence along which spreading occurs, then the ocean floor should become progressively older outward from the ridge crests. With the advent of the *Glomar Challenger*, the ocean-going drilling ship, it was possible to test that hypoth-

esis. One of the first lines of drill sites chosen was in the South Alantic Ocean, at about 30° south latitude, just south of Rio de Janeiro. The ship drilled eight sites, each progressively outward from the Mid-Atlantic Ridge. Drilling at each site was halted when the drill bit had penetrated the sedimentary sequence and had begun to bite into the hard basaltic rocks beneath it. The sediments recovered from each coring site were described in detail, and fossils contained within them were used to date each core throughout its length. The age of the sediments directly overlying the basaltic rocks was taken as a rough estimate of the age of the basement rocks beneath.

The results (Fig. 19-12) are exceedingly consistent and, much to the elation of the advocates of drift, show a continuous increase in age outward from the ridge axis. The slope of the curve, essentially a straight line, indicates uniform spreading from the Mid-Atlantic Ridge

at a rate of about 2 cm (0.79 in.) per year. That is just about the rate determinable from study of the linear magnetic anomaly pattern in that part of the South Atlantic Ocean.

Continued deep-sea drilling turned up another significant fact which at first seemed astonishing—in none of the world's oceans were basaltic rocks encountered beneath the sediments that were older than about 150 million years. The ocean basins are all geologically young features! Yet the continents all contained rocks 2500 to 3000 million years old or even older. Somehow the continents have floated about, like flotsam and jetsam, and

have lasted for billions of years, whereas the oceans appear to be in a continual state of renewal. If there is no old oceanic crust, what has happened to it to make it disappear? That is a question that will be pondered a little later in the chapter.

Continental drift in the 1960s

By the 1960s a whole new era had dawned in geology. Geologists had realized that the earth was not a static but rather a dynamic planet whose surface configuration was continually changing. The idea of continental drift was supported so strongly that most earth scientists considered the event to be a certainty. It is unfortunate that Wegener did not live long enough to witness that complete turnabout in geologic thinking.

With the acceptance of the continental drift theory, earth scientists turned their attentions during the 1960s and 70s toward finding a unified theory that could account not only for the drifting of continents, but for all of the other internally generated phenomena, such as mountain building, genesis of magma, and metamorphism as well. Through the recent development of the plate-tectonic theory we have come a long way toward that goal. Yet there is still a long way to go. It is an exciting time in geology, for almost each day brings new facts to light that require modifications in the way we look at the earth and its processes of formation.

Search for a unified theory Investigations of the ocean basins have indicated that new crust is continually being generated along the oceanic ridges and rises, but what happens to it after it is formed? If the crust is not being consumed at a rate equal to its generation then the earth would have to be expanding to accommodate it. And some geologists have proposed that this is so. However, opponents to the idea of expansion have pointed out that if the rate of expansion in the geologic past had been anywhere near the rate of sea-floor spreading observable in the recent past, the

Fig. 19-10. Pattern of bilaterally symmetric linear magnetic anomalies along the Mid-Atlantic Ridge. The areas of higher-than-background intensity are in black; those of lower-than-normal intensity are in white. Normally and reversely magnetized rocks beneath the ridge are youngest along the axis and increase in age laterally away from the axis.

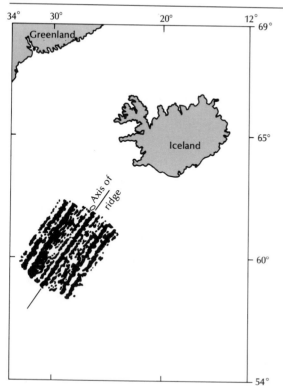

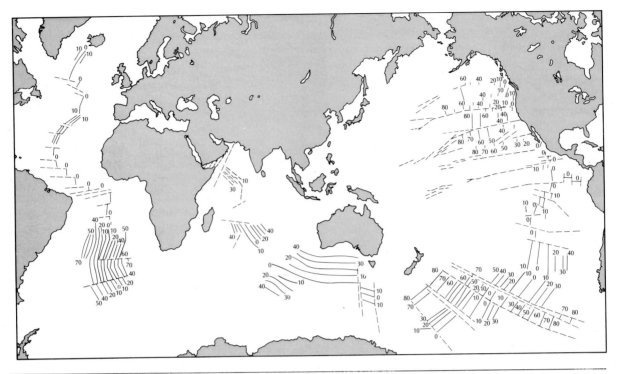

Fig. 19-11. Lines of equal age (in millions of years) on the ocean floor as determined from the linear magnetic anomaly pattern. The youngest rocks are found along oceanic ridges and rises, where the earth is splitting apart.

Fig. 19-12. The age of ocean sediments immediately above basaltic rock plotted as a function of distance from the Mid-Atlantic Ridge axis.

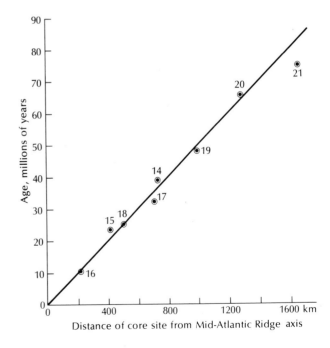

earth would have been an extremely small body initially, so small as to make that idea an unrealistic possibility. Moreover, the generation of new crustal material is not uniform, and if the crust is not consumed then the earth's surface should become distorted. But its shape is a fairly regular ellipsoid of rotation.

So most geologists have come to accept the idea that crustal material is being destroyed somewhere—but where?

In 1961, the American oceanographer Robert Dietz, followed in 1962 by Harry Hess, a geologist, proposed that the crust was likely to be consumed at the island arcs of the world. As you recall, island arcs are linear features and the sites of a variety of geologic phenomena (Fig. 19-13). The islands themselves generally consist of a string of volcanoes that erupt andesitic lavas and pyroclastics. Oceanward from the chain of volcanoes is a linear trench. Earthquakes are common to island arcs, and occur down to depths of about 700 km (438 mi). When the foci of all of the quakes are plotted on a cross section drawn at right angles to the arc system, they define an inclined seismic zone, called the *Benioff zone*, that begins at the surface near the trench and dips beneath the island arc at an angle of about 45° (Fig. 19-14). The zone was named after Hugo Benioff (1899–1968), an American seismologist who conducted pioneering studies into the distribution of earthquakes in continental margins. Studies of first motions recorded on seismographs reveal that, in most cases, the

Fig. 19-13. Kamchatka Peninsula and the Kuril island arc, showing the relation of earthquakes, volcanoes, and a deep-sea trench.

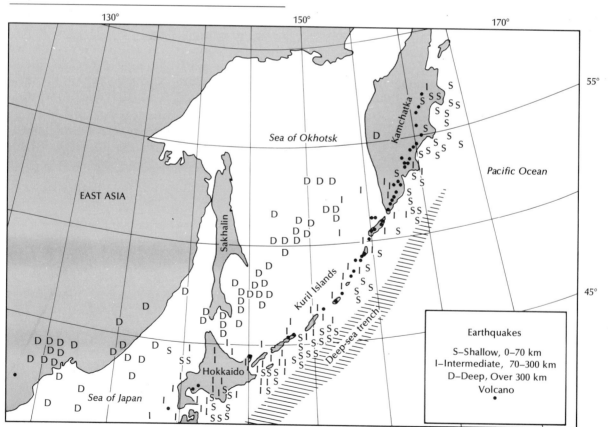

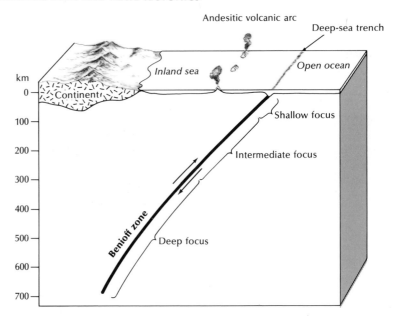

Fig. 19-14. The relation of a deep-sea trench, volcanic island arc, and the Benioff zone.

Benioff zone can be likened to a large thrust-fault zone along which the oceanward block pushes under, or underthrusts, the continental-ward block.

New global tectonics—A forerunner hypothesis
In the late 1960s, Jack Oliver and Bryan Isacks, two U.S. geophysicists, decided to investigate the possibility of the disappearance of crustal material at island arcs by studying the behavior

of seismic waves generated in the Tonga Island arc system in the southwest Pacific. They set up seismographs on islands both east and west of the deep Tonga trench and found that the nature of the seismic waves, especially the S-waves, were different, depending on the location of the stations. The differences, Oliver and Isacks reasoned, are the result of differences in the rocks beneath the island arc through which the waves travel. Their interpretation of the

Fig. 19-15. Hypothetical cross section A-A' extending from the Fiji Islands through the Tonga Islands to Rarotonga, illustrating the disposition and strength of the rock layers beneath the islands, as determined from interpretations of seismic waves.

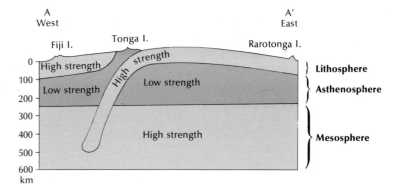

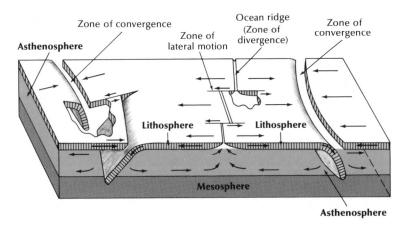

Fig. 19-16. The kinds of movement possible in the lithosphere and the relation of such movements with respect to compensating flow in the asthenosphere.

character of that rock substrate is shown in Figure 19-15.

Apparently a lobe of high-strength (brittle) material about 100 km (62 mi) thick parallels the Benioff zone and extends downward nearly 500 km (313 mi). They concluded that the brittle material, which they called the *lithosphere*, is being pushed down through a layer of lower-strength material, which they termed the *asthenosphere*. The latter extends from a depth of about 100 km down to about 250 km (155 mi) and corresponds to the zone around the world called the *low-velocity layer* (see Chap. 17). Beneath the asthenosphere, the mantle again becomes stronger, and they called that zone the *mesosphere*.

In 1968, Isacks, Oliver, and another geophysicist, Lynn Sykes, incorporated their findings into a hypothesis which they termed *new global tectonics*. They postulated that new lithospheric material is being formed at the ocean rises which, once formed, moves laterally away from the ridge axes and toward the island arcs (Fig. 19-16). They went on to speculate that in the vicinity of the trench off an island arc the lithosphere makes a sharp bend downward and sinks or is pushed into the softer asthenosphere below. It is much like a giant treadmill, with upwelling and generation of new lithospheric material along one zone and downwarping and consuming of the lithosphere along another zone.

PLATE TECTONICS—THE LAST WORD

Most earthquakes occur in the vicinities of the oceanic ridges and rises and along the Benioff zones of island arcs. In Figure 19-17 are plotted all of the foci for quakes that occurred in the world between 1961 and 1967. If one looks at the figure from a distance, a striking pattern is revealed. Essentially all of the world's seismic activity is confined to very narrow linear zones, and almost none occurs between those zones. Geologists have postulated from the pattern that the earth's crust is divided into a relatively small number of rigid blocks, or *plates*, as they have been formally named, that move and jostle one with respect to another (Fig. 19-18). Seismic activity takes place at the edges of the plates, but the plates themselves are nearly devoid of activity. The concept has been given the name *plate-tectonic theory*, and at the moment it is the dominant theory used by earth scientists to explain the internally generated features of the earth. It, being the most recent theory, embodies all of the concepts and features of continental drift and new global tectonics that went before.

Zones of divergence

Where plates move apart (pull-apart zones), such as along the Mid-Atlantic Ridge, new lithospheric material is formed (Fig. 19-19). They are zones of basaltic volcanism, and

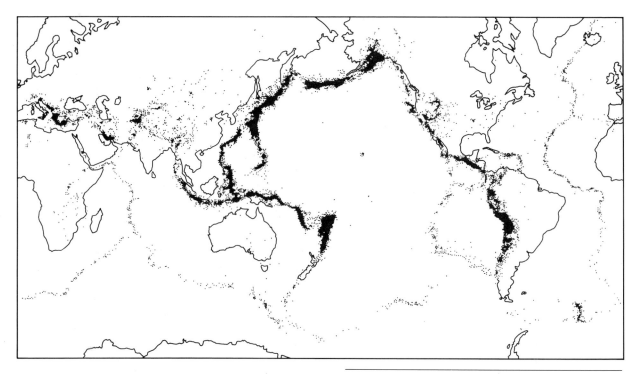

Fig. 19-17. Distribution of earthquake epicenters, 1961 to 1967. Most earthquakes occur in narrow belts which are now considered to mark the edges of moving lithospheric plates.

Fig. 19-18. The earth's lithospheric plates, based on occurrence of earthquake foci.

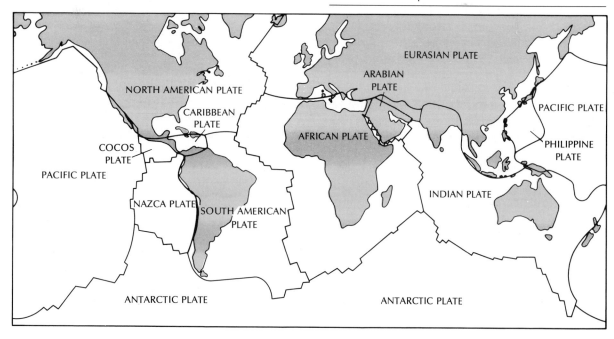

EURASIAN PLATE

ARABIAN
PLATE

NORTH AMERICAN PLATE

CARIBBEAN
PLATE

PACIFIC PLATE

AFRICAN PLATE

PHILIPPINE
PLATE

COCOS
PLATE

PACIFIC PLATE

INDIAN PLATE

NAZCA PLATE

SOUTH AMERICAN
PLATE

ANTARCTIC PLATE

ANTARCTIC PLATE

shallow-focus earthquakes. Rocks, when heated, tend to expand. Therefore, the newly formed hotter rocks near the axis of the zones of divergence occupy more volume, which results in an elongate topographical high—an oceanic ridge or rise. As the lithosphere moves laterally outward from the ridge crest it cools and contracts, and the ridge system is reduced in elevation. Earthquakes in the vicinity of the divergence zones occur at depths of less than 20 km (12 mi) as a rule and are restricted to the ridge axes or the fracture zones that cross them nearly at right angles.

Zones of lateral movement—transform faults

Lynn Sykes has carefully plotted the epicenters for many of the fracture-zone earthquakes and has noted a curious fact (Fig. 19-20). Although the fracture zone crosses and even offsets the ocean ridges and extends beyond the offset ends of the ridge, in places for many hundreds of kilometers, the seismicity is restricted essen-

tially to the region between the truncated ridge ends. Another fact was revealed about such quakes when Sykes studied their first motions recorded on seismograph records. From a casual glance at a diagrammatic representation of fracture zones (Fig. 19-21) one would think that the ridges are offset—the plate on the other side of the fault moving relatively to the left. Yet the first motions indicated fault movement of exactly the opposite direction; that is, right lateral movement. That appears contradictory until one recalls that the material is moving outward from the ridge-crest divergence zones. If new material is moving outward it will be moving past the fracture zone in a right-lateral sense. Once the material in the upper segment has moved out on the flanks to a point opposite the ridge crest in the lower segment then the two pieces of crust move in the same direction. That, concluded Sykes, is why almost all of the fracture-zone earthquakes are restricted to the portion of the fault between the two truncated ridge axes.

Fig. 19-19. World pattern of zones of convergence and divergence. Arrows indicate the supposed movement of lithospheric plates.

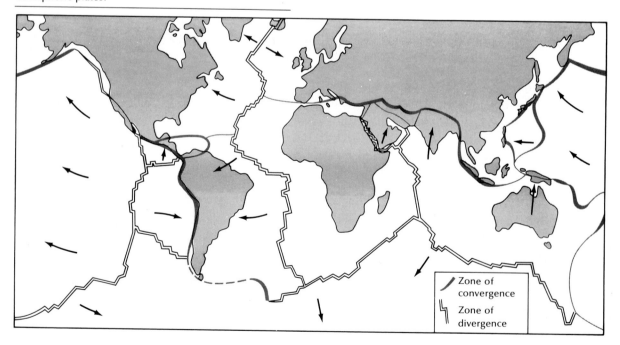

Zone of
convergence

Zone of
divergence

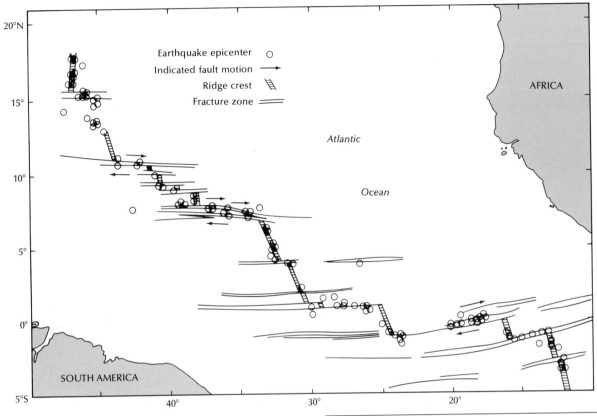

Fig. 19-20. Earthquake epicenters near the Mid-Atlantic Ridge and indicated fault motions along fracture zones.

Such an unusual type of faulting between segments of oceanic ridges and rises has been termed *transform faulting,* and the possibility of its existence was first postulated in 1965 by the Canadian geophysicist J. Tuzo Wilson. The seismicity along transform faults is but another proof that the oceanic ridges are active spreading centers.

Since transform faulting does not cause separation of ridge axes but stems from the fact that they are already offset, we still do not know what causes the initial offset in the ridge lines. Undoubtedly it is related to the generation of the spreading centers in the first place, but exactly when and how the offsets occur is still one of the mysteries of the earth.

Zones of convergence

The curvilinear island arcs of the world mark the zones where lithospheric plates push together and consuming of lithospheric material occurs (see Fig. 19-19). As a slab of lithosphere bends over and slowly descends into the asthenosphere below it is heated. Eventually, temperatures will rise to the point where the downgoing slab becomes indistinguishable from the mantle rocks that surround it (Fig. 19-22). How far down a plate edge can move depends on the rate of movement, but none have ever been found at depths greater than 700 km (435 mi). At those depths the slab is moving again through higher-strength mantle rocks.

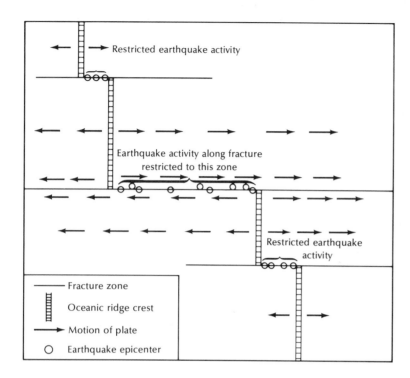

Fracture zone

Oceanic ridge crest

Motion of plate

Earthquake epicenter

Restricted earthquake activity

Earthquake activity along fracture restricted to this zone

Restricted earthquake activity

Fig. 19-21. Motion of blocks near the oceanic ridge system. Earthquakes along the transverse fracture zones are restricted largely to the region where two blocks are moving in opposite directions. Beyond the offset ends of the ridge crests, where the two crustal blocks are moving in the same direction, there are few earthquakes.

Fig. 19-22. Relation of a divergence zone to a convergence zone. New lithospheric material formed along oceanic ridges moves outward and bends sharply downward into a zone of convergence, where it is eventually consumed at depth. The axis of a deep-sea trench demarcates where the slab bends downward, and a line of andesitic volcanoes form where heating adjacent to the downgoing slab is sufficient to melt rocks. Shallow-, intermediate-, and deep-focus earthquakes of the Benioff zone are generated along the upper surface of the slab.

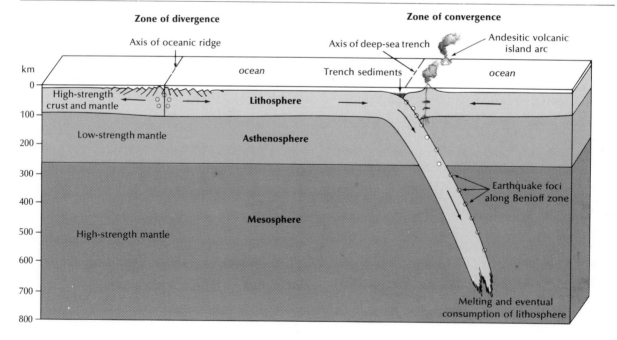

As the relatively cold lithospheric plate descends, frictional drag is developed. Sporadic release of the strain in the slab supposedly produces the shallow-, intermediate-, and deep-focus quakes characteristic of island arcs (Fig. 19-22).

Sediments derived from the nearby continental land masses and from erosion of the islands of the volcanic arc themselves can accumulate in the deep trench next to the arc. As the lithospheric plate moves downward, those stratified rocks are dragged down, or *subducted*. At first they will be folded and faulted; then as they move to greater depths, they will be heated—producing dynamothermal metamorphism and local melting. Eventually a linear mountain chain will be formed, as described in Chapter 16. Magma derived from melting of the sediments or parts of the downgoing slab can rise to the surface wherever cracks and fractures are present, to produce the linear chain of andesitic cones so common to island arcs. The exact nature of the material that is melted to produce the volcanic eruptions and the depth at which the melting takes place are still points of controversy among petrologists. Some even believe that some melting of the continental crustal rocks must be involved as well.

New wrinkles

The plate-tectonic model as first stated was relatively straightforward and elegant in its simplicity. Much of its appeal stemmed from those attributes. During the few years of its existence, however, it has become apparent that the simplified theory cannot explain all of the earth's geologic features. As a result, the theory has had to be modified to some degree. If exceptional phenomena become too numerous, however, there is a danger that the theory would become useless. But at the same time, modifications must not be ignored. A dominant theory, such as that of plate tectonics, tends to steamroll over facts and phenomena which may not be in accordance with it. A quote from the 1968 paper in which Isacks, Oliver, and

Sykes first formulated the forerunner theory of new global tectonics, is pertinent to the point: "At present there appears to be no evidence from seismology that cannot be reconciled with new global tectonics in some form."

Microplates The plate-tectonic model was initially meant to apply to the movements of a small number of large lithospheric plates. It has become apparent, however, that in some parts of the world, the lithosphere is composed of numerous small plates (called *microplates*) which can move in very complicated ways. The region around the Mediterranean Sea, for example, appears to be composed of microplates, and an understanding of its geologic structure is impeded by an inadequate knowledge of the complex movement of those plates.

Back-arc upwelling Oceanographic studies, mostly by the American geologist Daniel Karig during the last five years or so have shown that secondary zones of divergence commonly form between continents and island-arc systems, which leads to an oceanward migration, with time, of the arc and trench system. Such movement was a totally unknown phenomenon when the plate-tectonic theory was initially formulated. It is now the opinion of geophysicists that such *back-arc upwellings* result from the development of a convectional eddy current behind the island arc in response to heating of the mantle by the downgoing slab, as shown in Fig. 19-23.

Mantle hot spots and plumes Not all volcanism is restricted to the zones around plate margins; some, such as that of Hawaii, occurs out in the middle of the plates themselves (see Chap. 5). It is difficult to understand the origin of such volcanoes within the framework of the plate-tectonic model. J. Tuzo Wilson has speculated that somewhere at depth in the mantle, perhaps as far down as the core-mantle boundary, localized *hot spots* form. Heating leads to a reduction of the density of the mantle material, which then begins to rise toward the surface in a finger-like *plume*. The rising plume

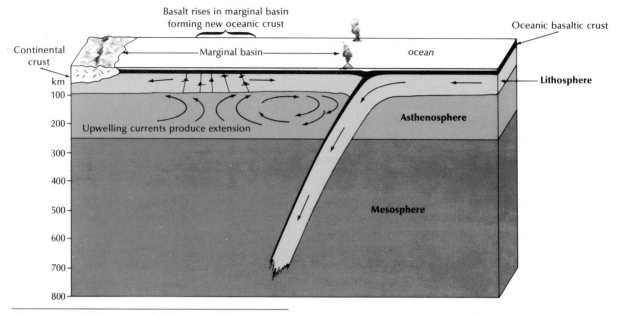

Fig. 19-23. Hypothetical model to account for the development of back-arc upwellings.

of heated rock can bring about the partial melting of parts of the asthenosphere or the lower parts of the lithosphere, and it is the eruption of that magma that builds volcanoes in the center of a plate (Fig. 19-24).

Some geophysicists have taken the idea a step further, speculating that the rising plumes bend over as they reach the asthenosphere. Where they strike the underside of the lithospheric plates they push, like moving fingers, and cause lateral movement of the plates.

Since the plumes supposedly are generated in the relatively static mantle beneath the moving lithosphere above, the volcanoes, once formed, are subject to being cut off from the magma source and carried away with the moving plate. The result is a formation of a line of volcanoes that increase in age away from the active hot spot (Fig. 19-24). The Hawaiian Islands form just such a chain of volcanoes, with the youngest ones over active vents on the island of Hawaii.

Not all geophysicists subscribe to Wilson's idea; some believe that the existence of

plumes in the mantle has not yet been demonstrated adequately.

Mechanism for plate tectonics

When Wegener first postulated his theory, many critics pointed out that he had failed to show how the continents might move from one place to another. In spite of recent progress in understanding the earth and its internal processes and the emergence of plate tectonics as a dominant theory, earth scientists are still no closer to explaining the mechanism than was Wegener.

One idea, introduced first by Arthur Holmes, is that the mantle of the earth is subject to inequalities in internal heating such that it can flow, or slowly convect, much like a pot of slowly boiling porridge (Fig. 19-25). Many advocates of plate tectonics have not only accepted the idea of convection but consider it as the force that drives plates over the earth's surface. Places where the currents rise and move apart correspond to zones of divergence,

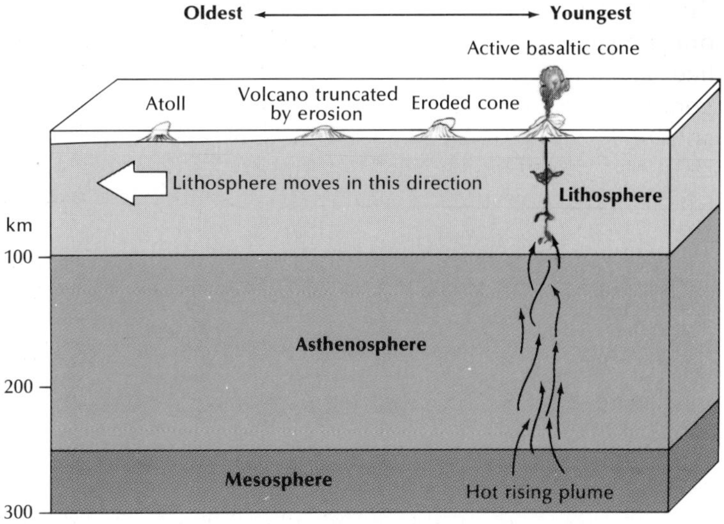

Oldest ←→ Youngest

Fig. 19-24. Diagram showing the postulated hot rising plume that melts rocks just below or within the lithosphere, leading to the formation of plate-center basaltic volcanoes. Once formed, the cones move laterally from the "hot spot," carried piggyback on the moving lithospheric plate.

whereas places where the cooler currents converge and descend correspond to the zones of convergence.

Yet we do not know whether the solid mantle is actually capable of convection as pictured, and even if it were, we do not know how far down in the mantle convection can operate. We are also at a loss to explain how an ephemeral phenomenon like convection can maintain itself in narrowly restricted locations for hundreds of millions of years.

Before it was known that a low-velocity layer in the depth zone between 100 and 250 km (62 and 155 mi) existed, those advocating convection imagined the cells to originate far down in the mantle (Fig. 19-25A). Today, most advocates restrict the process to the relatively soft as-

thenosphere—which would make particularly difficult the generation of long, linear zones of divergence that stretch continuously for tens of thousands of kilometers (Fig. 19-25B). If convection really is the primary driving force, one might wonder also how the upwelling currents at the zones of divergence can be offset along transform faults for distances up to 1000 km (621 mi) and more.

Since the convection theory seems to present insurmountable difficulties, many geophysicists have turned toward alternative hypotheses. One is based on the relative differences in density between the downgoing lithospheric slab and the mantle into which it sinks (Fig. 19-26). It is thought that the cold, sinking slab is more dense than the mantle beneath and that

Fig. 19-25. Cross section of the earth with (**A**) a deep convection model and (**B**) a shallow convection model.

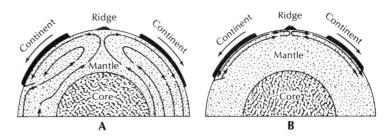

the slab sinks under its own weight, thereby dragging along the lithospheric plate behind it, like an immense, flattened tail. Yet some geophysicists do not believe that the lithosphere is more dense than the mantle, and certainly any differences that do exist would be small indeed. Moreover, in the plates diverging from the Atlantic and Indian oceans, there are no associated convergence boundaries and no sinking slabs, so it is hard to visualize how the mechanism could be responsible for plate motion in those two ocean basins.

Another theory that is now gaining support involves gravitational sliding of the lithospheric plates away from the topographically high oceanic ridge and rise crests. Opponents to that view, however, can point to several areas in the world where the downslope gradient apparently is inadequate for sliding to take place.

One other mechanism—expansion—deserves mention as a possible means of indirectly triggering plate movement, at least during the last 150 million years (Fig. 19-27). If the brittle lithosphere split apart as a result of expansion, then new material would have to flow from beneath and into the void to take up the space created by the splitting, and would, in a way, force some convection in the outer part of the mantle. But no one can really say whether or not the earth is expanding.

Almost all earth scientists subscribe to one or more of the above theories for driving the lithospheric plates. Yet there are questions and problems concerning all of them. It well may be that the actual mechanism is not a simple one but involves two or more processes working in unison or in tandem. Or perhaps the real mechanism is still unfathomed. The forces work so slowly and over such large regions that it probably will be some time before a realistic model can be formulated.

Fig. 19-26. The hypothetical sinking of the relatively cold and dense lithospheric slab into the asthenosphere and mesosphere. As the slab sinks, it pulls the oceanic lithosphere along behind it, causing opening at the ocean ridge.

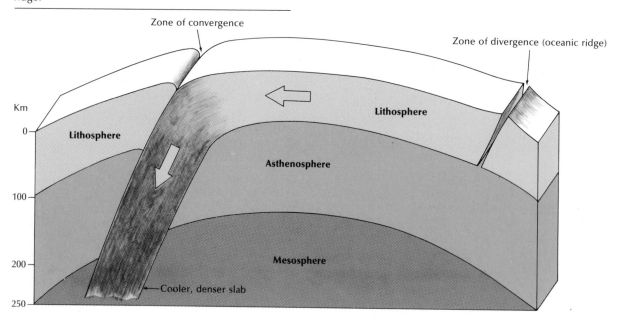

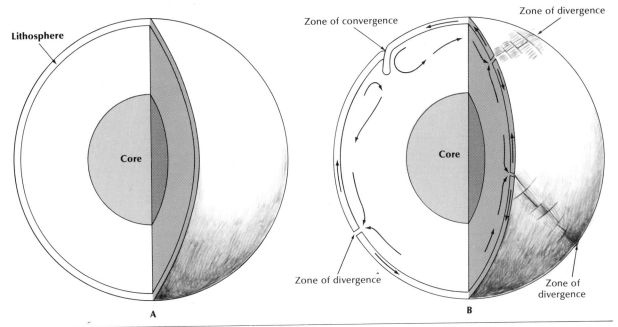

Fig. 19-27 A. Initial state of the earth. **B.** After expansion, the brittle lithosphere would crack apart with formation of zones of divergence; as material flowed up to fill in the voids developing at the zones of divergence, mantle convection would be generated, with attendant creation of subduction zones.

Fig. 19-28. Two deep-sea trenches showing undeformed sediments in (**A**) the Chile Trench, and (**B**) the Aleutian Trench.

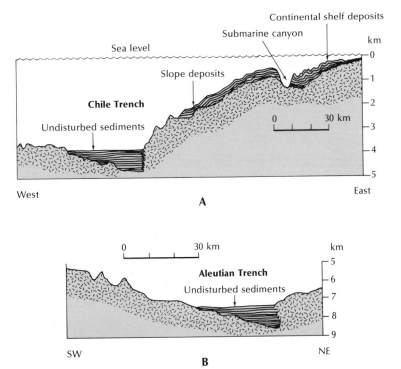

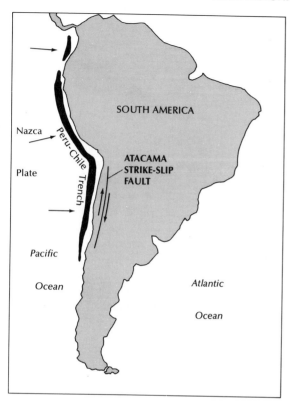

Fig. 19-29. South America, showing the Peru-Chile Trench and the Atacama strike-slip fault. The postulated movement of the Nazca Plate is shown by three heavy arrows. Notice that the direction of the plate motion is nearly at right angles to movement along the fault.

Assessment of the plate-tectonic theory

The theory of plate tectonics as it stands today is relatively simple and elegant. By means of one philosophical approach it has been possible for earth scientists to explain reasonably many of the earth's internally generated geologic features and processes. And that holds true both on a local or regional scale. As a consequence, during the last several years there has been an explosion of articles on plate tectonics in the earth-science literature.

In spite of the apparent success of the theory most proponents readily admit that it leaves some important questions unanswered.

John Maxwell, an American structural geologist, points out, for example, that there is a bilateral symmetry in the Atlantic and Indian oceans, with regard to the ocean ridge system and bordering ocean basin, but such is not the case in the Pacific. Why? And further, he asks, why does the formation of new lithosphere occur in all oceans whereas its consumption takes place almost exclusively around the margin of the Pacific? Another point that Maxwell makes is that many of the deep-sea trenches appear to have formed only recently and to be extensional, rather than compressional, in origin. Moreover, in some of the trenches the sediments are flat-lying, not folded and faulted as one might expect in the active zone of convergence (Fig. 19-28).

One structural feature which seems inadequately explained is a set of active major strike-slip faults around the margin of the Pacific Ocean such as the Alpine Fault in New Zealand, the Denali Fault in Alaska, and the Atacama Fault in South America. Strike-slip motion on those faults, which parallel the linear zones of convergence, is at right angles to the motion of the converging plates (Fig. 19-29).

The origin of certain mountain ranges also appears to be a bit cloudy. For example, the prominent east-west mountain belt that includes the Alps and Himalayas appears to have been formed by north-south convergence of plates during the last 50 to 60 million years. Yet the evidence from the linear magnetic anomalies for that same time interval indicates that essentially all plate motion has been in an east-west direction. And V. Beloussov, a well-known Russian structural geologist, notes that several mountain chains (such as the southern Rocky Mountains) have originated in the interiors of plates as a result of vertical motion. He contends that such vertical tectonics virtually are unexplained within the context of plate tectonics.

Some aspects of volcanism bring questions to mind. For example, andesitic volcanoes, according to the plate-tectonic model, are characteristically restricted to convergent zones. Yet active volcanoes in the Cascades of

the Pacific Northwest, which resemble convergent-zone volcanoes in all other respects, are not associated with an active subduction zone.

The existence of such questions and inconsistencies provides ample incitement for continued investigation of the nature of internal processes. Undoubtedly, as the theory is tested and reassessed, we can expect to see minor and perhaps even major modifications.

Perhaps the most significant test of the theory will come about as a result of precise measurements between far-removed points on the face of the earth. It appears that soon it will be possible to measure the distance between two points about 200 km (124 mi) apart with an error of 2 cm (0.79 in.). By measuring at yearly time intervals the long-range distances between points in a network on different plates, it will be possible within five to ten years to determine unequivocally the motions between and within plates. At that time we will know whether the theory is basically sound or in need of major modifications.

SUMMARY

1. Before the twentieth century several theorists had suggested that some continents had split and drifted apart. Alfred Wegener in 1911 uncovered evidence from several sources that supported his contention that such drift had occurred, and a great controversy over the theory of *continental drift* began.

2. The evidence that led to increased acceptance of the hypotheses included: fit of continental coastlines, stratigraphic and structural similarities between continents, paleoclimatic reconstructions and distribution of the late Paleozoic glacial deposits, distribution of similar fossil groups, paleomagnetic polar wander curves from different continents, the presence of linear mag-

netic anomalies over the oceanic ridges and rises, and dating of cores drilled from ocean sediments.

3. By the late 1960s most geologists believed in continental drift, but a unified theory was needed; one that would explain drift along with other internally generated phenomena such as volcanism, mountain building, metamorphism, and seismicity.

4. The model of *new global tectonics* was an attempt to provide a unified theory. The outer 100 km (62 mi) of the earth was defined as the *lithosphere*, the zone between 100 and 250 km (62 and 155 mi) the *asthenosphere*, and the zone beneath 250 km as the *mesosphere*. The lithosphere is strong and brittle and the asthenosphere is relatively weak. New lithospheric material supposedly forms at the oceanic ridges and moves laterally away from the ridge systems toward the island arcs. At trench-island arc systems, lithospheric plates bend sharply over and plunge into the asthenosphere.

5. The model of *plate tectonics* is the most recent unifying concept. It is thought that the lithosphere is subdivided into a number of large plates that are moving one with respect to another. Some converge, some diverge, and some slide laterally past one another.

Zones of divergence, where new lithospheric material is forming, are characterized by high heat flow, basaltic volcanism, and shallow-focus quakes.

Zones of lateral movement (*transform faults*) cross the ridges and rises nearly at right angles and are characterized by shallow-focus earthquakes.

Zones of convergence are marked by deep oceanic tranches, andesitic volcanism, and a seismic zone known as the *Benioff zone* that extends down to 700 km (435 mi). Shallow-, intermediate-, and deep-focus quakes occur along that zone.

6. Additions to the plate-tectonic model include the recognition of *microplates*, back-arc upwelling, and mantle hot spots (*plumes*).

7. The mechanism that causes plates to move is unclear. *Mantle convection,* a density difference between the lithosphere and asthenosphere, gravity sliding, and expansion of the earth have all been suggested as possible candidates.

8. Some data do not fit into the plate-tectonic model convincingly.

SELECTED REFERENCES

Bird, J. M., and Isacks, B., eds., 1972, Plate tectonics, selected papers from the Journal of Geophysical Research, American Geophysical Union, Washington, D.C.

Beloussov, V. V., 1974, Sea-floor spreading and geologic reality *in* Plate tectonics: assessments and reassessments, C. F. Kahle, ed., American Association of Petroleum Geologists, Tulsa, pp. 155–66.

Bullard, E., 1969, The origin of the oceans, Scientific American, vol. 221, no. 3, pp. 66–75.

Cox, A., 1973, Plate tectonics and geomagnetic reversals, W. H. Freeman and Co., San Francisco.

Heirtzler, J. R., 1968, Sea-floor spreading, Scientific American, vol. 219, no. 6, pp. 60–70.

Hurley, P. M., 1968, The confirmation of continental drift, Scientific American, vol. 218, no. 10, p. 52.

James, D. E., 1973, The evolution of the Andes, Scientific American, vol. 229, pp. 60–69.

Marvin, W. B., 1973, Continental drift, The evolution of a concept, Smithsonian Institution Press, Washington, D.C.

Maxwell, J. C., 1974, The new global tectonics: an assessment, C. F. Kahle, ed., American Association of Petroleum Geologists, Tulsa, pp. 24–42.

Orowan, E., 1969, The origin of the oceanic ridges, Scientific American, vol. 221, no. 5, pp. 102–19.

Seyfert, C. K., and Sickin, L. A., 1973, Earth history and plate tectonics, Harper and Row, New York.

Takeuchi, H., Uyeda, S., Kanamori, H., 1970, Debate about the earth, revised edition, Freeman, Cooper, and Co., San Francisco.

Tarling, D., and Tarling, M., 1971, Continental drift, Doubleday and Co., Garden City, N.Y.

Toksöz, M. N., 1975, The subduction of the lithosphere, Scientific American, vol. 233, pp. 88–101.

Vine, F. J., and Matthews, D. H., 1963, Magnetic anomalies over oceanic ridges, Nature, vol. 199, p. 947.

Wegener, A., 1966, The origin of continents and oceans, Dover Publications, New York.

Wilson, J. T., 1963, Continental drift, Scientific American, vol. 208, no. 4, pp. 86–100.

GLOSSARY

Aa A lava flow with a rough, blocky appearance.

ablation The processes by which snow or ice is lost from a glacier.

abrasion The mechanical wearing, grinding, or scraping of a rock surface by the friction and impact of moving rock particles.

absolute time The best estimate of the actual age, in years, of an object or event. Generally determined by the decay of radioactive elements.

abyss The deep ocean floor.

abyssal hills Hills that rise up to 1000 m above the deep ocean floor.

abyssal plains The relatively extensive low-relief plains of the deep ocean floor.

adobe A clayey deposit mixed with silt. Originally used to make sun-dried bricks in the Southwest.

aftershock A generally smaller earthquake that occurs after a larger earthquake and is closely related in origin.

aggradation (stream) Deposition of a stream's excess load on the channel bottom.

A horizon The dark-colored surface layer, or topsoil, of the soil profile.

A2 horizon A whitish layer in a podzol that lies between the A and B horizons and from which most of the iron oxides have been removed by downward-percolating water.

air-fall ash Volcanic ash that settles out of the air and can be deposited thousands of kilometers from the volcano's vent.

alluvial fan A fan-shaped deposit of sand and gravel deposited by a stream at the base of a mountain, usually in an area of arid or semi-arid climate.

alluvium Any clastic material deposited by a stream along its course.

alpine chain A complex linear mountain belt made up of batholithic igneous rocks, dynamothermal metamorphic rocks, and complexly folded and faulted sedimentary rocks.

alpine glacier A mountain glacier that moves downslope under gravity, often following a pre-existing stream valley. Also called a valley glacier.

amphibole group A large group of ferromagnesian silicates with similar physical properties. Their most distinctive property is the possession of two good cleavages intersecting at 124° and 56°.

andesite A gray to grayish-black volcanic rock composed mostly of intermediate plagioclase, augite, hornblende, and biotite. Volcanic equivalent of a diorite.

angle of repose The maximum slope at which relatively loose material will remain stationary without sliding downslope.

anion A negatively charged ion.

anorthosite A variety of gabbro consisting almost

entirely of coarse crystals of calcium-plagioclase.

antecedent stream A stream that has maintained its original course despite the occurrence of local uplift or movement along its course.

anticline A fold in which the limbs dip away from the axis; older beds are found toward the axis.

aphanitic texture A crystalline rock texture in which most of the crystals are too small to be seen by the unaided eye.

apparent polar wander curve The curve created by plotting on a map the succession of paleomagnetic poles for rocks of different ages from any one continent or lithospheric plate.

aquiclude A subsurface layer too impermeable or too tight to accept or transmit water.

aquifer A subsurface rock layer that readily yields water.

arête A jagged, thin rock wall or ridge separating two glaciated valleys, generally near their heads.

arkose A sandstone composed mainly of quartz and feldspar.

arroyo The steep-sided, flat-floored channel of an ephemeral or intermittent stream.

artesian well A well that taps a confined aquifer.

ash Volcanic fragments the size of dust blown into the air during an eruption.

ash flow See *glowing avalanche*.

asthenosphere The layer of the earth extending from 100 km below the surface to 250 km. It appears to be less strong than the zones above and below it and corresponds to the low-velocity earthquake zone, i.e., where *S* waves travel at reduced velocities.

atmosphere The gaseous envelope that surrounds the earth. It is composed largely of nitrogen (78 per cent by volume) and oxygen (21 per cent).

atoll A ring-shaped island with an interior lagoon and made up of the calcareous skeletons of marine animals.

atomic number A number characteristic of a chemical element and equal to the number of protons in an atom of that element.

atomic weight Of an element, a number that indicates, on the average, how heavy an atom of that element is compared to an atom of hydrogen.

augen Shear-resistant minerals in cataclastic metamorphic rocks; they appear as relatively large, rounded clots, or eyes.

augite A dark ferromagnesian mineral of the pyroxene group with two good cleavages that intersect at 87° and 93°.

aureole A halo of hydrothermally metamorphosed rocks that surrounds most batholiths.

axial plane A plane connecting the axial lines in successive beds in a fold.

axis (fold) A line drawn along the points of maximum curvature of a fold.

back-arc upwelling A secondary zone of divergence of lithospheric plates that commonly forms between a continent and an island arc system.

backswamp A section of low-lying ground between a natural levee and the river bluff.

backwasting A process by which valley slopes retreat essentially parallel to themselves.

badlands Rough, steeply gullied terrain usually found in dry areas.

bajada An apron of a desert mountain range formed from lateral coalescing of alluvial fans.

bar A spit that has grown almost long enough to close off a bay.

barchan A symmetrical, crescent-shaped dune that forms in deserts in which the wind direction is nearly constant, sand supply is limited, and the ground surface (bedrock) is hard.

barrier islands Long, narrow sand islands that are roughly parallel to the shore and are separated from it by a narrow body of water.

barrier reef A long, narrow coral reef roughly parallel to the shore and separated from it by a fairly deep and wide lagoon.

basal slip (glaciers) The sliding of glaciers over underlying material.

basalt A dark, volcanic rock, fine grained to aphanitic in texture and composed largely of pyroxene, calcium-plagioclase, and olivine. It is the most abundant volcanic rock.

base level (stream) The low point to which most streams flow, ultimately the ocean.

basin (structure) A saucer-shaped synclinal structure which lacks an axis and in which the beds dip toward the center.

batholith A pluton with an exposed surface area of at least 100 km; generally granitic in composition.

bauxite A chief ore of aluminum, formed as an end product of weathering in tropical climates.

bedding plane The surface that parallels the layers in sedimentary rock.

bed load The material moved along the bottom of a stream.

bedrock The rock that underlies soil or other unconsolidated surface material.

Benioff Zone   The planar dipping zone of shallow, intermediate, and deep earthquakes that is common to the Pacific Ocean margins.

B horizon   The subsurface horizon below the A horizon in the soil profile. Contains the most clay and is usually red to brown in color.

biotite   A common ferromagnesian mineral of the mica group. Sometimes called black mica.

birdsfoot delta   A delta formed by many distributaries of a river and resembling a bird's talons.

block faulting   A type of faulting in which the crust is broken into a number of sub-parallel blocks that are displaced with respect to each other, primarily by dip-slip motion.

body waves   Seismic waves (*P* and *S* waves) that travel through the interior of the earth.

bottomset bed   The horizontal strata at the bottom of a delta, deposited progressively in front of the advancing delta front.

boulder-clay   A glacial deposit consisting of rocks the size of boulders set in a clayey matrix.

Bowen's reaction series   The general sequence of crystallization in cooling magmas, going from minerals stable at higher temperatures to those stable at lower temperatures.

braided stream   A stream that flows in a network of wide, anastomosing channels separated by low bars or islands.

breccia   A clastic sedimentary rock composed of angular fragments cemented together.

calcite   A calcium-carbonate mineral with three directions of cleavage; effervesces in acid.

caldera   A large, basin-shaped depression formed by the inward collapse of a volcano after an eruption.

caliche   See *K horizon.*

calving   The process by which a chunk of ice breaks away from an ice sheet into water.

capacity   The potential load that a stream can carry.

capillary fringe   A relatively thin zone of water that migrates upward from the water table.

carbonates   Sedimentary rocks made up of calcium or magnesium carbonate.

carbon-14 dating (radiocarbon dating)   A method of radiometric dating in which decay of the isotope carbon-14 into nitrogen-14 is measured. The level of carbon-14 is continually readjusted in all organisms to an equilibrium level. Once they die, decay is not compensated for and steadily decreases, reaching an unmeasurable level in about 50,000 years.

cataclastic metamorphic rocks   Rocks formed by shearing and granulation during fault movement. The process is called cataclastic or dynamic metamorphism.

catastrophism   The belief that geologic history occurs as a sequence of sudden, catastrophic events.

catenary   A curve produced by a wire or chain suspended from two points. The cross section of a U-shaped valley approximates a catenary.

cation   A positively charged ion.

Cca horizon   In arid-region soils, a layer of calcium carbonate beneath the B horizon.

cementation   The process by which sediments consolidate: a cement (usually $CaCO_3$ or SiO_2) is deposited in the pore spaces separating the sedimentary particles.

cenote   In Yucatán, Mexico, a sinkhole formed by the collapse of the roof of an underlying cave.

Cenozic Era   The major division of geologic time following the Mesozoic Era and beginning about 65 million years ago. Characterized by the rapid evolution of mammals.

channeled scabland   The basalt plateau in eastern Washington crossed by deep, dry channels that formed during a huge prehistoric flood.

chemical weathering   The weathering of rocks and minerals by chemical reactions.

chert   A dense, hard, sedimentary rock made up of cryptocrystalline silica.

chlorite   A group of hydrous silicates containing magnesium, aluminum, and other metal ions. Occurs in thinly banded masses and is grass-green to blackish-green in color.

C horizon   The slightly weathered layer in a soil profile beneath the B horizon.

cinder cone   A relatively small volcanic vent commonly composed almost entirely of pyroclastic material.

cinders   Pebble-sized, reddened pyroclastic fragments blown from a volcano. Abundant in cinder cones.

cirque   A horseshoe-shaped, steep-walled glaciated valley head.

clast   A fragment produced from the breakdown of a pre-existing rock.

clastic sediment   Sediment made up of fragmental material derived from pre-existing rocks of any origin.

clastic texture   A fragmental texture usually associated with sedimentary rocks in which the angular-to-rounded grains (clasts) are broken fragments of pre-existing minerals and rocks.

clay-size A term used to designate the particle size of a sediment as very fine, and smaller than that of silt.

clay minerals Minerals with a sheet-like atomic structure similar to mica; generally the result of weathering.

cleavage A planar surface or set of parallel surfaces along which a mineral will tend to split when broken.

coal A partially metamorphosed sedimentary rock formed from decomposed and altered plant remains.

columnar jointing The joint pattern that forms when lava flows or tabular, shallowly buried magma bodies cool.

compaction The squeezing together of particles in a sediment.

competence (stream) The ability of a stream to transport sedimentary particles, numerically measured as the diameter of the largest particle transported.

composite volcano See *Stratovolcano*.

conchoidal fracture A type of fracture that resembles the markings on a conch shell; the surface has a number of concentric irregularities.

concretion The rounded bodies in sedimentary rock formed when cement preferentially collects in abundance around a small particle.

cone of depression A depression in the water table in the shape of an inverted cone; it develops around a heavily pumped well.

confined aquifer A layer of permeable material, such as sandstone, enclosed between layers of impermeable rock.

conglomerate A clastic sedimentary rock composed of rounded gravels cemented together.

contact The surface between two different types or ages of rocks.

contact metamorphic rocks Rocks recrystallized through heating marginal to a pluton.

continental drift The concept that the continents have moved apart relative to one another.

continental rise The area of coalescent sedimentary fans between the continental slope and the abyssal plain.

continental shelf The shallow, gradually sloping platform that surrounds almost all of the continents. Its width averages 80 km but ranges from 0 to 1500 km.

continental slope The slope that extends downward from the shelf break to the abyssal plain.

contour An imaginary line connecting points of the same value.

convection current The postulated circular movement of mantle material in a cell, as a result of local heating at the base of the cell.

coral Any of a large group of shallow-water, bottom-dwelling marine invertebrates having skeletons consisting of calcium carbonate. Common in warm seas.

core-mantle boundary The boundary, at a depth of about 2900 km, that marks a change from solid silicate rocks above to the fluid core below. The core is composed primarily of iron.

Coriolis force The apparent force that causes objects traveling over the earth's surface to deflect toward the right or left into a curved path.

country rock Pre-existing rocks into which plutonic rocks intrude.

covalent bonding A type of bonding in which the electron is shared by two atoms.

crater A funnel-shaped depression at the summit of a volcano marking the conduit through which volcanic products are erupted.

creep The slow movement of shallow soil material downslope.

crevasse A deep split, fissure, or crack near the surface of a glacier, caused by stresses during glacial movement.

cross-bedding The original layering of sedimentary rocks in which the layers are inclined, sometimes at steep angles to the horizontal.

crossing (river) The shallow part of a channel at the bend of a meander.

crust The rocks above the Moho (Mohorovicić discontinuity).

crystal A solid chemical compound or element having a regularly repeating atomic arrangement (crystal lattice) and commonly bounded by plane surfaces (crystal faces) that parallel prominent lattice planes.

crystal lattice See *crystal*.

crystalline texture A texture of interlocking mineral grains formed as a result of crystallization from a magma, precipitation of minerals, or metamorphic recrystallization.

crystallinity See *long-range ordering*.

cuesta A ridge formed by a resistant bed that dips at a low angle; has a gentle slope on one side and a steep slope on the other.

Curie temperature The temperature above which a rock loses its permanent magnetism and be-

comes virtually non-magnetic. The Curie temperature of pure magnetite is 580°C.

current ripples The corrugated surface made by a water current flowing across a sandy base or by the wind blowing across desert sands. The ripples are assymetric in cross section, with the more gentle slope facing the upcurrent or wind direction.

cut bank The outside bank in a meander, cut by lateral erosion of the stream.

cutoff A channel eroded through the neck of land between two bends, or meanders, of a river.

cut terrace A river terrace consisting of a thin layer of gravel resting on a fairly smooth bedrock surface.

daughter isotope The product of decay of a radioactive parent isotope.

décollement A complexly folded overthrust sheet common to alpine chains.

deep-focus earthquake Originating in the depth zone between 300 and 700 km.

deep-sea trench See *trench.*

deflation Erosion of earth materials by the wind.

degradation (stream) The downcutting of a stream's channel.

delta A low, nearly flat body of sediment deposited near the mouth of a river.

dendritic drainage A drainage system in which tributary streams join the main stream in an irregular pattern resembling the branching of a tree.

dendrochronology The study of tree-ring patterns for dating of the recent past (back to a few thousand years ago).

density current A gravity-induced underflow of relatively more dense water. Density differences may be affected by temperature, salinity, or sediment content.

desert pavement See *stone pavement.*

desert varnish A thin, shiny bluish-black coating composed largely of iron and manganese oxides formed in desert regions on stones and cliffs.

devitrification Conversion of volcanic glass to a fine, crystalline intergrowth, over a long period of time.

diapir A salt dome that has pushed through overlying sedimentary rock.

diastrophism Large-scale deformation of the earth's crust, such as folding, faulting, subsidence, or uplift.

diatom A marine plant that secretes silica. The remains form siliceous ooze.

differential weathering Variation in the rate of weathering in rocks. Ledges, recesses, and irregular forms are the result.

differentiation In cooling magma, the process that involves separation of earlier-formed minerals from the residual magma.

dike A discordant tabular intrusion of magma that cuts across the layering of the country rock.

dike swarm A group of dikes intruded at about the same time and possessing a geometrical relationship. Most commonly they occur in parallel alignment but also can occur concentrically or radially.

diorite A drab, gray coarse- to fine-grained plutonic rock with a composition midway between granite and gabbro. Composed mostly of calcium-plagioclase, amphibole, pyroxene, and biotite.

dip The angle between a dipping bed or surface and a horizontal plane, measured perpendicular to the line of strike.

dipolar field A relatively simple magnetic field configuration consisting of two magnetic poles called the north and south poles. Analogous to a bar magnet.

dipping bed A layer of sedimentary rock tilted at an angle to horizontal.

dip-slip fault A fault in which movement (slip) is parallel to the dip of a fault.

discharge (stream) The quantity of water that passes a designated point in a given interval of time.

disharmonic fold A complex fold in which the subsurface geometry has little resemblance to the surface geometry.

dissolved load (stream) The material a stream carries in solution.

distributaries The branches of a stream into which a river divides when it reaches a delta or alluvial fan.

dolomite A carbonate sedimentary rock in which the mineral dolomite predominates. Sometimes called dolostone.

dolostone See *dolomite.*

dome A shield-shaped anticlinal structure that lacks an axis and plunges nearly equally in all directions away from the dome crest.

doubly-plunging fold A fold in which the axis is either arched upward or bowed downward.

downwasting A process in which the slopes adja-

637

cent to a stream valley are gradually diminished by rock decomposition, soil creep, and other mass-movement processes.

drainage basin The entire area that gathers water and ultimately contributes it to a given river.

drumlin An elliptical, rounded low hill formed by glaciers and occurring in groups.

dynamic metamorphism See *cataclastic metamorphic rocks*.

dynamothermal metamorphic rocks Generally foliated rocks recrystallized by heat, pressure, and chemical solutions during alpine mountain-building processes. Also called regionally metamorphosed rocks.

earth flow A relatively slow earth movement, usually with a spoon-shaped sliding surface and a crescent-shaped cliff at the upper end. Commonly breaks up internally.

earth reference ellipsoid The ellipsoid of rotation closely approximating the size and shape of the earth.

echo sounder An instrument that records ocean depths by measuring the time required for a sound impulse to travel to the sea floor and back.

effluent stream A stream that flows at the level of the water table and derives some of its water from the latter.

elastic rebound theory The theory that movement along a fault results from an abrupt release of elastic strain energy that has accumulated in the rock masses on either side of a fault as a result of their deformation.

elastic strain energy The energy stored in a rock body during deformation, which can be regained during faulting.

electron A negatively charged particle of low mass outside the nucleus of an atom.

ellipsoid of rotation The shape that is generated when an ellipse is revolved about one of its axes.

end moraine See *terminal moraine*.

eolian Pertaining to the wind.

ephemeral stream A stream that does not flow continuously. Also called an intermittent stream.

epicenter The point on the earth's surface above the focus of an earthquake.

epoch A geologic time-unit and subdivision of a period.

era A major division in geologic time; eras are divided into lesser units called periods.

erg A broad, dune-covered area in a desert.

erratic (glacial) A rock or boulder carried from its source by glacier ice or an iceberg, and deposited when the ice melts.

escarpment A steep cliff formed either directly as a result of erosion or as an erosional modification of a previously existing step-like topographic feature, such as a fault scarp.

esker A narrow, sinuous ridge of stratified sediment probably deposited by streams flowing in ice tunnels at the bottom of a glacier.

estuary A funnel-shaped mouth of a coastal river valley formed as a result of a rise in sea level or land subsidence. Also called a tidal river.

eustatic change (sea level) A change in sea level resulting from changes in volume of water in the ocean.

evaporite A sedimentary rock that results primarily from the evaporation of water containing dissolved solids.

evapotranspiration The transfer of water to the atmosphere through evaporation and through transpiration from plants and animals.

exfoliation (sheeting) The process by which concentric layers form at the surface of bare rocks.

fabric The way in which grains or crystals in a rock fit together. Involves consideration of relative sizes and shapes.

facies changes (sedimentary) The lateral variation in sedimentary rock bodies due to lateral variation in the depositional environment.

fault A fracture along which significant movement has occurred.

fault block A segment of crust bounded by a fault on one side (tilted fault block) or by faults on both sides.

fault-block mountains Mountain ranges bounded at one or both lateral margins by large, high-angle normal or reverse faults.

fault gouge A clayey, soft material formed when the rocks adjacent to a fault are pulverized during slippage.

fault scarp A low, linear cliff resulting from displacement of the earth's surface during fault movement.

fault-zone breccia Rocks in a fault zone that are broken and sheared as a result of fault movement.

GLOSSARY

feldspar A silicate mineral group that includes both potassium feldspar and plagioclase (rich in sodium and calcium). Has two good cleavages that intersect at 90°, striations on one of the cleavage faces, and a hardness of 6.

felsic rocks Igneous rocks containing a large proportion of feldspar and silica quartz.

fenster (window) An erosional window through a thrust sheet that displays the rocks beneath the sheet.

ferromagnesian mineral A silicate mineral containing relatively abundant iron and magnesium and that tends to be dark in color.

fetch The distance or area in an open body of water over which wind friction can affect waves.

fill terrace A river terrace made up of relatively thick deposits of gravel resting on a bedrock surface.

fiord A narrow glacial valley partly submerged by the sea.

firn The gritty, granular snow formed when snowflakes melt and refreeze several times.

fissility The tendency of sedimentary rocks to split along well-developed and closely spaced planes.

fissure eruption An eruption that takes place through a fissure, or large crack, rather than through a localized volcanic vent.

flint A variety of chert, usually dark colored.

flood lavas (flood basalts) Flows of basaltic lavas that erupt from innumerable cracks or fissures and commonly cover vast areas to thicknesses of 1000 m or more.

floodplain The flat surface adjacent to a stream, over which streams spread in time of flood.

flow banding In an igneous rock, alternating layers of different texture and minerals; it is the result of the flowing of viscous magma.

focus The initiation point of an earthquake within the earth.

foliated rocks Metamorphic rocks characterized by parallel orientation of tabular minerals and varying degrees of banding, or color layering. Common to regionally metamorphosed rocks.

foliation Layering, or banding, in some metamorphic rocks caused by the subparallel alignment of platy minerals.

footwall The face of the block below an inclined fault.

foraminifera Single-celled marine animals that secrete a calcite shell. Some limestones consist almost entirely of foraminifera.

foreset bed The inclined layers that make up the front of a delta as it advances into a body of water.

fossils The preserved remains or traces of prehistoric life. Most fossils consist of the bones or exterior shells of organisms.

fringing reef A coral reef that is directly attached to the shore of an island or continent.

frost heave The uneven lifting of surface materials due to subsurface freezing of water.

frost wedging The mechanism by which jointed rocks are pried apart by ice acting as a wedge.

gabbro A dark, plutonic rock consisting typically of coarse-grained crystals of pyroxene, calcium-plagioclase, and olivine.

garnet A group of silicate minerals that lack cleavage and whose crystals are almost always well formed and equidimensional. They are common in some metamorphic rocks.

geoid A generalized earth shape in which high and low spots are represented but smoothed out and reduced. The attraction of gravity is everywhere perpendicular to the geoidal surface, which corresponds to the surface of mean sea level.

geologic time The vast period of time (about 4.5 billion years) covering the entire history of the earth.

geosyncline A huge, elongated trough at the earth's surface in which thousands of meters of sedimentary rocks and volcanic rocks accumulate over many tens to hundreds of millions of years. Thought to represent the site of subsequent alpine mountain building.

geothermal region A region of naturally occurring hot water and steam.

geyser A hot spring that erupts jets of hot water and steam, resulting from the heating of ground water by hot rocks and steam.

glacial abrasion The scouring or wearing down of the rocks over which a glacier moves.

glacial marine (sediment) A marine sediment containing stones carried out to sea by icebergs. Found around glaciated areas and areas of pack ice.

glacial quarrying The process by which a glacier prys rocks loose from the surface over which it moves.

glacial striations Long grooves and scratches made by a glacier on the rocks it carries or on the bedrock over which it moves.

glacial surge A rapid movement in which a glacier becomes decoupled from its base and advances downvalley at high velocities, some approaching 6000 m/yr.

glacier A large mass of ice formed by the compaction and recrystallization of snow, which moves slowly under the stress of its own weight.

glassy texture A texture in which a large part of the rock is composed of volcanic glass, as in obsidian.

glide A movement of a large mass of intact rock downslope along a planar surface, such as a bedding plane.

globigerina A single-celled surface-dwelling marine organism with a shell made of calcium carbonate.

glowing avalanche (ash flow) A turbulent mass of pyroclastic fragments and some high-temperature gas erupted from a volcano.

gneiss A coarse-grained metamorphic rock in which bands of light-colored minerals (quartz, feldspar) alternate with dark-colored ones (amphibole, biotite).

Gondwana One of two hypothetical land masses consisting of all the continents in the Southern Hemisphere.

graben A relatively down-dropped fault block with linear margins and generally lying between two horsts (uplifted blocks).

graded bedding A type of stratification in which the particles in each layer are graded—the larger particles at the bottom grade into smaller ones at the top.

graded stream A stream in equilibrium: one that has the velocity and channel characteristics required to transport the load from its drainage basin.

gradient (stream) The downvalley slope of a stream channel.

granite A coarse- to fine-grained plutonic rock made up largely of potassium feldspar, sodium-rich plagioclase, quartz, and micas.

granitization The theory that most granitic rocks are formed in place by solid-state crystallization of pre-existing rocks under the influence of chemical solutions.

graywacke A sandstone consisting mainly of detritus derived from mafic igneous rocks.

ground ice Ice in the thin soil layer overlying permafrost.

groundmass The fine material in a porphyritic rock that surrounds the phenocrysts.

ground moraine Till deposited beneath a glacier.

ground water Subsurface water, generally occurring in the pore spaces of rock and soil.

gumbotil Highly weathered glacial till.

guyot A submerged volcano with a planed-off, nearly level summit.

gypsum A common evaporite mineral composed of hydrous calcium sulfate.

gyres Large, rotating current cells in the major ocean basins generated by the deflective effect of the Coriolis force on moving water currents.

half-life The time required for a mass of radioactive isotope to decay to one-half of its initial amount.

halite Sodium chloride, or common table salt; a common evaporite mineral.

hanging valley A tributary valley whose floor is higher than the main valley; produced when erosion deepens the main valley more rapidly than the tributary.

hanging wall The face of the block above an inclined fault.

hematite A relatively non-magnetic oxide of iron having a red to red-brown streak.

high-grade metamorphic rocks Dynamothermal metamorphic rocks such as gneiss which are crystallized under high temperatures.

hogback A narrow, sharp-crested ridge formed by a resistant, steeply dipping bed.

horizon A soil layer with characteristic physical, chemical, and biological properties.

horn (glacial) A jagged peak formed in areas where mountain glaciers radiate away from a summit area and have removed most of the latter.

hornblende A dark ferromagnesian mineral of the amphibole group that resembles augite in color and luster, with two cleavages that intersect at 56° and 124°.

hornfels A dense, non-foliated contact metamorphic rock.

horst A fault block with linear margins. It rises above the blocks on either side of it.

hydration The volume expansion of salt minerals that occurs when water is added to their crystal structure. Pressures generated can break up a rock.

hydrologic cycle The complete cycling, or transfer process, of the earth's water.

hydrolysis A weathering process in which minerals are altered by chemical reaction with water and acids.

hydrosphere The waters of the earth.

GLOSSARY

hydrothermal metamorphic rocks Rocks formed by recrystallization through the action of hydrothermal solutions (hot fluids and gases) circulating marginal to large plutons.

iceberg A large chunk of ice that breaks away from an ice sheet and floats away.

ice dome The summit of a continental glacier or ice sheet.

ice sheet (continental) Large, irregular sheets of ice that cover a large area and are not usually guided in their flow by the underlying topography.

icing The freezing of ground water that reaches the surface in permafrost areas.

igneous rocks Rocks that have solidified from magma.

index fossil An easily identified fossil with a wide geographic range and limited span of time on earth. Useful in correlating and dating rocks.

influent stream A stream that flows above the water table and from which water flows in pores to the water table. Common in arid regions.

inselberg An isolated residual hill that rises abruptly above the surrounding plain in a dry region. Characteristic of the late stage of the erosion cycle.

interior drainage A pattern of streams that drain toward the center of a basin rather than toward the sea.

intermediate-focus earthquake An earthquake originating in the depth zone between 70 and 300 km.

intermediate rocks Igneous rocks intermediate in composition between felsic and mafic. Examples are andesite and diorite.

intermittent stream See *ephemeral stream*.

intrusive contact The surface of contact between a pluton and the surrounding rock (country rock).

ion An atom or group of atoms that carries a positive or negative charge.

ionic bond A bond that results from the electrostatic attraction between positively and negatively charged ions.

ionic-covalent bonding A type of bonding common in minerals; alternates between ionic and covalent.

island arc A volcanically active island archipelago.

isostasy The theory that blocks of the earth's crust are in a floating, gravitational equilibrium.

isotope A subdivision of an element. Each isotope of a specific element has the same atomic number (contains the same number of protons and electrons) but contains a different number of neutrons; therefore each would have a different atomic weight.

joint A fracture in a rock along which no movement has taken place.

joint set A group of joints with a similar geometry. In most cases a set consists of parallel, concentric, or radial joints.

juvenile water Water derived directly from magma that comes to the surface for the first time.

karst The name given to hummocky landscapes characterized by the features caused by the solution of rocks by ground water, such as sinkholes and caves.

kerogen The organic material usually found in shales which can be converted to petroleum products.

K horizon (caliche) A hard, thick calcareous crust that forms beneath the B horizon in arid-region soils (i.e., a strongly developed Cca horizon).

klippe An erosional remnant of a thrust sheet, or nappe.

laccolith A concordant igneous body more or less circular in outline, with a flat base and a dome-shaped top.

lag deposit See *stone pavement*.

lagoon The body of water separating a barrier island from the mainland.

lahar A mudflow on the flanks of a volcano.

laminae Sedimentary layers whose thickness is less than 1 cm.

landslide A relatively rapid movement of soil and rock downslope.

lapilli Pyroclastic fragments about 2 cm in diameter.

lateral fault A fault in which the relative displacement is primarily strike slip. Also called a strike-slip fault.

lateral moraine A ridge of debris that continually accumulates along the side of a glacier.

laterite A highly weathered tropical soil rich in oxides of iron and aluminum. Hardens upon drying and can be used to make bricks.

lateritic soil A deep, highly weathered reddish soil, widespread in tropical climates. Most soluble elements (calcium, sodium, potassium, and silicon) have been leached out, leaving a residuum enriched in oxides of iron and aluminum.

Laurasia One of two hypothetical land masses consisting of the continents in the Northern Hemisphere.

lava The molten material that erupts from a volcanic vent or the rock that results from the solidification of the molten material.

leaching The dissolving out or removal of soluble materials from a rock or soil horizon by percolating water.

levee A low embankment adjacent to and confining a river channel.

lichen dating See *lichenometry*.

lichenometry The study of lichen sizes on rocks for dating of very recent surface deposits.

limb (flank) One of the two sides of a fold.

limestone A sedimentary rock in which calcium carbonate predominates.

lithosphere The outer 100 km of the earth. It appears to be relatively strong and brittle.

lithostatic pressure The all-sided pressure at depth exerted by the rock load above. Also called confining pressure.

lodestone See *magnetite*.

loess Wind-deposited silt originating in glacial outwash plains or in desert regions.

longitudinal dune A dune aligned parallel to the prevailing wind.

long-range ordering An ordering of atoms in a mineral that extends throughout the mineral grain. Also referred to as atomic ordering or crystallinity.

longshore current A coastal current, parallel to the shore, produced by wave refraction

low-grade metamorphic rocks Dynamothermal metamorphic rocks, such as slate, crystallized under moderate to low temperatures.

low-velocity zone (earthquake) See *asthenosphere*.

L-wave (long wave) A seismic surface wave of relatively high amplitude and long wave length. It follows the P and S waves in arrival at a seismograph station.

mafic rocks Igneous rocks containing a large proportion of minerals rich in magnesium, iron, and calcium-plagioclase.

magma (melt) A liquid composed of molten material at high temperatures; generally rich in silicon and oxygen.

magnetic anomaly Any departure from the normal magnetic field of the earth. In the ocean, anomalous highs and lows occur in alternating ridges and rises.

magnetic field A region in which magnetic forces are exerted.

magnetic polarity time-sequence The dated sequence of changes in magnetic field polarity as determined from paleomagnetism of rocks and ocean-floor anomaly stripes.

magnetic reversal The 180° reversal of the earth's magnetic field that has occurred intermittently throughout geologic history.

magnetite A black, strongly magnetic iron oxide mineral, sometimes called lodestone.

magnetometer An instrument for measuring magnetism in rocks.

mantle The rocks below the Moho (Mohorovičić discontinuity) and above the core.

marble A fine- to coarse-grained non-foliated metamorphic rock recrystallized from limestone.

marine terrace (A) An accumulation of wave- and current-transported materials seaward of a wave-cut platform. (B) The term also applies to old marine beach deposits and wave-cut platforms now found above sea level.

mass movement The movement of rock material downslope through the direct pull of gravity.

mass wasting See *mass movement*.

meander A large, curving bend in a river.

mechanical weathering Weathering by physical forces such as frost action or absorption of water.

medial moraine A moraine in the middle of a glacier formed by the merging of the lateral moraines of two coalescing valley glaciers.

median valley The keystone-like depression, or rift, in the crest of an oceanic ridge or rise.

meltwater Melted ice and snow from a glacier.

mesosphere The mantle zone beneath the asthenosphere.

Mesozoic Era The era of geologic time between the Paleozoic and Cenozoic eras, lasting from about 225 to 65 million years ago. Characterized by the dominance of reptiles, especially dinosaurs.

metallic bonding A type of covalent bonding in which there are more metal atoms available than are necessary to satisfy bond requirements. Also called time-shared covalent bonding.

metallic luster A term applied to minerals that reflect light to about the same degree as a polished metal.

metamorphic rocks Rocks that form at depth through solid-state recrystallization of pre-existing rocks as a result of internal heat, pressure, and chemical activity of fluids.

metamorphism An isochemical process in which

rocks are crystallized from pre-existing rock under the influence of higher temperatures, confining and directed pressure, and interstitial fluids.

mica A mineral consisting of parallel sheets of silica tetrahedra strongly bonded at their bases but less strongly bonded across the sheets; the result is a well-developed cleavage in one direction.

microplate A relatively small lithospheric plate.

Mid-Atlantic ridge A submarine mountain range that stretches the length of the Atlantic Ocean.

mid-ocean ridge See *oceanic ridges and rises*.

migmatite A layered zone of igneous and metamorphic rocks between a batholith and the surrounding rocks. Generally found near the deeper parts of the batholith.

mineral A crystalline chemical compound (or element) that occurs naturally.

modified Mercalli scale (earthquake) A scale used to measure earthquake intensity based on damage caused.

Mohorovičić discontinuity A seismic-velocity boundary used to demarcate the earth's crust above from the mantle below. A relatively large increase in *P*- and *S*-wave velocities takes place at the boundary, which is also called the Moho or M-discontinuity.

Mohs hardness scale A scale used by mineralogists to judge the hardness of a mineral; ranges from 1 to 10.

monadnock An isolated hill of resistant rock that stands above the level of a peneplain.

monocline A one-limbed fold with horizontal strata on either side.

mud cracks The polygonal pattern produced by the drying and shrinking of wet, clayey mud.

mudflow A mass movement, usually of fine-grained earth materials containing a relatively high water content.

mudstone A shale without distinct bedding.

muscovite (white mica) A mineral of the mica group; usually colorless, gray, or transparent. Common in metamorphic rocks and in many sedimentary rocks, especially sandstone.

mylonite A dark, hard, fine-grained cataclastic rock.

nappe A complex, large-scale recumbent anticlinal fold.

neutron An uncharged particle in the nucleus of an atom.

new global tectonics The hypothesis that new lithospheric material, formed at the ocean ridges, moves away from the divergent zones and toward convergent zones, where the lithosphere bends sharply downward into the asthenosphere and disappears below.

non-plunging fold A fold in which the axis is horizontal.

normal fault A fault in which the footwall moves up relative to the hanging wall.

nuée ardente A turbulent gaseous cloud erupted from a volcano and containing primarily ash; accompanies eruption of a glowing avalanche.

oblique slip Fault movement (slip) oblique to the dip of the fault.

obsidian See *volcanic glass*.

oceanic ridges and rises Elongate suboceanic mountain ranges. See also *zones of divergence*.

oil shale Shale rich in kerogen, which on distillation and refining yields oil.

olivine A ferromagnesian mineral that usually occurs as rounded and glassy green crystals.

oölite Small, round grains of calcium carbonate that form as layers of the mineral build up around a nucleus.

ooze The remains of microscopic free-floating organisms that drift down and cover the deep ocean floor.

original horizontality, law of The law stating that layers of sediments are deposited horizontally, or nearly so.

oscillation ripples The corrugated surface made in the bed of a shallow body of water. The ripples are symmetrical in cross section, in contrast to current ripples.

outwash plain Floodplains formed by streams draining from the front of a glacier.

overthrust fault A low-angle thrust fault in which the dip angle is less than 10°.

overturned fold A fold in which one limb has been rotated past 90°.

oxbow lake A crescent-shaped lake formed when a stream bend, or meander, is cut off from the main stream.

oxidation A process of chemical weathering in which minerals or elements within the minerals combine with oxygen.

pahoehoe A lava flow with a ropy or corrugated surface and a glassy outer rind.

paleobotany The study of fossil plants.

paleoclimatology The study of past climates and the causes of their variations.

paleomagnetism   The study of fossil magnetism in rocks.

paleontology   The study of past geologic life based on plant and animal fossils.

Paleozoic Era   The geologic time-period between the Precambrian and the Mesozoic eras, lasting from about 600 to 225 million years ago. Characterized by relatively simple invertebrates and backboned animals.

Pangaea   The hypothetical single continent postulated by A. Wegener that split into fragments and began to drift apart during the Jurassic period.

parent isotope   The initial, or starting, radioactive isotope that will decay spontaneously and progressively to a daughter isotope.

parent material   The material from which a soil forms.

patterned ground   Polygonal patterns formed in surface material subject to intensive frost action.

pedalfers   Soils with a fairly high content of organic matter in the A horizon, and no calcium carbonate accumulation beneath the B horizon. Common in humid temperate regions.

pediment   An erosional bedrock surface that slopes away from a desert mountain range.

pediplane   An extensive erosion surface in deserts formed by the coalescence of two or more pediments.

pedocal   An arid-region soil characterized by a thin A horizon and a layer of calcium carbonate beneath the B horizon known as the Cca horizon or the K horizon.

peneplain   A broad, nearly featureless plain that has been eroded to nearly sea level by mass wasting and stream erosion.

period   The fundamental unit of the geologic time-scale and the subdivision of an era.

permafrost   Ground that remains below 0°C and usually contains small to large quantities of ice.

permafrost table   The upper surface of permafrost.

permeability   The measure of the capability of a rock to transmit a liquid.

phenocryst   The larger, usually well-formed crystals in a porphyritic igneous rock.

phreatic explosion   A volcanic explosion caused by the heating and expansion of ground water.

pillow lava   Lobes of lava resembling a bed of pillows. Results from extrusion into water or a water-rich environment.

plain   A broad area of low relief occurring generally at low elevations.

plateau   A broad, relatively flat or rolling region occurring at relatively high elevations.

plate-tectonic theory   The theory that the earth is divided into rigid blocks, or plates, that move relative to one another.

plastic flow (glacial)   The flow of the lower part of a glacier, caused by deformation of ice crystals under the pressure of the overlying ice.

playa   A dried-up playa lake consisting of clay and silt or sand and deposits of soluble salts.

playa lake   A seasonal lake in the center of a desert basin.

Pleistocene   A recent epoch of the Cenozoic Era (and part of the Quarternary Period) in which 90–100 per cent of the fossil shells still exist today. Usually thought of as coincident with the ice age of the Cenozoic.

plume   A column of heated rock rising from the mantle. It can bring about partial melting in the asthenosphere and the lithosphere.

plunging breaker   A wave formed in shallow depths near the shore and whose crest takes the shape of a half-cylinder that curls over and breaks suddenly with a crash.

pluton   A body of plutonic rock of any size or shape.

plutonic rocks   Rocks formed from magma that cools and solidifies underground.

podzol   A variety of pedalfer that forms in cooler climates toward the northern limit of trees. Podzols are characterized by a whitish layer, the A2 horizon.

point bar   The bar on the inside of a meander.

porosity   The percentage of the total volume of a rock that is occupied by open spaces.

porphyritic texture   See *porphyry*.

porphyry   An igneous rock containing two grain sizes: relatively large, well-formed phenocrysts imbedded in a finer-grained crystalline or glassy groundmass. Texture said to be porphyrytic.

Precambrian   All geologic time before the beginning of the Paleozoic Era. Characterized in the record by scant primitive life or no life at all.

pressure surface   The level to which water rises in a confined or unconfined aquifer.

proton   A positively charged particle in the nucleus of an atom.

pull-apart zones   See *zones of divergence*.

pumice   A frothy, porous volcanic glass that is generally rhyolitic in composition.

P **wave (primary wave)**   A relatively low-amplitude seismic body wave, the first to arrive at a seismograph station after an earthquake.

pyroclastic rocks Coarse- to fine-grained rocks formed from material hurled into the air during a volcanic eruption.

pyroxene group A group of two rock-forming silicate minerals characterized by good cleavages parallel to the crystal face and intersecting at angles of 87° and 93°; includes the common ferromagnesian mineral augite.

quartz A silicate mineral with a hardness of 7 that has a vitreous luster and commonly fractures conchoidally. Composed almost exclusively of silicon dioxide.

quartzite Unbanded metamorphosed sandstone or unmetamorphosed silica-cemented sandstone consisting of quartz grains. In the former, the rock breaks across the grains.

Quarternary The period of geologic time including the Pleistocene and Holocene epochs and covering the last 2–3 million years of earth history, up to the present.

quick clay Clay deposits with a high water content that become fluid when jarred, for example, by an earthquake.

radioactivity The spontaneous decay of an atom of one isotope into an entirely different isotope.

radiocarbon dating See *carbon-14 dating*.

radiolaria A microscopic marine organism that secretes a shell of silica. The remains form siliceous ooze.

radiometric date The age of a material as determined through measurement of radioactive decay.

recumbent fold An overturned fold; one in which one limb is overturned and roughly parallel to the normal limb and both limbs are nearly horizontal.

reef A ridge of layered sedimentary rock built by the secretions and remains of marine organisms, usually coral.

regionally metamorphosed rocks See *dynamothermal metamorphic rocks*.

relative geologic time-scale A time-scale in which geologic events, as determined by field relations and fossil evidence, are placed in chronological order.

remanent magnetization The permanent magnetization acquired by rocks. In most cases it parallels the earth's magnetic field lines at the time of the rock's origin.

reverse fault A fault in which the footwall moves down relative to the hanging wall. Also called a thrust fault.

rhyolite A light-colored fine-grained-to-glassy volcanic rock similar to granite in composition and commonly characterized by flow-banding.

ria coast A submerged coastline in which the sea extends inland, sometimes for long distances, in stream valleys.

Richter scale A scale of earthquake magnitudes developed by the seismologist C.F. Richter. The magnitudes can be determined from seismographs and are directly related to the amount of energy released during an earthquake.

rift valley A large, elongate trough, or graben.

river terrace Along a stream, an elevated flat area underlain by river deposits. Such a terrace is an abandoned floodplain.

rock avalanche A large mass of rock that slides very rapidly as a unit (100 km/hr) downhill, perhaps on an air cushion if the terrain traversed is flat.

rock cleavage The tendency for a rock to break along relatively smooth, closely spaced parallel surfaces. Slaty cleavage is an example.

rockfall The relatively free-falling movement of rock material from a cliff or other steep slope.

rock flour A fine silt that covers much of the surface of an outwash plain.

rock glacier A tongue-shaped slow-moving mass of rocks and ice in alpine areas.

salt A general chemical term that includes many compounds, the most familiar of which is table salt (sodium chloride).

saltation A mode of sediment transport in which the particles bounce along the floor of a stream or along a desert surface.

salt dome A structure dome produced by the upward movement of a body of salt through enclosing sediments.

salt playa A playa consisting of saline residues.

sandstone A clastic sedimentary rock consisting mainly of sand-size grains cemented together.

scarp A cliff or line of cliffs produced by faulting, erosion, or landsliding.

schist A foliated metamorphic rock of intermediate grain size. Individual folia are relatively thin; platy minerals commonly make up one-half of the rock; and color banding is not well developed.

sea cliff A cliff or slope produced by wave erosion.

sea floor spreading The hypothesis that continen-

tal drift occurs through cracking and spreading at the oceanic ridges and rises.

seamount A submerged volcano of basaltic composition.

sedimentary rock Rocks formed by the accumulation of layers of clastic and organic material or precipitated salts.

seiche A wave of oscillation set up in a lake, harbor, or bay and that is initiated by the motion of an earthquake or by local changes in atmospheric pressure.

seismic sea wave See *tsunami*.

seismic wave A wave or a vibration produced by an earthquake.

seismograph An instrument used to measure the vibrations, or waves, generated during an earthquake.

self-exciting dynamo A mechanism consisting of interrelated moving electronic currents and varying magnetic fields. Once primed, it is able to continually regenerate a magnetic field.

serpentine group A group of hydrous rock-forming minerals derived by the alteration of magnesium-rich silicate minerals (e.g., olivine) in water-rich environments at low temperatures.

serpentinization The process of forming serpentine from magnesium-rich silicates.

shadow zone A region from about 102° to 143° from the epicenter of an earthquake in which there is no reception of direct seismic waves.

shale A fine-grained laminated sedimentary rock composed of clay and silt.

shallow-focus earthquake An earthquake originating between the earth's surface and 70 km in depth.

sheet flood Muddy, turbulent water that fills and overflows the banks of an arroyo and eventually spreads out over the desert floor.

sheeting See *exfoliation*.

shelf break The outer edge of the continental shelf, characterized by a sudden steepening of the slope.

shield A large area of exposed igneous and metamorphic rocks, usually Precambrian, surrounded by sediment-platforms; e.g., the Canadian Shield.

shield volcano A shield-like volcanic cone built almost entirely of fluid lava flows.

silicate A mineral consisting of silicon, oxygen, and varying proportions of one or more metals. Most common minerals are silicates.

siliceous rocks Rocks made up largely of silica.

siliceous sinter A hot-spring or geyser deposit composed of silica.

sill A concordant tabular intrusion of magma that more or less parallels layers of the country rock.

sinkhole A pit in karst topography caused by the solution of surficial limestone or the collapse of a cave roof.

sinter A spongy, porous sedimentary rock formed from the chemical precipitation of silica from springs.

slate A common low-grade dynamothermal metamorphic rock, usually derived from fine-grained sedimentary rocks. Possesses slaty cleavage.

slickensides A polished and smoothly striated rock surface that results from movement along a fault plane.

slip The term used to denote actual relative displacement along a fault.

slip face The steep face of an asymmetrical dune.

slump The movement of rock material downslope as a unit along a concave-upward slip plane. Characterized by backward tilting of the mass.

snow Frozen water vapor, crystallized directly from the water vapor in the atmosphere.

snowline The line or altitude on a glacier separating the area where snow remains from year to year from the area where the snow of the previous season melts.

sodium chloride Common salt.

soil profile The three basic layers, or horizons, of most soils.

solar day The time between two consecutive passages of the sun at its zenith past a particular spot on earth. A mean solar day is an arbitrary unit of time, one that is always the same length, regardless of the season.

solid load The suspended load and bed load of a stream, taken together.

solid solution A mixed-crystal mineral composed of varying amounts of certain ions which can substitute for one another in the crystal lattice.

solifluction The slow downslope movement of water-saturated surface material occurring in permafrost areas.

solution A process of chemical weathering in which soluble minerals are dissolved.

sonar (Sound Navigation Ranging) A device that sends and receives sound signals underwater.

specific gravity The weight of a specified volume of a mineral divided by the weight of an equal volume of water at 4°C.

spilling breaker A wave whose crest collapses

gradually over a relatively long distance as water spills continuously down the wave front.

spit A curved embankment formed by a longshore current that trails down-current from the land.

stack A pillar-like rocky island or mass near a cliffy shore separated from the headland by wave erosion.

stalactites Icicle-like pendants of travertine that hang from the roof of a cave.

stalagmites Deposits of travertine built upward from a cave floor.

stellar day The time between two consecutive passages of a distant star past a particular spot on earth.

steppe An extensive dry region characterized by grass vegetation.

stock A large pluton, generally of granitic rock, less than 100 km² in exposed surface area.

stone pavement A thin veneer of stones several stones thick mantling a desert surface. Also called desert pavement or lag deposit.

stoping Enlarging of a magma chamber by the prying loose of small blocks or rocks from the roof and walls.

strata Layers of sedimentary rock.

stratovolcano A steep-sided volcano consisting of alternating layers of lava and pyroclastic materials. Also called a composite volcano.

strike The line of intersection made by a dipping bed or surface with an imaginary horizontal plane.

strike slip Fault movement (slip) parallel to the strike of the fault.

strike-slip fault See *lateral fault.*

subaerial erosion Erosion that takes place in the open air (compare subterranean and submarine erosion).

subdelta A small delta forming a part of a complex of deltas.

subduction The pulling down, or sinking, of lithospheric plates into the asthenosphere at the convergent zone.

sublimation The process by which a solid substance vaporizes without passing through a liquid stage.

submarine canyon A steep-sided valley or canyon cut into the continental shelf or slope.

superposition, law of The law stating that in a layered sequence of rocks the age of any one layer will be greater than the age of that above it, and less than the layer below it.

surface waves Relatively slow seismic waves that travel close to or at the earth's surface. Also called *L* waves.

suspended load (stream) The material a stream carries in suspension, buoyed up by the moving water.

S wave (secondary wave) A seismic body wave, the second to arrive at a seismograph station. Characterized by sudden onset and is normally greater in amplitude than a *P* wave.

swell A regular, somewhat flat-crested wave that has traveled far from its generating area; made up of long-period waves.

syncline A fold in which the limbs dip toward the axis; younger beds are found toward the axis.

talus A loose pile of angular rocks at the base of a cliff.

tectonic dam Refers to the origin of one type of continental shelf, in which shelf sediments are deposited behind a geologic uplift or lava; both act as dams.

terminal moraine A mass of debris that accumulates as a hummocky, rocky ridge around the snout of a glacier. Also called an end moraine.

Tertiary A period of the Cenozoic Era covering the time-span between 65 and about 2 million years ago.

Tethys sea The hypothetical oceanic waterway separating the super-continents, Gondwana and Laurasia.

texture The interrelations of the size, shape, and arrangement of the particles in a rock.

thrust fault See *reverse fault.*

till Material laid down directly by glacier ice.

tillite A consolidated glacial deposit consisting of sand, gravel, boulders, and clay.

time-shared covalent bonding See *metallic bonding.*

tombolo A strip of sand connecting a near-shore island to the mainland.

transform fault (zone of lateral movement) One of the numerous fracture zones in the ocean basin, along which the ridges and rises have been offset and along which lateral movement occurs.

transverse dune A dune aligned at right angles to the wind.

travertine A limy cave or spring deposit formed by chemical precipitation of calcium carbonate from solution in surface and ground water. Can also be precipitated by calcareous algae.

tree-ring dating See *dendrochronology.*

trench A narrow, elongate depression on the

deep-sea floor paralleling the trend of an island arc or continental margin.

tsunami A destructive wave generated by disturbances on the ocean floor. Also called a seismic sea wave.

tufa A spongy or porous sedimentary rock formed by the precipitation of calcium carbonate around the mouth of a hot or cold spring or in a stream or lake.

tuff A fine-grained rock composed of pyroclastic fragments, primarily ash.

turbidity current Density currents of sediment-laden water triggered by the slumping of oversteepened and unconsolidated material.

uniformitarianism The doctrine that the geologic processes now modifying the earth's surface have acted in essentially the same way throughout geologic time, although possibly at different rates.

U-shaped valley A valley suggesting the shape of the letter "U," with steep sides and a flat floor carved by a glacier.

valley glacier See *alpine glacier.*

varve A pair of thin, sedimentary layers made up of a coarse, silty lower layer and a fine-grained upper layer. Thought to represent seasonal variation in deposition.

velocity (stream) The direction and magnitude of displacement of a portion of a stream per unit of time.

ventifact A rock faceted by wind-driven particles.

vesicles Small, rounded cavities in lava formed by trapped gas bubbles.

viscosity Resistance of a liquid to flow.

vitreous luster A term applied to a mineral that reflects light to about the same degree as glass.

volcanic agglomerate An unsorted deposit of volcanic bombs, cinders, lapilli, and ash in crude layers. Sometimes called volcanic breccia.

volcanic bomb Irregular to spindle-shaped airborne blocks of lava hurled from a volcanic vent during eruption.

volcanic breccia Any rock composed of angular fragments of volcanic rock. See also *volcanic agglomerate.*

volcanic cone (volcano) The accumulated eruptive products around a volcanic vent, generally in a steep- to flat-sided cone.

volcanic glass Magma that has cooled so quickly that the liquid quenched to glass without crystallization taking place. Most commonly associated with rhyolitic magma.

volcanic neck A pipe-like pluton of solidified lava that once connected a magma reservoir with a volcanic vent at the surface.

volcanic rocks Rocks formed from magma that erupts at the surface and cools and solidifies.

V-shaped valley A narrow valley with steep, sloping sides resulting from downcutting by a stream and mass movement of material down the side slopes.

water table The surface at which water stands in wells, the upper surface of ground water.

wave base The lower effective limit of wave transportation and erosion.

wave-cut platform A planed-off rock bench cut by wave erosion at the base of a sea cliff.

wavelength The horizontal distance separating two equivalent wave phases, such as two crests or two troughs.

wave of oscillation A water wave in which the individual particles move in orbits with little or no change in position, although the wave form itself advances.

wave period The length of time required for two crests or two troughs of a wave to pass a fixed point.

wave refraction The bending of a wave as it approaches the shore.

weathering The mechanical disintegration and chemical decomposition of rocks.

weight The force that gravity exerts on a body.

welded tuff A pyroclastic rock whose particles have been fused together by heat still contained in the deposit after it has come to rest; generally associated with large-scale caldera-forming events.

xerophyte A plant adapted to dry conditions.

yazoo A tributary river that runs parallel to the mainstream between a natural levee and the river bluff.

zone of aeration A subsurface zone above the water table where pore spaces in the ground may range from completely to partially full of water.

zone of convergence The zone where plates push together and lithospheric material is subducted, or pulled down into the asthenosphere.

zone of divergence The zone where plates move apart and new lithospheric material is formed; equivalent to active oceanic ridges and rises. Also called the pull-apart zone.

INDEX

Boldface page numbers refer to charts and illustrations

and ocean deep, 482
and sea level fluctuations, 449
Glacier National Park, Montana, **399,** 407, **532,** 533, **534**
glaciers, 171, 307, **378,** 379–81. *See also* deposition; erosion; glaciation
budget, 388–91
climatic changes following, 423–26
and conversion of snow to ice, 385–86
deposits and depositional landforms, 171, 207, 256, 401–7, **401–7,** 416–17, 480
distribution and formation, 383–86
and graded stream equilibrium, 307
mechanism of movement, 386–88
rock, 407–9, **408**
surges, 386–88
velocity, 386
glass, volcanic, 95–**96,** 98, 100–101, **110**
glides, land, 267, **268**
Globigerina, 472, **478**
gneiss, **116, 219,** 222, **222, 224,** 242, 252
Gobi Desert, 366
Gondwana (supercontinent), 607, 613
Gore Range, 542
gouge, fault, 520
graben, 529, **529**
grade, 307
graded bed, 179, **180, 191**
gradient, and stream flow, 298–99
Grand Canyon, Arizona, **4, 8, 12,** 180, 182, 263, **328–29,** 330–31, **348,** 466–**67**
Grand Coulee, Washington, **166**
Grand Coulee Dam, Washington, 273, 297
granite, 86, 92, 98–**99, 124**–26, **125, 216, 235–37, 240,** 303, **362–63, 389, 401**
weathering, **243,** 245, 256
granitization, 126
graphite, 74
gravels, **176, 181,** 193, **193**
gravitation, law of, 32, 35
gravitational acceleration, 35–37
gravity meters, 37
graywacke, 194
Great Aletsch Glacier, Switzerland, 393
Great Barrier Reef, Australia, **199,** 200, 474
Great Basin, U.S., 353, 358, **370,** 538–40
Great Dike, Rhodesia, 116
Great Glen Fault, Scotland, 531
Great Ice Age, 382, 417
Great Lakes, 415
Great Plains, U.S., 491
Great Rift Zone, Pacific, 157
Great Salt Lake, Utah, 353, **370,** 371, 380
Great Sandy Desert, Australia, 368
Great Valley, California, **287,** 307, 487, 493
Great Whin Sill, England, 117–18
Greenland ice cap, 391, 409–11
Green River Formation, Wyoming, **206**

groins, sea, 445–46, **447**
ground ice, 283, 285
groundmass, 96
ground moraines, **414**–15
ground water
and aquifers, **490,** 491
as cementing agent, 497
contaminating, 493–94
and geothermal energy, 504–5
geysers and hot springs, 502–4, **503**
and karst, **500,** 501–2
and landslides, 271
occurrence and movement, 487–89
origin, 487
and permeability, 491
and porosity, 489–90
prospecting, 494–97
in underground caverns, 497–500
and wells, 491–93, **492**
and wind erosion, 364
gullies, **257–58,** 347, **375**
gumbotil, 419
guyots, **471,** 472
gypsum, 169, 196, **243,** 497
gyres, **46,** 48

Halemaumau fire pit, Hawaii, 160, 162, **163**
half-life, isotope, 19
halite, **68,** 195, 196. *See also* sodium chloride
halos, 122, 214
hanging valleys, 397, **397**
hanging walls, **527,** 528
hardness scale, mineral, 76
Hawaiian Volcano Observatory, 162
headlands, 452
heat, and metamorphism, 212–13
heat balance, earth, **40**
Hebgen, Lake, Montana, **525,** 565
Hekla Volcano, Iceland, **140**
Hells Canyon, Idaho, 330, **337,** 338
hematite, 85, 180
herring, armored, **206**
Hidden Glacier, Alaska, **415**
high-angle faults, 529, 543
high-grade metamorphic rocks, 219, **219,** 222
"Highland Controversy" on overthrust faults, 531–33
Himalaya Mountains, **36,** 37–38, 53, 280, 387, 393
Holocene Epoch, **16, 21**
Hood (Mount), Oregon, 136, **539**
Hoover Dam, 179, 468
horizontality, original, 9
hornblende, **82**–84, **83,** 93, 99, 101–**2,** 213
hornfels, 214, **215**
horns (glacial), 395, **400**
Horse Latitudes, 344, 345
deserts, 343, 344
horsts, 529, **529**
host rocks. *See* country rocks

hothouse effect, 428
"hot spots," 133, **219**
Hudson Canyon, Atlantic, 467
Hueneme Submarine Canyon, Pacific, 445
Humboldt Current, Pacific, 345
Hutton, James, 4–5, 8, 10, 91–92
hydration, 238
hydrocarbons, 77
hydrogen, 40, 67, **67,** 70
hydrologic cycle, 296–97
hydrolysis, 243–45
hydrostatic pressure, 213
hydrothermal metamorphic rocks, 214, 215–16

ice, 385–**86, 387,** 394. *See also* glaciers; permafrost
ice ages, 39, 143, 379–80, 382, 417, 423, **426**–27, 463
icebergs, 410, **410,** 480, **481**
ice domes, 428
ice sheets, continental, 391, 409–17, **409–13,** 427–28, 449
ice wedges, 285–86, **285–87**
Ichthyosaurus, **14**
icing, 289
idocrase, **73**
igneous processes and plutonic bodies, 111–128
igneous rocks, 87
classification, 98–111, **99–110**
mineral composition, 92–94
porosity, 489, **489**
texture, 94–98, **95–97**
Illinoian Glaciation, 419, 421, **422, 423, 424**
index fossils, 15
Indian Plate, **620**
indicators, glacial rock, 416
indistinct cleavage, 76
influent stream, 488
injection of magma, forceful, 113
inner core, earth, 584
inselbergs, 357
intensity, earthquake, **583**
interglaciations, 419, 482, **482**
intermediate aeration zone, 488
intermediate focus earthquake, 567, **618**
intermediate grade metamorphic rocks, 221
intermediate rocks, 101–2
intrusive contact, 113
intrusive rocks, 113–26, **114–25**
Inugsuin Fiord, Canada, **175**
invertebrates, **460,** 509, 510. *See also* fossils
ion(s), 46, 67–**68,** 69
ionic bond, 68, **69**
ionic-covalent bonds, 68
iron-oxides, **103,** 190, 203
iron meteorites, 587
Isfallsglaciären, Lapland, **425**

minerals vs. rocks, 245
rock, 229–30
and soil formation, 247–56
Wegener, Alfred, 606–8, 613, 615, 625
weight and gravity, 35
wells, 486–87, 489, **490,** 491–93, **492,**
504–5
Whitney (Mount), California, 37, 53,
237, 526
wind, 48
deposition, 171, 366–**69**
erosion, 358–66
global patterns, 40–**41, 345**

and waves, 434
Wind River Mountains, Wyoming, **335,**
398, 401
Wisconsin Glaciation, 415, 419–23, **419–**
24, 427, 449
wollastonite, 223
wrench faults, 530

xerophytes, 342
X-ray crystallography, 73–74, **73**

Yarmouth Interglaciation, 419
yazoos, 317

Yellow River, China, 366
Yellowstone National Park, Wyoming,
108, 112, 121, 150–**52,** 205, 421,
502–3, **504**
Yosemite Valley, California, 395, **396,**
397, 402
National Park, **240,** 394
Yucatán Peninsula, Mexico, 502

Zagros Mountains, Iran, **340**
Zion Canyon, Utah, 182
Zion National Park, 183, **184**

SOURCES FOR LINE DRAWINGS

CHAPTER 1

Fig. 1-16. After W. M. Wendland and D. C. Donley, 1971, "Radiocarbon Calendar Age Relationships," *Earth and Planetary Science Letter*, v. 11, pp. 135-39. **Fig. 1-19.** From J. T. Andrews and P. J. Webber, 1969, "Lichenometry to Evaluate Changes in Glacial Mass Budgets," *Arctic and Alpine Research*, v. 1, no. 3, Fig. 7. By permission of J. T. Andrews. **Fig. 1-20.** From J. T. Andrews and P. J. Webber, 1973, "Lichenometry: a Commentary," *Arctic and Alpine Research*, v. 5, no. 4., Fig. 1. By permission of J. T. Andrews.

CHAPTER 2

Fig. 2-17. After G. Dietrich, 1963, *General Oceanography*, John Wiley and Sons. **Fig. 2-26.** From B. C. Heezen and M. Tharpe, 1961, *Physiographic Diagram of the North Atlantic Ocean*, Geological Society of America. Reprinted by permission of B. C. Heezen and M. Tharpe. **Fig. 2-29.** From B. C. Heezen and M. Tharpe, 1976, *The Floor of the Oceans*, map painted by Tangy de Remur, American Geographical Society. Reprinted by permission of B. C. Heezen and M. Tharpe. **Fig. 2-32.** From B. C. Heezen and M. Tharpe, 1961, *Physiographic Diagram of the South Atlantic Ocean*, Geological Society of America. Reprinted by permission of B. C. Heezen and M. Tharpe.

CHAPTER 4

Fig. 4-26. After G. A. MacDonald, 1972, *Volcanoes*, Prentice-Hall, Inc. **Fig. 4-38.** After T. A. Steven, 1975, "Middle Tertiary Volcanic Field in the Southern Rocky Mountains," *Geological Society of America Mem. 144*, pp. 75-94. **Fig. 4-41.** After H. Cloos, 1931, "Neuer Jahrbuch für Mineralogia," Geologie und Paläontologie, Band 66, Abt. B.

CHAPTER 5

Figs. 5-3. After G. A. MacDonald, 1972, *Volcanoes*, Prentice-Hall, Inc. **Fig. 5-13.** After H. William, 1942, *Geology of Crater Lake National Park, Oregon*. By permission of the Carnegie Institution of Washington. **Fig. 5-16.** From John Shelton, 1966, *Geology Illustrated*, W. H. Freeman and Co., Copyright © 1966. **Fig. 5-20.** After R. G. Luedke and W. S. Burbank, 1968, "Volcanism and Cauldron Development in the Western San Juan Mountains, Colorado," *Colorado School of Mines Quarterly*, v. 63, pp. 175-208. **Fig. 5-21.** After P. W. Lipman and others, "Volcanic History of the San Juan Mountains, Colorado, as Indicated by Potassium-Argon Dating," *Geological Society of America Bulletin*, v. 81, pp. 2329-2352. **Fig. 5-29.** After G. A. MacDonald and D. H. Hubbard, 1970, *Volcanoes of the National Parks in Hawaii*, 5th ed., Hawaii Natural History Association. **Fig. 5-30 A.** After A. K. Baksi and N. D. Watkins, 1973, "Volcanic Production Rates: Comparison of Ocean Ridges, Islands, and Columbia Plateau Basalts," *Science*, v. 180, pp. 493-96. **B.** After A. C. Waters, 1955, *Volcanic Rocks and the Tectonic Cycle*, Geological Society of America Special Paper 62, pp. 703-22.

CHAPTER 6

Fig. 6-18. From R. Trümpy, 1960, "Paleotectonic Evolution of the Central and Western Alps," *Geological Society of American Bulletin*, v. 71, pp. 843-908, p. 2. **Fig. 6-26 A.** After C. O. Dunbar, 1960, *Historical Geology*, John Wiley and Sons, Fig. 215. **Fig. 6-32.** After A. J. Eardley, 1951, *Structural Geology of North America*, Harper and Bros., Fig. 21.

CHAPTER 7

Fig. 7-8. After H. L. James, 1955, "Zones of Regional Metamorphism in the Precambrian of Northern Michigan," *Geological Society of America Bulletin*, v. 66, pp. 1455-1487. **Figs. 7-12 and 7-14.** After W. W. Moorhouse, 1959, *Study of Rocks in Thin Section*, Harper & Row.

CHAPTER 9

Figs. 9-2 and 9-14. After G. A. Kiersch, "The Vaiont Reservoir Disaster," *Mineral Information Service*, v. 18, no. 7. **Fig. 9-16.** From D. R. Crandell, and D. R. Mullineaux, 1967, "Volcanic Hazards at Mount Rainier, Washington," *U.S. Geological Survey Bulletin 1238*. **Fig. 9-24 A.** From O. J. Ferrians, Jr., and others, 1969, *U.S. Geological Survey Professional Paper 678*, Fig. 1. **B.** From R. J. E. Brown, 1970, *Permafrost in Canada*, University of Toronto Press, Fig. 4. By permission of the University of Toronto Press. **Fig. 9-25.** After C. F. S. Sharpe, 1938, *Landslides and Related Phenomena*, Columbia University Press, Fig. 5.

CHAPTER 10

Fig. 10-2. After A. L. Bloom, 1969, *The Surface of the Earth*, Prentice-Hall, Inc. **Fig. 10-4.** After L. B. Leopold and others, 1964, *Fluvial Processes in Geomor-phology*, W. H. Freeman and Co., Figs. 6-1 and 6-9. **Fig. 10-8.** After Judson and Ritter, 1964, "Rates of Regional Denudation in the United States," *Journal of Geophysical Research*, v. 69, pp. 3395-3401. **Fig. 10-10.** After F. Hjulstrom, 1935, "Studies on the Morphological Activity of Rivers as Illustrated by the River Fryis," *University of Upsala Geol. Institute Bulletin 25*, pp. 221-527. **Fig. 10-14.** From L. B. Leopold, 1968, *Hydrology for Urban Land Planning – a Guidebook on the Hydrologic Effects of Urban Land Use*, U.S. Geological Survey Circ. 554, Fig. 1. **Fig. 10-21.** After D. J. Easterbrook, 1969, *Principles of Geomorphology*, McGraw-Hill Book Co., Fig. 6-10. **Fig. 10-28.** From L. B. Leopold and J. P. Miller, 1954, "A Postglacial Chronology for Some Alluvial Valleys in Wyoming," *U.S. Geological Survey Water Supply Paper 1261*.

CHAPTER 11

Fig. 11-3. After P. Meigs, 1956, *Future of Arid Lands*. By permission of the American Association for the Advancement of Science. **Fig. 11-15 A.** After T. M. Oberlander, 1972, "Morphogenesis of Granitic Boulder Slopes in the Mojave Desert, California," *Journal of Geology*, v. 80, pp. 1-20, Fig. 14. **Figs. 11-19 and 11-23.** After R. A. Bagnold, 1941, *The Physics of Blown Sand and Desert Dunes*, Metheun and Co. **Fig. 11-29.** From R. B. Morrison, 1968, "Pluvial Lakes," in R. W. Fairbridge, ed., *Encyclopedia of Geomorphology*, Dowden, Hutchinson & Ross, Stroudsburg, Pa. By permission of Dowden, Hutchinson & Ross. **Fig. 11-32.** From M. J. Bovis and R. G. Barry, 1974, "A Climatological Analysis of North Polar Desert Areas," in T. L. Smiley and J. H. Zumberge, eds., *Polar Deserts and Modern Man*, University of Arizona Press, Fig. 2-4. By permission of the University of Arizona Press.

CHAPTER 12

Fig. 12-27. From R. F. Flint, 1971, *Glacial and Quaternary Geology*, John Wiley & Sons. By permission of John Wiley & Sons, Inc. **Fig. 12-30.** After V. K. Prest, 1970, *Quaternary Geology of Canada*, Canada Department of Energy, Mines, and Resources; and Bryson and others, 1969, "Radiocarbon Isochromes on the Disintegration of the Laurentide Ice Sheet," *Arctic and Alpine Research*, v. 1, pp. 1-13. **Fig. 12-37.** After D. J. Easterbrook, 1969, *Glacial Map of the United States East of the Rocky Mountains*, Geological Society of America. **Fig. 12-40.** After the *National Atlas of the United States of America*, U.S. Geological Survey, 1970.

CHAPTER 13

Fig. 13-7. After W. Bascom, 1964, *Waves and Beaches*, Doubleday & Co., Inc., Figs. 40 and 41. **Fig. 13-11.** After W. C. Bradley, 1957, "Origin of Marine Terrace Deposits in the Santa Cruz Area, California," Geological Society of America, v. 68, pp. 421-44, Fig. 2. **Fig. 13-17.** After A. L. Bloom, and others, 1974, "Quaternary Sea-Level Fluctuations on a Tectonic Coast, New 230 Th 234 Dates from the Huon Peninsula, New Guinea," *Quaternary Research*, v. 4, pp. 185-205, Fig. 5. **Fig. 13-19.** From J. T. Andrews, 1975, *Glacial Systems*, Duxbury Press, Fig. 7-7. By permission of Duxbury Press. **Fig. 13-24.** After D. W. Johnson, 1919, *Shore Processes and Shoreline Development*, John Wiley and Sons, Fig. 88. **Figs. 13-29 and 13-30.** From J. H. Hoyt, 1967, "Barrier Island Formation," *Geological Society of America Bulletin*, v. 78, pp. 1125-36. By permission of the Geological Society of America and J. H. Hoyt.

CHAPTER 14

Fig. 14-2. From B. C. Heezen and M. Tharpe, 1976, *The Floor of the Oceans*, map painted by Tangy de Remur, American Geographical Society. Reprinted by permission of B. C. Heezen and M. Tharpe. **Fig. 14-3.** After K. O. Emery, 1969, *The Continental Shelves in the Ocean*, W. H. Freeman and Co., pp. 44-45. **Fig. 14-5.** From R. P. Shepard, 1963, *Submarine Geology*, 2nd ed., Harper & Row, Fig. 150. **Fig. 14-7.** From B. C. Heezen and M. Tharpe, 1976, *The Floor of the Oceans*, map painted by Tangy de Remur, American Geographical Society. Reprinted by permission of B. C. Heezen and M. Tharpe. **Fig. 14-10.** From H. W. Menard, 1964, *Marine Geology of the Pacific*, McGraw-Hill Book Co., Fig. 4-8. By permission of McGraw-Hill Book Co. **Fig. 14-12.** After W. M. Davis, 1928, *The Coral Reef Problem*, Special Pub. no. 9, American Geographical Society. By permission of the American Geographical Society. **Fig. 14-14.** From B. C. Heezen and C. D. Hollister, 1971, *The Face of the Deep*, Oxford University Press, Figs. 7-44, 45, and 46. By permission of B. C. Heezen. **Fig. 14-19.** After W. S. Broecker and J. van Donk, "Insolation Changes, Ice Volumes, and the 0^{18} Record in Deep-Sea Cores," *Reviews of Geophysics and Space Physics*, v. 8, no. 1, Fig. 3.

CHAPTER 15

Fig. 15-6. After A. N. Strahler, 1969, *Physical Geography*, John Wiley & Sons, Inc., Fig. 32-29. **Fig. 15-8.** From R. C. Heath and others, 1966, *The Changing Pattern of Ground-Water Development on Long Island, New York*, U.S. Geological Survey Circular 524. **Fig. 15-9.** After K. Cartwright and F. B. Sherman, 1969, *Evaluating Sanitary Landfill Sites in Illinois*, Illinois State Geological Survey Environmental Notes no. 27. **Fig. 15-11.** After W. J. Wayne, 1956, *Thickness of Drift and Bedrock Physiography of Indiana North of the Wisconsin Glacial Boundary*, Indiana Dept. of Conservation, Geological Survey, Report of Progress no. 7, Figs. 2 and 8.

CHAPTER 17

Fig. 17-6. After O. Nuttli, 1973, "The New Madrid Earthquake of 1811 and 1812, Intensities, Ground Motion, and Magnitudes," *Bulletin of the Seismological Society of America*, v. 63, pp. 227-48, Fig. 1. **Fig. 17-8.** After G. Plafker and J. C. Savage, 1970, "Mechanism of the Chilean Earthquake of May 21 and 22, 1960," *Geological Society of America Bulletin*, v. 81, pp. 1001-1030. **Fig. 17-10.** After R. Siever, 1973, "The Earth," *Scientific American*, v. 233, pp. 82-91. **Fig. 17-11.** After Gilluly and others, 1968, *Principles of Geology*, 3rd ed., W. H. Freeman and Co. **Fig. 17-14.** From S. T. Algermissen, 1969, "Seismic Risk Studies in the U.S.," *Proceedings of the Fourth World Conference on Earthquake Engineering*, v. 1, pp. 14-27. **Fig. 17-15.** From C. Kissingler, 1974, *Earthquake Prediction*. By permission of C. Kissingler. **Fig. 17-16.** After D. M. Evans, 1966, *Geotimes*, v. 10, pp. 11-18. **Figs. 17-18 and 17-19.** After A. N. Strahler, 1971, *Principles of Physical Geology*, John Wiley & Sons, Inc. **Figs. 17-22 and 17-23.** After O. M. Phillips, 1968, *The Heart of the Earth*, W. H. Freeman and Co. **Fig. 17-25.** After C. F. Richter, 1958, *Elementary Seismology*, W. H. Freeman and Co. **Fig. 17-31.** After F. Press and R. Siever, 1974, *Earth*, W. H. Freeman and Co. **Fig. 17-33.** After A. N. Strahler, 1971, *Principles of Physical Geology*, John Wiley & Sons, Inc. **Fig. 17-37.** After P. J. Wyllie, 1975, "The Earth's Mantle," *Scientific American*, v. 232, no. 3, pp. 50-63.

CHAPTER 18

Fig. 18-7. After Takeuchi, Uyeda and Kanamori, 1970, *Debate About the Earth*, W. H. Freeman and Co. **Figs. 18-8 and 18-9.** After A. Cox, 1969, "Geomagnetic Reversals," *Science*, v. 163, pp. 237 and 202. **Fig. 18-10 A.** After J. R. Heirtzler and others, 1965, "Magnetic Anomalies Over the Reykjanes Ridge," *Deep-Sea Research*, v. 13, p. 427. **Fig. 18-10 B.** After F. J. Vine, 1960," Magnetic Anomalies Associated with Mid-Ocean Ridges," in R. A. Phinney, ed., *History of the Earth's Crust, A Symposium*, Princeton University, p. 73, Fig. 6. By permission of Princeton University Press.

CHAPTER 19

Fig. 19-4. After S. W. Carey, 1958, "A Tectonic Approach to Continental Drift," in *Continental Drift, A Symposium*, Geology Dept., Univ. of Tasmania, pp. 177-355. **Fig. 19-5.** After E. C. Bullard and others, 1965, "The Fit of Continents Around the Atlantic," in P.M.S. Blackett and others, eds., *A Symposium on Continental Drift*, Phil. Tran. Royal Soc., London, v. 100. **Fig. 19-6.** After Tarling and Tarling, 1971, *Continental Drift*, Doubleday and Co. **Fig. 19-7.** After P. M. Hurley, 1968, "The Confirmation of Continental Drift," *Scientific American*, v. 229, pp. 60-69. **Fig. 19-8.** After Takeuchi, Uyeda, and Kanamori, 1970, *Debate About the Earth*, W. H. Freeman and Co. **Fig. 19-9.** After S. K. Runcorn, ed., 1962, *Continental Drift*, Fig. 20, Academic Press, Inc. **Fig. 19-10.** After E. Orowan, 1969, "The Origin of the Ocean Ridges," *Scientific American*, v. 221, no. 5. **Fig. 19-11.** From W. C. Pitman and others, 1974, *Age of the Ocean Basins*, Geological Society of America map. **Fig. 19-12.** After A. E. Maxwell and others, 1970, "Deep-Sea Drilling in the South Atlantic," *Science*, v. 108, pp. 1047-59. © 1970 by the Amer. Assc. for the Advancement of Science. By permission of A. E. Maxwell. **Fig. 19-13.** After B. Gutenberg and C. F. Richler, 1954, *Seismicity of the Earth*, 2nd ed., Princeton University Press. **Fig. 19-15.** After J. Oliver and B. Isacks, 1967, "Deep Earthquake Zones, Anomalous Structures in the Upper Mantle and the Lithosphere," *Journal of Geophysical Research*, v. 72, pp. 4259-4275. **Fig. 19-16.** After B. Isacks and others, 1968, "Seismology and the New Global Tectonics," *Journal of Geophysical Research*, v. 73, pp. 5855-5899. **Fig. 19-18.** After M. N. Toksoz, 1976, "The Subduction of the Lithosphere," in *Continents, Aground, Readings from Scientific American*, W. H. Freeman and Co., pp. 112-22. **Fig. 19-25.** After E. Orowan, 1969, "The Origin of the Oceanic Ridges," *Scientific American*, v. 221, no. 5. **Fig. 19-28 A.** After D. W. Scholl and others, 1970, "Peru-Chile Trench Sediments and Sea-Floor Spreading," *Geological Society of America Bulletin*, v. 81, pp. 1339-1360. **Fig. 19-28 B.** After M. L. Holmes and others, 1972, "Seismic Reflection Evidence Supporting Underthrusting Beneath the Aleutian Arc Near Amchitka Island," *Journal of Geophysical Research*, v. 77, pp. 959-64.